DATE DUE

Demco, Inc. 38-293

MAR 1 6 2009

Listening on
the Short Waves,
1945 to Today

Listening on the Short Waves, 1945 to Today

JEROME S. BERG

McFarland & Company, Inc., Publishers
Jefferson, North Carolina, and London

LIBRARY OF CONGRESS CATALOGUING-IN-PUBLICATION DATA

Berg, Jerome S., 1943–
Listening on the short waves,
1945 to today / Jerome S. Berg.

p.   cm.
Includes bibliographical references and index.

ISBN 978-0-7864-3996-6
illustrated case binding : 50# alkaline paper ∞

1. Shortwave radio — History.   2. International
broadcasting — History.   I. Title
HE8697.4.B476   2008   384.54 — dc22   2008024675

British Library cataloguing data are available

Cover photographs ©2008 Shutterstock

Manufactured in the United States of America

*McFarland & Company, Inc., Publishers
Box 611, Jefferson, North Carolina 28640
www.mcfarlandpub.com*

To Ken Boord,
for his many years of service to shortwave listeners,

and to P.V.

Contents

Preface

After the 1999 publication of *On the Short Waves, 1923–1945: Broadcast Listening in the Pioneer Days of Radio*, a number of readers urged me to write a volume covering the years 1945 to the present. While I was flattered, I never intended to do so. Other obligations prevented me from devoting any more time to radio, and while my own listening activities date to 1958, I felt there were many who had a better understanding of the period.

Things changed for two reasons. First, my retirement in 2003 meant that I had the time such a project would demand. And second, no one else had done it. Hence this next installment of the story, and its companion volume, *Broadcasting on the Short Waves, 1945 to Today*. The present book covers the listening side (clubs, publications, the audience, etc.) and *Broadcasting*, the stations. This is the first attempt to chronicle in one place post–World War II shortwave broadcast listening in the United States. I have tried to be true to the events as I understand them (and, in many cases, experienced them), recognizing that others may have a different perspective or may wish that certain subjects received more or less emphasis.

As was the case with *On the Shortwaves, 1923–1945*, the present work is written from the perspective of the listener, not the broadcasting professional. It is about those who enjoyed shortwave listening enough to join clubs, write to magazines, send reception reports, and otherwise share their experiences—"serious" listeners. Closet SWLs will have to tell their own story.

It is also about listening to shortwave broadcasting stations. Amateur radio, broadcast band, utility and ham band listening are not addressed except where necessary to round out the shortwave broadcast story. In addition, I have treated the subject from an American perspective. Much could be written about the development of shortwave listening in other parts of the world, and I encourage others to do so. And while the Canadian experience parallels the American in many ways, I have not presumed to cover it in any detail save for the histories of the Canadian shortwave clubs.

Since this book will likely be read mainly by those with an interest in the subject, a basic understanding of shortwave concepts and nomenclature is presumed. The titles, abbreviations, acronyms, etc., will be familiar to shortwave listeners (a full list follows). In quoting material, I have made occasional minor changes in punctuation, case and spelling to make it easier on the eye, and in a few places I have converted megacycles or megahertz to kilocycles or kilohertz for the sake of uniformity. I have also changed some abbreviations to others that are more understandable, and I trust that James Watt will forgive my use of kw. rather than kW.

While the book is not intended as an academic work, this road is not likely to be traveled again any time soon, and thus I have included many citations to references from inside and outside the shortwave listening community so that worthy resources are not forgotten and so that others will be aided in doing further reading and research. Citations are mainly to North American print materials. Website references were correct as of August 1, 2008. Readers are reminded that defunct websites can often be resurrected through the "Wayback Machine" at http://www.archive.org. Occasionally I have included some summary history for the years before 1945 to set the scene for later events. For more extensive treatment of the pre–1945 years, the reader is respectfully referred to *On the Short Waves, 1923–1945*.

Consistent with the modern practice, shortwave is spelled as one word throughout the book. It is two words in the title in order to preserve continuity with *On the Short Waves, 1923–1945*, which covered a period when shortwave was typically spelled as two words rather than one.

Preparing the two new volumes, *Listening* and *Broadcasting*, proved the truth of my comment in the preface to *On the Short Waves, 1923–1945* that much of the history of shortwave broadcast listening is to be found in the basements and attics of its practitioners. This is especially true of the topics covered in this volume. I cannot express enough my appreciation to those who have willingly searched through troves of long forgotten material, packed it up and sent it to me, sometimes at considerable inconvenience to themselves. In this regard a thousand thanks are due in particular to Rich D'Angelo, Reuben Dagold, Mickey Delmage, John Herkimer, Alan Johnson, Bob LaRose, Adrian Peterson, Jack Rugg, Harold Sellers, and George Zeller. I am sure that the completion of this project is as much a relief to them as it is to me — maybe more.

Don Jensen deserves special thanks for his encouragement, his perspective and his help in reconstructing events. His ongoing advice and counsel have been nearly as important to me as is our friendship of half a century.

Numerous others provided me with material, helped me locate information, suggested additional sources, offered their insights, read preliminary drafts, and let me plumb their memories. Many are friends of long standing, and those friendships appear to have survived intact the last several years of pestering. I also met many new people along the way, including some whose names I knew but with whom I had not previously been in contact. New or old, and with apologies to any I have forgotten, I extend my thanks to them all: "Skip" Arey, John T. Arthur, Guy Atkins, Ray Barfield, Michel Baron, Mike Barraclough, Rachel Baughn, Kirk Baxter, John Berenyi, Ralph Brandi, Walter Brodowsky, John Bryant, Joe Buch, Wolfgang Bueschel, Bill Butuk, John Callarman, John Campbell, Neil Carleton, Bryan Clark, Harold Cones, Mark Connolly, Bob Cooper, Pete Costello, Wendel Craighead, Richard Cuff, the late Neville Denetto, Gerry Dexter, Torre Ekblom, Kim Andrew Elliott, Dan Ferguson, John Figliozzi, John Fisher (Massachusetts), Bill Fisher, Karl Forth, David Foster, Bernd Friedewald, Harold Frodge, Tom Gavaras, Victor Goonetilleke, Bill Graham, Manosij Guha, Dave Hammer, Chris Hansen, Nicholas Hardyman, Sheldon Harvey, Harry Helms, Dan Henderson, Eric Hitchcock, Jim Howard, Bob Hill, Günter Jacob,

John Kapinos, Dave Kenny, Tony King, Henrik Klemetz, Joerg Klingenfuss, Erik Køie, Kraig Krist, Finn Krone, Mathias Kropf, Marie Lamb, Russell Lay, Chris Lobdell, Ken MacHarg, Stew MacKenzie, Larry Magne, Jonathan Marks, George Maroti, David Martin, Judy Massa, Ed Mayberry, Paul McDonough, Ian McFarland, Rich McVicar, Kevin Mikell, Don Moman, Don Moore, Ray Moore, Richard Moseson, Graham Mytton, Serge Neumann, Mark Nicholls, Mike Nikolich, Tim Noonan, Toshi Ohtake, Bill Oliver, Harold Ort, Fred Osterman, Kathy Otto, Bob Padula, Gene Parta, Jack Perolo, Anker Petersen, Nandor Petrov, Bill Plum, Christos Rigas, Tony Rogers, Ron Pokatiloff, Rob Rabe, Jim Ronda, Andreas Schmid, Andy Sennitt, Ed Shaw, Ken Short, the late Morris Sorensen, Simon Spanswick, Larry Steckler, Tom Sundstrom, "Tex" Swann, Jan Tunér, Rob Wagner, Dave Walcutt, Andy Wallace, Tom Walsh, Brent Weeks, Gerhard Werdin, Jeff White, Bob Wilkner, Barry Williams, John Wright, Larry Yamron, Andy Yoder, Bob Zanotti, Ken Zichi, and Oliver Zöllner.

My interest in shortwave history has paralleled my chairmanship of the Committee to Preserve Radio Verifications (CPRV), a committee of the former Association of North American Radio Clubs from 1986 to 2005 and a standalone project since. My current partners in that effort have been named above — Gerry Dexter, Tom Gavaras, Dan Henderson, John Herkimer, and Don Jensen. Their dedication to keeping the flames of DX history alive has been important to me during my writing, as has the assistance of curator Chuck Howell and reference specialist Michael Henry of the Library of American Broadcasting, the University of Maryland facility which is home to the CPRV collection.

My thanks as well to Jean Williams and Steffie Lowder of the Interlibrary Loan Department of Cary Memorial Library, Lexington, Massachusetts, who proved themselves without peer in their ability to locate and retrieve numerous publications, no matter how distant or obscure. They aided my research immeasurably.

And finally, thanks to my wife, Ruth, who has acquiesced, usually willingly, in the expenditure of practically all my waking hours for three years on a project that non-shortwave types could never appreciate. She has been a DX widow for 43 years, so this is just the latest, albeit the biggest, in a long line of shortwave frolics to which she has been witness. But it still takes understanding and forbearance, traits which, thankfully, she has in abundance.

Graphics in this book are from the CPRV collection, my own collection, and the collections of various individuals. I have made every effort to contact copyright holders of original material when permission to utilize it appears necessary, and any errors in this regard are unintentional.

No doubt some gremlins have crept in to the vast amount of factual material amassed for this volume. Errors are mine alone, and corrections are welcome.

As this is written, shortwave receivers are better than ever, and information about shortwave, at least on the internet, is more widely available than ever. There is even experimentation with "hi fi" quality digital shortwave signals (DRM). That shortwave broadcasting is both more accessible and potentially more attractive to the non-technical public at the same time that it is being supplanted by other media may seem ironic, but time marches on and events are what they are. For those of us who have loved long distance listening over the years, it has been a great run. As NNRC broadcast band editor Carroll H. Weyrich once said about a horse named Radio Rome that had come in second at Pimlico, "It at least shows that someone in the Sport of Kings is interested in the Hobby of Kings."[1]

Titles, Abbreviations and Acronyms

ACE, A*C*E . . . . Association of Clandestine Radio Enthusiasts
ADXR . . . . Association of DX Reporters
AIR . . . . All India Radio
ANARC . . . . Association of North American Radio Clubs
ARDXC . . . . Australian Radio DX Club
AWA Journal . . . . bulletin of the Antique Wireless Association, Inc. (since January 2005)
AWA Review . . . . annual anthology of the Antique Wireless Association, Inc.
ASWLC . . . . American Shortwave Listeners Club
AWR . . . . Adventist World Radio
BADX . . . . Boston Area DXers, or British Association of DXers
BBC . . . . British Broadcasting Corp.
BBG . . . . Broadcasting Board of Governors
BCB . . . . standard broadcast band, medium wave
BDXC . . . . British DX Club
BPL . . . . Broadcasting Over Power Lines
CADX . . . . Chicago Area DX Club
CADEX . . . . bulletin of the Canadian DX Club
CANDX . . . . bulletin of Canadian SWL International
CB . . . . Citizens Band
CBC . . . . Canadian Broadcasting Corp.
CDXC . . . . Canadian DX Club
CDXQ . . . . Club des DXers du Quebec
CHAP . . . . Canadian Handicapped Aid Program
CIDX . . . . Canadian International DX Club

Communication . . . . . . bulletin of the British DX Club
CONTACT . . . . bulletin of the World DX Club
CRI . . . . China Radio International
CPRV . . . . Committee to Preserve Radio Verifications
CSWLI . . . . Canadian SWL International (club)
CW . . . . continuous wave
CW . . . . *Communications World*
DSWCI . . . . Danish Shortwave Club(s) International
DUDXS . . . . *"Down Under" DX Survey*
DW . . . . Deutsche Welle
DX, DXing . . . . distance, listening to distant stations
DX Ontario . . . . bulletin of the Ontario DX Association (through December 1999)
DXLD . . . . *DX Listening Digest*
DXPL . . . . DX Partyline (HCJB program)
DX Reporter . . . . bulletin of the Association of DX Reporters
DXSF . . . . *DX South Florida*
EDXC . . . . European DX Council
EI . . . . *Electronics Illustrated*
FBIS . . . . Foreign Broadcast Information [Intelligence] Service
FEBA . . . . Far East Broadcasting Association
FEBC . . . . Far East Broadcasting Co.
FRENDX . . . . bulletin of the North American Shortwave Association (through December 1989)
GCSS . . . . Great Circle Shortwave Society
HAP . . . . Handicapped Aid Program
HFCC . . . . High Frequency Coordination Conference [Committee]
HTL . . . . *How to Listen to the World*
IBB . . . . International Broadcasting Bureau
ID . . . . identification
IDXC . . . . International DXer's Club of San Diego
International Short
   Wave Radio . . . . . . bulletin of the International Short Wave Club
ISWC . . . . International Short Wave Club
ISWL . . . . International Short Wave League
ITU . . . . International Telecommunication Union
kc., kHz. . . . . kilocycle(s), kilohertz
kw. . . . . kilowatt(s)
Listening In . . . . bulletin of the Ontario DX Association (from January 2000)

LSB . . . . lower sideband

mc., MHz. . . . . megacycle(s), megahertz

MDXC . . . . Minnesota DX Club

Messenger . . . . bulletin of the Canadian International DX Club

MARE . . . . Michigan Area Radio Enthusiasts

MVDXC . . . . Miami Valley DX Club

MT . . . . *Monitoring Times*

MW . . . . medium wave, standard broadcast band

NASB . . . . National Association of Shortwave Broadcasters

NASWA . . . . North American Shortwave Association

NASWA Journal . . . . . . bulletin of the North American Shortwave Association (from January 1990)

New Zealand DX Times . . bulletin of the New Zealand Radio DX League

NNRC . . . . Newark News Radio Club

NNRC Bulletin . . . . . . bulletin of the Newark News Radio Club

NU . . . . *Numero Uno*

NZRDXL . . . . New Zealand Radio DX League

ODXA . . . . Ontario DX Association

Old Timer's Bulletin . . . . bulletin of the Antique Wireless Association, Inc. (through October 2004)

ORF . . . . Österreichischer Rundfunk

ORTF . . . . Office de Radiodiffusion Television Francaise

Passport . . . . *Passport to World Band Radio*

PE . . . . *Popular Electronics*

PLL . . . . phase locked loop

PTT . . . . Post, Telephone & Telegraph

QRM . . . . interference

QRN . . . . noise, static

QSL . . . . verification of reception

RBI . . . . Radio Berlin International

RCI . . . . Radio Canada International

RFE-RL . . . . Radio Free Europe-Radio Liberty

RFPI . . . . Radio for Peace International

RIB . . . . *Review of International Broadcasting*

RNM . . . . *Radio Nuevo Mundo*

RRI . . . . Radio Republik Indonesia

SABC . . . . South African Broadcasting Corp.

SCADS . . . . Southern California Area DXers

SCDX . . . . Sweden Calling DXers

Short Wave News . . . . . . bulletin of the Danish Shortwave Club(s) International

Short Wave News [ISWL] . bulletin of the International Short Wave League (U.K.) or its predecessor magazine

SPEEDX . . . . Society for the Preservation of the Engrossing Enjoyment of DXing

SSB . . . . single sideband

SWBC . . . . shortwave broadcast

SWC . . . . Short Wave Center, a *FRENDX* column

SWL . . . . shortwave listener

SWL . . . . bulletin of the American Shortwave Listeners Club

TDF . . . . Télédiffusion de France

TWR . . . . Trans World Radio

UADX . . . . Union of Asian DXers

USB . . . . upper sideband

Universalite . . . . bulletin of the Universal Radio DX Club

URDXC . . . . Universal Radio DX Club

USIA . . . . United States Information Agency

VOA . . . . Voice of America

VOFC . . . . Voice of Free China

WDXC . . . . World DX Club

WPE Call Letter . . . . . bulletin of the Great Circle Shortwave Society

WRH, WRTH . . . . *World Radio Handbook, World Radio TV Handbook*

WT . . . . *The World at a Twirl*

1

Prelude to 1945

Broadcasting experiences during World War II have clearly demonstrated the effectiveness of short-wave radio and that it should prove an ideal medium in helping to achieve and maintain better understanding between the nations of the world. In no other way can the spoken word reach so many people in so short a time. In no other way can one tell one's story so quickly either to or from the more remote places on the globe. I, for one, feel strongly that international short-wave radio will assume its rightful place and responsibility in the era just beginning.

Kenneth R. Boord, International Short-Wave Editor,
Radio News, January 1947[1]

Although experimentation with shortwave transmission had commenced soon after the turn of the century, its long-distance properties were not discovered for another 20 years. The first transatlantic amateur contacts (1923), and the discovery that shortwave frequencies, long thought worthless, could "bounce" in multiple hops between earth and the ionosphere (then called the "Kennelly-Heaviside" layer), led to the recognition of shortwave as a medium with worldwide reach.

Shortwave broadcasting entered the public consciousness in the early 1920s as standard broadcast stations like KDKA, themselves a novelty at the time, began transmitting their programs on parallel shortwave channels.[2] Although at that time shortwave transmission and reception was mainly the province of experimenters, it became part of the public debate on "national broadcasting stations." As early as 1923 there was optimism that stations could be heard nationwide by relaying their programs on shortwave to local stations for rebroadcast on standard broadcast band frequencies.[3] Once superheterodyne shortwave receivers became available to individual consumers in the early 1930s, it was a short leap to the notion of broadcasting on shortwave direct to the listener.

Early radio publisher Hugo Gernsback was a booster of shortwave broadcasting. In 1928 he predicted that shortwave would open a new era in radio. Soon many countries, including England, Holland, Germany, France, Spain, Australia and Russia, were experimenting with broadcast programming on shortwave. Receiving equipment became available to the general public. Gernsback was careful to note, as he often would in the future, that you needed a "really good set" to get good long distance reception, and that manufacturers were not being sufficiently candid with the public in this regard. Listeners were interested in the thrill of distant reception. "It would not surprise me at all," he said, "if, during the next five years, the broadcasting of both sound and sight will be done completely on short waves; and the upper wave channels from 200 to 600 meters [the standard broadcast band] gradually abandoned, as fast as we learn more about the short waves."[4] A few months later he predicted that "soon every important station will have its short-wave simultaneous transmissions."[5]

Others, like NBC General Engineer C. W. Horn, were hopeful but more cautious. There were natural difficulties to contend with, such as static, fading and distortion. Shortwave transmitters were more complex than their broadcast band counterparts, and they required larger aerials.[6] And there was a "dead zone" surrounding the transmitter. But there were many more frequencies available in the shortwave bands than in the standard broadcast band. This held promise for relieving broadcast band congestion, and placing within a station's reach listeners at hitherto unachievable distances, and with less power (although it would soon be learned that listenable reception would demand substantial power).

Listening to shortwave was not simple. Even in the 1930s, Charles A. Morrison, president of an early shortwave club, the International DXers Alliance, noted the congestion in the 49 meter band and urged a system of international frequency allocation. He felt that receivers with better signal-to-noise ratios were needed, local electrical interference being the greatest single handicap to shortwave reception. And he wanted better visual signal peaking devices, and better frequency calibration. "[N]ot far in the future we shall see all-wave receivers that will possess such a degree of exact calibration that any known frequency will be located almost instantly with a very small percentage of error! Easy location of important short-wave stations on a receiver will do much towards making for a universal popularity for the high frequencies."[7] (This advance would not be widely available until the 1970s.)

Shortwave broadcasting progressed, with more and more stations coming on the air. There were two main types of stations: international stations, and domestic broadcasters. The former targeted other countries, usually in their own languages, while the domestic shortwave stations covered areas of a country that would otherwise have required many separate AM stations (which were susceptible to high static levels in tropical areas). With few exceptions, international shortwave stations operated with a maximum of 50 kw., domestic shortwave stations with much less. Receivers improved and shortwave tuning became easier, although it was never as simple as tuning a regular radio.[8] Said Westinghouse Vice-President H. P. Davis in 1931, "[I]t is almost a matter-of-fact statement for me to say that international broadcasting is an established accomplishment, and eventually will be as reliable and as perfect as local broadcasting."[9]

In the 1930s there were magazines devoted exclusively to many aspects of shortwave, and there were enough stations on the air for one such publication, *Short Wave Radio*, to pick its ten best. They were: VK2ME, Sydney, Australia, and VK3ME, Melbourne, "The Voice of Australia"; HJ1ABB, La Voz de Barranquilla, Colombia; the BBC; the Deutsche Kurzwellensender; PCJ's successor, PHI, Hilversum, Holland (at that time already carrying the voice of Eddie Startz, who would be known to shortwave listeners for over 40 years); I2RO, Rome; EAQ, Madrid; Radio Nations, the League of Nations station in Geneva, Switzerland; and Venezuelan stations YV3BC, Radiodifusora Venezuela, and YV5BMO, Ecos del Caribe, both in Caracas.[10] In the United States, the number of standard broadcast

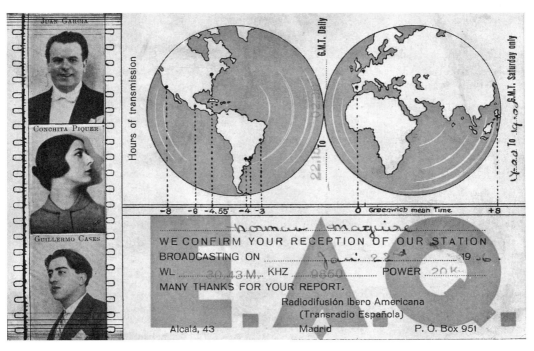

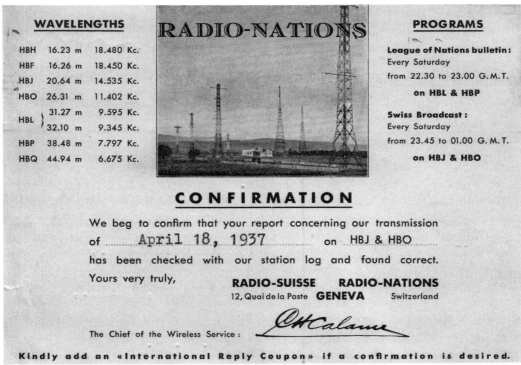

Top: Spanish station EAQ, Radiodiffusion Ibero Americana, was widely heard. Its monthly Spanish-language program guide contained cultural articles and a summary of each day's program. *Bottom:* League of Nations station Radio-Nations had two 20 kw. transmitters located in Prangins, Switzerland. It was also used for communicating with member states and League delegations.

stations simulcasting on shortwave was growing, and included W2XAF, 9530 kc. and W2XAD, 15530 kc. (relaying WGY, Schenectady); W3XAL, Bound Brook, New Jersey, 6100 and 17780 kc. (relaying WJZ, New York); W2XE, 6120 and 11830 kc. (WABC, New York); W8XAL, 6060 kc. (WLW, Cincinnati); W9XF, 6100 kc. (WENR, Chicago); and W8XK, 6140, 11870, 15210 and 21540 kc. (relaying KDKA, Pittsburgh). Canadian station VE9GW at Bowmanville, Ontario was well heard on 6095 kc.[11]

By the mid–1930s, all-wave radios equipped with the shortwave bands were in common use, and the newspapers were playing up shortwave.[12] There was great optimism surrounding shortwave broadcasting. Morrison predicted that every country, including the United States, would have a national shortwave broadcasting service such as those of England, France and Germany.

Shortwave broadcasting was also seen as an important element in the search for world peace. Morrison felt that as we learned more about the people of other lands, "we will find fear, trust and jealousy vanishing."[13] Some leaders in the radio industry, including Frank Conrad, echoed these sentiments.

> Today every nation in Europe displays a keen appreciation of the importance of short wave transmission in inter-country mass communication. The short wave knows no borders and passes freely from one country to another. Some nations are using this method to spread their particular ideologies, it is true, but the day will come when short waves will find their rightful use as bonds of international understanding and appreciation.... ¶[W]ho knows but that the day may come when the short wave will bring forth a new and better understanding between the great nations of the earth.[14]

It was in war time that shortwave would come into its own, however. World War II brought with it major, competing shortwave efforts between the Allied and the Axis powers, as well as the start of clandestine broadcasting of various types. *National Geographic* observed:

> [Radio] has made words into weapons as vital as bullets and bombs. Modern war, it almost could be said, is fought at the speed of light, 186,000 miles per second, for that is the terrific pace at which words travel on the lightning wings of radio waves. * * * Guns cease firing now and then, and soldiers take time to sleep, but the War of Words never stops.[15]

Shortwave broadcasting also made it possible for countries like England to stay in touch with colonies under siege, and for colonial stations like FZI, Brazzaville, French Equatorial Africa, and Radio Nationale Belge, Leopoldville, the Belgian Congo, to serve as free national voices during the occupation of their home countries.

The Allies usually came in second in the shortwave radio war. Reporting to the Newark News Radio Club, an American staff sergeant serving in North Africa noted: "As we have known for years, the Axis (mostly Germany) transmits much more propaganda than the Allies, even if it isn't as effective. And to date I have heard only one station that claims to be the 'Voice of America' broadcasting anti–Nazi programs."[16]

During the war, *RADEX,* a popular shortwave magazine of the day, considered dropping coverage of stations in Germany, Italy and Japan, but settled for reminding its readers that broadcasts from these stations should be viewed as propaganda. Radio club members commented on the broadcasts:

> [W]e used to like to listen to the Nazi and Jap propaganda broadcasts and many a laugh we had over them. At times while on Guadalcanal, I used to help carry a radio that was portable and which was handled by three men. We used to go on two-to-seven day patrols out in the jungles.

And when we bivouacked we set up our radio. Twice we heard via Tokyo that we had been anni-hilated and that Guadalcanal was in the hands of the Japs. Did we get a laugh at that![17]

———

An amusing situation was created last month when the Berlin short wave announcer offered to accept cabled reports from America collect, asking short wave listeners to tell how the programs were being heard. A New York newspaper suggested that everyone send a cable in order to make Germany reach down in the old pocket-book. For one week Germany received thousands of cables daily, many of which were in no way related to short wave radio. Many persons took advantage of the opportunity to tell what they thought of the Fuehrer in particular and Ger-many in general. Telegraph companies reported that many cables were not sent because of the language used. The German announcers took everything in a very liberal manner.[18]

Organized shortwave listening on the home front declined greatly during the war years. Some clubs closed down. Limitations on the availability of paper, and interruptions in mail service to some countries, made bulletin publishing difficult. Moreover, many club mem-bers were in the service, and those who weren't were working overtime in the war effort. But even though their time for radio was limited, they still liked to stay in touch with their club's headquarters.

My DXing last year wasn't so good but my excuse is the same as that of other members—too much war work. I work from 9½ to 13 hours a day, six days a week at our local porcelain plant. And for at least three hours every night I repair radios (goodness knows there sure are a lot of them to be repaired).[19]

———

Since my last couple of reports way back in '41, I have been working and working, building the Norden bombsight. That kept me up nearly eleven hours a day (and sometimes at night). As I was doing precision work it was required that I get plenty of sleep. Therefore, no DX.[20]

———

Although a bomb shattered his radio gear early in May, our good friend Arthur E. Bear of Lon-don reports that he and his family as [sic] still safe and sound. For which we are thankful.[21]

———

Am making out as well as can be expected. My plane is the P-51 and it is plenty good! I have been just about everywhere over Germany including Berlin, Leipzig, Munich, Kiel, Bremen and dozens of other places. Have the Air Medal with three oak leaf clusters for knocking down a couple of Jerry Focke-Wulfs. Now I am leading a flight and should be a captain in about three weeks. What's more important to me is to get back home as quickly as possible to start teach-ing my youngster how to DX.[22]

Contacts with the belligerent countries were renewed after the war. In November 1941 the Newark News Radio Club received a letter from Shokichi Yoshimura in Japan, together with enough Japanese mint stamps to cover the cost of NNRC membership. Then came the war. The club kept his letter on file, and contacted him in 1946. He was still interested in renewing his DXing, although his QSLs, as well as his typewriter and household items, had been lost in a fire during the war.[23]

Many familiar shortwave voices disappeared after the war: DJB, DJC, DJD and the rest of the "D" call letters of the powerful Nazi "Zeesen" station; JZI, JZJ and the other Japanese "J" call letters; the European clandestine stations; etc. But other shortwave voices would soon take their place, and the level of shortwave broadcasting would increase.

Gone also were the messages from prisoners of war that were carried by stations in Germany, Japan, and other Axis countries. Many SWLs listened attentively to the mes-

Listening Post!

ALEX. E. GORDON, Indianapolis, Ind.— legislative representative of the Brotherhood of Locomotive Firemen and Enginemen— brings good news to hundreds of mothers of missing servicemen, through his hobby and his faithful 16-tube, 1940 model

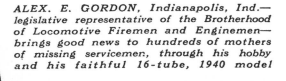

MIDWEST RADIO

ALWAYS a short-wave radio enthusiast, Alex. E. Gordon has spent many a night listening over his 16-tube Midwest Radio to foreign broadcasts. Several months ago he noticed that the Nazis, along with their propaganda, were mentioning the names of a few American prisoners each night. Mr. Gordon began to jot down the names and sent postcards to the parents of the men named. The response to these cards was so instantaneous and gratifying that Mr. Gordon induced others to join with him in a Short Wave Listeners Club—each member of which is allotted a definite time at his listening post.

Mr. Gordon feels that he is amply repaid for his trouble by such grateful expressions of appreciation he has received: "It is a patriotic service for which I cannot thank you enough" . . . God bless you for your kindness" . . . and other similar statements received by this Midwest Radio owner.

Just another case where a Midwest Radio, famous for its ability to pull in long distance stations even under the most adverse conditions, is doing yeoman duty, until Victory will permit us to turn from our production of radio and electronic devices for our Armed Forces and resume the manufacture of finer radio receivers—at lowest Factory - To - You prices and at savings up to 50%.

BUY MORE WAR BONDS

SEND FOR FREE INTERNATIONAL TIME CHART CALENDAR

An attractive 4 - color calendar with International Time Calculator will be sent FREE on your request if accompanied by 10c in stamps or coin for which we will send you a War Savings Stamp.

MIDWEST RADIO CORPORATION
DEPT.   11-G   ESTABLISHED 1920 CINCINNATI, OHIO

The POW messaging activities of a satisfied customer were featured in this Midwest Radio Corp. advertisement.

sages, which were read in the station's studio. They copied down the prisoner's name and address, and notified families by letter, postcard or telephone. A few, who were equipped to record such messages on discs, sent the discs. Sometimes a family would hear from as many as 100 people who had heard their loved one's message. Although the government provided like monitoring and notification, the news usually came first from a shortwave listener, and the initial contacts were often touching. "I will never forget the scene when I called to deliver [a message in Newark] in person," wrote James J. Hart of Irvington, New Jersey. "It seems that the boy, before entering our armed forces, lived with his mother and grandmother, both of whom had believed him dead previous to my visit."[24] An NNRC member in Bloomfield, New Jersey observed:

> In a radius of 400 miles I have been delivering messages via phone. One recent call to Ohio was the first direct contact from an Air Corps man to his family in more than three years. He was one of the first Jap captives. This family had two phones in its home, so it was a three-way conversation. Such is the satisfaction we derive from doing this work. When this family learned I was calling from New Jersey, the phone charges were promptly reversed. And I'm afraid we didn't adhere to the five-minute period requested by the phone company, but in cases like this I am sure it can be overlooked. ¶Many cannot believe we are doing this work without compensation. When I had to apply to an official of the WPB [War Production Board] for a priority to purchase a set of crystal earphones to replace by 22-year-old ones, and after explaining why I wanted a new set, he said: "You surely are entitled to a priority, as this is a damn fine service you listeners are doing." He then spent a half-hour reading some of the letters I received from parents and relatives of our men now prisoners of war.[25]

Listeners in other countries monitored POW messages as well, and the practice was resumed by New Zealand DXers during the Korean War when POW messages were broadcast over Radio Peking and the voices were those of the POWs themselves. Within a year of the war's start, New Zealanders had handled upwards of 1,000 messages to families in many countries, especially the U.S.[26]

World War II ended in 1945. There was great hope among broadcasters. As William R. Reid, Acting Director of the BBC's North American Service, put it, "[i]f the radio organizations of the world will take advantage of the technical advances in short-wave broadcasting and news reporting brought on by the war, we should embark on a postwar era in which radio will play as vital a role for the maintenance of peace as it has in the war against fascism."[27]

Listeners were similarly hopeful. The NNRC opened its October 1945 bulletin with this message:

> The advent of October marks the beginning of another DX season, the first one in peacetime after nearly four years of war. Already there is manifested a quickening of interest in matters DX and the consensus is that the 1945-46 season will be one of the best since the inception of that great utility, radio. The boys who have served so valiantly in the armed forces are returning home in increasing numbers and the clubs that have carried on so well under adverse conditions will soon be back to their normal active strength.... To add to the picture is the certainty that before long those new and more efficient receivers we have heard about will be available to the dialer of distant stations.[28]

While advances in receiver design would lag, the best days of shortwave listening were just ahead.

2

The Shortwave Audience

Q: What was your favorite place to DX from?
A: New Zealand, I think, because of the general enthusiasm of the DXers I met
there, as well as the standard of the DX itself. Finland (especially north of the Arc-
tic Circle) comes a good second for the same reasons. It's probably no coincidence
that these are about the only two countries in the world where you can say at an
average dinner party that you're a DXer and find at least one other person who
knows what that is (and doesn't confuse DX reception with ham radio) because he
or she has a friend or relation involved in the hobby or has read about it in news-
papers.

John Campbell, 1985[1]

All shortwave listeners in the United States are familiar with a basic shortwave anom-
aly. If the large shortwave stations have tens of millions—sometimes over 100 million—
listeners, why don't we ever run across one? Shortwave listeners know from experience
that, outside of shortwave circles, the chances of finding someone with knowledge of shortwave broadcasting is slim. Explaining the medium invariably leads to its confusion with
ham radio. While in modern times some of this can be attributed to the presence of competing delivery vehicles in an increasingly rich media environment, people's reactions were
not much different a half century ago.

Identifying the shortwave audience and answering basic questions, such as how large
it is, and what stations and programs people listen to on shortwave, is at best an inexact
process. Even the most reliable of the available techniques are convincing only in degrees,
and at the end of the day we are left still wondering.

Understanding the distinction between shortwave broadcasting and international
broadcasting, and the division of the organized shortwave listening community into program listeners and DXers, is important. For many years, shortwave broadcasting and international broadcasting were basically synonymous. Except for some cross-border medium
wave broadcasting, all international broadcasting was done via shortwave. International
broadcasting *meant* shortwave. Further, among shortwave listeners the terms typically
included both those stations whose programs were intended for foreign audiences, and
domestic shortwave stations which served mainly local listeners but whose signals were
heard internationally due to the long-distance properties of shortwave transmission.

Today, shortwave broadcasting and international broadcasting mean different things, shortwave being but one vehicle for reaching international radio audiences. Others include satellite broadcasting, internet broadcasting, and the increased use of local AM and FM stations, and cable, to rebroadcast programs from other countries. The availability of these alternate platforms, together with the increased number of local radio stations worldwide and the improvement in the quality of their programming, have had a dramatic influence on the size and makeup of the general shortwave audience. As one observer has put it, "international radio has been taken out of the ghetto ... of shortwave broadcasting and into mainstream society, mainstream life.... [A] majority of the developed world's population has never switched on a shortwave radio service. If you walked down the street in Washington, D.C., or London, and stopped ten people ... I wonder how many of them would actually say: 'I've got a shortwave radio and have listened.' I suspect the answer is one out of ten, if you're lucky."[2]

Among shortwave enthusiasts, a distinction between program listeners and DXers grew up over the years and bears on questions of audience size and composition in various ways. It originated in the 1920s, soon after the relatively homogeneous body of tinkers called "radio amateurs" divided into "transmitting amateurs," or ham radio operators, and "BCLs," or broadcast listeners. Program listeners are those "BCLs" who listen to shortwave for content and who find shortwave programs distinctive and enjoyable. For DXers it is the thrill of the hunt, the romance of hearing distant signals, weak and battered as they may be, that is the main attraction.

The two groups are not mutually exclusive. All DXers are program listeners to some degree, and most program listeners at least dabble in DX. In a major survey of the NASWA membership conducted in 1976, about 40 percent viewed themselves more as program listeners than DXers. Another NASWA survey, in 1995, put the DXer/program listener dichotomy among members at 50–50. In a series of annual surveys of the readers of *Review of International Broadcasting* conducted between 1981 and 1986, roughly two-thirds indicated that they spent 80 percent or more of their time listening to content rather than DXing, a result broadly echoed in a survey of the attendees at the 1994 Winter SWL Fest.

However, the largest body of "shortwave listeners" consists of those around the world who belong to no club and who listen to shortwave because it fills a need that is not being satisfied by their local media. They may be motivated by many factors: the desire to hear a different point of view; to be informed; to be entertained; to be inspired by faith; to be confirmed in their political beliefs; or to escape the increasingly homogenized world of domestic broadcasting.

Over the years, there have been some who have felt that shortwave broadcasting could serve even grander purposes.

> Short wave radio, by bringing the different parts of the world closer together, offers the greatest means yet known to man to bring our people closer together, to break up the habits that bring on war and to create a better understanding between the various races of the world. Nothing before was ever so potent as a war-preventing, peace-building hobby. By getting a better idea of how the others in this world of ours get along, we become more tolerant to their faults, more understanding to their habits. And by becoming more tolerant and more understanding we become better friends.[3]

However, within the overall shortwave listening audience, it is the organized shortwave listeners—club members, magazine readers, and other "active listeners"—who have

Late-night radio sessions were common among DXers and frequently the target of radio humor.

defined and organized the listening side of the shortwave experience and been the most visible part of the audience. And among all of these armchair adventurers, it is the DXers who have formed the core.

To the DXer, shortwave usually has little to do with programs. It is about stations—hearing them, and (often) QSLing them. The further the distance and the lower the power, the greater the challenge. While reception reporting requires that the program be described in some fashion, understanding the content is not important, and so language makes little difference (save for the ability to identify the language, a skill which many DXers develop). The object is to hear the station, and, for most DXers, to QSL it and put another notch in one's belt. The following poem, written in 1939, captures the essence of DXing the shortwave bands, and still rings true.

"The Lament of the Sleepless Knight"

by Capt. E. N. Massey[4]

The foreign foemen will not stand; they flee my trusty lance.
  The cowards lurk 'neath QRM, as witness Fort de France.
I had a go at SPD — we only had one fall,
  But lo! When back to him I turned, he wasn't there at all!

The Cuban dastards are the worst; change places every week;
  I hunt the old familiar spots and do not hear a squeak,
But when I think that I've discovered something that is new,
  I strain my ears to listen, and I hear "COCQ."

The knights of Daventry go by, contemptuous of my skill,
  They sneer at all my challenges and swear they always will.

I bid them stand and fight with me but still they pass me by
   With heads erect they calmly say, "We do not verify."

On JZJ in Tokyo I've squandered quite a heap
   Of dimes and patience, to say nothing of my morning sleep.
I've penned them hopeful cartels, just itching for a fight,
    How can my honor be increased when they won't even write?

HH3W as well, is a thorn within my side;
   Challenging all listeners, they later run and hide.
I hammer on their portals, but silence then ensues,
   What wonder that I am beset by symptoms of the Blues?

I'm told that Delhi and Bombay are worthy of my steel,
   Invading both their frequencies, I've asked them how they feel.
I'd stake my DX honor on just one fight with them,
   Alas! The only answer is a lot of QRM!

ZIK2 is lost to view, or rather, to the ear,
   He signaled loudly, but of late, has hushed through starkest fear,
Some messenger told him of me and made him change his mind,
   It's enough to make a valiant knight lose faith in all mankind.

I'm not like Alexander, with no more worlds to whip,
   The world is full of foemen but they all give me the slip,
There are many foreign warriors who should doughty champions be
   But if they just ignore reports, how can I VAC?

While active shortwave listeners, whether interested in programs or DX, are a small fraternity, and nearly invisible outside the shortwave community,[5] they never cease won-

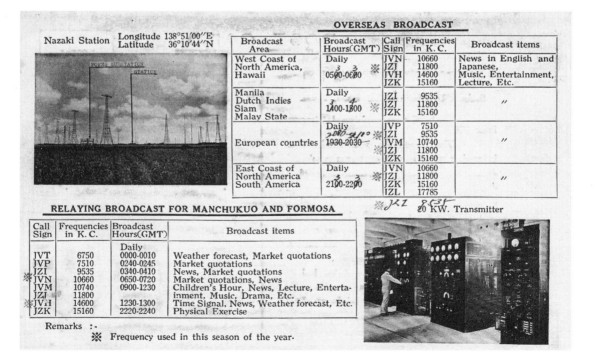

JZJ, a channel of the Broadcasting Corp. of Japan, was one of many stations of the 1930s mentioned in the poem "The Lament of the Sleepless Knight."

dering why others do not share their enthusiasm. Publicizing shortwave was a major goal of the Association of North American Radio Clubs, an SWL umbrella organization, from its inception in 1964 until it closed in 2005. The Handicapped Aid Program was formed in 1972 specifically to promote shortwave listening among handicapped persons burdened with a sedentary lifestyle. And Tiare Publications proclaimed a "Shortwave Radio Week" for March 12–18, 1990 and several years thereafter, albeit without much effect.

There have been some successes in interesting others in shortwave listening. The Ontario DX Association reaped large membership increases as a result of promoting short-wave listening at hobby fairs and similar events, and in making its monthly bulletin available at radio stores. Some other clubs had limited success with similar activities. Overall, however, publicity efforts had little effect, and wiser heads accepted the anonymity of short-wave listening as the natural order of things. One knowledgeable observer pointed out that SWLs were connected mainly through club bulletins and lacked the feedback and interaction normally attendant to ham radio or citizens band radio. "Can you seriously imagine a group of people, say a family or some friends, sitting around and listening to Radio Moscow? Radio Peking? Or even Radio Nederland or the BBC? ... [P]resent SWLing to the masses and the reaction will be a big 'so what?'..."[6]

Professional Audience Research

Historically, the principal efforts to define the international broadcasting audience have been conducted by the BBC and the U.S. government. The BBC began domestic audience research in 1936.[7] Informal research for the external services began in the mid–1940s. It became more comprehensive in the late 1960s, with weekly global audience figures added in 1974. VOA audience research was done first by the VOA itself, then by the Office of Research at USIA, VOA's parent organization for many years (1953–1994), and, since 1994, by the International Broadcasting Bureau. Radio Free Europe–Radio Liberty (RFE-RL), Radio Free Asia, Deutsche Welle, and Radio France International also do research, as do the Australian Broadcasting Commission and the Canadian Broadcasting Corp. for Radio Australia and Radio Canada International respectively.

In general, professional audience research attempts to determine how many listeners a station has, how often they listen, what they listen to, and, if possible, what political, cultural or opinion groups they represent. The information is intended to be used to improve the focus or the quality of a station's broadcasting. Methodologies common in measuring domestic listenership, such as diaries and electronic devices, typically are unavailable in international audience research. Quantitative research is usually conducted through prompted questions in face-to-face surveys, while qualitative research uses focus groups, focused interviews, questionnaires, listener panels and the like. The research is conducted by the broadcaster or by a contractor such as Intermedia, a Washington, D.C., firm with roots in RFE-RL that performs much of the audience research for U.S. government broadcasters. Surveys may be targeted to media issues or they may be part of a broader survey.[8]

Research targeted to listening on shortwave as opposed to other media has become increasingly hampered by the changing dynamics of international broadcasting. When shortwave broadcasting was the only means of reaching foreign audiences it was the sole target of the research. Today, in addition to shortwave, international broadcasting encompasses rebroadcasting over local AM and FM stations, satellite radio, internet broadcasting, cable radio, and television. Stations are interested in knowing about all the vehicles by which they reach their audiences. However, distinguishing the listenership for each is sometimes problematic.[9] Listeners may not recall by what means they heard a particular station, and in some parts of the world, especially underdeveloped countries, listeners may not even know the difference between, for example, shortwave and medium wave, and may be unfamiliar with the particular bands to which they are tuned.

Graham Mytton, for 14 years Head of Audience Research and Audience Relations (and later Controller of Marketing) at the BBC World Service, and a widely recognized expert on audience research in international broadcasting, working from surveys encompassing 77 percent of the world's population, has calculated that approximately 2.2 billion people live in the over half-billion households with a shortwave radio, and that about 250 million people over age 15 listen to international shortwave at least once a week. An additional, undeterminable number listen to domestic shortwave. In 2002, the most recently available worldwide data indicated that the percentage of radio households that have shortwave radios is 60–65 percent in Africa and the Middle East and 30–40 percent elsewhere (except North America and the Caribbean, where it was said to be about 10 percent).[10] Traditionally, many radios in Europe have had shortwave bands that have largely gone unused. And quaere how many radios with a shortwave band are truly capable of dependable reception.

The central finding of surveys taken over the years is that listening to shortwave, particularly international (as opposed to domestic) shortwave stations, is most popular in countries with a deficient media environment — few stations, or heavy government control and lack of credibility — or where there is a special reason to listen, as in times of civil disturbance. In politically stable countries with media environments that are free and that offer choice, and with better reception via local transmitters, listeners prefer domestic radio, and the audience for international broadcasting is small. In short, people in the general population who listen to international radio do so because they have to.

Other basic findings on shortwave broadcasting are as follows:

- The large stations claim huge audiences. In 2002, the BBC World Service reported that it reached over 150 million people at least once a week when all languages and all media (not just shortwave) are combined. That is about twice the 1980 estimate. Its English-language audience was about a third of the total. The VOA's reach was over 100 million, a figure that has more than doubled since 1967. For Deutsche Welle it was 90 million. The BBC and VOA figures for 2007 were 183 and 116 million respectively (BBC and VOA use different methods in making their calculations). Audiences for other stations are much lower, but often still large by commercial radio standards, e.g. Radio Canada International, 4.1 million.[11] There is little station-specific research data for most stations, however, because most do not conduct professional audience research.

- Listenership tends to be higher among opinion leaders and the better educated. Whether the cause or the result of this phenomenon, international broadcasting emphasizes news and current affairs rather than entertainment.[12]

- The level of listenership is low or very low in developed countries, higher in less developed areas. In the United States it is under 1 percent, probably considerably under.

- More men than women listen. (This is reflected in shortwave club membership, which is overwhelmingly male.)
- Shortwave listening tends to be higher in countries where shortwave is used for domestic broadcasting. As domestic shortwave is replaced by FM and other media, shortwave listening drops off.
- The shortwave audience has been declining over all. The main reason for this has been the reduced regulation of the media environment and the growth of private broadcasting in many countries, particularly since the fall of the Soviet Union. As local listening options increase, there is less shortwave listening.
- Shortwave audiences for all but the largest broadcasters tend to be scattered and small, and are often located in rural, low-priority areas. In their total numbers, however, they still constitute a large number. If a station reaches only one person in 10,000, on a global scale that translates into hundreds of thousands of people. This is difficult to prove because many listeners are in areas that tend to be under researched. The tendency to conduct research in urban environments may slant the results against shortwave.[13]
- In recent years, shortwave has had its broadest use in Africa and Asia. Although the situation is changing, for a long time Africa has had a relatively large shortwave audience, in part because domestic stations in Africa began using shortwave many years ago and shortwave sets were common. Based on the latest available data in 2002, Mytton put the number of shortwave households in Kenya at 90 percent.[14] He estimated the BBC audience in Nigeria alone as 17 million in 1998, the VOA's at 14 million.[15]

Listener Letters

Those who write to stations are atypical of the general audience in what is basically a one-way medium. Thus, estimating the size of the audience by counting letters is not reliable, and often a mix of speculation and optimism. However, because most stations lack the resources for professional audience research (save for the occasional listener questionnaire), reliance on listener letters is often the main vehicle available for gauging success.[16] Within nine months of its founding in February 1942, the Voice of America was broadcasting 24 hours a day, seven days a week, in 24 languages, via direct shortwave, recorded disc, and relay arrangements with stations in allied and neutral countries (mostly medium wave but some shortwave as well, including such DX sites as the Belgian Congo). The size and nature of the VOA audience was estimated principally from letters, the comments of travelers from overseas, and reactive comments to VOA programs by news media in the targeted countries.[17]

Letters do provide good data on reception conditions, and they also provide information about those listeners who take their radio listening seriously enough to write. Listener letters tend to be of value more in a qualitative than a quantitative sense, since they may include comments about particular programs. Whatever their value, many stations track at least the gross number of letters received, and their numbers can be huge. A small but well-known shortwave (and medium wave) station in Costa Rica, TI4NRH, claimed in 1942 to have received 860,000 listener letters during its 13½ years of operation. In 1969,

Radio RSA in South Africa received almost 24,000 letters. By 1988 the number had grown to 120,000. The figure for the BBC External Services was 300,000 in 1974,[18] 340,000 in 1982. Radio Moscow reported receiving 300,000 cards and letters in 1980.[19] Deutsche Welle received the same number in 1982. In 1983, HCJB received 75,000 letters, and the following year Radio France International received 127,000 letters from 174 countries. Radio Japan received 80,652 letters in 1986–87, of which a modest 4,019 were from the United States. The year 1986 brought Deutsche Welle 390,775 letters, including 15,000 from North America. And in 1993, China Radio International received over half a million letters, the great bulk of them from Asia. In more recent years, e-mail has become a major point of listener contact.

The American Shortwave Audience

The shortwave audience in the United States is one of the smallest. Accurately establishing its size, or that of a surrogate indicator, the number of shortwave receivers, has always been more art than science, and a very rough art at that. Figures, seldom explained or authenticated, or easily comparable, have been all over the lot.

In 1942, the NNRC reported that 35 out of 50 million radios in the United States included shortwave bands.[20] In 1970, a "recently conducted survey" was cited as authority for the proposition that there were more than 25 million shortwave receivers in the country.[21] The following year, two knowledgeable observers estimated that there were two million shortwave receivers in the United States.[22] In 1973, a Gallup survey conducted for Radio Canada International concluded that about 13 percent of adults in the United States, or 18 million people, reported owning a shortwave receiver.[23] A second RCI–sponsored study put the figure at 11 percent in 1975, the same year that the BBC was reported to have estimated that 16 million Americans owned shortwave sets.[24] In 1977, BBC audience research was cited for the proposition that around 10 million adults in the U.S. had a shortwave receiver in their home or car.[25] The extent to which any of these figures include ham radio receivers that are not used principally for listening to shortwave broadcasts is not known.

Whatever the number of shortwave receivers, it is certainly far larger than the number of people who listen to broadcasts via shortwave. In 1971, the editor of *Popular Electronics* summed it up. "Statisticians can tag the number of smokers or boat owners," he said, "but no one can hazard a guess as to the 'depth' of SWLing. Is SWLing of interest to only 5000 of the 200,000,000 people in the U.S., or do 750,000 or even 1,000,000 take to their shortwave receivers at some time each day?"[26] Another knowledgeable observer thought the figure was probably less than 500,000 in 1990.[27] No one knew for sure, although it is now universally agreed that the number of North American shortwave listeners is very small.

There was some early academic research on the subject which expresses some audience-related themes that still echo.

The first published examination of the question appears to have been that of Harwood L. Childs, Director of Research at the Princeton Listening Center,[28] who in 1941 examined five then-current studies on the subject: two nationwide studies— one done in 1940 for the Columbia Broadcasting System and another done in 1941 for the Princeton Listening Center by the American Institute of Public Opinion — and three local studies done in Baton Rouge, Louisiana (1940), Princeton, New Jersey (1941), and Erie County, Ohio (1941).[29] Childs' goal was to determine who was listening to shortwave broadcasts from Europe and why, and to analyze the impact of the broadcasts on American public opinion. While the research methods were somewhat rough, and not always comparable across studies, Childs felt some concrete conclusions could be made, including the following:

- The number of people who listened to European shortwave at least once per month was roughly 5–10 percent of the population. However, the number of people who listened "seriously and regularly" was thought not to exceed one percent.[30]
- Roughly twice as many people listened to shortwave in New England as in other parts of the country (perhaps owing to the comparatively high quality of reception from Europe).
- The higher one goes on the economic and educational scales, the greater the number of shortwave listeners. Thus the significance of shortwave listening is not just a matter of number of listeners, but also the influence of the listeners, who tended to be opinion leaders.
- The number of listeners decreased after age 50, and there were many more male than female listeners.
- Listeners did not tune in very often, and they tended to be haphazard in their selection of programs. Generally they tuned to whatever they could hear rather than seek out particular programs. The bulk of listeners listened mostly to news, preferring local stations for entertainment or educational programming. The BBC was the most often listened-to station, with the daily audience in the U.S. thought to be about 438,000.
- There appeared to be considerable turnover in shortwave listeners, partly due to the difficulty in obtaining information about shortwave programs. Many returned to domestic radio after the novelty of shortwave wore off or as they realized the propagandistic nature of wartime shortwave programming. Many who stuck with shortwave were "radio fans" who also listened to a good deal of domestic radio.

What was the impact of shortwave *programming*? The Baton Rouge study found listeners to be skeptical of the accuracy and impartiality of news heard on shortwave. "Nine times out of ten they think the [shortwave] program is essentially a propaganda device, and they therefore probably listen to the news programs critically and distrustfully if not actually with negative attitudes."[31] Respondents in a small study conducted in 1961 expressed similar sentiments about Radio Moscow but felt that exposure to such views helped them develop a more balanced perspective.[32]

Another interesting study, conducted in 1941, focused on the "North End" of Boston, then a poor community composed almost entirely of first and second generation Italian immigrants. The researchers expected to find that shortwave listeners were distinguishable based on their interest in news or current events, their use of Italian in reading and speaking, or the degree to which they were "serious radio listeners." But these factors did not seem to matter much. What the study found was that regular shortwave listening often correlated with several ethnic-related factors. One was "militant Italianism" — strong identification with Italy and the Italian side of public issues, resentment toward America over job discrimination and the like — not an insignificant problem at the time — and general insecurity over ethnic marginalization. Another factor was skepticism about the accuracy of the news in the domestic media. A third factor was the willingness among shortwave listeners to criticize domestic radio and to point out faults and suggestions. Another was the enhanced prestige in the community that came with being, or at least appearing to be, better informed through sources of information from home rather than from the U.S.[33]

The "North End" study was unique in its focus on a small population with distinct ethnic characteristics, and probably not replicable in 21st century America. However, its findings do suggest the interesting question of whether factors similar to the hostility and discontent of that time are behind today's use of private U.S. shortwave stations by domestic right-wing groups. Is there an attraction to shortwave broadcasting among the disaffected?

Of perhaps greater interest to shortwave enthusiasts than ethnic considerations was the "North End" study's observations about the "casual" shortwave listeners it encountered.

> [T]he casual listener does not have a fixed preference for a particular nation's programs to nearly the same extent as the regular listener. Unlike the regular listener who tunes to the same station even when conditions are poor, he may listen to the BBC, Berlin, Rome, South America, or Boston's WRUL — depending upon atmospheric conditions. The gratification he gets, moreover, will not be the same as that gained by the "regulars," for it is really amusement rather than confirmation of beliefs or bolstering of the ego which he seeks from his pursuits. The "casual's" attitude toward foreign programs is best compared with the attitude of radio listeners during the 1920's toward programs from out-of-town. It's "fun" to get Europe, just as, in the old days, it used to be "fun" to tune in San Francisco from the East.[34]

The American shortwave audience was the focus of a study undertaken in 1966–67 by Don D. Smith, then Associate Professor of Sociology at Florida State University, and published in 1970,[35] by which time 150 hours of shortwave broadcasting were being beamed to North America daily and the annual sales of shortwave receivers was said to have reached two million. This study targeted adults who had listened to shortwave for news and discussions of world events during the previous year. (It specifically excluded hams and "shortwave listeners.") It found that 6 percent of the sample fell into this category, 9 percent if programming other than news and world events was included. The 6 percent dropped to 2 percent if limited to those who listened either once a week or more, or one to three times a month. The 2 percent projected to a national audience of two million, one million of whom listened once a week or more. As to the composition of the audience, the results were similar to those reported by Childs in 1941, although the west was by now better represented than New England, and declines in listenership based on increasing age and lower income were not noticed.

Excluding the roughly one-third who listened to shortwave as a result of their personal ties with a particular country, Smith found that the rest were characterized by high interest in current events. And they were a steady, consistent audience (two thirds had been listening for at least three years) which focused on a limited number of deliberately-selected stations but usually listened for only brief periods, and most often listened alone. And the listeners often felt that while shortwave broadcasts were sometimes propagandistic, they nonetheless broadened their understanding of world events. The study concluded that "the American international political broadcast audience is still there, in notable numbers and with notable consistency," and that "[d]espite the importance attached to the ideological struggle between nations, it is a mass communications audience that has been virtually ignored for over a generation."

Most of the above studies were intended to gauge the size and composition of the American shortwave audience in general rather than the subset of known shortwave aficionados. However, a 1975 study of the membership of the Newark News Radio Club, clearly a group predisposed to long distance radio listening, made an interesting point regarding the importance of these "professional" shortwave listeners.

> The appeal of DXing, or tuning the shortwave bands for distant broadcasting services, was high among these respondents but the desire to acquire news and information from international broadcasting services was ranked significantly higher than previous research would lead us to believe. Thus, the hobbyist listener (the SWL) should not be dismissed as a mere technician. Interestingly, the primary difference between the hobbyist and the non-hobbyist shortwave listener of international broadcasts apparently was in the amount of listening. The hobbyist listened to over two hours per day while the non-hobbyist often listened less than once per month. ¶This study, then, demonstrates that the regular shortwave listener is an important target audience for international broadcasts. Educated, modestly affluent, and *interested* in international news and information sources, the SWL may widely disseminate program content to an even larger audience.[36]

Less charitable to shortwave listeners have been some in the ham radio community who have always been protective of their frequency allocations, and vocal about the importance of ham radio and about interference from other stations, including shortwave broadcasters. Said Wayne Green, editor of *CQ*, commenting on the results of a poll taken by the magazine in 1957:

> These broadcasters are fighting real hard for frequencies, but who is really listening? Sure, there are a few hundred high school kids who tune the short wave bands in order to get verifications from the shortwave stations, but does anyone ever turn on a foreign broadcast station and sit down and listen? I have never heard of anyone who did.... Perhaps, if the small size of the audiences for these broadcasts were known, we could open the 19 and 25 meter bands for ham use. ¶Shame on us. Here are a dozen or so countries who are spending millions of dollars and fighting desperately for more and more frequencies to send political broadcasts to us that we aren't listening to.[37]

Although NNRC shortwave editor Hank Bennett, himself a ham, suggested that Greene may have written this just to stir up some response from SWBC listeners, he acknowledged that the number of SWBC devotees was small compared to the number of hams, but pointed out some of the benefits of SWBC listening, such as learning more about other cultures, getting news first hand, and keeping in touch with other lands. NNRC members added their views. Didn't the hams have enough frequencies already? And as to listening only to get QSLs, how about those ham contacts that lasted but a few seconds and consisted of nothing more than an exchange of signal reports and "please QSL"?

Green countered that the SWL receiver market was almost extinct — a situation which, if it was true, would change radically in the coming years — and that propaganda broadcasting — "what other purpose can the foreign governments justify for such expense?" — was taking up much more spectrum space than the number of listeners warranted.[38]

Green returned to the subject two years later.

> Who do you know, in the United States for example, who listens regularly to commercial shortwave broadcasts in preference to looking at TV or listening to the local medium wave BC station? We bet, not many! ¶Those blind SW broadcasts emanating from Moscow, Paris, Cairo, Stockholm, Tokyo, Rome and other countries on a 24-hour per day basis have few listeners

except the monitoring stations of the news services (if a few facts are considered). ¶Some of the facts? So-called all-wave receivers cost *money*, even in the United States where the production of electronic items surpasses every country in the world. Even if the average family throughout the world could afford one they usually only use the medium wave band. In certain countries, consumer type radio receivers are designed and built about local frequency groups; that is, so as not to receive certain international shortwave bands. The same countries with jamming equipment manage to keep those lucky enough to own or operate an all-wave receiver from hearing a large percentage of overseas broadcasts. (They also manage to transmit harmonics into our international hambands.)[39]

Green's conclusion was that there are more hams than SWLs, that the number of hams was growing, and that in the fight for more frequencies, the hams were more important than either the SWLs or the shortwave broadcasters. "We firmly believe that it is better to have a large group of people within a country contributing directly to a nation's welfare (technically and culturally) than a lot of blind broadcasting SW BC stations to which the majority of the world's population do not listen."[40]

The American Radio Relay League argued similarly in 1978, referring to the regular audience for shortwave broadcasting as "vanishingly small" and citing an ARRL–commissioned SRI International (formerly the Stanford Research Institute) study. "Reduced to a single comprehensive statement," said the report, "this study clearly shows that any demands made by HF broadcasters for increased spectrum *due to increased audience demand* simply cannot be supported by the information now available."[41] Larry Magne, who would later found *Passport to World Band Radio,* a basic shortwave reference, offered a strong counterpoint, challenging both the methodology and the conclusions of the SRI report.[42]

Over the years the debate became less contentious. By 1991, hams were being urged to give shortwave listening a try. Said one author in the ARRL journal, *QST*: "Now, in the 1990s, shortwave listening is worth rediscovering as a medium for obtaining legitimate news and views *directly* from countries around the world.... * * * [S]hortwave broadcasting can provide news of world events as no other medium can. * * * Now, even state-controlled shortwave voices may speak with a candor that astonishes and refreshes."[43]

Station Popularity Polls

For many years, radio clubs and radio publications have conducted shortwave station popularity polls among their readers and members. While not scientific, they do offer snapshots of opinion from groups of committed listeners. One of the first was done in 1949 when the U.K. publication *Short Wave News* polled its program-listening members. Among the most popular stations were Radio Netherlands, Radio Australia, HCJB, WRUL, and OTC in the Belgian Congo, a station which, like Radio Netherlands, cultivated its audience.[44] In August, the magazine started carrying a monthly column called "...From the Month's Short Wave Broadcast Programmes." It concentrated on the programming that could be heard over various stations. (It was re-named "Guide to Short Wave Broadcast Programmes" and made quarterly in June 1950.)

Other polls followed, including many in North America. The results of a selection of polls spanning more than half a century are shown in the table below. All the polling organizations are North American save for the International Shortwave Club (ISWC), which was British.

LONDON
CALLS
THE
WORLD

Strong signals and good programming contributed to the BBC's consistently high popularity among shortwave listeners.

Popularity Polls, 1950–2004[45]

		No.	Most Popular	Second	Third	Fourth	Fifth	Sixth
1950	ISWC	—	OTC-Congo	Australia	Switzerland	CBC	BBC	Netherlands
1953	ISWC	—	BBC	Australia	Switzerland	Belgium	CBC	Netherlands
1956	ISWC	—	Australia	Switzerland	BBC	CBC	Netherlands	VOA
1959	ISWC	—	Australia	Switzerland	Netherlands	BBC	CBC	VOA
1962	ISWC	—	Australia	Switzerland	Netherlands	BBC	VOA	CBC
1965	ISWC	—	Netherlands	Australia	BBC	Switzerland	CBC	VOA
1968	ISWC	—	Netherlands	BBC	DW	Switzerland	CBC	VOA
1971	ISWC	—	Australia	BBC	Netherlands	VOA	DW	R. Japan
1974	ISWC	—	Netherlands	BBC	VOA	Australia	DW	RCI
1976	NASWA	754	BBC	Netherlands	RCI	Australia	HCJB	R. RSA
1977	ISWC	—	Netherlands	BBC	VOA	DW	Australia	Sweden
1979	RIB	—	BBC	CBC No. Svc	–	–	–	–
1980	RIB	132	BBC	CBC No. Svc	New Zealand	–	–	–
1982	RIB	104	BBC	RCI	Netherlands	Switzerland	VOA	Australia
1983	RIB	207	BBC	RCI	Netherlands	Australia	VOA	R. Moscow
1983	CIDX	94	BBC	RCI	Netherlands	Australia	HCJB	WRNO
1984	RIB	236	BBC	RCI	Netherlands	Australia	VOA	R. Earth
1984	SPEEDX	93	Australia	BBC	Netherlands	HCJB	R. RSA	DW
1985	RIB	395	BBC	RCI	[Tied: Australia-Netherlands-VOA]			[DW-R. Moscow]
1986	HCJB	–	HCJB	BBC	Netherlands	VOA	Australia	RCI
1987	ASWLC	25	BBC	Netherlands	Australia	RCI	HCJB	–
1989	CIDX	104	BBC	RCI	Australia	Netherlands	VOA	WRNO
1992	CIDX	83	BBC	Netherlands	Australia	DW	HCJB	RCI
1994	Fest	73	BBC	Netherlands	DW	RCI	Australia	HCJB
2000	CIDX	56	BBC	Netherlands	RCI	–	–	–
2001	CIDX	63	BBC	Netherlands	RCI	–	–	–
2004	NASB	96	BBC	WBCQ	Netherlands	RCI	VOA	DW

It will come as no surprise to experienced shortwave listeners that, with few exceptions, the most popular stations have been those which project a friendly, informal, western style, and have a strong signal.[46] Some listeners may spend much of their time hunting elusive stations, but for ordinary listening they want an interesting program that can be heard without difficulty. This was evident in the comments that sometimes accompanied people's votes in some of the early polls.

OTC, Radiodiffusion Nationale Belge, Belgian Congo: "The International Goodwill Station had kept closely to the recipe that for an enjoyable radio dish, two parts of fantasy should be mixed with two parts of unaffected simplicity, no commercial or propaganda being allowed to offend our ears." (1950)

Radio Australia: "Excellent news coverage of world-wide events without the large doses of obvious propaganda and political advertisement. Excellent musical programmes with variety. The 'Mail Bag.' A word from the 'Children's Australian Diary,' and above all the cheerful friendliness of their announcers, in place of the formal and (to a listener) 'stand-offishness' of the announcers at so many British stations." "They cater for all tastes. They give encouragement to both new and old DXers alike and the programmes are free from propaganda." (1956)

BBC: "I vote for the BBC for fine entertaining programmes, presenting the news and with a high degree of fine reception. Brings us the best in radio reception over 12,000 miles, impartially presented, with a sense that short wave broadcasting can be made to make friends with everyone." (1956)

Radio Nederland: "I vote for the 'Happy Station,' Hilversum because it is so pleasant to listen to, has a style and presentation all its own, broadcasts something for everyone and a wonderful friendly atmosphere prevails throughout." "Virtually a 'One Man Show,' Eddie Startz imparts sheer spirits, exuberance and care-free happiness, with real genuine friendliness." (1956)

Swiss Short Wave Service: "For its friendly and interesting programmes." "Programmes are always interesting, instructive and perfectly reflect the way of life, people and the country of Switzerland." (1956)

Radio Canada: "I like Radio Canada because it tells me in a very short time what I want to hear about Canada. Ninety-nine times out of a hundred reception is good and rarely does it not get through at all. The schedule keeps me in touch with the wavelengths so that even if I am unable to listen for a while I do not lose the station." (1956)

Listener polls were not without some controversy. A minor storm over the ISWC rankings arose in 1968 when Radio Portugal came in in eighth place, the first year it had placed among the top ten. During the polling, the station had announced that it would offer a "gold caravel" to the listener who placed the station as No. 1 on their list and gave the best reason for doing so, an action which some participants felt should have voided the ranking. The station said it was just following the ISWC's instructions, which encouraged stations to publicize the contest and offer a prize for the best reason underlying one's No. 1 choice. Although the implication of the instructions was that a station's decision to award a prize should be announced only after it had been named No. 1, this was by no means clear. The poll results stood.[47]

In 1980, the European DX Council adopted a resolution that polls not be given on-air publicity because, while they reflect the preferences of a select group of people (those who were active enough in shortwave to subscribe to publications where the polls were announced), this was a small part of the listening audience. Radio Netherlands opined that basing a conclusion on such polls was like concluding from a bus station survey that 98 percent of people travel by bus.[48] Some stations agreed, others did not.

Programs

There is some very good programming on shortwave. Overall, however, programming has been the Achilles heal of international shortwave broadcasting. This is due to several factors: the propagandistic nature of much shortwave broadcasting during World War II and the cold war; shortwave's focus on news rather than entertainment; and the problem of producing special programming for foreign listeners, often in foreign languages, with the minimal staffing that most shortwave services enjoy. In addition, during the cold war many shortwave services enjoyed a presumptive political value, and thus security within their country's broadcasting administration. As a result, audience-related factors that would typically influence program development in a competitive, commercial media environment received scant attention. As the editor of *Voices*, a short-lived TV-guide style publication for shortwave, noted in 1981: "Program production in too many countries has been relegated to the status of a relic. The remoteness of the audience has been used to cover a multitude of sins.... * * * Today's shortwave broadcaster ... can no longer presume a tolerance for the drivel of yesteryear."[49] Around the same time, the well-known author on international broadcasting, Donald R. Browne, observed that few stations differentiated among listeners in ways that might assist in reaching them.[50]

In general, shortwave programming is divided into news; features, current affairs or a "magazine" format; music and other entertainment; and listener programs, i.e. DX and

The North American program of the Swiss Shortwave Service was one of the strongest signals from Europe.

mailbag programs. In countless member surveys taken by North American shortwave clubs over many decades, the DX programs and similar shows have always been the most popular (even among those who consider themselves mainly program listeners), as have the stations that carry them and the personalities who host them. This is no surprise, given that the subject matter of DX programs— shortwave listening — is the *raison d'etre* of shortwave clubs and the bond shared by all members.

After listener programs, news is usually the most popular, with the BBC typically getting the highest marks. The rankings of music programs are seldom better than middling. "Propaganda" and religious programs are the least popular, although well-delivered on-air propaganda has been found capable of impacting political beliefs under certain circumstances.[51]

Save for the examination of wartime propaganda broadcasts,[52] and some academic studies of the content of international broadcasting (usually news) and the differing approaches of a few of the large broadcasters,[53] there has been comparatively little organized information about the programming that emanates from international shortwave stations. In general, however, listeners know that, historically, much of it has been relatively unimaginative. As one observer has put it:

> Too many countries have spent large sums of money on broadcasting equipment that will let you hear them without improving the programs enough so that people will want to listen.... ¶Instead of complaining that too many listeners are only interested in QSLs, broadcasters should be asking themselves why their programming fails to attract listeners other than QSL seekers. Listeners are not failing shortwave broadcasters; instead, shortwave broadcasters are failing listeners.[54]

To be sure there are exceptions to the rule, and they become immediately evident to the listener. VOA, BBC, Radio Netherlands, Radio Canada International, Deutsche Welle, Radio Australia, Radio New Zealand, and Swiss Radio International (when it was on shortwave) have delivered excellent programming.

Jonathan Marks, formerly the head of the English service and Director of Programmes for Radio Netherlands, observes:

> Programming on shortwave is a mixed bag because stations have traditionally been remote from their audience and the format, for the most part, was "news and features" about the country of origin.... For the most part, international broadcasters are not really trying to share information across cultural boundaries. They make programmes in one language, translate them and chuck them in the direction of a particular continent. Their editorial charter is very limited; they have a brief to cover events in their own country.[55]

The dynamic is different in domestic shortwave broadcasting, which is usually an offshoot of a local medium wave or FM operation. Even if it does not meet high production standards, it does not suffer from the same kinds of deficiencies as international programming. However, domestic shortwave broadcasting is usually in a language that the foreign SWL does not understand, and thus a station is likely to be of interest mainly for its DX value, or its music content.

Kim Andrew Elliott, who later worked in audience research at VOA and the U.S. International Broadcasting Bureau, put the matter in perspective in 1980 and 1981 when he proposed several alternative approaches to international broadcast programming.[56] He described the traditional approach as one focused on political or politico-economic topics, and delivered in an impersonal, script-based, third-person, non-impromptu format. To

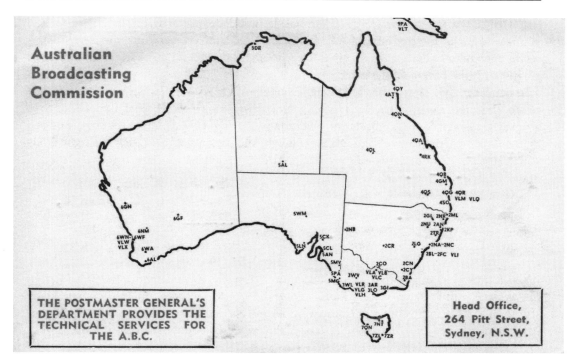

In addition to the overseas programs of Radio Australia, some of the ABC's domestic shortwave stations were regularly heard in the United States.

increase listener interest among both casual and serious listeners, he proposed greater use of extemporaneous and first-person formats, i.e. "personality-based" programs; less world news and more domestic news; more information about the broadcaster's home country; more interesting programs, including readings of short stories and separate music programs (with the focus on domestic music); fewer talks (particularly political talks), which are usually more suitable for the print media; and fewer interviews and press reviews.

While Elliott's ideas may not have had much impact on international shortwave broadcasters, their echos can be heard in post–cold war improvements in shortwave programming. Competition and existential economic and administrative threats to many shortwave services have, it seems, done what good broadcasting theory could not.

The Absence of Programming Information

Within the listening community itself, shortwave programming received some limited attention during the 1930s when shortwave was still thought to have potential as a practical communications medium that might be broadly accepted by the radio listening public. By the time shortwave listening regained its footing after the war, however, its consumer potential had been largely forgotten and the focus of the organized shortwave community was mainly on DX.

For the listener who was interested in programming, the availability of information on specific shortwave programs— times, frequencies, and content — has long been a problem. Only a few stations offered program guides with advance program information, and you had to be on their mailing list to get them. Most stations were unresponsive in supplying such material to listeners or to those who might publicize it. The situation improved greatly with the advent of the internet.

For several years starting in 1995, Ian McFarland's Marbian Productions International offered listeners a plan whereby, twice a year, registrants would receive a packet of program schedules of cooperating stations. Stations were asked to help defer the costs, and less than a dozen participated. Although at its height the packet was being sent to some 2,500 people, the plan foundered as stations dropped out, increasing the cost to others and requiring a subscription fee which few recipients were willing to pay. The coming of the internet, with its free resources, was probably also a factor.

Now many stations offer program details on their websites, if one is willing to look for them, and the larger broadcasters provide advance program details by e-mail. Prior to the internet, a number of periodicals, which are described in more detail in Chapter 4, attempted to fill the gap. The first post-war effort was in 1965 with the *SWL Program Guide*. It was followed by three other short-lived efforts: *The Shortwave Reporter* in 1973, *International Listener* in 1976, and *World Radio Report* in 1987. More successful in the 1980s, albeit still published for only short periods, were a Finnish publication, *Voices* (1980), and the American *International Shortwave Listener's Program Guide*, published from 1984 until its merger with *Monitoring Times* in 1986. Two other one-time, program-focused publications of the era were *Tune In the World* (1983), and *Muzzled Media* (1986), the latter devoted to international news.

Several "lists" focused on times and frequencies of individual programs with various levels of description but no attempt to evaluate content. These included Kraig Krist's *The Shortwave Listener's Program Guide* (1989), and John Figliozzi's *The Shortwave Radioguide*, first published in 1990 (it became *The Worldwide Shortwave Listening Guide* and in some years was sold in Radio Shack stores). A similar *Monitoring Times* publication, issued annually from 1992 to 1994, was Kannon Shanmugam's *Guide to Shortwave Programs*. *MT* had included some shortwave programming coverage even before that, and expanded it in 1998 when Figliozzi began penning a monthly column on the topic.

Two major program-oriented publications were Glenn Hauser's periodic *Review of International Broadcasting* and Larry Magne's annual *Radio Database International*. *RIB* was inaugurated in 1977 and focused on opinion and discussion about programming during its 20 year life. Varied in content, and somewhat idiosyncratic, it had many contributors and contained a wealth of information not usually available in club bulletins. *Radio Database International*, inaugurated in 1984 and soon renamed *Passport to World Band Radio*, has been a major attempt to serve, and help expand, a consumer audience for shortwave. Its coverage of shortwave programming developed into *Passport's* present hour-by-hour review of the programming of many international shortwave stations. Each year since 1991 it has also tagged "ten of the best" programs, usually emanating from the large western broadcasters but with some entries from Russia and other countries.

Shortwave clubs gave limited attention to programing in their monthly bulletins starting in the mid–1970s, with varying results. Some clubs offered just the times and frequencies of particular programs. Others reviewed programs that were part of a station's regular lineup, provided advance information in the limited instances where it was available, and printed listener comments. Glenn Hauser included some program information in NASWA's "Listener's Notebook" column from the time he assumed its editorship in 1975. By 1976, when the Ontario DX Association was but one year old, it had a program column which continues to the present day. The American Shortwave Listener's Club and SPEEDX followed with their columns in 1978. NASWA began the forerunner of the present "Easy Listening" column in the early 1980s. The Canadian International DX Club soon followed with its program offering.

However, unlike much of the rest of a club bulletin, which relies on listener contributions, the "program" editor usually had to generate his or her own material. Thus the quality of these columns was directly proportional to the talent of the editor. In recent years the club and magazine "programming" editors have become among the best sources of listener-generated information about shortwave programming.

3

Clumbs

Radio broadcasting was a mere youngster of almost six years of age and this writer about double that number of years the bitter cold evening of December 8, 1927 when this DXer started out by streetcar for a history-making rendezvous. This was the evening when Rider of the Air Waves would meet for the first time with such stalwart DXers as the Dialist, Dial Twirler, Knob Twister, Old Man Low Notes and the many others whose DXing exploits had filled the columns of "Broadcasts Winnowed" in letters to the Dialist, Charlotte Geer, distinguished editor of that column in the Newark Evening News. * * * ¶*As I stepped off the elevator into the auditorium of the* Newark Evening News *building, I found a considerable number of DXers already there.... In all, 55 enthusiasts assembled in that hall on that memorable evening. We began to know each other by our proper names, rather than the pen names of old.... We were all so enthusiastic about the excellent turnout that organizing ourselves into some kind of club seemed to be of prime importance.*

Eugene C. Bataille, 1975[1]

While the international radio stations may measure their audiences in the millions, the organized shortwave audience is much smaller.

Since the 1920s, the principal organizational vehicle for shortwave listeners has been the shortwave club. At first clubs were local. Soon, however, there were regional, national, and even international clubs. The main benefit of membership was a printed bulletin, usually sent monthly, which consisted of material submitted by members or developed by the column editor.

All clubs suffered from the same maladies. A recurring problem was their volunteer nature. Although some editors remained in their posts for decades, more often there was a need to find new volunteer talent on a regular basis as editorial responsibilities were displaced by the pressures of work, school and family, or as new editors discovered that the job required more effort than they thought.

A constant problem was member contributions. Typically only a small percentage of the membership contributed to the club bulletin. Efforts to improve member participation were constant, and many strategies were tried, usually with only limited success. This did not necessarily prevent the production of a good bulletin, however, for the small contributor base usually included some of the best-informed and the most prolific. A much larger contributor base probably would have exceeded the capabilities of most clubs. Starting in the 1970s, the most experienced DXers often found their needs better met by smaller newsletters which could get information out faster, and this robbed the clubs of the participation of some valuable contributors.

While clubs usually aspired to a democratic structure, nearly all operated on a key man basis, with one, or a handful, of people controlling events. Forays into democracy were usually unsuccessful and short lived. And while successors were sometimes waiting in the wings, the departure of a key person was usually cause for concern, save for those occasional instances where the change offered more opportunity than risk.

Many clubs with small memberships came and went, especially in the 1950s and 1960s. Among the larger clubs, in the postwar years none was larger than the North American Shortwave Association, whose membership reached about 2,000 in early 1981 and again in 1991. The Ontario DX Association had nearly 1,300 members in 1991. For the other major clubs, membership was typically smaller. And many members belonged to more than one club.

SWLs of the 1950s through the 1970s will probably be surprised to learn that the leading club for decades, the Newark News Radio Club, was not larger than it was. Prewar NNRC membership was around 2,000, about the same number as in 1934,[2] but dropped to a tenth of that during the war years. In the 1950s and 1960s, the NNRC grew to be the principal DX club in the United States, the one to which almost every SWL belonged at one time or another. While there is little hard data, references to membership figures in the club bulletin in 1963 and 1980 put the number both times at around 600, and an academic study of the membership conducted in 1973 revealed it to be 675 at the time.[3] Given the club's downhill path in the years before its demise in 1982, it is doubtful that postwar membership ever exceeded that figure by very much.

The attraction of youth to wireless experimentation and long-distance radio was widely acknowledged as early as the 1920s, and continued during most of the decades since. Many listeners active in club affairs in the 1950s would probably point to a youth orientation in the clubs of that era. The main source of new members were the news stand electronics magazines, many of which carried shortwave news. Many persons first learned of shortwave, and of shortwave clubs, through these vehicles, whose focus on electronic experimentation and on careers in the rapidly developing electronics industry attracted young readers. As a result, the clubs did enjoy an influx of young members, at least for a time. Many of today's shortwave leaders trace their interest in shortwave back to their teens.

In 1963, the ASWLC reported that 49 of 71 respondents to a club survey were under 20 years of age. Probably not much can be extrapolated from this, however, because of the small sample and the ASWLC's reputation as a club for beginners. A more valid figure may be the one quarter of 199 respondents to a NASWA survey in 1971 who were below age 20. In 1973, 14 percent of the NNRC membership was under 19, with another 8 percent aged 19–22. In 1976, 14 percent of 754 NASWA respondents were under 17, with another 18 percent aged 19–24.

The following figures from later surveys suggest an overall advance in the average age of club members.

Organization	Year	Avg. Age of Respondents
ADXR	1982	50
SPEEDX	1983	38

Organization	Year	Avg. Age of Respondents
CIDX	1983	42
ASWLC	1987	36
Fine Tuning	1987	37 (median)
NASWA	1988	42
ODXA	1993	51
NASWA	1995	48

Other available figures show a similar trend. The average age of the attendees at the 1994 Winter SWL Fest was 44. At the 2004 Fest, of 96 attendees answering a survey conducted by the National Association of Shortwave Broadcasters, 31 percent were between 40 and 49, and 29 percent were between 49 and 50. Only 13 percent were under 40.

In some quarters it was the absence of youth that was a problem, even as long ago as the 1960s. In 1965, the Newark News Radio Club was seeking more young people to participate in the club's management. "More and younger hands are needed, and they are needed now," warned the club leadership.[4] But the NNRC was an old line club, with many senior members, and it probably appeared intimidating to younger members. (The seniority of its membership is suggested by the high average age, 50, reported in 1982 for members of the Association of DX Reporters [ADXR], the club that attracted many NNRC members when the NNRC closed in 1982.)

As time went on, the graying of the club membership was clearly evident, and confirmed by the data. In 1992, the ADXR had one member under 21 years of age. In 1993, the ODXA had one under 20.

There have been a small number of women involved in shortwave clubs over the years. Overall, however, shortwave listening has been a male activity.

North American Clubs of the 1940s and 1950s

There had been a winnowing of radio clubs in the years before the war. The Transcontinental Radio Club closed in 1933. The International Short Wave Radio League in Boston began publishing *International Short Wave Radio News* in 1930, but closed thereafter. The Atlantic Radio Club was absorbed by the National Radio Club in 1934, as was the Universal DX Club in 1938. The Quixote Radio Club appears to have ceased operation in 1938. Others that disappeared in the 1930s included the Chicago Short Wave Radio Club, the Globe Circlers' DX Club, and the United States Radio DX Club.

Club activity declined greatly during the war, the result of members' military service or overtime work in defense-related industries. Defense needs meant that new receivers for consumer use were in very short supply. There was also a need to conserve paper and supplies. There wasn't much in the way of time or resources for hobbies.

Many clubs stopped functioning during the war. The International DXers Alliance was founded in 1932 and published the popular *Globe Circler* DX bulletin. It began as a broadcast band–only club and expanded to shortwave in 1934. Two years later it had 1,200 members in 50 countries. It closed in July 1943, a victim of wartime conditions.

Only two major American shortwave clubs survived the war years: the Universal Radio DX Club and the Newark News Radio Club. A third, the International Short Wave Club,

MEMBERSHIP CARD
of

No. 865 · Date 11/12/35.

'International Dx'ers Alliance'

Having Qualified Admits to full Membership

Name _____ S. R. Steele

Address _____ 3606 Homewood Avenue,

City _____ TOLEDO, _____ State _____ Ohio.

Country _____ United States of America.

District No. _____ Six.

Class _____ Official.

_____ Director

_____ Pres.

The International DXers Alliance (IDXA) was one of the best-known pre-war clubs. Its bulletin, the *Globe Circler,* concentrated on medium wave stations at first, but the focus shifted to shortwave broadcast and ham band DXing.

ceased operation in 1942 but reappeared, in the U.K., in 1946. The National Radio Club is the only American club that was operating during the war years and that has survived to the present. However, it dropped shortwave in 1944 and has been broadcast band–only ever since.

UNIVERSAL RADIO DX CLUB (1933–1961)

Founded in 1933 and headquartered in California, the Universal Radio DX Club (not to be confused with the Universal DX Club) was for years the only DX club west of the Mississippi. At first it covered the broadcast band only, eventually expanding to an all-band format. During most of its existence the club was headed by Charles C. Norton of Hayward (later Vallejo), California. The *Universalite* was published twice monthly during the months of October through April, other times monthly.

During the postwar years, the *Universalite* was 10–20 pages in length, depending on the season. By this time, broadcast band coverage had been dropped and the bulletin was devoted roughly 60 percent to SWBC (including utility stations) and 40 percent to amateur listening. Over time, the ham band coverage declined and the SWBC coverage increased, and by the late 1950s the URDXC was principally a SWBC club. (There were occasional attempts to renew broadcast band coverage and expand to FM and utility DXing, but these never caught on.)

Members' logs made up most of the SWBC section, to which was added material by country. The organization of the bulletin varied over time. At one point it included a cer-

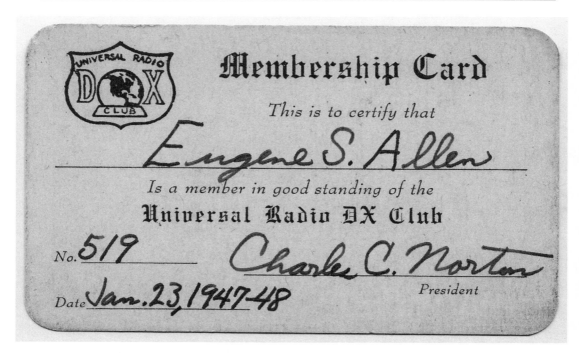

Like the IDXA, the Universal Radio DX Club changed from a medium wave club to a mainly-SWBC club. Begun in 1933, it remained in operation for almost 30 years.

tificates and awards section where members were urged to list their stations and countries verified and talk about their DXing accomplishments. At another time, in order to encourage participation, points were awarded to contributors—10 points for each bulletin in which a reporter's material appeared, and two points for each item used. A list of contributors, arranged by points earned, was published periodically.

As of 1951, the club had members in 33 states and more than a dozen foreign countries, and it counted many experienced hobbyists among its members. The club hit its stride in the mid–1950s. In 1955 it added 24 new members in one month alone. The content of the bulletin on the SWBC side was high in both quantity and quality. In addition, the URDXC featured narrative comments from members. While this was standard practice in medium wave circles, most shortwave bulletins focused on logs and other station news. The *Universalite's* "Letters to the Editor" were a nice touch, and some members had interesting stories to tell.

> I started as amateur 2nd class in 1914 or 1915 as 8JN.... [I] am very proud of the fact that I was broadcasting records on 'phone some months before KDKA came on the air. Only had 15 watts. I got the circuit from Doc. Conrad (then 8XK) when I went out to his shack and played a sax duet with Doc. Thomas on his first version of the electric organ. We played "Home Sweet Home." Although most of my interests then were in SW CW on the 160, 80, 40 and 20 meter bands, I did a little experimenting with 'phone on 5 meters.... In receivers I went through all stages of the crystal detector, DeForest Audion, Paragon, Grebe, and finally home made receivers. After I gave up ham radio, I fooled around once in a while on short waves until my wife gave me a Zenith Super Trans-Oceanic receiver for Christmas in 1953. Since then I have gone "hog wild" on SWL.[5]

Among the club's shortwave editors were such well-known DXers as Marvin E. Robbins, Al Niblack, Robert J. Hill, David Morgan, John A. Callarman, William F. Flynn, Ernest R. Behr, and C. M. Stanbury II. However, as the 1950s progressed, editors changed frequently, and at the end of 1959 a series of personal circumstances compelled Charles Norton's reduced involvement in the club. Thereafter the bulletin, much-reduced in size and content, appeared irregularly, and the club ceased operations in 1961. Charles Norton died in 1991.

INTERNATIONAL SHORT WAVE CLUB (1929–1980)

Although a British club in its postwar incarnation, the International Short Wave Club had its origins in the United States. It was founded in Klondyke, Ohio in October 1929 and operated from there (and later from East Liverpool, Ohio) until its close in 1942. The ISWC had the distinction of being the only major club devoted solely to shortwave broadcast listening. Its small but information-packed bulletin, *International Short Wave Radio,* edited by club leader Arthur J. Green, contained extensive logs reported by members, as well as other information about shortwave broadcasting stations. In May 1942, wartime conditions, including the unavailability of new receivers, the induction of many members into the military, and a drop in advertising revenue, caused the ISWC to abandon monthly publication for what it anticipated would be an irregular schedule.

The club was dormant until 1946 when *International Short Wave Radio* began publish-

Certificate of Membership

INTERNATIONAL SHORT WAVE CLUB

THIS is to Certify That ___ Mr Lawrence Lundberg ___ *is an active member of the International Short Wave Radio Club, East Liverpool, Ohio, U. S. A., and is entitled to all the privileges accorded members, including the regular monthly magazines.*

ARTHUR J. GREEN, President

Expires ___ December 1940 ___

A LIVE CLUB FOR THE SHORT WAVE BROADCAST FAN

This International Short Wave Club membership certificate dates back to the days when the club operated from its original location in Ohio. It closed there in 1942, and was reborn four years later in the U.K.

ing from London, England under the tutelage of Arthur E. Bear whose name (and memorable London address of 100 Adams Gardens Estate) would be firmly associated with the club for the balance of its existence. Bear had joined the club in 1930 and had been its European and Colonial Representative. Membership in 1952 cost $1 per year. The post–1946 bulletin had a different look from the 5 × 7", professionally-printed bulletin of its U.S. incarnation. It was now a four-page, legal-sized, mimeographed bulletin. Typically it had a page of news devoted to the doings of members and comments about club activities, followed by two pages of SWBC loggings and a page of amateur band loggings.

The postwar *International Short Wave Radio* never matched its prewar ancestor in either quality or quantity. Although the club bulletin was widely read, and during the Bear years had a significant worldwide membership (including some of the best DXers in the United States and elsewhere), it is best remembered not for its depth but for its periodic shortwave station popularity polls and its anti-jamming campaign, which began in 1956. A 100-page *ISWC Yearbook* was advertised in 1958 but never materialized.

Bear suffered a home accident in 1972. Following a six-month hiatus, the club soldiered on under the new management of Reginald Mason. Bear died in 1973 at the age of 71, and the club never fully recovered. Jim Malone succeeded Mason in 1976. By late in the decade, bulletin content had become increasingly spotty, and many of the foreign members had dropped their memberships, greatly reducing the club's visibility outside Britain. The bulletin did not appear for four months during late 1979, and by the following year the ISWC was gone.

Newark News Radio Club (1927–1982)

The Newark News Radio Club was the major American shortwave club during the 1940s and 1950s. It traced its roots back to 1927. But all clubs struggled during the war years, and the NNRC was no exception. Membership, once around 2000, dwindled to 200. During the 1940s, NNRC bulletins often began with a plea for more member contributions. There had been a total of 288 pages of material in the NNRC bulletins for 1936, the first full year of NNRC mimeographed bulletin production. In 1942 the page count was 94.

The bulletin was generally published weekly during October through April, and monthly during the remaining months. However, immediately after Pearl Harbor, the board of directors adopted a twice monthly schedule. Six months later the bulletin went monthly year round. The summer picnic-convention for 1942 was cancelled as well, "[d]ue to restrictions on auto travel by reason of gasoline (and tire) rationing." (The next such event would be in 1947.) The extensive activities of the club's "Courtesy Programs Committee" in setting up special programs from BCB stations, a staple among BCB DXers, was suspended in March 1942 when the War Censor Board banned such request programs.[6]

Although an amateur band column (as well as a statistics column) had begun in 1946, the soul of the NNRC was the broadcast band, where its charter members had started out. In depicting the "average" NNRC member after a questionnaire survey in 1947, Carleton Lord observed: "His principal interest is DXing on the BCB, which he prefers because it offers more of a challenge to tuning efforts than the easier SW and ham bands. If he ever did switch allegiance to the higher frequencies, it would be the result of difficulty in crawling out of a warm bed on a cold early morning."[7]

There had been a shortwave column in the NNRC bulletin since December 1935, soon after the bulletin's inception. It was called "High Frequencies." It disappeared in August 1942 because shortwave editor Earl R. Roberts of Indianapolis, Indiana had become Pvt. Roberts and was serving at Chanute Field, Illinois, awaiting training as a radio instructor. "In the meantime," he reported, "I am getting KP four times a week." By May 1943 he was in Wyoming giving instrument instructions to B-17 pilots.

Many members were busy with war-related activities. For servicemen who retained their memberships it sometimes took months for the bulletins to reach them. "Haven't been doing much in the DX line lately as I have been very busy with my work, a large portion of which is connected with government contracts," wrote Jake Field of San Diego, California in 1943. Don Hill of Broken Bow, Nebraska reported: "Have no time for DX or anything else except farm work. I had 135 acres of corn to harvest this season and was unable to get any help until just lately...." BCB DXer Ida Van Nostrand of Flushing, New York, observed: "I'd never ask for a verification these days from stations heard. It seems most unpatriotic when men are scarce and time is so valuable in our radio stations.... Our chief concern now is that the war will soon end and our men in uniform return safely." Corporal Elwood Borowski wrote from Camp Crowder, Missouri: "Just to let you know I am still following club activities and still actively interested in radio. Have been here for quite some time now and expect to continue with further study of radar, which is quite the thing these days."

The war years required the club to make some adjustments. One member suggested that, since many NNRCers could DX only sporadically, one week a month should be designated a "DX Week" during which members could focus some effort on listening. Club elections were postponed, and dues were reduced from $2 per year to $1. Some members volunteered to pay the dues for members who were in the service, and members were urged to buy war bonds and war stamps. Members in the service wrote in about their travels, their duties and their promotions. Their addresses were listed, and it was suggested that members write to them.

There was no let up in the good-natured ribbing among the old-timers, who still used the "handles" they had adopted over the years— Phil "Old Potato" Nichols of East Hartford, Connecticut, Lloyd "The Rooster" Hahn of Baltimore, Maryland, Carleton "Count DeVeries" Lord of Corning, New York, Howard "Brass City Night Owl" Kemp of Laconia, New Hampshire, and many others. They were devoted to the NNRC, which would acknowledge in the pages of its bulletin for decades to come the birthdays, marriages, anniversaries, illnesses, and other family events of its members. And members who died in combat were remembered.

The shortwave column reappeared in March 1943 with James J. Hart of Irvington, New Jersey as editor. Among the stations reported in his first column were Finland on 15190 kc., requesting that reports be sent to the Finnish Legation in Washington; Aden on 12115 kc., IDing at 1:30 EST ("ZNR is using a wave of 24.76 meters and is now closing down"); CR7BE, Lourenço Marques, Mozambique, on 9843 kc., "one of the best Africans"; and CR6RC, Luanda, Angola, 9470 kc., "another good African." Also reported, on 9590 and 11470 kc., was Azad Moslem Radio which "tries to sell a free India idea on its broadcasts. This station must be either in Germany or in the Near East because of its volume at 11:30 A.M. when a program is broadcast in English. Best on 11.47, the quality of this one is good

Vol. 8, No. 11.   - Page 5 -   August 1, 1943.

HIGH FREQUENCIES (EWT) - By James J. Hart

Mcs.   **Station, Location, Operating Hours, Etc.**

18.115   LSY, Buenos Aires, Argentina, program for CBS- "Pan-America Calls", from 4:30 to 5 P.M. Saturdays.

16.035   AFHQ, Allied Force Headquarters, Algiers, with press news at 8:45 A.M. Also makes schedule for CBS at 8:10 A.M. daily.

12.000   Shonan (formerly Singapore), Malaya, heard well at 11:15 A.M. July 19th with Japanese program. According to IDA, this station has news periods in English at 6:05, 9:25 and 10:15 A.M.

11.970   FZI, "Radio National Francaise" at Brazzaville, Free French Africa, is now using a new high-power transmitter and is on the air daily from 1 to 8:45 P.M. with broadcasts for Canada, North and Central and South America in English, French, Spanish and Portuguese. News periods are as follows: 4:45 P.M. in Portuguese for Portugal; 6 and 7 P.M. in French for Frenchmen in America; 7:45 P.M. in English for the United States and Canada; 8:15 P.M. in Portuguese for Brazil and at 8:30 P.M. in Spanish for Latin-America. They are received like a local station during the entire emission and verify 100 per cent. Jose Carriazo, Jr., DX Editor of Radiomania, Havana, Cuba. Best signal on the air nightly with news and comment in English at 7:45 P.M. (JJH)

11.950   ZPA5, Encarnacion, Paraguay. Good signal and announcements every 15 minutes. Easy to identify.

11.948   Radio Centre, Moscow, comes on the air at 6:48 P.M. and announces that the following frequencies will carry its program: 15.230, 15.110, 11.948, 9.880, 9.860, 9.480 and 5.440 mcs. All of these frequencies are heard nicely with 11.948 mcs. being the best. At signoff, announcer says that Radio Centre will be back at 8 P.M. on 11.948 and 9.480 mcs. with an all-English program. The 9.480 mcs. frequency is badly interfered with; likewise 9.860 for about 20 minutes from 6:55 to 7:15 P.M. when Radio Madrid is on the air.

11.755   PZX3, Free Netherlands Radio, Paranaribo, Netherlands Guiana, new SW broadcast station first heard at 7:20 P.M. July 10. There was quite a crowd in the studio at the time and each speech was applauded. Apparently it was a dedicatory program. They announce 20 minutes after the hour and 10 minutes before the next hour and in English at beginning of the program. PZX3 is between GSD on 11.750 and DXR, 11.760 mcs. On the air daily, except Saturday and Sunday, at 7:20 P.M. Earlier programs are broadcast on these two days.

11.740   HVJ, Vatican City, at 7:45 P.M. with ticking of metronome until 8 P.M. Then the sign-on with a program for Polish-speaking people.

11.670   "Radio National Belge", Leopoldville, Belgian Congo, very good from 3:15 to 5:15 P.M. Also very good from 12:30 to 3:15 P.M. on 17.770 mcs. They have two

(Continued on page 6)

HIGH FREQUENCIES (ENT) - Cont. from Page 5

Mcs.   Station, Location, Operating Hours, Etc.

{ transmitters and one may hear both carriers at 3:15
P.M. Broadcasts are in French and Flemish. (Jose
Carriazo, Havana)

11.805   COGF, Matanzas, Cuba, signed on at 6:30 P.M. July 19th
with very good signal.

11.780   HP5G, Panama City, Panama, has newscast in English at 8
P.M.

10.055   SUV, Cairo, Egypt, is now used for relay to CBS at 6:50
P.M.

9.890   KROJ, San Francisco, "Voice of America" station. News in
English at 8 A.M.

9.843   CR7BE, Lourenco Marques, Mozambique, has news in English
at 3:50 P.M. Usually signs off at about 4:45 P.M.
Good signals!

9.840   VIS, Sydney, on at 8 A.M. for CBS with story of action
in the South Pacific. This station is on nearly
every day.

9.760   DEUTSCHE SENDER ATLANTIC, with German program, signed off
at 8 P.M. with "Auf Wiedersehn" and a weird bugle
call. Good volume. This station had not been heard
previously and is not listed. However, it is easy
to identify.

9.640   WBAW, (no location), calling Army No. --- at 4 P.M. R6-7.
LRY, Buenos Aires, signing on at 5:57 A.M. daily, plays
a march before announcing. Tied up with the Argentine chain, it announces several calls for member
stations.

9.590   WLWK is the new call for WLWO, Cincinnati and is on the
air every day at 8:45 A.M.

9.570   WBOS, Boston, now using this frequency, signs off at 6 A.M.

9.540   VLG2, Melbourne, broadcasts to the U.S.A. at 8 A.M.

9.470   ---- A station (unable to get its call) gives call and
or   announces in English at 3:25 P.M. It gives the news
9.465   in English for three minutes, stating it is using
3.169 meters and will be back next day, same time.
Plays music until 3:37 P.M. and then signs off. No
national anthem. Possibly TAP but when last heard
that station used several languages and played the
Turkish national anthem. Can anyone help on this
one?

7.280   VLI9, Sydney, broadcasts to U.S.A. at 8 A.M.

6.370   WBAW, (no location given), announces as one of the "Voices
of America!" This station seems to be on the air
all night long.

6.040   WRUW, Boston, also a "Voice of America" station, all-night-
er, so it seems at the present time.

New BBC frequencies and calls when known: 6.125 mcs.; (GRW-
6.150): 9.490, 9.550, 11.700, (GVV-11.730): (GVU-11.780): 11.930,
11.955, 15.060, (GRP-11.787). GRE is now on 15.420 mcs. instead of
15.385. (IDA).

Don't forget to send your short wave tips and news to Editor
James J. Hart, 617 Grove Street, Irvington, New Jersey.

Opposite and above: A medium wave club at first, the Newark News Radio Club inaugurated a short-wave section in 1935. During the war years, members were busy with military and home front obligations, and the section, called "High Frequencies," seldom ran more than two pages.

and Axis propaganda is employed." In the same issue of the bulletin it was reported that, nationwide, there were 37 commercial FM stations then on the air.

The shortwave column was typically one or two pages of loggings, plus occasional country-specific or time-specific surveys—"Let's Listen to London," "Let's Listen to Australia," "Let's Listen to the U.S.S.R.," "Let's Listen at 8 A.M. EWT"—and brief features on topics such as when to listen for signals from particular parts of the world. All loggings were shown in Eastern Standard Time (or "Eastern War Time," the equivalent of EDT), probably for the sake of consistency with the broadcast band loggings for which EST was arguably more appropriate.

Member participation began picking up in 1944. June was an important month.

> Due to circumstances beyond your officers' control, issuance of the June 1st bulletin has been delayed until this rather late date. ¶It goes out to you on one of the most critical days in our history. This is D-Day and our American boys and Allies have landed in France. Our unified prayers are said for the success of this, the most stupendous military undertaking since the beginning of the world. It is quite possible that some Newark News Radio Club members who have been in England for many months now are taking part in the invasion. Their fellow-members wish them God speed with the fervent hope that this is the beginning of the end and that they, and all those who are fighting so valiantly for the cause of the United Nations, will soon be able to return to their homes and loved ones.[8]

By September, optimism was in the air. At year's end, the shortwave news from members required four to five pages, and it was around this time that there began appearing in the NNRC new member rolls the names of a "new generation" of DXers who would become well known in listening circles. These included Paul Dilg, Charles S. Sutton, Gilbert L. Harris, Paul Kary, John J. Oskay, and many others. Some of these members were associated with other clubs as well. But whatever other club affiliations one might have, the NNRC was *the* club to belong to.

The year 1945 was a turnaround year for the club. May 7 was V-E Day, August 15 V-J Day. In October the bulletin resumed twice-monthly publication during the BCB DX season. Effective November 15, amateur radio operators were authorized to resume transmitting activities, halted after Pearl Harbor. In December, annual dues were increased from $1 to $2 ($3 in 1947, $4 in 1957). Annual elections were reinstated the following April (later they would become biennial). And it was in 1945 that the name of Hank Bennett, then in the Seventh Army's Signal Corps in Europe, began appearing as a bulletin contributor.

The shortwave column grew under Jimmy Hart. Both BCB and shortwave contributors were given "tip cards" for reporting unusual loggings or information on new stations. Roger Legge of Forest Hills, New York was appointed Assistant Shortwave Editor in May 1947, although it appears that his contributions were more in content than in the editorial area. As for the club in general, things got back to normal. An NNRC picnic-convention was held in 1947 for the first time since 1941. Eighty-five people attended, and, except for one year (1969), it continued to be held annually until 1974.

In order to give its editors some relief, in November 1949 the club dropped its practice of issuing the bulletin twice-monthly during the broadcast band DX season and adopted the policy of a monthly bulletin year round. In December of the same year the editorship of the shortwave section passed to Hank Bennett, then of Woodlynne, New Jersey, who would retain the post until the NNRC closed its doors in 1982.[9] Bennett had become interested in shortwave listening in 1938, and obtained his ham license eight years later.

The Korean War broke out in 1950, and while it did not have as great an impact on the club as did World War II, the absence of some NNRC members was noticeable. In addition to urging NNRCers to write to fellow members who were in the military, the club reminded NNRCers to be alert to what they were hearing:

> While the part that our club can play must of necessity be a small one, nevertheless we are enlisting the aid of our members in a monitoring service to detect the source of any subversive transmissions or broadcasts of a detrimental nature which might be harmful to our security or the cause to which we are now committed. ¶In this connection, we are not concerned so much with those countries behind the iron curtain but any transmissions that might originate in this country or in countries in close proximity to us. Therefore, if you should at any time be successful in picking up such a subversive broadcast, try to get complete information on the station with regard to frequency, operating hours, possible location, etc., and forward the information to your respective editors without delay.[10]

Socializing was a big part of the NNRC. A 25th anniversary dinner party was held in 1953 at the Crystal Brook Inn in Eatontown, New Jersey. Ladies were gowned and corsages were provided by the club.[11] Subsequent annual dinners were held through 1962. By the mid–1950s, anywhere from four to eight pages of a typical 50-page bulletin were devoted to birthdays, marriages, illnesses and other personal events, lists of new members and re-enrollments, members' travels, updates on old members, etc. It is ironic that in these days of rapid travel and instant communication, clubs have not matched the level of camaraderie achieved by the NNRC in the days of manual typewriters, ponderous auto travel, and expensive long-distance calling.

The club also benefitted from a highly developed organizational structure that was unusual for radio clubs. By the mid 1950s, in addition to the various column editors, there was a president (Irving R. Potts from 1928 to 1962), treasurer, executive secretary and assistant executive secretary, 12 U.S. vice-presidents, a Canadian vice president, 13 honorary vice-presidents, a bulletin manager, 50 directors, 11 local directors, and a clerk to the board of directors. While many of these posts were honorific, they were a way of rewarding the club's supporters and encouraging their continued involvement. This structure remained largely untouched long after it had outgrown its usefulness.

In addition, each year starting in 1950 the club named a "Man of the Year." At various times it awarded memorial plaques named after various late distinguished members: club vice-presidents Jacob C. Goldstein, Adolph J. Koempel, and Peter J. McKenna, Treasurer Walter L. Townley, and President Irving R. Potts. A James J. Hart Memorial Plaque was awarded annually following Hart's death in 1972. And every month the board of directors named a "Member of the Month."

The shortwave column grew under Hank Bennett. It consisted of members' loggings in frequency order, along with by-country listings of schedules and other station news, and members' comments. Some limited QSL information, including new station addresses, was included as well, as were utility station logs. In October 1955, a two-page "Short Wave Odds & Ends" department was set up under Richard Labate of Yardley, Pennsylvania. It focused on SWL-card swapping, tapesponding, shack descriptions, letters from members, QSLs, equipment and antennas, and miscellaneous material. By the end of 1956, Labate's column often occupied four to six pages.

Today's DXers may be surprised to learn how popular SWL-card swapping was in the 1950s. It had fallen into disuse in the war years, but thereafter it grew until by the mid 1950s

SWL-W2

LeRoy Waite   Ballston Spa
39 Hannum Street   New York - U.S.A.

AMATEUR EDITOR

NEWARK NEWS RADIO CLUB

Radio......*Rex*....Talking to....*Rog*.....Sigs R....S.....
...........195....At...........G.M.T. on....Mc.

Rcvr............. Ant.............

Remarks. *Would Appreciate a Report*........

WISQF   Pse QSL, Tnx. 73, *Roy*

LeRoy Waite was NNRC amateur section editor from 1952 to 1969.

it was a major shortwave-related activity. "Short Wave Odds & Ends" was one of the best sources for names of SWLs who were interested in swapping, and pages of "good" names and addresses (SWLs who would swap), and comments on members' swapping accomplishments, appeared each month. Serious swappers would attempt to swap as many states and countries as possible, list those they had and those they wanted, etc. Most SWLs engaged in card swapping to some extent, although some thought it a pastime with only a tangential relationship to DXing and little to commend it.[12]

An FM column was added in 1952. The shortwave column was reorganized in 1956. Shortwave broadcast stations were placed in one of three lists: new stations and new frequencies, unidentified stations, and general loggings, a format that would be followed for 20 years. Next appeared country-specific schedule information. Some band surveys began appearing in the shortwave column in 1956 as well. These were usually prepared by Roger Legge, then living in McLean, Virginia, and Stew West of Union, New Jersey, and appeared regularly until mid–1961.

"Short Wave Odds & Ends" was preserved in the reorganization. The daily sunspot counts for the previous month were also an NNRC shortwave section staple, provided by member Grady Ferguson of Charlotte, North Carolina. Utility station logs and related items continued to have their own space within the shortwave section of the bulletin, as they had for some time, but they had their own editor, Charles McCormick, Jr., of Baltimore, Maryland. Soon, utility information occupied four to six pages monthly. McCormick would edit the utilities column until his death in 1969 when he was succeeded by Robert H. French of Bellaire, Ohio, who edited the column for the rest of the club's life.

Bennett experimented with various ideas to improve the column. In 1961 he expressed the intention to inaugurate a reporters' "honor roll," with points awarded for each useable item reported. (The system was never put into effect.) The following year a "saturation coverage" project was begun whereby members were asked to focus on a particular band each month. Although this resulted in more loggings for the band, because they were integrated into the regular loggings column they lost the significance they might have had if placed in a band survey-type format. "Saturation coverage" faded from view.

In the NNRC, more youth participation was sought. In commenting on the NNRC 1958 annual dinner, one member observed: "While the dinner was a great success ... it could have been a lot better if some of the youngsters could have attended and lent a hand to the festivities. It must be remembered that we will not always have the oldtimers around to keep the club going and if the present younger membership doesn't become more active in the club there may be no NNRC...."[13] They were prophetic words.

Although originally it did not appear in every bulletin, the bulletin's statistics section, called "Status of NNRC Members' Logs" and showing in various formats the states, countries and zones that members had logged and verified, had been around since 1946. Beginning in 1958 and for many years thereafter, the statistics column ran an annual receiver survey where members were encouraged to list their receiving equipment. Another popular column (although it dealt mainly with medium wave and, for a time, TV DX) was Carleton Lord's "Leaves From A DX-er's Scrapbook." It ran most years from 1947 to 1961. And one of the club's most experienced members, Carroll H. Weyrich of Baltimore, Maryland, wrote a monthly column looking back on the club's history and reviewing the contents of past bulletins. It was called "The Pioneer Days," and it appeared from June 1966 right up to the bulletin's last issue.

Irving R. Potts, the man who was the only club president most members had ever known, died in November 1962.[14] The transition to a new president, William Schultz of North Arlington, New Jersey, was smooth. Schultz was appointed by the board of directors, and elected by the full club membership in the club elections of 1964. NNRC membership was said to be 600 by mid–1963.

By 1966, Assistant Shortwave Editor Dick Labate's popular "Short Wave Odds & Ends" column was usually two to four pages, sometimes more. In May 1971 he was succeeded by a father-son team, George and Bob Smith of Grand Rapids, Michigan, who served until the column was phased out in 1973. (Labate died in 1975.) A card swapping column was resumed in 1975, but it lasted only a year, swapping having gone out of fashion by then. Labate's coverage of tapesponding had fallen off by the mid–60s, but the subject was re-established in 1967 under assistant editor Bob Fowler of Bloomfield, New Jersey. Fowler passed away early in 1973 and the column appeared under two successor editors through June 1978. It was tried again for a few months in 1979, and then dropped.

Although in June 1968 it was observed in the bulletin that the club was showing a steady increase in membership, careful NNRC watchers sensed that things were not the same. There had been a major decline in participation in BCB activities, leading in 1967 to a scaling back of the BCB section. Broadcast band DXers were the heart of the NNRC, and now the club was facing competition from not just one specialized BCB club, the longstanding National Radio Club (NRC), but also, since 1964, from the International Radio Club of America (IRCA). The BCB environment was changing., and special BCB DX broadcasts were more difficult to arrange.

The club was facing stiff competition in other areas as well. FM-TV DXers were attracted by a specialized FM-TV club, the Worldwide TV-FM DX Association, born of the Worldwide Monitors Radio Club in 1967. And on the shortwave side, the decision of the North American Shortwave Association (NASWA) to go all-shortwave broadcast in 1966, and its rapid ascendancy to the pre-eminent position on the North American SWBC scene, also hurt the NNRC. In addition, while in 1969 the club could still count among its members five who had joined when the club was founded in 1927, the "old timers" were much less in evidence and the social side of the club was declining. The 1968 annual picnic drew about 60 attendees, 40 of them club members. Only 14 percent of the membership voted in the 1970 club elections. There were frequent pleas for more contributions from members and greater involvement of "younger blood" in the preparation and production of the bulletin. At the end of the 1960s, NASWA had 50 monthly shortwave reporters, twice the number of those reporting to Hank Bennett's column.

Schultz resigned as club president in January 1971. The Executive Board appointed Eugene G. Vonderembse of Astoria, New York to fill the unexpired portion of Schultz's term, which ended in April 1972, and named Schultz President Emeritus. Vonderembse was elected president in 1972. He realized that the club was in need of renewal and would have to adjust to changing times in order to keep members and attract new ones. Various things were tried. Occasionally the monthly Executive Board business meetings were followed by a discussion of some aspect of DXing, with all members invited to attend and participate. A club picture page, containing photos of the 1971 picnic, was published. A members' correspondence column was instituted in July 1972 and appeared from time to time when the mail warranted it. A program to distribute audio cassettes of part of the bulletin to sight-impaired members was begun in June 1973 but ended after two years due to lack of interest. Member ads ("NNRC Radio Exchange") started appearing in small number in 1975. And in some years the club was able to offer the *World Radio TV Handbook* to members at a reduced price.

Although the club had been independent of the *Newark Evening News* for decades, it was a sad day, and perhaps a bad omen, when the paper closed down on August 31, 1972 after 89 years of publication. It never regained the circulation and the advertising lost during a debilitating 10-month strike. Although club members had long used the home addresses of the individual editors for their contributions, right up to the end the club's official masthead address—215 Market Street, Newark, New Jersey—was the address of the newspaper, which forwarded mail received.

Where's the Club?

"Back in the summer of 1958 I was visiting an aunt in beautiful downtown Newark and decided to visit NNRC World Headquarters at 215 Market Street. I walked in and innocently asked the receptionist if I could see someone from the club, even though I did not have a formal appointment. She had no idea what I was talking about and finally called someone to either cart me away or help me, I know not which. A gentleman came out and informed me in a rather stuffy manner that 'Oh, they only use this as a mail drop. Someone picks up the mail every week or so. There hasn't been a direct connection between the Newark Evening News and the radio club for years.' Away went disillusioned me." John Kapinos, "Memories," *WPE Call Letter* (Great Circle Shortwave Society), January 1988, p. 8.

An important step was taken in January 1974 when Hank Holbrook of Chevy Chase, Maryland, who had been doing the statistics column since 1969, was succeeded by Sam Barto of Morris, Connecticut. Barto converted the column into two columns, one on statistics and the other on QSLs. This was the club's first dedicated QSL column. Although the column covered all bands, the emphasis was on SWBC. It grew quickly, and Barto leavened the details of members' QSLs with station profiles and other newsy items, turning it into a mini-bulletin all its own and filling a void, QSLs and features, that had long existed in the club's shortwave coverage. "QSL Report" covered the subject admirably. But after the May 1978 column, "QSL Report" disappeared without a trace. (A separate statistics editor had been appointed in November 1977.)

In addition to serving as shortwave editor, Hank Bennett, together with his wife, became the publishers of the entire bulletin in November 1975. A test bulletin in offset format was tried in October of that year, but the mimeograph style was retained until July 1978, when the club finally went offset and retired the mimeograph machine which had produced over 7 million pages of club material. The offset format was an economy measure and a means to better handle graphics.

When the club went to offset it was already exhibiting the symptoms typical of a club in trouble: reduced quality and quantity of content, reduction in the overall number of pages (approximately 40 pages per month), frequent changes in editors, reduced member participation, missing columns, a decline in bulletin appearance, and — the key element — insufficient members to maintain financial viability.

Other difficulties dogged the club as well. In December 1976 it stopped issuing "Member of the Month" awards because it was running out of candidates. The award was replaced with a monthly recognition of NNRC's "Cavalcade of Stars," members who had provided outstanding service to the club. The club's membership had dropped from 675 in mid–1973 to around 575 at the start of 1979. That year, only 35 members participated in a major club contest, and ten listed their totals in the statistics column. Ten percent of the members participated in the 1980 elections. And while a 50th Anniversary celebration held at the Bloomfield Civic center, Bloomfield, New Jersey, in October 1977 attracted nearly 100 people, a major general membership meeting in September 1981 drew fewer than hoped.

In the changeover to offset, an essential step for economic and management reasons, the bulletin lost its distinct appearance, an anchor to the past from which the club had always benefitted and that gave the NNRC a special status among listener clubs. Now that the NNRC bulletin looked like all the others, it lent itself to easier comparison, which was not always favorable. Anchor columns had largely run out of gas. The strongest column for most of the club's life, the broadcast band column, was now but a few pages. The amateur column, though still having a small core of regular reporters, was also much reduced in size. A favorite for many years, the heard-verified statistics list, was supported by only a handful of members.

One of the club's difficulties was the relative rigidity of the bulletin structure. While some new columns were tried (and usually abandoned due to of lack of sustained interest by either the members or the editor), the basic structure and organization of the bulletin had not changed greatly since the club's early days. A particular problem was the relative absence of feature articles or other interesting material to give the bulletin some variety. Save for technical topics, where an "Information Please" column did a commendable job, there was no space regularly set aside to accommodate feature material. The gap was most noticeable in the shortwave broadcasting area, where content was limited to loggings and whatever limited station-related information members might submit. This resulted in a sameness of content from month to month.

There were some bright spots. The club held major, week-long DXing contests in the month of January each year from 1979 to 1982. Although Bennett increasingly was drawn to minor conflicts of one kind or another with other clubs and with some contributors, his shortwave column, while never the source of breaking news or rare DX, had enjoyed something of a renaissance and was still substantial in size and respectable in content. In November 1979, John Santosuosso of Lakeland, Florida, who had been editing a CB column for the club, began a new column, "Pirates, Spies and Clandestines." This was the first time this subject had received separate treatment in the bulletin. Although not everyone agreed with the notion of highlighting illegal transmissions, Santosuosso's column was meaty and well-written, and included both loggings and other material. (In later years he would write a similar column for *Monitoring Times*.) And in July 1981, a quarterly QSL column was reinstated under editor Charles C. Boehnke of Los Gatos, California.

Requests from club president Eugene Vonderembse for greater participation of younger members in club administration went unheeded. In 1981, presaging what was to come a year hence, Vonderembse expressed concern over the difficulties of conducting club business. The major organ of the club, its board of directors, had been established in radio's early days when the club was as much a social group as a service to radio listeners. Then there were many members nearby to help with club administration, and personal relationships were strong. But now most of the old timers had retired, left the Newark area or passed away, and those remaining were no longer able to devote their time and energy to club affairs. Quorums were sometimes difficult to obtain, and it was becoming increasingly difficult to find members willing to serve on committees. The club's many non-local members who had not experienced the years of NNRC fellowship and whose main contact with the club was by mail viewed the NNRC as the publisher of a magazine rather than a club to which they owed support and participation. Vonderembse observed: "The past can be a comfortable shelter to hide in, and it also can be a cavern of self-delusion to shut out reality. But the future will not go away. We cannot ignore it if we want to remain alive and vigorous."[15]

This was a repeat of past calls and it did not mention the critical problems: membership and finances. Those came to the fore a year later, in March 1982, when Vonderembse advised members that the club was in danger of not meeting its monthly expenses, which were about $600. Increased costs had eroded the club's financial reserves and it was living month to month. Although increased costs, mainly postage, always resulted in the loss of some members and the corresponding revenue, the club usually built up a financial cushion in the fall and winter which tided it over during the leaner summer months. But this had not happened in 1981–82.

The end came in April 1982, when members were advised that continued operation would only produce debts that could not be paid. Various alternatives had been considered — a smaller bulletin, bi-monthly publication, perhaps a single-page bulletin for a few months to improve the club's cash position. But these were not viable alternatives. As Vonderembse put it, "To decide to dissolve the club was a difficult and melancholy duty. But to have the NNRC face a disastrous ending was even more disturbing. We chose to cease operations while we could inter the club with reasonable dignity. Our group has built a fine reputation through the years, and we wanted to retain as much of it as was possible."[16]

At the April 24, 1982 meeting of the Executive Board, Honorary Vice President Eugene C. Bataille, the last remaining member who had been present at the club's birth, made the final motion to dissolve. After a brief remembrance of the club's first meeting in 1927, the motion was unanimously approved. In the face of changing times and a changing membership, the club's greatest strength over much of its lifetime, the personalized culture of long term relationships among senior members, could not save the day.

An NNRC personality worth noting was Jack R. Poppele of South Orange, New Jersey. Jack earned his federal wireless license at the Marconi wireless school in New York in 1915. He was a shipboard operator in World War I, and in 1922 he and another engineer built the original 250-watt transmitter for New York's WOR, then owned by L. Bamberger & Co. of Newark. He became the station's chief engineer and vice president. He was an early NNRC supporter (and honorary president) whose counsel was always valued in NNRC circles. In the late 1930s, WOR hosted a weekly 15-minute NNRC radio program over its "shortwave" outlet on 25.3 mc., W2XJI. On December 7, 1940 the program was moved to the station's FM outlet, W2XOR, 43.4 mc. (later W71NY, 47.1 mc.). (WOR also carried an annual NNRC DX Program on its BCB channel.) Poppele served as President of the Television Broadcasters Association in the mid 1940s, and in 1953 President Eisenhower appointed him Director of the Voice of America. In 1992 the VOA transmitter site in Delano, California was named the Jack R. Poppele Transmitting Station.[17] In 1960, after leaving the VOA, he engaged in radio and TV manufacturing and consulting.

A Safe Landing

"[T]o my mind the greatest thing Jack ever did occurred on the stormy night that the dirigible 'Shenandoah' broke loose from its mooring and flew at random all over the eastern seaboard. Jack stood on the roof of the Bamberger store in Newark and kept calling the giant ship, advising them to change their course when he saw them flying over the Prudential tower in Newark and headed out to sea. With Jack's help, the Shenandoah's commander was able to get his bearings, plot his course and return safely through the dense fog to Lakehurst, N.J. After an all-night vigil, Jack finally went home when he heard that all on board the dirigible had landed safely. For his work that night, Jack was cited by the U.S. Navy, a well-deserved reward." James J. Hart, NNRC Shortwave Editor, NNRC Bulletin, February 15, 1947, p. 25.

OTHER CLUBS

There were numerous other clubs during the 1940s and 1950s, and while a few were groups with significant membership and organization, others represented little more than the good intentions of a few enthusiastic listeners.

In the former category were the Victory Radio Club and the United 49ers. The Victory Radio Club was started in 1942 by the editors and reporters of *Radio Index (RADEX)*, a popular DX magazine which had ceased publishing that year. The club was based in

Worcester, Massachusetts and it published the monthly *Victory News* as well as a weekly *Victory Flash Sheet*. During the war it operated a War Prisoner Message Bureau to coordinate the delivery of POW messages heard by members. It was absorbed by the Universal Radio DX Club in 1946.

The United 49ers Radio Society was in operation from 1949 to 1954, and counted some well-known hobbyists among its members. Edward I. Broome of Medford, New Jersey was club president and host of an annual club meeting at his home, and James R. Pickering of Hightstown, New Jersey was club treasurer. Although the editor and publisher of the club's official bulletin, both at the time of the club's formation and at its end, was William C. Peters of Collingdale, Pennsylvania, during most of the club's life those roles were filled by well-known DXer Anson Boice of New Britain, Connecticut. "Ans" was also the SWBC editor, and was assisted in his duties by his wife, Julia. True to his roots, he was fond of reminding people to "keep in touch with the Dutch," a phrase he borrowed from Eddie Startz. The United 49ers bulletin covered SWBC and ham listening. It also included news about members, reminders of member birthdays and anniversaries, and occasional inspirational messages from the club chaplain.

Among the many clubs that came and went largely unnoticed after the war were the World Wide DX Society, based in Texas, and the Dial Spinners Club in Rapid City, South Dakota (the two merged in 1949); the modestly named United Short Wave Listeners of the World, formed in Piketon, Ohio in 1951; the Minnesota Short Wave Listeners Association, Cloquet, Minnesota, also organized in 1951; the Empire City Short-Wave Listeners Club, New York City, established in 1952; the Southwestern Short Wave Club, Dallas, Texas, organized in 1954; the World Shortwave Club of San Bernardino, California, founded in 1956; and the Connecticut DX Club, established in Wallingford, Connecticut in 1958. Others included the Association of Mid-West SWLs, Chicago, Illinois, with its publication *A-SWL*; the Nationwide Short Wave League in Gloucester, Massachusetts; the Northern Valley DX Club of Tenafly, New Jersey; the Tri-State SWL Club, Vincentown, New Jersey; and the Knobtwisters of Waterloo, Iowa.

Major Shortwave Broadcast Clubs of the 1960s and Later

AMERICAN SHORTWAVE LISTENERS CLUB (1959–1998)[18]

The ASWLC was founded in 1959 by Ken MacNeilage of Cranford, New Jersey and Maxey Irwin of Sparta, Tennessee. In November of that year they printed a two-page introductory bulletin and sent it to 15 prominent DXers selected from the ranks of the Universal Radio DX Club and 15 from the NNRC and the National Radio Club. Twenty of the 30 agreed to join, and the first regular bulletin was issued in January. Called *SWL*, it would be published for almost 40 years. The club's philosophy was reflected in its motto, "World Friendship Through Shortwave," adopted in July 1960.

MacNeilage was the general editor. Irwin was the SWBC editor until October 1960, when Robert Newhart of Merchantville, New Jersey assumed the role. C. M. Stanbury II of Crystal Beach, Ontario became shortwave editor in December 1961 when Newhart entered military service.

Anson Boice, shortwave editor for the United 49ers Radio Society, at the dials of his Hallicrafters receiver.

The club got off to a good start. The mimeographed bulletin was substantive and good looking, and contained material for both beginners and experienced listeners. Many well-known listeners were among the contributors and early club patrons. In addition to short-wave broadcast material, *SWL* contained broadcast band and amateur sections, some TV-FM DX information, card swapper lists, a reporters' scoreboard, technical features, and other items. There were frequent listening contests for ham band and SWBC listeners, and the club had an operating style that was commendably democratic by shortwave club standards. In little more than six months it had 100 members, within two years nearly 250.

In early 1962 the club suffered the first of several organizational crises when MacNeilage relinquished his duties as Executive Editor in advance of relocating. Several difficult years followed. In March, James J. Howard of Kansas City, Missouri, an early ASWLC member and editor of the "Technical Digest" section, became Executive Editor. Howard was a strong hand when the club needed one, and a hard worker. He started a section on space communications, and also prepared a weekly ASWLC DX program, "Calling SWLs" (later "DXing Worldwide"), which was carried on WRUL, a Boston-based shortwave station. An extensive "Friendship Center" column covering SWL-card swapping, tape swapping and pen pals was edited by William P. Eddings of Altoona, Pennsylvania. Eddings resigned effective September 1, 1962. He had been editing a similar column for the North American Shortwave Association since May of the same year, and would become NASWA Executive Editor and Publisher the following year. A QSL column edited by Alan Roth of Bridgeport, Connecticut was begun in April 1963.

Although bulletin content remained fairly high, it became increasingly difficult to find editors, and more of the burden fell on Howard, whose "free speech" editorial policy ensured that *SWL* reflected many personality-based conflicts. A club constitution appeared in the March 1963 bulletin, and in September a board of directors was elected by the membership, which by that time numbered 475. As its chairman the board named another early ASLWC member and well-known senior shortwave DXer, 73-year-old August Balbi of Los Angeles, California.

A second crisis occurred when Howard resigned. Although it appeared that a good transition plan was in place, many issues of the bulletin went unpublished, and membership suffered, as the club went through two executive editors before reconstituting itself in July 1964 under new Executive Editor and Publisher Jerry Klinck of Buffalo, New York. Klinck gave new life to the club, which would now be referred to informally as "the new ASWLC" or "ASWLC II."

The new board of directors was almost identical to the old. There were some new editors— Irwin Belofsky of Brooklyn, New York, Tom Cardullo of Emerson, New Jersey, Robert E. LaRose of Binghamton, New York, Serge C. P. Neumann of Santa Monica, California, and Stewart MacKenzie of Huntington Beach, California, among others— and some holdovers. C. M. Stanbury remained shortwave editor, and Jim Howard was back with the "Technical Digest." Although the club was restarting largely from scratch, with no carry-over membership, it was soon on a strong footing both organizationally and financially, and the bulletin's appearance was much improved. There was a renewed focus on DXing rather than extraneous topics. A new constitution was published in September 1964, with August Balbi continuing as Chairman of the Board, a post he would hold until 1969. Consistent with the interests of the overwhelming number of members, the club became principally shortwave broadcast in December 1964. This decision was honored in the breach, however, as the bulletin continued to contain a utilities column and an international medium wave column, and would inaugurate an FM column in 1983.

In March 1965, the club weathered another change in leadership when Klinck relinquished the executive editorship to Stanbury, who also continued serving as *SWL* shortwave editor. Klinck stayed on as publisher-treasurer and occasional commentator (and editor of a short-lived "Novice Log" column). Soon Stanbury was the mainstay of the club. Sharing the shortwave editorial duties were Associate SWBC Editor Robert LaRose and "QSL Corner" editor Serge Neumann. An elected board of directors composed mainly of club regulars took office in September 1965, and Balbi was again elected Chairman and club President. The DX content of *SWL* was by this time at its peak.

In January 1966, the duties of publisher and treasurer were assumed by the club's utilities editor, Stewart MacKenzie. MacKenzie had been a member of the club since its earliest days and a board of directors member since 1964. Klinck stayed on as an Associate SWBC Editor. Soon *SWL* had a much-improved appearance.

Through it all, the ASWLC bulletin remained well-balanced, with much material for newcomers and experienced listeners alike, notwithstanding occasional complaints from each group that the bulletin favored the other. There was, however, a persistent concern that the culture of the club had changed. When it was new, the ASWLC was the club "for the rest of us." Now there was a feeling that some of the club's leaders were taking their DX a little too seriously.

In November 1967, MacKenzie became Executive Editor and club headquarters moved officially to Huntington Beach. Two months later, Stanbury was replaced as shortwave editor by Don W. Billingsley of Sacramento. Stanbury was a colorful character and an excellent DX editor. However, he symbolized the changes about which some members were concerned. His love of the controversial and the conspiratorial led to unnecessary criticism of other clubs and club editors, as well as discourses and theories on matters of limited interest to the membership at large (e.g. the "real" stories behind Radio Swan and Radio Americas, the origins of the airborne "Blue Eagle" stations of the Vietnam war, the American role in the establishment of the anti–Rhodesian BBC Francistown relay station, etc.).[19]

Although ASWLC editors and columns changed from time to time, the bulletin remained strong. By the late 1960s, *SWL* was carrying SWBC news, an ANARC report (ANARC was the Association of North American Radio Clubs), information on utilities, QSLs and international medium wave DXing, "Tricks of the Trade" (for newcomers), SWL-card swapping, and various other topics. There was an excellent propagation column authored by Jack White of Gresham, Oregon, and in May and November, when the major international frequency changes occurred, the "Calling North America" section detailed the times and frequencies of the broadcasts to North America. The club also had an awards program and a country list, a translation service, and periodic listening contests for members.

In 1969, the club issued the *Proper Reporting Guide,* a detailed manual for preparing reception reports in English, Spanish, Portuguese and French. (Subsequent editions were published in 1974 and 1981.) In the same year, the club constitution was revised by A. E. (Art) Glover of Port Angeles, Washington, and Serge Neumann. Although it was never put to the membership for ratification, it was unanimously approved by a new board of directors in January 1970 and amended from time to time thereafter. August Balbi, while remaining on the board, was succeeded as club president by Glover and granted life membership in the club. The new board began publishing quarterly reports of its activities in the bulletin.

Then followed yet another difficult period for the club. The board had been attempting to bring about a series of what it felt were necessary improvements in the club's administration. MacKenzie, who handled most club operations and was by this time, as Publisher-Treasurer, the club's key man, was in agreement with some, including the appointment of Bill Ray of Mill Valley, California as ASWLC Executive Editor (Ray was editor of a ham radio bulletin), and David W. Thorne of Santa Ana, California as club Secretary. He was not in agreement with some other proposed management changes, however, and an impasse ensued. As a result, in July 1971 most of the club officers, directors, and editors resigned and formed a new club, the "Society for the Preservation of the Engrossing Enjoyment of DXing," or SPEEDX.

The immediate result within the ASWLC was the consolidation of authority in MacKenzie and an influx of new editors, most of whom were less experienced than their predecessors. The bulletin continued to be published without interruption, however, and membership surpassed 500 in 1971. From a longer-term perspective, the departure, at one time, of so many well known and well respected staff members deprived the club of some of its most experienced talent. Although MacKenzie was supported by many loyal and talented editors through the years,[20] the priority which he attached to supporting new listeners resulted in a "beginner's culture" that persisted during the life of the club. The ASWLC was also viewed as a west coast club in a DX world dominated by east coasters.

In 1972, the club's dormant awards program was reinvigorated. In 1973 a new countries list was adopted which was geared for BCB and utility listeners as well as SWBC types. On June 17, 1976, Balbi died at the age of 85. The club's man of the year award was renamed the August Balbi Award.

The bulletin went to an offset-printed pamphlet style in June 1972. Although columns were sometimes absent, the bulletin was consistent in its organization. At the heart of *SWL* were the three loggings sections: Asia-Oceania, the Americas, and Europe-Africa, with loggings listed by country, together with a separate "Time Index" column cross-referencing the previous month's logs by time. Countries would be reassigned within the loggings sections, and the number of loggings sections expanded and consolidated from time to time, but they retained their basic format. Each loggings column had its own editor who exercised considerable discretion in adding non-logging material. As for the time index, at first it covered all loggings. It was later modified to cover only loggings of English-language programs, then it returned to the all-loggings format, and still later it adopted a "selected loggings" approach.

There was also a QSL column; "DX-Showcase/Shortwave Review," containing station schedules and a "DX-Hotline" subsection covering new stations, frequency changes, information about the smaller stations, etc.; "DX-Showcase/Spotlight" featuring station profiles and the like; and "Pot Pourri," as its name implies, a mix of ANARC news, Handicapped Aid Program (HAP) news (see p. 119), hobby satire, clippings, and a members' comments subsection called "DX Forum." "Pot Pourri" was edited by Woody Seymour, Jr., of Sanford, North Carolina. There was an introductory column by Bill Ray called "Eyes, Ears and Knows" containing general observations, as well as some times and frequencies of English-language programs.

Stew MacKenzie presented a "Headquarters" section covering upcoming events, club doings, etc. Each month MacKenzie would acknowledge the "publishing committee," those who helped get the bulletin out at a monthly club meeting normally held at his home. Other club events would occur at these monthly meetings as well, including the presentation of awards to club members and the welcoming of hobby personalities who might be in the area. The bulletin also contained an international BCB column (retired in July 1975), a utilities section, and a propagation column which would run until May 1980. In January 1979, a central distribution editor was appointed in order to facilitate reporting loggings to the various column editors.

The content of *SWL* changed from time to time. "DX-Showcase/Spotlight" did not appear after May 1974. However, "DX-Showcase/Shortwave Review" was expanded.[21] Some advertising for companies like Drake, Kenwood and Radio West (and later Grove Enterprises and Universal Radio) began appearing in 1975. A scoreboard column was started in 1976 and appeared intermittently until July 1977. Starting in January 1978, *SWL* carried the humorous "Clara Listensprechen Report," a take off on the Charlie Loudenboomer reports in *FRENDX*. (Clara appeared through 1979 and a few times in 1980 and 1981.) The last of Woody Seymour's "Pot Pourri" columns appeared in May 1978. (Seymour began a "Program Panorama" column in the *SPEEDX* bulletin the following month.)

Several columns appeared briefly or intermittently from the mid–70s to the mid–80s, including an awards committee column, a technical column, a novice column, a "Smorgasbord" column, and "They Said It About," which quoted what various stations were saying about important international issues. Sam Barto's HAP auction page, which was carried by a number of clubs, began appearing in January 1980. "Computer Corner" was introduced in September 1979. It was to be a series of articles authored by Stu Taylor of Pleasanton, California. However, only one article appeared. In October 1980 it became a regular column edited by Richard Varron of Wayne, New Jersey, who over the years had edited the propagation column and several shorter-term ASWLC columns. Although "Computer Corner" was to focus on how SWLs could use "microcomputers" as an aid to DXing, it quickly became more of a technical column for those interested in computers per se. It appeared through July 1982.

Club membership was approximately 600 by 1980, and typically the bulletin was over 50 pages long. An eight year run of a broadcast band column was begun in February 1981, the first time the bulletin had covered the broadcast band since 1975. A second HAP page was introduced in March 1981 to complement the HAP auction page, and lasted until January 1983. An FM-TV column was added in April 1983 and appeared through April 1985. In March 1985, Zichi added an informative and popular "Q & A" which was incorporated into a new "ASWLC Almanac" column that appeared from October 1986 to April 1988. "ASWLC Almanac" included member profiles, book reviews, station features, clippings, equipment information, and newsy items about shortwave. In September 1986, a telephone "DX Program Hotline" was set up to accept DX news for incorporation into an ASWLC segment on HCJB's "DX Partyline" and on Radio Discovery. October of that year saw the return of a computer column, "Computer Connection." From June 1987 to January 1993, the bulletin also carried a "Pacific Report" by Arthur Cushen, Invercargill, New Zealand. This column featured news of stations in the Pacific area and stations heard in the Pacific.

Although *SWL* never had the semi-professional appearance of some of the other club bulletins, and missing columns were commonplace, it had many contributors. With its headquarters on the west coast, and with many more west coast contributors than other clubs, it was an alternative to the east coast clubs. And while ASWLC was not the club to which the more experienced DXers gravitated, it had a loyal following, fostered in part by the close relationship of the club with SCADS, the Southern California Area DXers, a group which held frequent meetings and events for local listeners.

By 1984, membership had grown to an all-time high of approximately 1,100. However, a decline in club activity soon followed, and the club entered a long period of struggle. There was no May 1985 issue. Zichi departed his editorial post in December 1986, expressing concern about the club's month-to-month financial existence and the absence of information on club finances. MacKenzie observed that the reduction in the club's membership and the halving of club revenues was a normal product of reduced shortwave listening in a time of low sunspot activity, and would improve. Some editors who were concerned about the club organized a Staff Advisory Council to review various aspects of club operations.

Member contributions declined, and by 1988 the bulletin had dropped in size to about 40 pages. The October 1988 issue was skipped because funds were low and several columns were missing. The December issue was delayed for lack of postage, and bulletin publication was suspended for 90 days. Club membership had dropped to about 350, roughly a third of the 1984 figure, and the club had suffered a financial loss in connection with hosting the 1988 ANARC convention. A number of members made special contributions to help retire the club's printing debt. At $17 per member, the club was on the brink of bankruptcy. Survival would require dues of $25, which MacKenzie did not believe was feasible.[22]

In January 1990, as a result of a continuing decline in membership (then around 300), it was decided to publish SWL every other month. Columns were frequently missing, and even at the bi-monthly pace, issues were occasionally skipped altogether for lack of contributions. Editors changed often, there were frequent editorial vacancies, columns varied in style and setup, and the bulletin lacked an organized, uniform appearance. Starting in July 1990, the bulletin contained a special ASWLC edition of the Southern Cross DX Club (Australia) "DXpress" column focusing on African, Latin American and Asian DX. It did not last beyond year's end, however. Later a pirate column appeared a few times, but no long-term editor could be found. A promised propagation column never materialized.

Monthly publication resumed in January 1994, and the bulletin, then down to 24 pages, grew to 36. Increasingly, however, it relied on filler material, e.g. newspaper clippings, press releases, reprinted schedules, etc. The problem of minimal member contributions in most areas persisted, and editorial vacancies or other absences often dictated that MacKenzie edit columns himself. There were some improvements in content. The one-page "Arthur Cushen's Shortwave Review" appeared from January 1995 to Cushen's death in September 1997, and, after much searching, an editor was found for a BCB column, which was restarted in November 1995. In addition, in April 1995 Spence Naylor returned as editor of the utilities column, a post he had held from 1977 to 1991 when the utilities column had been one of the bulletin's strongest features.

By mid–1995, membership was down to around 250. Although members were offered a month's renewal extension for each new member brought into the club, only 27 new members joined in 1995, 18 in 1996, 26 in 1997. In September 1996, in an effort to attract more members, club dues were reduced from $24 to $18 per year, or $30 for two years. (The dues of the only other national-level, general SWL club in the United States at that time, NASWA, were $26.) There was no June 1997 issue because there was no new or renewal member income at the time and thus no funds on hand. The February 1998 bulletin was skipped as well. Financially, the club was living hand to mouth. On the editorial side, in September 1997, with the loss of the third of the three loggings editors and with MacKenzie already doing the other two loggings sections himself, the three sections were consolidated into one "Shortwave Review" column edited by MacKenzie.

MacKenzie's efforts to save the club had been herculean. However, it appears that the ASWLC's last bulletin was issued in April 1998. The club has lived on in the form of an ASWLC Yahoogroup and combined monthly meetings with the Southern California Area DXers.

NORTH AMERICAN SHORT WAVE ASSOCIATION (1961 TO THE PRESENT)

The old adage that great oaks from little acorns grow has no better exemplar among shortwave clubs than the North American Shortwave Association, or NASWA. (It was "NASA" until November 1966 when it was changed to avoid what could have been only the remotest possibility of confusion.) No one present at its birth could have predicted that it would grow into the largest SWL club in North America.[23]

The first issue of the NASWA bulletin, a one-page affair, appears to have been issued in September 1961. The club's president was Sterling Pike, a 19-year-old employee of Western Union in Heart's Content, Newfoundland, site of the landing of the first Atlantic cable in 1866.[24] The vice-president and treasurer was 18-year-old Richard D. Roll of Hamburg, New York.[25] Within a few months, Roll was also president (and publisher), and the bimonthly bulletin had grown to 20 pages.

The bulletin contained a headquarters column, columns covering shortwave broadcasting and citizens band, an all purpose "Broadcasting Highlights" column (soon called "NASA Newsroom"), a swap and sell column, and an awards program. Added within the club's first year were columns on medium wave DXing, ham band DXing, SWL-card swapping, and tapesponding (the latter two combined into a "Friendship Corner," which was later divided into two columns). "Tech Trails" and "Space World" (satellites and related) followed, while CB fell by the wayside.

Both the appearance and content of the columns were uneven. The medium wave and "Friendship Corner" columns were the most successful, while the shortwave broadcast column could seldom muster more than two pages. Incremental progress was apparent, however; the bulletin went monthly in September 1962, and membership reached 100 in February 1963. Still, the club's long term prognosis would have been poor save for the efforts of William P. Eddings of Altoona, Pennsylvania. Eddings had been editing the "Friendship Corner" from its inception in May 1962. More important, his experience as editor of the American Shortwave Listeners Club "Friendship Center" column made him familiar with club administration, and a familiar name to potential NASWA members. Eddings became NASWA Executive Editor around the middle of 1963, although from his distinctive (some would call it odd) writing style it was clear that he had been writing part of the "Headquarters" section from the start of that year. A contest won by member Bill Harris of the U.K. resulted in the choice of "FRENDX," a conflation of "friendship" and "DX," as the name of the bulletin.

Eddings was a man of modest means and lifestyle who projected a kindly, almost pastoral mien, different from the flintiness sometimes evident among club editors and administrators. He placed great value on "unity and friendship," a phrase which he began using informally in January 1963 and which became the club's motto when he took over. (The phrase was perhaps influenced by "World Friendship Through Shortwave," the motto adopted by the ASWLC in 1960.) In addition to his frequent exhortations to members to eschew apathy and contribute to the bulletin, Eddings would often editorialize about the meaning of peace, friendship and understanding, and cooperation among clubs. Most important for the club, he was completely devoted to NASWA's success, and willing to work hard to accomplish it (he once estimated that he spent 40 hours a week on NASWA business).

A crisis occurred in July 1963 when Roll resigned as publisher. There were no bulletins in July or August, and the September and October bulletins were published by member Don Erickson of Sunnymead, California. Soon Eddings became publisher. Roll's title of "President" disappeared, and from then on Bill Eddings was NASWA's key man.

An important factor that benefitted NASWA's prospects at the time was that the ASWLC had stopped publishing, and it did not look like it would survive. (It did.) NASWA benefitted from this temporary decline in competition.

In response to a writer asking about NASWA headquarters, Eddings described a well-equipped office on the third floor of an elevator building in Altoona's downtown business district, and then said:

> Truth/fact of the matter is that NASA hdq, although located on the third floor, is in a humble apartment building "on the other side of the tracks" here in the city of Altoona. The two rooms we occupy are filled with only the barest and the most basic of furniture, equipment. The heat is sometimes most uncomfortable in summer, many times rather chilly/lacking in winter. No elevators have we either; just plain old leg-power. As for the view, take your pick from one of two windows and you see naught but the house or building next door.[26]

Robert Newhart had done a good job with the ASWLC shortwave column during 1960–61, and hopes for an improved NASWA shortwave broadcast section increased when he became editor in 1963 while he was on active duty in Germany. However, his tenure was only a year and a half, and the SWBC column was still the weakest part of the bulletin. The turnaround came in March 1964 when the editorship passed to Doug Benson of Schenectady, New York, an announcer at WSNY radio. Soon the column (called "Shortwave Center" from June 1964) was ten pages long, and more substantive in content and newsy in format. In addition to loggings, there were station profiles, QSL information, members' comments, a popular question and answer section called "Shortwave Roundtable," feature articles, and utility coverage. Some of the more advanced SWBC DXers, like August Balbi of Los Angeles and William S. Sparks of San Francisco, plus many up and comers, joined the club and contributed to Benson's column.

The year 1964 was a year of progress in other areas as well. The bulletin, then published in 8½ × 11" mimeographed format, adopted a neater, more uniform appearance. Small graphics began appearing. "NASA Newsroom" developed into an eclectic mix of information on the doings of other clubs, radio-related press items, and other interesting miscellany. The SWL-card swapping column was popular; NASWA highlighted a "swapper of the month," and even had a swappers scoreboard, indicating how many states and countries a listener had swapped. Eddings was sensitive to the needs of newcomers, and for several years published brief bios of most new members. The club won in the third year of

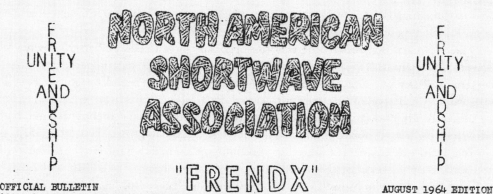

F R I E N D S H I P

UNITY AND

NORTH AMERICAN SHORTWAVE ASSOCIATION

F R I E N D S H I P

UNITY AND

"FRENDX"

OFFICIAL BULLETIN   AUGUST 1964 EDITION
****
Headquarters—NASA   1503 Fifth Avenue A2   Altoona, Penna. 16602
****

WHAT IS DX?

The answer to this question is as varied...well almost...as there are DX enthusiasts!
Here then are typical examples from a survey made.
"To me, DX is any low-powered station that I cannot hear unless conditions are right;
Choice DX is that elusive, low-powered station half round the world that I never seem
able to tune in." "DX includes stations that are audible only under freak conditions"
"A good criterion is "watts per mile"" "It depends on the type reciever, antenna,
distance, power of station, time, season, frequency and so on; Super DX is the long-
distance, low-powered reception on a channel not usually favorable for long-distance
reception at the time heard...rather a "miracle".! "I feel that any station more
3000 miles distant is DX, regardless of power, frequency and there is no doubt that
a relatively low-powered station considerably closer can also legitimately be labeled
DX." "DX is anything of interest to the particular listener and that is a new station
or logging or country, etc." "It is reception of a SW station that is difficult to
hear except during ideal conditions and one that is not heard regularly." "DX is a
station located outside your local range. Distance and power determine its class."
(Reprinted, in part from WT'56; Via August Balbi, California)

LATEST de ANARC

Most recent action before the ANARC is a
vote on whether a country list should be
adopted, whether it should be one already
in use by other clubs/organizations, or
whether it should be be one originally
made up and adopted by ANARC itself.

AMERICA'S PROMISE

"To every man his chance.
To every man, regardless of his birth,his
shining opportunity.
To every man the right to live, to work,
to be himself, and to become whatever
his manhood and his vision combine to
make him." de THIS WEEK by Thomas Wolfe

FM/TV QRM

"Sorry to notify you so late, but will not
be able to get section done this month due
the fact I now have a job, also am taking
two courses at MIT...have lotsa homework.
There will however, definately be action
on the section for September and I will
have it to you in time. Six reports are
now on hand and I hope to have more then."
Paul Gough.

NASA at CDXC CONVENTION

A tape recording containing segments of
many NASA staff/officials has been sent to
CDXC and will be hrd by those at Montreal
convention. Tape represents NASA/members.

*****   *****   *****   *****   *****   *****
Much of this edition was done at hdq this month and although we are willing it must
be pointed out that it does cause added work and additional burden. Do/ did you help?

Like most club bulletins at the time, in 1964 the NASWA bulletin, "FRENDX," was produced by mimeograph.

what had become a friendly annual DX competition between member teams of NASWA and the Canadian DX Club, and the contest trophy came south for the first time. And NASWA became a charter member of ANARC.

Like Newhart, Benson's tenure was also brief, cut short in September 1964 by his relocation to New Hampshire. Eddings took over "Shortwave Center" and prepared it from headquarters until January 1966 when former "Newsroom" editor Ronald Luyster of Flushing, Ohio was put in charge of the column. A "scoreboard" of members heard and verified totals was added in July 1965 (the ham and BCB columns had their own scoreboards), and by the end of the year the number of members listing their totals had grown from 13 to 55.

In an effort to democratize the club, elections were scheduled for 1964. Although there were some 20 nominees for the board of directors, only three were willing to be considered, and Eddings (who himself was unopposed in the election for Executive Editor) canceled the elections and declared the willing nominees the winners. They were Don Jensen of Racine, Wisconsin, ham band editor Lavoyd Kuney of Detroit, Michigan, and past BCB editor John T. Arthur of San Jose, California. There being no club constitution, the exact duties of the board were unclear. A constitution was drafted and distributed to members in December 1965. Following approval by all but 7 of 319 members voting, it was adopted.

The bulletin took on a slightly different appearance in February 1965 when an eagle logo began gracing its front page. It was modeled on a sketch of a plaque which NASWA briefly offered in 1965 to the North American member of the SWL-Certificate Hunters Club who scored highest in an annual ham band competition. The eagle would be a standard *FRENDX* symbol until the bulletin went offset.[27]

Eddings had first raised the issue of specialized clubs in an editorial in October 1965, concluding that NASWA's future as an all-band club was dependent on the level of member support for the various columns. A subsequent poll found that most members focused on shortwave broadcast listening rather than hams, medium wave, etc. Eddings continued to raise the topic, hinting that specialization was coming. An official vote of the members was taken and the result, with about 70 percent of the members voting, was 192 in favor, 102 opposed. In July 1966 it was announced that NASWA would henceforth cover solely shortwave broadcasting. The small number of members who wanted to leave the club received a pro rata refund of their dues or had their memberships transferred to either the International Radio Club of America (a medium wave club) or the Canadian DX Club (a club with a strong ham section).

The first SWBC-only issue of *FRENDX* was in August 1966. The citizens band, space and "Tech Trails" columns had been dropped during the club's first few years. They were now joined by the amateur and medium wave columns, and the "Friendship Corner." Don Jensen became overall shortwave editor, preparing "Shortwave Broadcast Center," which now became a features column. At the outset he was assisted by Gregg A. Calkin of St. John, New Brunswick, who did "QSL Report" (which included the members' scoreboard), and Ron Luyster, who did the "Loggings" column. "Newsroom" was prepared by Eddings, who also continued to do a headquarters section. Over the next few months, Del Hirst of Snyder, Texas became loggings editor and Dan Henderson of Laurel, Maryland took over "Newsroom."

The impact of the change to all SWBC was immediate and dramatic. The bulletin, which had been improving over time, now enjoyed a huge jump in both substance and

O F F I C I A L   B U L L E T I N

O F F I C I A L   B U L L E T I N

*************************************JULY 1966*************************************
HEADQUARTERS NASA   1503 FIFTH AVENUE, APT.2   ALTOONA, PA.16602:
****

$\frac{196}{102}$ NASA GOES ALL SWBC! $\frac{196}{102}$

SEVENTY PERCENT OF MEMBERSHIP VOTES

It is all over but the shouting! This is the final edition of Frendx to be published
by NASA as a general club; Beginning next month we go all SWBC! "You asked for it"!!!
By a plurality of ninetyfour votes, the action has been decided! Final tally has
been made; Majority favors the change. Apathy on part of approximately thirty per
cent of membership would seem to have figured in the tabulation, but we will never
know, can only speculate. Fortunately most eligible members did exercize their
right, did vote, and the outcome clearly shows a new course for NASA in future!!!

NEW EDITOR.............NEW FORMAT

So that we can get off to a worthy start, prove to the SWBC realm of DXers that we are
earnest in our hopes to have NASA become a mainstay, we take pleasure in announcing
appointment of Donald Jensen as our SWBC editor. He will take over beginning with the
August section and will be assisted by Ron Luyster and new/additional column editors
now being contacted by hdqs. The new format for the DX section will run along lines of
old, but will have "flavor" not heretofore found in a club bulletin. Frendx will not
be just another "news-sheet", but then we like to think it never has been! We expect
it will attract many new SWBC DXers to our ranks, both oldtimer-toppers, and novices.
With the cooperation of them, along with those regulars already aboard, our future is
to be shaped; Success will depend on team work! "Unity and Friendship" will be vital!

!MORE TO COME!

There will be other changes, both as to bulletin staff, and as to headquarters from
the operational functions. These and all future changes to be made purposefully with
want for more solidarity, more security as in preparation for time ahead. Improvements
in bulletins, mimeo-wise, will be taken to task; New club services will be offered....
each/all to make NASA the "new era club" for the realm of SWBC enthusiasts. Now, more
than ever before, all sincere members will need come forth and be recognized by giving
support, sending reports, showing faith, promoting our motto. If each will do his or
her part, then our course to the future is plotted, we will arrive! Harken and hear ye!
                    *****   *****   *****

FRENDX OF FUTURE WILL BECOME A JOURNAL FOR DXERS AND SWLS WHO ARE INTERESTED IN SWBC!

In July 1966, Bill Eddings announced that henceforth the club would be all shortwave broadcast.

member participation. If you were a SWBC DXer, NASWA membership was a must. By December, the loggings column was ten pages long and had almost 50 contributors, including both experienced DXers and newcomers, twice the number of NNRC SWBC contributors at the time. "Newsroom" focused on schedules, the doings at various stations, listening tips, radio-related treatment of politics and international affairs, as well as features from other clubs and from the public press.

Although QSL reports had long appeared in *FRENDX*, they were usually perfunctory and not very informative. Now they received more attention. Based on members' submissions, "QSL Report" was soon the best organized, most detailed body of information on the QSL practices of individual stations—what the QSLs looked like, time to verification, return postage sent, etc.—along with graphics. The column also had a countries heard and verified scoreboard (with over 100 participants by year's end, and peaking at 186 in August 1968). The scoreboard appeared monthly at first, then bi-monthly, then quarterly.[28] "QSL Report" also included occasional features on such topics as new QSLs, verifying Russian regional stations, "white lists" of reliable verifiers (an idea borrowed from the America Central Radio Club), lists of stations which sent pennants, etc.

In all, *FRENDX* was some 50 pages long. The club's first contest editor, Bob Hill, then living in the Washington, D.C., suburbs, conducted a two-week "Logging Marathon" contest on the 31 meter band with transistor radios and *WRTHs* as prizes. NASWA contests continued through 1969, although usually there were only a handful of participants.

The biggest change was in "Shortwave Broadcast Center," which now included a wide range of schedule information; write-ups about particular stations and countries ("DX Atlas"); profiles of radio personalities (including interviews with the likes of "Sweden Calling DXers" host Arne Skoog, journalist John Chancellor, and Senator Barry Goldwater); and features on various DXing techniques. In addition to attracting articles by experienced club members, the column also became the focus of original research into various shortwave topics, much of it done by Jensen, whose writings within the DXplorer Radio Association (see p. 215) now had a new outlet. His work included both historical research, much of it on clandestine radio, and research into current operations, the latter resulting in several exhaustive articles on the geography, politics and shortwave broadcasting activities of such places as Indonesia, China and the U.S.S.R. Material of this scope and quality had not appeared in the American DX press before.

The shortwave news was supplemented with headquarters information that included extensive ANARC reports (long a NASWA staple), new member bios, members' comments ("The 'Open Mike'"), and the humor of Charlie Loudenboomer, plus the NASWA "Flash Sheet." Inaugurated in September 1966, the "Flash Sheet" was a two-page column of "hot" information from both NASWA members and others. Prepared at the outset by Bill Eddings, it was assigned its own editor, Al Niblack of Vincennes, Indiana, in November 1967.

The changes at NASWA did not go unnoticed by other clubs. NNRC shortwave editor Hank Bennett, not one to throw undeserved bouquets, wrote:

> NASWA has, as a result of going to an all–SWBC format, become one of the leading clubs in that category anywhere in the world. Their monthly bulletin is really a masterpiece. There are far more interesting items in that bulletin than in any other that we've seen. And — and this is most important — they have a crack editorial staff.... Other clubs will do well to look at NASWA for a good example of producing a bulletin![29]

The editorial team largely eclipsed the board of directors, which faded from view (and soon was dropped from the monthly "staff roster").

NASWA had enjoyed two years of stability in its editorial team since the switch to all–SWBC. In August 1968, Richard E. Wood of Honolulu, Hawaii succeeded Jensen as chief NASWA shortwave editor. Wood's tenure was brief, however — eight months, after which "Shortwave Broadcast Center" was prepared by Eddings until the appointment of William A. Matthews of Columbus, Ohio in October 1969, the same month that Robert Hagerman of Hemlock, Michigan took over "Newsroom" from Dan Henderson. In November of 1968, after a two month stint by Rod Williams of Rossville, Georgia, Dan Ferguson, then of Coral Gables, Florida, assumed the editorship of "Log Report" from Del Hirst. Thus by the start of the new decade, a new editorial team was in charge.

The club responded well to an unsettled period in the early 1970s. On October 16, 1970, Eddings suffered a heart attack. Club member Edward C. Shaw of Norfolk, Virginia stepped in as publisher for three months, and was then appointed Corresponding Secretary in order to relieve Eddings of that burden.[30] Shaw picked up the publishing again in January and February 1972 when Eddings returned to the hospital. In March, publishing was assumed by Dan Ferguson, by then living in Charleston, West Virginia and editing "Listener's Notebook" (as "Newsroom" had been renamed), which he expanded to include by-country DX information in addition to schedules.

The club suffered a major loss when Eddings died on August 23, 1972 at age 51. Early in his tenure, in talking about his work on behalf of the club, he said: "...I hope to think I did a bit of good somehow, somewhere along the road and that in days to come there will be just one, at least, who will recall that [the club] was once run by an obscure, humble OM ["old man," a ham term] who wanted nothing but friendship and goodwill between those he came to know as brothers. Can anyone say I was wrong in preferring my number of friends over my number of DX catches?"[31]

Throughout his administration, Eddings had often suggested that it was time for someone else to take over, but he had taken few steps in that direction. Although it was fortuitous that Shaw and Ferguson were willing and able to step in when they did, the most fitting tribute to Eddings was that the club survived his absence.

The club had taken many initiatives since switching to all–SWBC. In 1967 it issued the NASWA Country List. Designed mainly for SWBC DXers, it contained some 212 countries.[32] A Country List Committee was set up to consider cases of new countries and instances where clarifications or interpretations were necessary, and revisions were issued from time to time. During 1967 and 1968 the club distributed reception report forms in French, Portuguese, Spanish, and Indonesian. These were the best forms commonly available at the time in the United States. The awards program was rejuvenated and expanded in 1968 and again in 1974. New awards were offered, and the names of award recipients were featured in the "NASWA Awards Reports" column. Club member stationery was made available in February 1971.

To fill the void left by Eddings' death, Ferguson stepped in as Executive Editor and applied a level of energy and organization that brought the club to still higher levels. He used his extensive contacts within the shortwave fraternity to enlist people for editorial posts or to work on club projects. His administration was a busy one.

In January 1973, *FRENDX* introduced columns on two specialized topics: Larry

Magne's "Clandestine Bulletin," and Glenn Hauser's "Harmonic Hunting with Hauser."[33] The harmonics column lasted until July 1973. "Clandestine Bulletin" ran until February 1975, when Magne concluded that the subject was receiving sufficient attention in the general shortwave press as to no longer warrant separate treatment.

Also in January 1973 a separately mailed, mid-month "Flash Sheet" was introduced to supplement the monthly in-bulletin "Flash Sheet." (The latter went to separate mailing as well in 1974 but returned to in-bulletin status in May 1975.) Called "DX Hotline," "Update" and "DXtra" at different times over the years, the "Flash Sheet" continued under various editors until 1989 (and was resurrected in electronic form in 2001). In May 1973, in order to relieve the workload on any single editor, "Log Report" was separated into four sections, each with its own editor. (It was variously consolidated and expanded over the years.) During the summer of 1973 a survey of members was taken, and for months afterward "Listener's Notebook" concentrated on the stations that members wanted most to hear. In June 1974, "Scoreboard" was separated from "QSL Report" and became a separate, quarterly feature.

The first (and only) NASWA convention was held in Columbus, Ohio in July 1974. The three-day event was hosted by Bill Matthews and Randy Minnehan, and 58 people attended. The banquet speaker was Radio Canada International's Ian MacFarland. Also in 1974, three special broadcasts for club members were arranged, from Radio Cook Islands, Radio Maldives, and the Polish Pathfinders Station. The first two were not heard and the third did not take place. Several members did hear a NASWA special from Turks Island Radio in January 1976.

A variety of NASWA special publications was issued during the 1970s. These included "QSL Verification Signers," a list of the signers of almost 300 stations (1972); "After the Sun Goes Down" by Ed Shaw (about the darkness pattern) (1973); the "Niblack Reprint Booklet" on receivers (1973); the NASWA Country List (1973); and a "Survey of Bolivian Shortwave Radio Stations Known to Be Active Between December 1973 and October 1974" by Tony Jones of Paraguay (1974). A number of band surveys were completed during the years 1970 to 1974, and the results of a survey of antennas and accessories used by club members ran in *FRENDX* from May through November 1975.

One of the club's most ambitious projects was Ed Shaw's *DXing According to NASWA*, published in 1975 (see p. 189). About six months later the club offered a 16-page pamphlet by Shaw, *Welcome to the World of Short Wave Listening!* Its purpose was to promote shortwave listening in as many venues as possible, and the thousand copies printed were gone in a few months.

In July 1975, Glenn Hauser took over the "Listener's Notebook" column from Dan Jamison of Richmond, Virginia. Hauser's LN was content rich. In addition to a huge quantity of by-country DX information, it contained, at various times, subsections on clandestine stations, pirates, broadcasts in English, harmonics, publications, program news, DX programs, future station plans, reader opinion and other topics—a mini-bulletin all its own. While Hauser was sometimes out of sync with NASWA administrators, his column was a major asset to the club.

Some disputes over club structure and editorial guidelines played out in the pages of *FRENDX* during 1976 and 1977. However, the club continued to prosper into the mid–1970s. In August 1976 it published its first membership directory, of which several revisions followed. The NASWA "Special Publications" series was published in 1976. It included primers on precision frequency measurement, propagation, and the elimination of reception noise. A fourth summarized a comprehensive NASWA membership survey which was the basis of a master's thesis by member Kim Andrew Elliott. A Boy Scout Merit Badge Committee was established in 1978 to assist the Boy Scouts of America in updating their radio merit badge requirements. The first "NASWA Radio Country List and Awards Program" booklet was published the same year (it would be revised and republished in 1980 and 1986). In 1978, Diane Lévesque of Carbondale, Illinois began editing the "Music Page," and the following year Mike Agner of Burlington, Massachusetts started an "Antenna/Accessories" column which would be published until 1984.

Organizationally, Ferguson was the mainstay of the club. However, he was seeking relief, which he finally got in May 1975 when Alan E. Mayer of Park Forest, Illinois took over the publishing, and M. R. (Mac) Leonhardt of Liberty, Indiana, then one of the "Log Report" editors, assumed the role of Executive Editor. The club had long been operating outside the terms of the constitution and by-laws adopted in 1966, and in December 1976, after a 10 to 1 vote of the membership, a new charter and constitution were adopted, replacing the former structure with an Executive Council composed of an Executive Director (Leonhardt) and not more than three councilors (increased to four in 1977). In the 1976 restructuring, the publisher (Mayer) also became the managing editor. In February 1978, Mayer was succeeded as publisher by William E. Oliver of Levittown, Pennsylvania, then the club's QSL editor. (Mayer became Correspondence Secretary, serving until August 1981.)

As the decade of the 1980s began, NASWA was a strong club, with a membership of approximately 2,000. *FRENDX*, which was by then 58 pages long, had four main components: "Shortwave Center," the features column; Glenn Hauser's "Listeners Notebook"; "QSL Report" (by far the best QSL section of any club bulletin, and then edited by Sam Barto of Watertown, Connecticut, who remains the column's editor today); and the multi-section "Log Report." There was also a headquarters report, the monthly in-bulletin "Update" (successor to the "Flash Sheet") and a separately mailed mid-month "Update," plus a quarterly "Scoreboard," a quarterly awards report (monthly from 1985 to 1989), and other columns.

During the first half of the decade the bulletin continued to improve. Although editors frequently bemoaned low member participation, typically in 1980 some 60–80 members contributed monthly loggings (the number had exceeded 100 a few times in the latter 1970s). The number of members submitting their totals to the "Scoreboard" reached 225. When John J. Moritz, Jr., of Youngstown, Ohio took over "Shortwave Center" in October 1980, he enlisted several experienced members to edit sub-columns. "It Sounds To Me" was a well-received opinion piece by journalist Alvin V. Sizer of North Haven, Connecticut, and drew considerable member response. Other sub-columns were "Vintage Vignettes" by Bill Taylor of Unionville, Pennsylvania, dealing with early receivers and early shortwave broadcasting; "DX Smorgasbord," in depth, by-country DX news by Dan Ferguson; and "Technical Topics," at first written by Ed Shaw, then by James G. Herkimer of Caledonia, New York, and, from 1985, by T. J. "Skip" Arey of Beverly, New Jersey.

The sub-columns were continued by Chris Hansen of the Bronx, New York, who succeeded Moritz in 1982, and by John C. Herkimer of Caledonia, New York, who took over from Hansen in 1983. Herkimer added a member discussion column called "Contact."

Edited at first by Andy Robins of St. Joseph, Michigan, and later by Stephen G. Moye of Cranston, Rhode Island, it did a good job of bringing forth member opinion on many topics during its five-year run. Also added by Herkimer was "Pappas On Programming" by Nick Pappas of Fairless Hills, Pennsylvania. It offered the best content analysis of shortwave programming of all the club "program" columns of the time. "Computer Corner" was added in 1984.

In 1980, QSL editor Sam Barto began featuring a monthly HAP auction page where he auctioned off stickers and other station memorabilia on behalf of the Handicapped Aid Program. In 1982 the club published the *NASWA Tropical Band Guide*. Developed by George Sherman of Minneapolis, Minnesota, it was an extensive listing of tropical band loggings reported by North American *FRENDX* reporters from February 1976 to February 1982. A second edition was published in 1983.

William E. Oliver became Executive Director (in addition to publisher) on May 1, 1981, with Mac Leonhardt continuing as Business Manager and Treasurer and handling most club financial and membership matters until January 1984 when he retired and Oliver assumed those duties as well. (Leonhardt died in 1988.) With all headquarters duties now in his portfolio, Oliver's work on behalf of the club became herculean, and but for his efforts the club most likely would not have survived, fort it was a particularly challenging time. Many of the top DXers who had actively supported *NASWA* when it went all–SWBC, and were largely responsible for its growth and reputation, were no longer in evidence. Some had become inactive, while others had shifted their attention to weekly resources such as the *Numero Uno* and *Fine Tuning* newsletters. In any event, whether it was the absence of the old guard that contributed to a reduction in the bulletin's hard core DX content, or vice versa, it was clear that now a great many of the club's members were more interested in enjoying shortwave listening for its own sake than in reaching the heights of the DX game.

Whatever the exact dynamics, after 20 years the club was showing its age. As early as 1981 only a few people were contributing to "Update," and by mid-decade there were months with no "Update" contributions at all. When John Herkimer's period as "Shortwave Center" editor ended in 1985, no successor was named. *FRENDX* continued to carry the various SWC sub-columns, but in the absence of a unifying editorial hand they became disjointed. Whether it was because *FRENDX* had become a less welcoming venue for shortwave authors (the more prolific of whom were the hard core DXers), or simply because no one was working on content development, the number of feature articles, long a *FRENDX* staple, diminished. In addition, the infrequent appearance of the headquarters page from 1985 deprived the club of a unifying element and contributed to a sense of organizational drift.

A "NASWA Novice" column was tried in 1985 but soon disappeared. The last in-bulletin "Update" was in October 1986, after which it became solely a mid-month publication. In 1987, an attempt was made to reprise the successful 1980–83 opinion column, "It Sounds to Me." Its successor, "Kap's Korner," edited by John M. Kapinos of Shrewsbury, Massachusetts, lasted only a few months. The need for content was met more and more with filler material—full-page announcements, press releases, and random station schedules.

Some initiatives were more successful. In 1985 the Pappas programming column was replaced with two columns: "Easy Listening," describing some of the programs that could be heard on shortwave, and "Soundwaves," which covered musical programs. ("Soundwaves" was dropped in May 1987.) Also in 1987, in an attempt to return some traditional DX feature coverage to the bulletin, "DXers Forum," a column devoted exclusively to the techniques of hard core DXing, was begun. It was edited by R. Charles (Chuck) Rippel of Virginia Beach, Virginia, who would soon become a very important person in NASWA.

Also successful was the computer printout of English-language broadcasts which appeared in the bulletin every month starting in August 1986. These lists were prepared by Thomas R. Sundstrom of Vincentown, New Jersey, and while their graphic appeal left something to be desired, they were handy, informative and accurate (as were Sundstrom's comments introducing the material each month). They had the added benefit of filling some 20 pages in the seasonal changeover months, fewer in other months. The Sundstrom lists replaced Roger Legge's shorter listings of English broadcasts to North America which had appeared in the *FRENDX* "Shortwave Center" column in the changeover months between November 1977 and July 1986.[34]

But by 1988 it was clear that NASWA needed a restructuring. Membership had dropped to about 1,000, half the number at the club's height. The number of members receiving the mid-month "Update" was down by nearly half, and the number of "Log Report" contributors was less than half what it had been. Those listing their totals in "Scoreboard" had dropped from a high of 225 to 75. Most importantly, the quality of the bulletin was suffering, in part because the burdens of club administration and publishing had left little time for content development and editorial management.

In order to give these functions more attention, in April 1988 the Executive Council embarked on the first phase of what would turn out to be a two phase effort to improve the club. It established an Editorial Committee under the chairmanship of Rippel. Joining him on the committee were Rowland Archer of Raleigh, North Carolina; Bob Brown of Lansdale, Pennsylvania; John H. Bryant of Stillwater, Oklahoma; and Harold N. Cones of Newport News, Virginia. In September 1988, with Oliver's approbation, Rippel was named NASWA Executive Director, returning the club's leadership to an arrangement similar to that which had existed from 1978 to 1981. Cones became chairman of the Editorial Committee (and soon *FRENDX* Editor-in-Chief). Bryant headed the group's Graphics Task Force.

The enthusiasm of this group, whose members had worked together on other shortwave projects, was evident as it undertook the mission of improving the quality and appearance of *FRENDX*. An editor's manual was created. "DXers Forum" became the lead section. Now edited by Archer, it covered the how-and-why of DXing and drew upon the authoring talents of many well-known DXers. "Computer Corner" and "Contact" were dropped, as was "DXtra," which had been receiving little support. "FRENDX Portraits," showcasing NASWA members, was added, along with a monthly "From the Executive Director" page (called "NASWA Notes" from March 1990). "The CPRV Page," a monthly column of historic QSLs prepared by the ANARC Committee to Preserve Radio Verifications and appearing in several club bulletins, was carried in *FRENDX* from its inception in January 1989. And in June of that year, Bruce MacGibbon of Gresham, Oregon succeeded Glenn Hauser as "Listener's Notebook" editor.

Increased use of computers by editors improved the appearance of the bulletin, and by the end of 1989 membership had increased to around 1,300. Much of the increase was a result of a November 1988 arrangement whereby, in exchange for a monthly page of advertising space in *FRENDX,* Universal Radio, a major national SWL and ham vendor, began enclosing a NASWA promotional flyer with its orders. (A similar arrangement would be made with Gilfer Shortwave from 1994 to 1997.)

The second phase of the NASWA makeover occurred in January 1990 when the bulletin underwent a major facelift. It was given a plain cover (some said it was boring; decorative covers eventually returned), and the name was changed to *The Journal of the North American Shortwave Association.* The appearance of the bulletin improved greatly, and professionalism was the watchword. Expanded use of desktop publishing led to a more uniform appearance, with such touches as photos of the column editors and better graphics. There were new columns: "Listener's Library," dealing with new shortwave-related publications and software; "Listener's Classroom," with answers to members' questions; "Destinations," where a team of editors focused on a different part of the world each month; "Equipment Review," covering new receivers and related gear; and feature articles on other topics. "Log Report" was condensed from four columns into two, one for the tropical bands and the other for international band loggings. Following a club-wide contest, a new logo was adopted and soon appeared on club mugs, hats, T-shirts, sweatshirts, jackets, etc., offered by the new "Company Store," which soon added shortwave publications to its inventory. A reprint service was begun, and a new-member packet developed. Decorative covers returned in January 1992. It was all part of the promotional effort dubbed "NASWA for the Nineties."

Editors changed from time to time, as did the club leadership. In February 1990, Bob Brown, who had been serving as Assistant Executive Director, succeeded Rippel as Executive Director. (Kris Field of Ambler, Pennsylvania became Assistant Executive Director.) Two years later, Brown, who by then had established a close working relationship with Oliver (who was still the publisher), also assumed the newly created job of Managing Editor (the position of Editor-in-Chief was dropped). These steps reduced the burden on Oliver, who could now concentrate on publishing and on membership matters.

The early 1990s were growth years for the club. The Gulf War brought many new people into shortwave listening. The sale of shortwave receivers jumped markedly, and many purchasers became interested in clubs. The NASWA membership in 1991 was approaching the record levels of a decade earlier, with as many as 100 new names added to the roles in some months. Regional chapters were active in Boston, Philadelphia, Ohio and Washington, D.C.

In the *Journal,* member letters and "musings" were encouraged and grew in number. For those more interested in program listening than DXing, Richard D. Cuff of Allentown, Pennsylvania, who took over the "Easy Listening" column in December 1990, gave the program side of shortwave new life. Several new awards were introduced in 1991. A periodic "Radio Stamps" column ran during 1992 and 1993. The format for the English schedules was improved. In August 1994, "DXers's Forum," which had not appeared since June 1993, was reborn as a features column called "Shortwave Center" (the name used from 1966 to 1985). And the Country List Committee issued periodic reports on its doings. Some columns disappeared, including "Listener's Classroom" (in 1991), "NASWA Portraits" (1994), and "Destinations" (1996).

In September 1993 the Executive Director's baton was passed to Richard A. D'Angelo of Wyomissing, Pennsylvania, who retains the post today. Brown stayed on as Managing

Editor until August 1995, when he was succeeded by Harold N. Cones. In June 1994, Ralph Brandi of Middletown, New Jersey, began assisting with desktop publishing and graphics, and in December 1995 the club opened a website with Brandi as webmaster. The bulletin's appearance was greatly enhanced, and Brandi became an increasingly important part of the club's administration. In November 1994, the Executive Council, which had become relatively inactive following the 1988–90 reorganization, was itself recast as an advisory body, with members appointed by the Executive Director.

By the mid–1990s, the growth of the previous five years had abated. Many of the post–Gulf War memberships turned out to be short term, and by 1997 membership was back to around 1,200. Changes in the *Journal* continued. A pirate radio column was added in July 1995. The following year, "Adrian Peterson's Diary," based on the radio history scripts of Petersen's "Wavescan" program over AWR, began. (A long time DXer, Peterson was AWR's Coordinator for International Relations and the former host of AWR's "Radio Monitors International" DX program.) In May 1997, Mark J. Fine took over preparation of the biennial English-language schedules from Tom Sundstrom. This continued until 2005 when the schedules were supplied by Mike Barraclough of the World DX Club (U.K.), who provided a like service to that club. Sundstrom continued with a new column, "Net Notes," which focused on internet resources. It appeared until 2005.

In 2000, the *Journal* received another facelift. The same year saw the establishment of a member of the year award. Called the William P. Eddings Award, the first recipient was Bill Oliver.[35] In 2000 the club also assumed responsibility for the Winter SWL Fest which since 1988 had been held annually near Philadelphia and had been ably run by a group of SWLs who were in the NASWA leadership and were, by 2000, ready to pass the baton to others. In 2003, the club filed extensive formal comments with the FCC in opposition to the authorization of broadband over power lines, a potentially serious source of interference to shortwave reception.

From 2000 to 2004, the *Journal* carried a major, twice yearly section called "SWL Program Guide." In it, John Figliozzi of Clifton Park, New York, provided a by-time listing of programs broadcast to North America in the morning and evening prime time hours. Published for the spring-summer and fall-winter seasons, it included specific program titles, days and times. In December 2001, the "Flash Sheet" was reintroduced, this time on a weekly basis and in electronic form. News of updates to the www.ontheshortwaves.com website, which covered shortwave broadcast and DX history, were carried in the *Journal* starting in April 2003. And in 2005, Kim Andrew Elliott, by now a long time employee of VOA and IBB and a well-known member of the SWL community, joined the *Journal* editorial team. "Kim's Column" covered developments in international broadcasting and public diplomacy, and provided the bulletin with some new breadth. Through it all, D'Angelo continued to serve as Executive Director and contributed numerous station profiles to the *Journal,* as he had been doing since 1984.

By 2007 the club was experiencing both the benefits and the burdens of its senior status. As the sole survivor among the national shortwave clubs in the United States, it had a heightened responsibility to America's audience of SWLs. Now, however, that audience was much smaller. NASWA membership was about 650, far below that of the halcyon years but still more than twice the size of the next largest North American club, the Ontario DX Association. ODXA had gone all electronic, and the question of whether a paper bulletin like the *Journal* could survive in the contracting world of shortwave was an open question. The club's membership was also older, and just a few dozen stalwarts carried the torch in terms of contributing to the *Journal*. But while the high-energy days were over, and the environment had changed, the determination to chart a successful path to the future remained strong.

CANADIAN DX CLUB (1961–1968)

Although it is largely lost to history, the CDXC appears to have been the first national Canadian shortwave listeners club. It was formed in 1961 by Bill Graham and Fred Woodley of London, Ontario, and Dave Bennett of Richmond, British Colombia. Bennett was 18, Graham and Woodley were in their early 20s. The first issue of the bulletin, soon to be known as *CADEX,* was published in April 1961. It outlined the aims of the club: to bring DXers a complete report of the hobby in Canada; to publicize the hobby in order to attract new members; to help DXers solve their problems, offer assistance and help them further enjoy the hobby; and to "help bring about world peace through international understanding via the hobby." Whatever may have been the club's contribution to the last of these, it did make an impact on the others.

The club's constitution articulated the objects of the club a little differently: "to create a stronger unity among Canadians of all races and creeds through a mutual interest in Radio and TV monitoring"; inform DXers in other countries of radio and TV activities in Canada; and assist radio enthusiasts in better enjoying the hobby. The constitution provided for an elected three-man council, the members of which were to be Canadians. For most of its life the club would operate with little regard for these niceties, however. A system of full membership for Canadians and non-voting associate memberships for others was soon replaced with one class of membership. Duties and titles of club leaders changed from time to time. Woodley, as "Chairman of the Board," was the key man during most of the club's life. However, both Bennett and Graham also played important roles.

By the start of 1962, after some experimentation, the contents of *CADEX* had stabilized. The bulletin covered SWBC, medium wave, ham, utilities, and FM-TV. Bennett was the chief editor. Because he typed the entire bulletin, *CADEX* had a uniform look and feel. A system of regional editors was tried but scrapped in favor of a more conventional system of topic editors, of which there were many. In addition to coverage of the various DX bands, the bulletin contained extensive member correspondence, member ads, card swapper and tape swapper names, statistics on members' states, provinces and countries heard and verified, reviews and brief owners' reports on receivers ("Buyer's Guide"), limited news about QSLs received, and short features, including some technical and construction articles. There was also an awards program, a column devoted to those who collected listening awards (mostly ham), and a contest program. Contests were held annually for ham band and BCB DXers, less often for others. One year there was even an SWL-card swapping contest. Year-long, all-band "DX-Athons" were held in 1965–66 and 1966–67, with points equaling the number of stations verified times the power of each station. The prize was a choice of one of several books. And for several years a friendly rivalry with NASWA existed in the form of an annual listening contest, each year featuring a different band, each club choosing a team to compete against the other. Various books, club stationery, club seals and other items were available through club headquarters, which also arranged for the sale of Canadian and U.S. stamps to DXers on the opposite sides of the border.

In January 1962 the bulletin reached 15 mimeographed, legal-sized pages, and membership was at 56. By mid-year Woodley had received some 2,500 pieces of mail about the club. A membership drive brought membership to 161 by the end of the year. The bulletin was 34 pages long. The ham band column had grown especially quickly, followed by medium wave and card swapping, with respectable showings in the utility and FM categories as well. (A few clearly erroneous reports were inserted each month to see if the bulletin was being read, and at least one member received a six-month extension on his membership for questioning the "camel stampedes in Alaska" item in the utilities section.) From August 1963, the club's important BCB column was edited by Andrew F. Rugg of Pointe Claire, Quebec, whom Woodley described as "the best all-around DXer in this country."[36]

Shortwave broadcast coverage lagged behind the other sections. The SWBC column, with charter editor Jim Tomkins of Regina, Saskatchewan, consisted of members' logs arranged by time, supplemented by country-specific schedule information. The bulletin's SWBC content expanded when Tomkins was succeeded in December 1962 by Dave Clark of Toronto, Ontario, who would become a well-known and highly respected DXer in the decades to come. Clark supplemented the regular shortwave news with information about QSLs, but the column had few contributors. Shortwave broadcasting received additional attention in July 1963 when a new column, "World Wide Review," was begun. It concentrated on individual stations, SWBC and BCB (mostly the former), and it was edited by another DXer who would soon make his mark, Gregg A. Calkin of St. John, New Brunswick. The column lasted only for a year, however. The club's SWBC award, the A6Z Award ("A to Z in Six Bands"), was based on the total number of countries, up to 25 per band, that the applicant had heard on the six SWBC bands. A parallel award was available for ham listeners.

One of the club's strengths was the socializing among its members. The first unofficial get together of seven club members was held in August 1961 at Woodley's home. Local chapters soon formed in London, Ontario, Toronto, and Montreal. A club convention was held at the Westbury Hotel in Toronto in July 1962 and drew about 30 people, roughly a quarter of the club's membership. The plan was to have another national convention five years thence, with three regional gatherings each year in between. Thirty club members attended the regional meeting in Kingston in 1963, but proposed meetings in Vancouver and Halifax that year never took place. There were many informal get togethers among members, however, and a few DXpeditions. Twenty-five members attended a club convention in Montreal in 1964 (a train coach was chartered for the trip from Toronto to Montreal), and a convention in London, Ontario in 1965 drew 21. To facilitate member contact, the club's full membership list was published annually, a practice that would be adopted by the Canadian International DX Club during most years between 1982 and 1997 and by the Association of DX Reporters during its entire 13-year existence. (The CDXC dropped it in 1965.)

In May 1963 the club weathered its first crisis when the bulletin's chief editor, Dave Bennett, resigned. He took over as club contest chairman and utilities editor, but his duties as *CADEX* chief editor were assumed jointly by Fred Woodley and by Dave Noon of London, Ontario, who were assisted by Bill Graham and by club treasurer Jim Warner of

Toronto. However, club membership had now grown to approximately 200, and the work involved in running the club — it was estimated that it took some 70 hours monthly just to produce the bulletin — had become too great, and both Woodley and Noon resigned in October. Woodley wanted to focus on his own DXing and on special projects, and Noon was returning to university.

Graham picked up the cudgel, but the club proved no less a burden for him. With the club close to collapse, Woodley agreed to assume all club functions save for the time-consuming task of typing the stencils, which would now be done by the individual editors. Money donations were requested of members in order to improve the club's troubled finances.

Although membership dropped to about 175 by mid–1964, the reincarnated CDXC continued to function relatively well. The bulletin was chatty and interesting. It featured many letters from members, along with DX content of good quality and good-natured ribbing among club "old timers." A column on space-related radio appeared for a time starting in May 1964. However, the strongest parts of the bulletin continued to be the ham and BCB sections. Although he had long before left SWBC DXing in order to focus on the ham bands, Bill Graham took over the SWBC section in November 1964. Despite his requests for increased member contributions, SWBC content never reached the level of the other columns.

The SWBC column disappeared altogether after February 1965, but was reinstituted in June under a new editor, Jerry Klinck of Buffalo, New York, who had recently brought the American Shortwave Listeners Club back from virtual extinction (and still served as ASWLC publisher-treasurer). Klinck's reorganization of the SWBC section did little to breath new life into the subject matter, however, and the club dropped SWBC coverage altogether, along with utilities, within a few months. SWBC reappeared briefly in March 1966, and a SWBC contest was held in the spring of 1966. However, it was planned that this would be the end of the club's SWBC coverage, and members interested in SWBC were given the opportunity to transfer their memberships to NASWA when it went all SWBC (NASWA gave its ham banders the chance to switch to CDXC). It was thus a surprise when a SWBC column reappeared in February 1967. It was edited by Mike Scott of Roxboro, Quebec.

Woodley's illness in September 1964 brought Dave Noon to the editor's chair. Woodley remained club chairman, but Noon handled most of the bulletin production. This again proved a daunting task, and in February 1965, observing that they had gotten the club back on its feet, both Noon and Woodley resigned. Although there was some movement to have the club run by a group of Montreal members, in the end Dave Bennett, the bulletin's original editor, became the club's leader, with Woodley soon reappearing to replace Noon as ham band editor. In 1965, Woodley was also named CDXC "Man of the Year."

News from other clubs gained more prominent coverage in 1965 and 1966. In August 1965 CDXC membership stood at 168. A French page that was added to the bulletin in September lasted for a year.

The year 1966 was one of change for the club, and while some events were positive, it was the start of the club's decline. In February, Andy Rugg was succeeded as BCB editor by Sam Simmons of Montreal. BCB contributions were dropping, however. A strong bimonthly utilities column, edited by Glenn Hauser, then of Albuquerque, New Mexico, was reinstituted in May. In December the FM-TV column was split into two, and both columns received good support from the membership. On the negative side, Woodley quit his ham column in September, and while *CADEX* contained other ham coverage, Woodley's column was missed.

In March 1967, Bennett indicated that the burden of publishing the bulletin and handling club business was too great and that he wished to hand over the reins to someone else. Renewals and new memberships had declined drastically. Only 102 copies of the March 1967 bulletin were distributed. In April, Andy Rugg again took over the BCB section when Sam Simmons was drafted. In August he summed up the situation saying, "Only two reports were received. When I had the column two years ago it was common to have 20." The SWBC section showed some promise, although building a SWBC section at a time when NASWA was a high-performing, all-SWBC club was a challenge. For those who had been around early in the club's life, the bulletin suffered by comparison with early editions.

Bennett agreed to stay on as publisher, and in June 1967 a new member, Ralph J. Irace, Jr., of Avon, Connecticut, a ham and a listener both, stepped forward to lead the club. He was the first non–Canadian to hold a club office. Although a Canadian, SWBC editor Mike Scott, became club Secretary, the club was losing its Canadian identity. A newsy column, "CADEXtras," edited by Doug Benson, who previously had edited for NASWA, was introduced in November 1967. Benson was from Laconia, New Hampshire. Three new editors appointed in 1966, and two in 1967, had been from the United States. DX Inter-Nationale, a small club in Newton, Massachusetts, was absorbed by the CDXC in September 1967, and some of its members transferred to the CDXC rolls.

As 1968 began, Irace, whose appreciation for club democracy had come late, was seeking four persons to run for an elected board of governors. The headquarters section of the bulletin reflected a good deal of activity on his part as several of the columns, including SWBC, appeared to be getting their second wind, and the bulletin's appearance was improved. Irace moved to Riviera Beach, Florida, early in the year and appeared to be managing the club acceptably from there. However, amid rumors that C. M. Stanbury might take on the role of CDXC Executive Editor, it all came crashing down. The March 1968 CDXC bulletin was the last. The club dropped from sight and went out of business.

Especially in its early and mid life, the Canadian DX Club had shown that there was sufficient critical mass among Canadian DXers to make a national shortwave club possible in Canada. It would be two other clubs, the Canadian International DX Club and the Ontario DX Association, which would prove that it could be viable over the long term.

CANADIAN INTERNATIONAL DX CLUB (1962 TO THE PRESENT)[37]

At first called the Canadian International DX *Radio* Club, the Canadian International DX Club, or CIDX, was founded on May 11, 1962 by seven young Winnipeg radio aficionados, some of whom had come to know each other from running an illegal radio station with a range of six blocks. Its locus, a family attic, would soon double as club headquarters.

The club started out as a not-too-serious project. Awards made under the club's "awards program" were practically given away, and the club presidency was rotated among members on a monthly basis to give everyone a chance. Gerry Porter was chosen "permanent" president in August 1962. (Porter passed away in 1981 at age 36.) Several people would serve in that role during the next few years. August also saw the first publication of the club's one-page bulletin, the *Messenger*. The club pulled through a near-collapse in November 1962, but by January 1963 it was back on track, with a bulletin masthead and a growing, if small, membership. In July a "convention" was held in a Winnipeg church hall, with 30 persons attending. Honorary memberships, and a member of the year award, were established. To raise money the club held a series of dances, such socials being a favorite fundraising technique in Manitoba at the time.

The year 1964 was not a good year for the club, with member participation down and the bulletin issued tri-monthly at one point due to the competing interests of its leaders and a lack of member support. Things looked up in 1965, however. Jim Rzadkiewicz took over as president, and in addition to holding another convention (this one in Hamilton, Ontario), the club published two "*Messengerettes,*" professionally printed bulletin supplements with photos of members and other club information. Membership increased; in November the first club contest was held (medium wave only); and the club was admitted to ANARC.

Elections were being held annually, and in 1966, 17-year-old R. Lorne Jennings was elected CIDX president. He would remain the club's key man for almost 15 years. In May 1967 another "*Messengerette*" was issued, and in July another CIDX convention was held in Winnipeg, this time in conjunction with the Pan American Games. The convention again drew 30 club members, some from as far away as Seattle, Quebec, and Massachusetts.

The next few years were a period of experimentation with the *Messenger*. Columns were begun and dropped depending on the editorial manpower available and the level of member contributions. A monthly "*Messengerette* award" was made to the editor of the best column, and "active member contests" were held in order to encourage contributions. In these six-month contests, members accumulated points for various club activities, such as reporting to the bulletin and signing up new members. Member support was still lagging, however, and there was no CIDX convention in 1968 nor in the four years thereafter.

One of the club's most active members in terms of bulletin contributions, obtaining new members, and publicizing the club was Curt Webber of Leavenworth, Kansas. Although he had been a club member for only four years, the death of the "grand old gentleman" in April 1969 at age 77 was a blow to the club. A memorial certificate was issued to him posthumously, and for many years its inscription was displayed in each issue of the *Messenger*— "In loving memory of Curt Richard Webber for his never ending work to promote the short wave listening hobby for the betterment of all."

Webber's death was a bad omen, for while membership had topped 300 by mid–1969 and the mimeographed bulletin was typically over 40 legal-sized pages, the club was plagued with problems. Bulletins were delayed, finances were shaky, and administrative work languished. Both the publisher and treasurer resigned, and no bulletins were issued during most of 1970. The club regrouped toward the end of the year, however, and bulletins resumed. In 1971 an elected board of directors was established. Members held two-year terms, and the board issued monthly reports of its activities. By this time the club covered shortwave broadcasting, the broadcast band, utilities, the ham bands, QSLs, even "Music of the 70s." The main shortwave column was "Logbook," which had appeared since the club's early days. A TV-FM column, and a members' opinion column, were added in 1972, the club's tenth anniversary, an event celebrated by various activities including on-air birthday greetings from several stations.

CIDX member Harold Sellers of North Bay, Ontario was named Member of the Year in both 1972 and 1973. Resigning as editor of the bulletin's "Shortwave Corner" column in 1973 after three years, 20-year-old Sellers was elected to the board of directors and designated chairman the following year. By 1973, club membership had grown to 470. The first club convention since 1967 was held in Kingston, Ontario, and once again it drew about 30 people. In 1974, a club ham band contest was held. It would become an annual affair, and would be held through 1999 (and once again in 2003). A few utility and medium wave contests were held in the mid–1980s as well.

Another successful CIDX convention was held in August 1974. It was sponsored by the University of Manitoba DX-SWL Club, and 40 people attended. However, that summer saw the start of several contentious years in the club's history. Several members of the board, including Sellers and Ivan Grishin of Toronto, Ontario, resigned over disagreements with Jennings, who was then the club's Executive Secretary. (The Ontario DX Association would be formed within six months, with Sellers and Grishin two of its founders.) There was a constant shuffling of editors, most of whom quickly lost interest in the work. With a declining membership, and decreasing support from those who stayed, the *Messenger* grew smaller and less informative. Jennings was deluged with complaints. On the positive side, there was some improvement in the appearance of the bulletin. But this was no panacea for the club's problems. Both a newly found publisher and the appointee to the new position of Chief Editor stayed in their posts for only brief periods.

Better days were ahead, however. Donald E. Moman of Edmonton, Alberta joined the club in February 1978, and within a year was co-editing "CIDX Logbook," the club's main shortwave column, assuming sole responsibility for it in July 1979. In September of that year, Rob Gerardi of Benton, Illinois was named club president by the Executive Council (as the board of directors was called for a short time). Following a long-term fund raising effort, purchase of an electronic scanner led to a better looking bulletin.

By this time there had emerged a group of strong editors and board members who were dedicated to moving the club forward and avoiding the pitfalls of the past. The club reached a turning point in December 1980 when Moman assumed the Executive Secretary's post and CIDX headquarters moved from Winnipeg to Edmonton, where a critical mass of active members could facilitate club operations. Lorne Jennings had served the club well, but he was inactive in the hobby and the growth and complexity of the club had brought unanticipated burdens. Moman brought new blood and a new spirit. He continued to edit "Don Moman's Logbook," a listing of DX programs on shortwave (later expanded to include station schedules as well), and soon he was also publishing the bulletin. Moman also introduced small electronic construction projects ("Project of the Month") to the pages of the bulletin. Soon he would enter the shortwave equipment mail order business under the name "Shortwave Horizons."

The 1980s were good years for the club. It hosted a successful 1981 ANARC convention in Thunder Bay, Ontario, and 25 members attended the club's 20th anniversary convention in Edmonton in 1982. In December 1981, Brian Pimblett took over as club President (the highest vote getter among the board members became President). In May 1982 the *Messenger* went to a 40-page offset booklet format. The bulletin's shortwave broadcasting coverage was contained in two columns: "Logbook," which included members' logs and station schedules, and "International Broadcast Memo," written by well-known shortwave author C. M. Stanbury II, who focused on clandestine stations, pirates and the various oddities and perceived conspiracies to which he was attracted.

The bulletin also included medium wave and FM, ham band, and utility coverage (including, from October 1978, a unique "Arctic DX" column focusing on utility stations in the antipodes); a headquarters column with announcements, new member bios, and member ads ("Shortwave Marketplace"); and a catchall column called "CIDX Forum," which covered ANARC news (including Harold Sellers' highly regarded ANARC "Marketplace Report"), news about other clubs, equipment information, members' comments, etc. There was a monthly list of DX programs ("DX Program Guide"), plus Sam Barto's "Handicapped Aid Program" auction column, and, from time to time, equipment projects and news of new products and members' DXpeditions. (The first of a series of club DXpeditions to Alberta's Jasper National Park and other mountain venues was held in 1979.)

A QSL column ("Verie Interesting") covering shortwave broadcast and broadcast band QSLs, returned in July 1982 after several years' absence. The editor was John Fisher of Medicine Hat, Alberta, who in December added a short-lived, heard-verified "scoreboard" to the column. Ham band QSL coverage was added in June 1983.

In October 1982 the "Loggings" column was divided into east and west sections, with Moman retaining the west and Sheldon Harvey of Greenfield Park, Quebec taking the east.[38] Moman and Harvey had met at the 1982 ANARC convention in Montreal where they had come up with the "east-west" concept of CIDX organization. One factor underlying this idea was the decision to offer CIDX membership for the unexpired terms of members of Canadian SWL International, a club which closed toward the end of the year. Most of the incoming CSWLI members were from the eastern part of the country. The addition of the former CSWLI members was an important step in making CIDX a national club.

In April 1983, station schedules were placed in a separate column, and beginning in July 1984 this column, now named "CIDX Guide," was expanded to include other by-country shortwave information. It was edited by Mickey Delmage of Edmonton, Alberta, who had joined the club in January 1982 and would become one of its strongest supporters. In August 1984, Moman instituted a "Technical Talks" column. In March 1985, Stanbury closed his "International Broadcast Memo" column. (He passed away the following year.)

During the 1970s and 1980s the club created several small publications, including an introduction to long distance listening by Brian Pimblett called *Radio as a Hobby;* the *CIDX Station Address Guide;* the *CIDX Reception Report Guide;* the *CIDX Shortwave Antenna Handbook;* the *CIDX List of Utility Stations* (four editions); the *CIDX Utility Guide* (also four editions); and the *CIDX Morse Code Lessons* (a 90-minute cassette). The club's "New Member Booklet" was updated and published in two installments in the August and September 1985 issues of the *Messenger* and was made available thereafter to new members.

The club took some new directions when Sheldon Harvey became President in June 1986. Harvey had joined the club in May 1981 and become a board member in 1983. Although publishing would remain in Edmonton until 1990, Moman, whose radio interests had expanded into other areas, welcomed the relief from the day to day responsibilities of club operations. Harvey would be supported by an active group of core Montreal members.

Under Harvey, the publication of monthly board of directors reports in the *Messenger*

Canadian International DX Club

MESSENGER

NOVEMBER 1984

CANADIAN INTERNATIONAL DX CLUB
MESSENGER

November 1984

In this issue:

ITEMS OF SPECIAL INTEREST:

The *Messenger,* bulletin of the Canadian International DX Club, at 22 years.

was reinstated. Nigel Pimblett of Medicine Hat took over the QSL column from John Fisher, and two new columns, "Window On the World," dealing with shortwave programming, and "Radio Stamps," covering radio-related philately, verification stamps, and the like, were begun. ("Radio Stamps" soon would be a monthly feature on HCJB's "DX Partyline.") A column for beginners, "Groundwaves," was added (it was combined with "CIDX Forum" in 1989), as well as a column on computers, and a "What's News" column containing radio-related newspaper and magazine clippings. The club also carried "The CPRV Page." During this time, Harvey continued to edit the "Logbook East" column.

In 1987, Harvey, together with Bill Westenhaver of Montreal and other club members, began an "International Radio Report" program over McGill University FM station CKUT. Hosting was passed to Janice and Steve Karlock of Dorval, Quebec in 2004, and the program is still on the air.

Publications and bulletin reprints were made available to members via the club. These included the annual sale to club members of the *WRTH* and *Passport to World Band Radio*. There were more announcements of radio meetings in the bulletin, and greater club participation in hobby shows, hamfests and other gatherings where the club could be promoted. The club was a regular exhibitor at the ANARC conventions and the Winter SWL Fest. In July 1990, in order to facilitate attendance at events, a monthly calendar of radio gatherings was instituted. The *Messenger* became available in some Canadian retail outlets, and mailings to amateur radio clubs brought in ham members interested in learning more about other parts of the radio spectrum. There were more promotions of the club over the air, and in 1988 the club began participating in the periodic report on Canadian radio clubs broadcast over RCI's "SWL Digest" program.

CIDX membership included the opportunity to socialize. Fifty members attended the club's 25th anniversary convention in Edmonton in August 1987. Thirty were present for a club gathering in Calgary, Alberta a few months later. There were club Christmas parties (still an annual tradition for Montreal members), monthly meetings in Montreal and Vancouver, club DXpeditions, and many barbeques, both east and west. In 1989 the club sponsored the First Annual Montreal Shortwave Radio Festival. The day-long event attracted 200 attendees and was conducted by the club's Montreal chapter every year during the next five years. And during the 1980s and 1990s, in order to facilitate member contact, the *Messenger* would periodically contain the full list of member names and addresses.

In January 1990, the publishing of the bulletin was moved from Edmonton to Montreal. Over time a number of steps to improve the appearance of the bulletin were taken. New printers and new printing methods were tried, usually to good effect. More use was made of the reproduction of articles, and the look of individual columns improved as more editors used computers and uniform formats were adopted.

In June 1990 the club's computer column was dropped, as was "Radio Stamps" in March 1991. In August 1990, Larry Shewchuk of Winnipeg, Manitoba began a column called "SW News Journal." Well written, informative and entertaining, it was unique in its coverage of the *content* of the news, current affairs reporting and editorial opinion of the world's shortwave stations. Shewchuk's timing was good because he was able to capture the drama of several major "shortwave events"—German reunification, the Gulf War, and the collapse of the Soviet Union and the eastern bloc. The column ended in February 1992, but Shewchuk returned in September 1994, taking over the "CIDX Guide" column from Mickey

Delmage, renaming it "Prime Time SW" and presenting as its centerpiece a monthly listing of the stations and frequencies that could be best heard during the North American evening period, and supplementing it with station schedules. (Delmage updated "DX Programme Guide," the monthly list of DX programs, until Glenn Hauser assumed the job in 2007.) Shewchuk's illness led to the appointment of Richard Hankison of Kansas City, Missouri as "Prime Time SW" editor in December 1996. Hankison maintained the column's high standard.[39]

The first club Lifetime Achievement Award was presented in March 1991. The recipient was Ian McFarland of RCI. In 1992 it would be presented to Moman and to Robert A. Curtis, originator of the "Arctic DX" column and a long time club member and supporter.

A scanner column was added in January 1992, followed in June by the reinstatement of a computer column. A 30th anniversary convention was held in Montreal in August 1992. Golf shirts and caps bearing the club logo became available that year as well. The most significant event of 1992, however, was the decision by RCI to allow CIDX volunteers in Montreal to handle the station's QSLing when it looked like budgetary constraints would lead to a cessation of all RCI QSLing. Some 2,500 reception reports were responded to in the first year alone. This service brought the club valuable publicity.[40] The club also increased its visibility through some reciprocal, radio-related advertising with *Monitoring Times* and the Ontario DX Association, and limited paid advertising from others, which began appearing in the *Messenger* in 1993.

Harvey established "Sheldon Harvey Radio Books" in 1992 and had some 50 titles available by mail order. In 1995 he opened a store under the name "Radio H.F.," expanding his business to include the sale and servicing of equipment, antenna installation, and other shortwave-related services. He moved the business to his home in 2000, maintaining business contact with the DX community via the "Radio H.F. Internet Newsletter" which continues to serve as both a sales vehicle and a source of interesting URLs.

Mickey Delmage took over the "Verie Interesting" column in June 1994 when Nigel Pimblett moved over to the broadcast band column. In November 1995, after 13 years of editing "Logbook East," Harvey, wishing to concentrate on publishing and on overall club management, relinquished the post to Bob Poirier of Dollard-des-Ormeaux, Quebec. Poirier took over "Logbook West" in January 1998. The two sections were combined into a single "Logbook" when Poirier was succeeded by Sheryl Paszkiewicz of Manitowoc, Wisconsin in October 2000. David Ross of Hamilton, Ontario took over the column in September 2002.

The number of club members was around 370 in 1994, double what it had been in 1982. However, membership declined to a level of approximately 250 by 1997 as the club felt the effects of a confluence of factors that were impacting all clubs: fewer shortwave stations, the decline in interest in radio generally, the growth of alternative media, and reduced leisure time. Other reasons to which shortwave club doldrums were often attributed included the low sunspot count, which affected shortwave reception; the absence of events like the Gulf War, which had sparked public interest in shortwave; and a weak economy which made spending on leisure pursuits a lower priority.

But it was competition from the internet, which had become a pastime all its own and a powerful vehicle for the distribution of shortwave information, that was widely blamed for the decline of the clubs and the shrinking DX community. While computers were not as ubiquitous as they are today, even then some believed that the volume of information on the internet and the speed with which it circulated would soon make paper bulletins obsolete. On the plus side, most of the CIDX column editors were now accessible via e-mail, which facilitated communication with both members and club administrators.

The reality for the club was that member participation had declined dramatically. Only a handful of CIDXers were contributing regularly to the *Messenger,* and in the 1999–2000 elections only 15 ballots were cast.[41] The reduction in contributions and the absence of original material had a negative impact on the content of the bulletin as well. Editors changed more often, necessitating the ongoing recruitment of new, often less-experienced people. Ironically, as with all the clubs, it was the availability of information from open internet sources that permitted clubs to offset the decline in member contributions and club-originated content. However, this led to a homogenization of the information in club bulletins across the entire DX community.

In the CIDX there was some drop in local esprit as well. The Montreal chapter meetings stopped in 2000 when Harvey closed the Radio H.F. facility which had served as the group's meeting place. The Vancouver chapter stopped meeting around the same time. The DX community had matured, and the energy level, especially among those who had been at it for a long time, had declined.

Notwithstanding these challenges, the club's 35th anniversary in 1997 was a happy occasion, celebrated with an October convention in Edmonton which was preceded by several months of selling raffle tickets that brought the winner a South African Baygen Freeplay "wind up" shortwave radio, one of the novelties of the day.

In July 1995, in recognition of the growing importance of the internet, Larry Shewchuk began a "SW On the Internet" column. Although the early columns, which focused on using the internet, were supposed to have been followed by others addressing radio-related internet resources, Shewchuk's unsuccessful battle with cancer meant that only four columns would be published. (Sewchuk died in 2001.) A column focusing on internet resources, "Signals Unlimited," was begun in December 1997. Several other new columns were instituted as well. "Captain's Log," covering pirate radio, was added in December 1997, and in January 1998 the *Messenger* began carrying a column by Glenn Hauser called "Shortwave/DX Report," consisting of material from Hauser's publication of the same name (later "DX Listening Digest"). This was a boon because it provided a large volume of excellent DX material, often in excess of 20 pages.

Also in 1998 the club began its bi-weekly "Canadian International DX Report" in the RCI mailbag program. This would run until 2004. The "Over 25" column, covering the largely overlooked 25–30 MHz. band, was started in February 1999, and monthly propagation charts from the Geological Survey of Canada were instituted in October of the same year. And the *Messenger* once again carried regular ANARC reports. The bulletin varied in size but was usually 50–60 pages in length.

In January 1998, Don Moman, who had been the club's Executive Secretary since December 1980, stepped down and Fred Van Driel of Montreal took his place. (Moman continued to edit the "Technical Talks" column, and in 2001 was part of a group of hams that purchased and ran a store for two-way radio enthusiasts, Alfa Radio Ltd.) Van Driel, who would serve until 2004, was also the club's webmaster. In December 1999, the club set up a Yahoo "CIDX Club On Line" where members could exchange messages, post photos or links, and meet in a club chat room. The year 2000 brought a survey of listeners' favorite stations and programs, conducted in conjunction with RCI's "Maple Leaf Mailbag" program. This survey was repeated in 2001. In 2002 the club celebrated its 40th anniversary with a convention in Montreal. And in 2007, a new column, "Free Radio Scene," was added to the *Messenger*.

In January 2005, Harvey announced that the club was going all electronic with the next issue. Henceforth, the *Messenger* would be published as a .pdf file and distributed as an e-mail attachment, with paper printouts available to members who were not on line. Since Glenn Hauser's material was available elsewhere on the internet, his column was dropped (but material from Hauser's "DX Listening Digest" supplemented members' reports in "Logbook"). The *Messenger* retained its regular content and appearance, but with the addition of color, photos and other graphics, and active links. Although many would have preferred retaining the paper bulletin, the electronic version looked much better and was well received by the membership. And the annual membership fee dropped from U.S. $27 for paper to U.S. $10 for electronic.

The change to all-electronic should not have been unexpected, for Harvey had often discussed the future of the club in a shrinking and increasingly electronic DX world. By the end of 2004, club membership had fallen below 100 and economics dictated the decision. The other major Canadian club, the Ontario DX Association, would go all-electronic a year later.

The CIDX was the first Canadian club to establish a strong, long-term presence within the DX community. Through able leadership it has built an admirable record of success in an environment dominated by larger clubs.

SWL International (1972–1976)

SWL International, an all-SWBC club, was founded in December 1972. Its monthly bulletin, which typically ran some 16 pages, was called *SWL News*.

For most of the club's life its Board of Directors consisted of Marc Fisher of the Bronx, New York, Chairman; Charles F. Edwards of Maryville, Missouri; and Marc Champagne of Montreal, Quebec. Champagne had succeeded Ken Boyd of Lethridge, Alberta during the club's second year, and was the bulletin's Executive Editor as well as publisher and treasurer, and the club's key man.

Once bulletin organization settled down, the regular columns in *SWL News* were "DXer's Dialogue," consisting of member comments, want ads, meeting notices, surveys and the like; "Bits 'n Pieces," newsy items about stations, publications, etc.; "St. Data," which presented station schedules; "QSL Corner"; "Electronics Corner, Late Loggings and Listener's Letterbox," an amalgam of those subjects; "SW Digest," articles on equipment, stations and other topics; and "SW Log," the member loggings section. Early on there was a mid-month *SWL News,* but it was dropped during the club's first year. The bulletin also covered the activities of the Canadian Handicapped Aid Program, and of ANARC, SWL International having become an associate member in July 1973 and a full member the following year.

The bulletin improved over time. The quality of the sections varied, the strongest being the loggings and QSL sections, both of which were quite substantive. An excellent

QSL survey tabulated useful data on all the QSLs reported for a two-year period. Columns were sometimes missing, however, and, overall, the club lacked sufficient critical mass to meet the standards of some of the national-level Canadian and American clubs.

Editors changed chairs from time to time. At various points editorial duties were performed by, among others, Champagne, Edwards and Boyd, plus Kenneth Compton of Brighton, Massachusetts; Doug Copeland of Winnipeg, Manitoba; Neville Denetto, Kakabeka Falls, Ontario; David Elkjar, Vancouver, British Colombia; Joseph Gannon, Chicago, Illinois; Randy J. Malko of Carignan, Quebec; Craig Maskell, Chatham Head, New Brunswick; Don Moore, then of Milesburg, Pennsylvania; Robert Payne, St. Eleanors, Prince Edward Island; and Chuck Rippel, Virginia Beach, Virginia.

There was a club constitution, but this did not prevent the occasional conflicts and sniping in matters of governance which seemed to plague clubs in those days. SWL International's bid to host the ANARC convention in August 1975 was not accepted, which was just as well because, while noting in March of that year that the club had come a long way, SWL International abruptly ceased operations when Champagne resigned due to other commitments. Although there was an attempt by Christian P. Hansen of New York, together with club officials, to keep the club alive, by January 1, 1976, it was evident that their efforts would not be successful.

CANADIAN SWL INTERNATIONAL (1977–1982)

Canadian SWL International (sometimes written as "Canadian S-W-L International") was founded on May 21, 1977. The founding members were Neville Denetto, age 43, of Thunder Bay, Ontario (later Kakabeka Falls), and John J. Garner and Wesley Rogers, both of Thunder Bay. (The club's headquarters address was in Kakabeka Falls, some 23 miles from Thunder Bay, only because post office boxes were not available in Thunder Bay when the club was formed.) Although the club had no relationship with the similarly named and by-then defunct club, SWL International, Denetto, who had DXed for 15 years, had been an editor for SWL International. He had also served as an editor and board member of ODXA, whose chairman, Harold T. Sellers, encouraged the formation of Canadian SWL International and assisted it in getting started. Garner and Rogers, both age 44, had been DXing for about five years.

The club's lofty goals were to promote the shortwave hobby, QSL collecting, etc.; to promote knowledge of Canada and things Canadian (membership was not restricted to Canadians, however); to promote international friendship through letters, visits, etc. and by helping members obtain shortwave equipment; to keep members informed of developments in radio communications; to provide information on construction projects; and to assist international broadcasters by reporting interference and frequency alternatives, providing summarized reports, and commenting on program content.

For most of its life the club's board of directors consisted of Denetto as the club administrator, Garner as secretary-treasurer, and Bill Butuk of Thunder Bay, who succeeded Brian Baumgartner of Ottawa, Ontario as bulletin managing editor in 1979 and also became the club's technical editor. In 1982, Nandor Petrov of Winnipeg, Manitoba joined the group as membership secretary.

The first issue of the club's bulletin, *CANDX*, was published in October 1977 following a canvas of the members' wishes with regard to bulletin content. Although CSWLI was not a SWBC-only club, SWBC was its focus. Among the contents of *CANDX* were a headquarters report and a managing editor's report, brief biographies of new members, an ANARC report (the club soon achieved full membership in ANARC), HAP and CHAP reports, member loggings, station schedules ("On the Air"), other station information ("DX-Casts," later "DX Spectrum"), SWBC QSLs, station profiles, press clippings, features, a technical column, general announcements, a utilities column, and occasional other items. A medium wave column was added in 1982. *CANDX* was originally in mimeographed form, and usually exceeded 40 pages. In 1982 it was restyled as an offset booklet which eventually reached 20–24 pages. The club also had a three-minute news spot over RCI's weekly "DX Digest" program.

Denetto, Garner and Rogers shared in the editing while they were active in the club. However, many others edited various columns over the years. By May 1978, a year after the club's formation, CSWLI passed the 100-member mark. A year later membership leveled off at almost 150, with members in 18 countries, all Canadian provinces, and 21 U.S. states.

As was the pattern in many of the smaller clubs, member contributions declined over time and increasingly the bulletin consisted of material originating from outside the club. In 1981, CSWLI hosted the ANARC convention in Thunder Bay, Ontario, which, despite strikes at Radio Canada International and Canada Post, was the second-largest ANARC convention at the time, with attendance nearing 200. This successes notwithstanding, the club closed its doors in September 1982 due to financial difficulties.

Society to Preserve the Engrossing Enjoyment of DXing (SPEEDX) (1971–1995)

The Society to Preserve the Engrossing Enjoyment of DXing, commonly known as SPEEDX (originally "SPEE DX"), was born in July 1971, the product of a dispute within the American Shortwave Listeners Club that resulted in the resignation of nearly all the officers, directors and editors and the simultaneous creation of this new club.

Although created out of conflict, SPEEDX earned success quickly as a result of the energy and experience of its leadership, all of whom were well-known in the shortwave community. The pro tem President was Art Glover of Port Angeles, Washington, the Vice-President Jack White of Gresham, Oregon, and the Secretary David Thorne of Santa Ana, California. The founding members appointed a board of directors, and annual financial reports were issued. Among the club's early editors were Bill Flynn of Mountain View, California; Carroll R. (Pat) Patterson of Decatur, Georgia; Christopher Lobdell of Reading, Massachusetts; Steven P. d'Adolf, Woburn, Mass.; Serge C. P. Neumann, Anderson, Indiana; Robert E. LaRose, Rochester, New York; and William Sparks of San Francisco, California. Flynn soon took on the job of Managing Editor and Thorne the duties of Business Manager.

Aside from the bulletin's construction as an offset booklet (a style which the ASWLC would adopt in June 1972), its content and organization tracked the ASWLC bulletin in many ways. It covered shortwave broadcast loggings, station schedules, QSLs, technical topics, and propagation, and it contained feature articles and a monthly ANARC report.

SPEE DX

THE DX RADIO BULLETIN FOR ACTIVE SHORTWAVE LISTENERS

Volume 1   July 1971   Newstand price = **$1.**

contents

For an inaugural issue, the *SPEEDX* bulletin looked good and was high in content, thanks to the experienced editorial staff, most of whom had just departed the ASWLC.

It also contained extensive coverage of the utility bands. A uniquely informative column was "Propagation" by Jack White. "DX Montage" was an informative catch-all column of material that did not fit elsewhere. Edited by Serge Neumann until November 1972 when he left to join the staff of Radio Netherlands, "DX Montage" was taken over by Alan W. Brooks who was unmatched in his application of wit to shortwave topics during the column's remaining six years. (It was resurrected in 1983 with David Sharp of Lutz, Florida as editor.)

SPEEDX also initiated the *SPEEDX-Gram,* a two to four page, mid-month supplement containing the most current station information. It was available to all members who kept a supply of stamped, self-addressed envelopes on file.

A novel idea to encourage member participation was "SPEEDXtras," a department of the club (and a monthly column) which sponsored special projects of various kinds, including supplemental publications which would be available only (or at a reduced price) to contributors. These included reference leaflets of various kinds, reception surveys of particular stations, multi-lingual reception report forms, and sunrise-sunset tables, as well as larger publications such as the *SPEEDX Utilities Handbook,* first issued in November 1971 (and several editions thereafter); the *U.S. Coast Guard DX Handbook* (1972); the *World Utility DX Handbook* (1973); the *Foreign Language Reporting Guide* (1976); the *Guide to Soviet Broadcasting* (1980); and the *SPEEDX Reference Guide to the Utilities* (1983). A logo was selected by club-wide vote. Contests were held. A constitution and bylaws were approved by the membership in 1972. Periodic elections for the SPEEDX board of directors were held. An awards program was established, the awards having such exotic titles as Wayfarer, Expeditionist, Globe Trotter and Global Voyager. March 1972 saw the start of the Handicapped Aid Program (HAP), a project that would become widely known among international listeners.

Within three months, membership was at 150. By January 1972 it stood at 200. In order to encourage participation, SPEEDX adopted a two-level membership structure, with full membership for those who earned a certain number of points for their contributions to the bulletin or to the club's administration, associate membership for all others. (The ASWLC had experimented with a similar point system but had dropped it.) Only full members could vote in club elections.

The bulletin's Western Hemisphere editor, E. J. (Jim) Whitehead of Reading, Massachusetts, was elected President of the club in 1974. In March 1976 he was succeeded by Don Johnson of Elsinore, California. John Trautschold of Menomonee Falls, Wisconsin, became President in January 1980. A succession of members served in other headquarters capacities. In June 1978, a new column, "Program Panorama," edited by Woody Seymour, Jr., of Sanford, North Carolina, was added. The column focused on the programming available on shortwave rather than on DX.

By 1980, SPEEDX was enjoying a successful middle age. Membership reached 255 "full members" in 1983. The total number of members reached 1,200 the following year. Although column editors changed from time to time, the club continued to benefit from experienced editors and strong member support. A number of SPEEDX publications were issued, including the *SPEEDX Guide to African Listening, The First Shack,* and *Pitfalls and Problems.*

Some of the energy of the early years was missing, however. In particular, the "SPEEDXtras" concept, which had produced many interesting club initiatives while being handled by Jack White, fell into disuse after his departure, and efforts to reinvigorate it were unsuccessful. But the club kept up with the times, and featured many user reviews of new entries in the expanding shortwave receiver market.

A major effort to improve the bulletin was undertaken in 1985. A "min/max" page allocation system was instituted wherein bulletin size would be driven by the volume of member contributions rather than strict page allocations. Some new columns were instituted, including an "Antennas and Accessories" column, "Product Profiles," and a "Contest Page" within "DX Montage" which ran monthly quizzes with small prizes. The awards program was upgraded, and a three-day All-Band DX Contest was held. The establishment of local chapters was encouraged, and SPEEDX published a 36-page *SPEEDX Guide to Latin American DXing,* as well as a revised *Guide to Soviet Radio* (previously *Guide to Soviet Broadcasting*).

None of this brought fundamental change to the bulletin, or the club, however. When Jack Sanderson became President in 1986, SPEEDX was suffering the same membership decline as other clubs. Technology, the replacement of DXers with more casual listeners, and competition from computers and video games were widely perceived as the culprits. By 1987 the club no longer had enough members to support a 56–60 page bulletin. It went to a maximum of 52 pages, and the "min/max" system was dropped. The contest page went bi-monthly, and publication of the *SPEEDX-Gram,* which was suffering declining member participation, was suspended. In August,"DX Montage" was temporarily incorporated into an expanded introductory "Keynotes" column, but was reinstated as a separate column the following year.

In 1988, club Vice-President Peggy Thompson of La Crescenta, California, became President, Bob Thunberg of Dubois, Pennsylvania, became Business Manager, and Ed Janusz of Bricktown, New Jersey, took over the publishing. There was a further reduction in the size of the bulletin. The few local chapters that had been established were disbanded.

The "Hawaii Report," begun in 1986 by Dr. Richard E. Wood, lasted only a few months. Since 1984, member Ralph Famularo in Japan had been presenting a one-page "Japan Report" of stations heard there. It was not widely read, however, and was cancelled in 1987. In the years that followed, numerous other columns and editors came and went. Woody Seymour's "Program Panorama" column was dropped. On the plus side, an interesting "Latin Notebook" column edited by Carl Huffaker in Mexico was instituted.

Seymour became Managing Editor in October 1989, and William Westenhaver took over as President the following year. To encourage member contributions, a Contributor Incentive Plan was adopted whereby a member could get a free one-year membership renewal by contributing to the greatest number of columns during the remainder of the year, or by contributing to the bulletin in at least ten of the remaining 11 months. There would be only one winner in each category, however, and the plan had little effect.

Peggy Thompson returned as President in 1991. Although the club benefitted financially from increased membership thanks to the enhanced interest in shortwave that accompanied the Gulf War and publicity about the club in the "SPEEDX Report" over HCJB's "DX Partyline," the decline in bulletin content continued. Don Thornton of Belle Mead, New Jersey took over as President in 1992 and some new initiatives were undertaken. "The SPEEDX Supermarket" was set up for the sale of publications and T-shirts, and some special publications were issued, including *The Literature of Shortwave Radio* by Ed Janusz, which consisted of over 50 reviews of shortwave-related publications, and *Carl Huffaker's Latin Notebooks,* an interesting compilation of all his columns. In March 1992, the bulletin was renamed *Shortwave Radio Today.*

In 1993, the board of directors dissolved itself, the stated reason being its inability to assist the publisher or the business manager in their responsibilities. Don Thornton became the publisher, and for all the club's problems, the bulletin still had a very professional appearance. However, the club's month-to-month existence eventually caught up with it, and SPEEDX closed its doors after the January 1995 issue. As Thunberg explained in a subsequent letter to the membership, in the past the club had been able to rely on new and renewing memberships to get it through bleak financial periods. Now the club's funds were completely depleted and membership revenues were not enough to meet increasing postal and production costs.

Ontario DX Association (1974 to the Present)

ODXA was formed in late 1974 and early 1975 by three Ontario DXers: Harold T. Sellers of North Bay, Ivan Grishin of Toronto, and Harry A. Dyer of Kingston. The idea was Dyer's. Following a poll of interested DXers in the Ontario area, the name Ontario DX Association was chosen, and elections were held for a board of directors to replace the group of three. The board members elected were Sellers, Grishin, and Neville Denetto of Kakabeka Falls. Sellers was named board Chairman and Publisher, and his home was designated as club headquarters.

The club bulletin was called *DX Ontario,* and the first mimeographed issue appeared in February 1975. The message on the cover of the second issue captured the optimism of the founders: "Thanks to you ... ODXA is on its way ... to the top!" The plan was for the club to be all-wave. Editors were appointed, and a modest bulletin began to take shape. Within a few months the club had 40 members.

Like all radio clubs in their formative stages, ODXA had lofty goals. Unlike some, however, its leadership had imagination and know how, plus a businesslike approach and a flair for public relations that would become club hallmarks for years to come. ODXA was to be for Ontarians only, with the national and international scene left to the Canadian International DX Club and Canadian SWL International. The club knew how to capitalize on the benefits of its regional structure. It held regular club meetings, "DX rallies," barbeques, open houses, DXpeditions, and similar social events. A members' telephone directory (later a full membership directory) was published, and the names and contact information for new members, along with brief personal data, was included in the bulletin. Members' activities were often highlighted, and in the mid–1980s "Meet an ODXer," with a photo and a write-up about a member, was a regular *DX Ontario* feature. In later years, ODXA chapters were established in various cities throughout Ontario and beyond, and their meeting dates listed in the *DX Ontario* calendar.

The bulletin focused on members' interests, which were mainly shortwave and medium wave loggings and QSLs. By the club's first birthday, the bulletin was 24 pages long (except in the summer). It contained a headquarters section with general news about the club and DX-related topics; a "ragchewing" column called "Ontario DX Observer" (later "Shack Talk"); "Listener's Corner," a column devoted to the programming available on shortwave

DX Ontario, bulletin of the Ontario DX Association, was humble in look, if not message, in this issue of June 1975, a few months after the club was founded. The bulletin would become the best looking of all the shortwave club bulletins in North America.

(later called "Listening Inn," then "Listening In," and, in 2000, "Programming Matters"); plus columns for shortwave and medium wave loggings, station news and schedules, a heard-verified scoreboard, and articles on technical topics and other aspects of DXing.

In its second year the club adopted the NASWA country list for ODXA use, published an equipment survey of its members, commenced publication of a quarterly list of English broadcasts to North America (later published monthly and expanded to include other English broadcasts), and initiated a quarterly by-time cross-reference table of the previous quarter's loggings. The club also gained membership in ANARC and published news of ANARC activities. The bulletin continued to improve in appearance and grow in size. By the end of 1976 it was 42 pages long, and membership had reached 70. It doubled in the next two years.

An awards program was announced in 1977. A reprint service was established the following year, as was the "DXer's Hotline Service," a bulletin section devoted to loggings made between the column deadlines and the time of publication (it was later called "DXpress"). An "ODXA DX Report" became part of Radio Canada International's "DX Digest" program. (In later years, "ODXA Perspective" would be heard monthly over HCJB's "DX Partyline.") Biennial board of directors elections were held, and reports of board activities, as well as club financial reports, were published. Questionnaires seeking member opinion on the bulletin and on club operations were distributed periodically.

As the club grew, the status of non–Ontarians kept arising. In 1978, the club decided to permit non–Ontario clubs and organizations (but not individuals) to subscribe to *DX Ontario* in order for their members to benefit from its content. In June 1979, subscriptions to *DX Ontario* (later called associate memberships) became available to individuals outside Ontario. However, non–Ontarians could not vote, hold club office or submit loggings. By 1985 there were 50 such non–Ontarian subscribers. The policy was further relaxed, and associate members were allowed to vote in club elections. The associate member classification was discontinued in 1998 and all benefits of membership (save for holding club office) were opened to non–Ontarians.

In February 1980, the full-page mimeo bulletin format was replaced by an 80-page, two-column booklet-style. Although the content remained the same, the bulletin's appearance improved greatly, thanks in large part to club member Ron Hopkins, a professional printer who was generous in his assistance to the club. *DX Ontario* was soon the best looking of all the North American club bulletins. It was upgraded further in 1981 through the use of a commercial printer and a professionally designed cover. Radio-related advertising began appearing the same year.

In 1983, a monthly "Equipment Review" column was instituted, as was a monthly chart showing the times and frequencies on which the BBC could be heard around the clock. (Years later it would be expanded to include some specific BBC program recommendations for the month.) A similar chart, "U.S.A. Survey," showing times and frequencies for broadcasts over the private U.S. shortwave stations, was begun in 1988. "Alternate Airwaves," covering pirate broadcasting, first appeared in 1984, as did a regular listing of DX programs. Columns covering utility transmissions, computers, and DX programs were added in 1985 (the computer column was short lived). Also in the mid–1980s, a "Program Calendar" was added showing some recommended programs for various dates throughout the month.

The ODXA issued various special publications over the years, among them *Introduction to the Tropical Bands* by Cedric Marshall, Great Circle Sunrise-Sunset Maps centered on Toronto (also produced by Marshall), multiple editions of the *Pennant Guide for Collectors,* and *QSL-Summary-Africa* and *QSL Summary-Asia,* which compiled QSL information from many sources.

A major expansion of ODXA membership took place in the mid–1980s. By 1986, membership numbered 450, with an additional 97 non–Ontarian subscribers. By this time the typical *DX Ontario* was 90 pages in length.

The active outreach of the early years continued. Although *DX Ontario* did not cover amateur radio until 1994, displays at hamfests, ham flea markets and hobby shows proved to be a fruitful source of new members. In addition, *DX Ontario* became available in selected retail stores, and a joint project with Sony provided an ODXA-Sony promotional poster to all Sony distributorships in Canada. In 1993, Radio Shack agreed to place ODXA brochures in all Radio Shack stores in Ontario.

The club hosted successful ANARC conventions in 1984 and 1987 in Toronto and Mississauga respectively. The first annual ODXA convention was held on three days in October 1985, with 84 members and visitors in attendance. ODXA conventions (later called "Radio Fests") would be held almost every year thereafter. In some years, tapes of convention presentations were made available to members who could not be present, and later conventions included a silent auction and a raffle (prizes sometimes included receivers donated by dealers). An essay contest, "Radio In My Life," was held in connection with the 1999 convention and the essays compiled into an ODXA "Radio In My Life" essay collection.

Many less formal member get togethers were held as well. Weekend DX camps have been held once or twice a year since 1991. Descriptions of these camp experiences were printed in *DX Ontario* in order to encourage attendance. Several Rochester, New York-area DX Camps were held in the early 1990s, and for the next ten years these were carried on in other New York venues by a group calling itself the Mohawk Valley Shortwave Listeners Club. Twice a year from 1980 to the early 1990s, ODXA members Terry and Judy Ferguson hosted the Gravenhurst DX Weekend at their Muskokan Hotel in Gravenhurst, Ontario. SWL flea markets (later expanded to include an auction) were held from 1993 to 1999. ODXA picnics were annual affairs from 1994 to 1998. In 1994 an ODXA Air-Waves Summer Weekend was conducted in conjunction with the 1994 Trenton Air Show, with the itinerary including, in addition to the air show, tours of a radio station and of a Department of Defense communications base, a DX camp, and a day at the beach. ODXA members knew how to have a good time.

The club also conducted contests, including the month-long ODXA DX Challenge, which has been held most years since 1986, the shorter "DX Sprint" in 2000, and from 1994 to 1999 the ODXA 2-Down Challenge (limited to stations below 2.0 MHz). There was an ODXA ham net, and the First Annual Ontario QSO Party, which sought to maximize contacts by and with Ontario hams, was held in 1997. Subsequent QSO parties were held annually.

For ten years beginning in 1986 the club offered courses on shortwave radio in several educational venues around Ontario. The instructor was usually either ODXA member (and later Chairman of the board of directors) Stephen Canney, or club propagation expert Jacques d'Avignon. In 1991, the club took over the QSLing duties for CFRX.[42] A day of listening post tours, where members welcomed other members to their homes to view their listening setups, was held in August 1999.

In 1990 the club's key man, Harold Sellers, relinquished his membership on the board, and his post as Chairman, in order to focus on headquarters business and on the publication of *DX Ontario*. He was named club General Manager and *DX Ontario* Managing Editor. In recognition of his central role in the creation and development of ODXA he was also named Founding Chairman. Following club-wide elections, highly regarded DXer and long-time club contributor David M. Clark became Chairman of the board. He was succeeded by Joe Robinson in 1994. Robinson took over as General Manager in 1996 and Jack Henshaw became Chairman. Sellers remained Managing Editor of the bulletin, a post he still holds. He returned to the board from 1998 to 2004 and again from February 2006 to the present. He served as Vice-Chairman before taking over as Chairman for 2000–2002. Stephen Canney served as Chairman from 1998 to 2000, Brian Smith from 2002 to the present. (The board was renamed the Executive Committee in 2006.)

The club made the most of advances in technology. In 1989 it started the DX-Change, a 24-hour telephone information service where members could leave messages. They could also listen to a bi-weekly message about interesting stations or programs that were being heard, and a calendar of upcoming ODXA activities. The ODXA "Listening Post" Bulletin Board System for computer users was set up in 1992 and handled more than 40,000 calls during its four-year existence. An ODXA website appeared the following year, and an ODXA e-group was established in 1999.

In 1989 and 1990, in order to purchase a desktop publishing system, the club conducted a fund raising campaign that yielded $10,000, followed by almost $5,000 more in a second fund raising effort. Desktop publishing contributed further to *DX Ontario's* professional appearance. The bulletin continued to develop in content as well, with some new columns added and others retired. A "Scanning VHF/UHF" column was added in 1990. During 1991, "DXer's Classroom," intended for beginners, went from being an occasional bulletin entry to a full-fledged column. A "Station Profile" column, focusing on a different station each month, was begun in 1991. It was edited by Thomas B. Williamson of Peterborough, Ontario, whose extensive DXing experience reached back to 1935.[43] The "Propagation Forecast" column was begun in 1995, "FM/TV Report" in 1996, and "On the Net" in 1997. The latter column, whose purpose was to explain the equipment and software needed to access the internet, review software, and point members to interesting URLs, was in response to the rapidly growing role of the internet in the shortwave community.

In 1994 the club began experimenting with a column called "Target Listening" which each month showed the current time and frequency information for stations in most radio countries. The by-country listing was followed by a table cross-referencing the items by time so that a user could see what might be heard at each hour. (The country list and the cross-referencing table were later separated and presented on an alternating basis.) The information was based on contributions from members and from other sources. Originally a separate publication available by mail, "Target Listening" became a monthly bulletin feature in February 1995, whereupon the international and tropical loggings columns, which had been separate, were consolidated into a single, abbreviated "Shortwave Loggings" section which focused on the most interesting of the members' loggings. "Shortwave Loggings" was later renamed "Your Reports."

June 1991
ISSN 1183-0344

DX ONTARIO

MONITORING THE WORLD VIA RADIO SINCE 1975

$2.50

RADIO VATICANA 1931-1991

Professionalism in both content and appearance was the watchword for the bulletin of the Ontario DX Association. *DX Ontario* was restyled many times.

Also in 1994, an amateur radio column, "QRZ," was instituted, and the club reprint service was reinvigorated. Through it and an annual bulletin index, one could obtain all the significant articles ever published in *DX Ontario.*

Although most shortwave clubs enjoyed increases in membership during the late 1980s and early 1990s, the growth of the ODXA was especially noteworthy because of the club's regional character. By 1991 it had nearly 1,300 members, three-quarters of them in Ontario. Membership remained above 1,000 for five years. High membership notwithstanding, as was the case in other clubs, the number of members who actually contributed to the bulletin was low — between 3.4 to 6 percent per month in 1993. In addition, the club suffered the same decline in membership that all clubs were experiencing in the latter part of the 1990s, and by 1999 total ODXA membership was below 700, with two-thirds being non–Ontarians, most of whom considered ODXA a national club.

The club moved to a 40-page, full-size magazine format in 1996. Four years later, *DX Ontario* changed its name to *Listening In* so as to reflect the diversity of members' interests — many were more interested in program listening than DX — and the club's considerable non–Ontarian membership and activities.

In 2000, a .pdf version of *Listening In,* with color graphics, became available to members in lieu of or in addition to the print version. A new column covered receivers and computer hardware and software. A "Nostalgia" column was begun. Coverage of scanning, VHF-UHF and related computer topics was consolidated into the club's long-running utilities column, "Monitoring Services" (most of this content was later placed in a new "Scanning Report" column that was published from February to December 2005). In 2006, "Looking Back," a shortwave history column by Tom Williamson, replaced Williamson's "Station Profile" column, and "Click!," a column featuring station websites, was begun.

With the increased interest in program listening, several editions of *Listening In* between 2000 and 2002 featured John A. Figliozzi's "SWL Program Guide," a detailed by-time and by-station listing of many of the programs of the major shortwave broadcasters. (ODXA and NASWA had jointly published Figliozzi's shortwave program publication, *Shortwave Radioguide,* from 1991 to 1995.)

The club celebrated its 30th anniversary in 2004. Total membership in May was 500. The following month, as a cost-saving measure, the bulletin returned to its earlier 5½" × 8½" two-column size with the promise that no content, columns or articles would be missing. The page count would vary from month to month, but usually was around 50 pages.

In 2005 the club established an innovative arrangement with two venerable South Pacific clubs, the Australian Radio DX Club and the New Zealand Radio DX League, both of which published .pdf versions of their bulletins. In an effort to provide their members with more value, members of any of the three clubs who received their bulletin electronically could also obtain the e-bulletins of one or both of the other clubs at no additional charge.

Sellers observed in June 2005: "Although we presently have some stability, the future of the club is always in flux with changes in membership, magazine sales, supply sales, etc. Your support is crucial to continuing things like *Listening In* magazine." In January 2006, such uncertainties led the club to discontinue the print version of *Listening In* in favor of distribution exclusively over the internet. Photo copies were sent to those without internet capability. Said Sellers: "[D]eclining numbers in the club membership are reaching the point where the board had considered going to an all electronic version. Many members have already switched to the electronic edition and there are not enough print members to justify the costly run of magazines we were currently printing."[44] By year's end, almost half the members were receiving the electronic edition.

Although smaller (about 300 members) and less Ontarian-centric than in earlier times, ODXA confronts today's club challenges with the formula that has served it well for more than 30 years: businesslike management, continuous outreach, and a democratic structure, all bottomed in close relationships among staff and core members.

ASSOCIATION OF DX REPORTERS (1982–1995)

The demise of the NNRC in April 1982 may have been anticipated by a few insiders, but it was a surprise to the vast majority of the members, and the editors. A core group wanted to carry on, and on May 1, 1982, the first edition of the spirit-duplicated *DX Reporter*, an eight-page bulletin of loggings on hand from a few ex-NNRCers, went out to 35 people. The "Baltimore Publishing Committee" consisted of Reuben Dagold, Carroll Weyrich, Bob Colgrove, Matt Zahner, Charles Wackerman, and honorary member Eugene Vonderembse (former NNRC president). It emphasized that it was trying to fill the void, but that it was not a continuation of the NNRC.

The *DX Reporter* hit the ground running. By June the group had a name—the Association of DX Reporters—and dues—$15 per year. Three of the NNRC's experienced editors immediately reestablished their old columns: Hank Bennett of Cherry Hill, New Jersey, NNRC shortwave editor since 1949; Ray L. Fansler of Fairfield, Illinois, NNRC ham band editor since 1977; and Robert H. French, of District Heights, Maryland, NNRC utilities editor since 1969. Bob Colgrove of Columbia, Maryland, who had edited the NNRC's "Information Please!" column from 1974 to 1979, started "Underneath the Headsets," covering long wave, medium wave, TV and FM; and Larry Ledlow of Sykesville, Maryland, who had edited "Information Please!" since July 1980 and the NNRC statistics column before that, took on "The Data Bank," which covered equipment, propagation, and technical topics. A mailbag column and an all-wave (except amateur) QSL column were soon added.

Reuben G. Dagold of Baltimore, Maryland was the publisher, and the club's key man. The bulletin moved to booklet style, and *DX Reporter* looked good. The ADXR received some publicity in other publications and on DX programs, and by year's end membership stood at 123. In March 1984, ADXR won probationary membership in ANARC. Full membership followed, and ANARC news began appearing in the bulletin regularly. The monthly ANARC "Marketplace Report" appeared occasionally starting in 1986, and "The CPRV Page" was carried monthly from January 1989.

Hank Bennett's last shortwave column for ADXR was in January 1984. He resigned following a conflict with the publishing committee over some critical comments about other editors and clubs which the committee felt were unwarranted. He was succeeded by Donald E. Stidwell of Baltimore. Bennett's column had been basically loggings and members' comments. Stidwell leavened it with country-specific station information from members and other sources, a time index, and miscellaneous shortwave-related news.

Other columns and editors changed from time to time as well. After January 1983, "Data Bank" was replaced with a mix of standalone articles, press clippings, book reviews, etc., which grew into one of the more interesting parts of the bulletin. A members' statistics column was added in October 1984 but dropped after June 1986.

From January 1984, "Underneath the Headsets" was edited by John Wilkins of Wheat Ridge, Colorado. (The name of the column was changed to "Broadcast Band DXing" in February 1987.) A separate TV-FM column had been tried unsuccessfully in May 1983, but emerged again in October of that year and continued until 1989, by which time it was clear that it was garnering little support. Rather than drop it altogether, it was consolidated into the BCB section in July 1989.

Mike Witkowski of Stevens Point, Wisconsin, who had edited the ham band column for the NNRC from 1972 to 1977, took over the ADXR ham band column from Ray Fansler in March 1986. It would return to Fansler in July 1994. A column covering pirates and clandestines was begun in February 1991. It was edited by Michael Goetsch of Lakewood, Ohio. In October 1991, Goetsch assumed responsibility for the SWBC column *in toto*, and the pirate and clandestine coverage was folded into it. However, Goetsch's final column was in March 1992, and Dagold pinched hit until Charles George of Dallas, Texas took over in April 1994.

The strongest section of the *DX Reporter* was utilities. Bob French yielded the column to Robert L. Jones of Eldersburg, Maryland, in January 1988. In February 1990 it was divided into voice and CW/RTTY/FAX sections, with a separate editor for each. It was recombined in October 1991 when the editorial duties were passed to Spence Naylor of Ventura, Florida, who had edited the utilities column in the ASWLC bulletin since 1977. A separate "LF Beacons" column was begun in November 1986, and in February 1991 it was expanded to include all long wave stations.

Although the heart of the club was the *DX Reporter,* there were other club activities as well. In October 1984 the club adopted a country list, which was a combination of the ARRL and NNRC lists. The first club contest, a two-day ham band listening event, was conducted in December 1984. It was repeated in 1985, the same year in which a two-day BCB contest was held. From 1987 to 1993, ham band "semi-contests" of several months' duration were also held. Members were awarded points for their level of activity as measured by their ten best loggings for the month. Winners earned books and certificates. And in 1985 the club published the *ADXR List of Limited Coastal Stations* ("company" utility stations), utilizing funds donated to the Matt Zahner Memorial Fund which had been established when long-time NNRCer Matt Zahner passed away in 1984.

The number of members remained relatively consistent throughout the life of the club. There were 154 members in 1985, 130 in 1993. Although member participation was strong at first, it dropped off over time, and by 1989 the trend of non-participation was apparent. Columns were shorter, and support was coming largely from a small group of members. In 1993, BCB editor John Wilkins calculated that the number of contributors reporting to his column, and the number of loggings, were well below half what they had been ten years earlier. The bulletin was relying increasingly on internet and other non member-originated material, and there were some months when columns went unpublished. Exhortations for member support generally went unheeded.

Beginning in 1990 the August issue was skipped each year. By the end of 1994 the bulletin was barely making 30 pages, compared to 40+ in the club's early days. In addition, Dagold had filled in for several editors when their posts were vacant — sometimes for long periods—and while all the editorial positions were filled by mid–1994, it was becoming increasingly difficult to find capable editors from within the relatively small ADXR membership. In early 1995 Dagold announced that the ADXR was closing, which it did in February 1995 (coincidentally a month after SPEEDX closed).

Regional and Specialized Clubs After 1960

AMERICA CENTRAL RADIO CLUB (1964–1965)

The America Central Radio Club was formed in the summer of 1964. It was a reflection of its creator, Richard E. Wood, who at the time was on the faculty of the University of Indiana in Bloomington. Wood began DXing from Newcastle-on-Tyne, England, in 1958. He was a linguist, and came to the United States to obtain his Ph.D. and teach. He was widely traveled and he had many contacts among DXers in Scandinavia, from whom he developed a love for the Latin American DXing that was popular there.

Wood's sole interest was broadcast stations. The *America Central Radio Bulletin* at first covered both shortwave broadcast and medium wave DXing. Long distance TV and FM DXing was added in 1965. However, Wood dropped medium wave, an area in which he had extensive experience, concluding that it was sufficiently specialized to require separate attention. The club was small, probably never exceeding 50 members. Keeping it small was a goal at first, although eventually he opened it to a broader membership and actively recruited new members. The club became an ANARC member early in its existence.

Wood eschewed organizational and democratic complexities. He "was" the club, and while he encouraged and welcomed contributions from members, he viewed the club in part as a vehicle for showcasing his own considerable DXing accomplishments, from which he felt others could and should benefit. Indeed, improving the quality of DXing in North America was one of his goals. He set high standards of accuracy in reporting, and stressed the importance of linguistic correctness in logging and reporting to stations. The focus of the bulletin's shortwave section was Latin America, and secondarily the smaller stations in other parts of the world, principally non English-speaking areas. Free club memberships were offered to Latin American DXers who agreed to contribute information on the shortwave doings in their parts of the world. After the 12-page bulletin adopted a booklet format, Wood was able to include illustrations of QSLs and pennants, of which he was an avid collector. The bulletin also contained articles by Wood on various facets of DXing.

The club closed late in 1965, and memberships were transferred to *SWL News*, a new DX "magazine" published by James J. Howard of Kansas City, Missouri for which Wood briefly wrote. Wood was well known and highly respected in the DX community, and contributed to many DX publications. He edited the NASWA "Shortwave Broadcast Center" from 1968 to 1969, and in the latter year wrote *Shortwave Voices of the World*. Around the same time he was named "International DXer of the Year" by the National Radio Club, an American medium wave club. In recent years he had again become active in medium wave DXing, this time from his home in Hawaii. He passed away in 2007.

AMERICA CENTRAL RADIO BULLETIN

America's most popular ILLUSTRATED Radio DX magazine

Organ of the AMERICA CENTRAL RADIO CLUB

New Series, Vol. 1, No. 1.   September 1965

207 E. 16th., Bloomington, Indiana, USA

Rádio Cultura de Campos, Campos SP, Brasil

PENNANTS FROM AROUND THE WORLD.
This issue of the AMERICA CENTRAL RADIO BULLETIN will feature illustrations of
pennants, banners and other flags received from all over the world by the many
DX-ers who have discovered this fascinating aspect of their hobby. Stations in
many areas of the world are willing to send these flags out to their listeners
in other countries. Among the countries which will be represented in the coll-
ection of the experienced pennant-hunter are Canada, Japan, France, Spain, the
Canary Islands, Morocco(a happy memory for the older DX-er, the pennant issued
some years ago by the now inactive Voice of Tangier), Upper Volta, Israel, and
China.
But these pennants from Europe, Africa and Asia will not dominate the display.
Names like Rádio Cultura de Campos, Rádio Difusora de Aquidauana, La Voz de la
Patria, Radio El Mundo...these will set the tone of the collection. For, as is
known to every DX pennant collector, the true home of the DX radio pennant is,
undoubtedly, Latin America. México, Guatemala, Honduras, Panamá...from the Rio
Grande to the tip of Tierra del Fuego...these are the lands which are paradise
to the pennant-collector. And it is these pennants which the present issue and
future issues of the AMERICA CENTRAL RADIO BULLETIN will feature.

LATIN AMERICA...LAND OF DX.
This BULLETIN of DX news and features devotes a large slice of its contents to
broadcasting stations in LATIN AMERICA. This is our primary interest. European
stations have the News, and a reliable all-round service to offer the listener
abroad. There is some interest in the African and Asian stations...but most of
these are Government-dominated, with only one national network in each country
or state. Many of these stations are following the trend towards high kilowatt
power, and their DX value is decreasing. But Latin American stations continue,
as always, to maintain their highly personal and characteristic DX value. This
BULLETIN makes no apology for devoting much of its space to these fine DX stations. We hope to contribute to the rising tide of their popularity.

The America Central Radio Club was the creation of well-known medium wave and shortwave DXer
Richard E. Wood. Wood relocated several times. The club was founded while he taught in Bloomington, Indiana.

Miami Valley DX Club (1973 to the Present)

The Miami Valley DX Club began publishing its monthly newsletter, *DX World*, in August 1973, the year the club was co-founded by Mark J. Brown of Minster, Ohio and Brad Lovett of Coldwater, Ohio. The two had met after Lovett heard Brown's name on the Radio Netherlands "Happy Station" program. The then-five member group was originally going to call itself the Tri-County DX Club.

DX World covered SWBC loggings, station news, QSLs, propagation, and other topics. Although school commitments of its founders almost caused the club's demise late in 1974, it continued into 1975 on a wing and a prayer. That year and the following year the club sponsored an Ohio Weekend DX Seminar. Membership topped 100 in 1977, and the first of many quarterly club meetings were held. The same year, publication of the bulletin was assumed by Dave Hammer of Columbus, Ohio, who remains the club's leader today. A separate *MVDXC Flashsheet* was also begun in 1977. Full ANARC membership was achieved in 1978, as was the introduction of a twice monthly MVDXC segment on HCJB's "DX Partyline."

The 1980s saw the expansion of *DX World* from mainly shortwave broadcasting to some non–SWBC coverage. An awards program was begun in 1981, and it was the MVDXC that in 1982 originated the SWL Monitors Forum at the annual Dayton Hamvention, one of the country's largest amateur radio gatherings. A possible merger with the St. Louis International DXers was considered in 1983 but did not take place.

The MVDXC went from a regional, beginners club to one with members at all experience levels and located throughout North America. The 12 to 20-page *DX World* covered all bands, with an emphasis on shortwave broadcast and utility loggings, but with news on propagation, station schedules, programming information and equipment reviews as well. The club has downsized in recent years. Although in 2004 it began referring to itself as The DX Club in order to de-emphasize its Ohio orientation, with an informal membership of about 40 it now confines its activities to occasional local get togethers and the distribution of a mini–*DX World*.

Chicago Area DX Club (1973 to the Present)

The group that would be known for more than 35 years as the Chicago Area DX Club met for the first time in January 1973. Mike Nikolich of Wheeling, Illinois, was the main force behind the effort, whose founding members included both shortwave listeners and medium wave enthusiasts. Many early participants had been members of the Midwest DX Club, which had been active during the preceding five years.

Things moved quickly, and by February Nikolich had produced and mailed the first issue of the *DX-Probe*. He did the main SWBC loggings column. Karl Forth handled BCB loggings, with Robert Kramer and Don Petravick covering other BCB matters, Eric Draut doing QSLs, and Steve Wayman editing a members' comments column. Ham and utilities columns appeared for a time as well.

The heart of the club was a dozen active members. The *DX-Probe* focused on shortwave and BCB loggings, along with station profiles and reports on club doings. The bulletin was newsy and substantive, and the typos and mistakes couldn't detract from the enthusiasm that CADX members shared. Club democracy occupied much time at first, but was soon dispensed with.

Its success notwithstanding, by late 1974 support for the club dropped off as many of its members headed off to college. The last issue of the *DX-Probe* was in December 1974. Many of the members stayed in touch, however, and six years later, again with Nikolich in the lead, the club was rejuvenated. In October 1980, the first issue of the new *DX-Probe* (later called *DX Chicago*) was published. Only four of the first 25 members traced their membership back to the 1970s; the rest were new. However, the enthusiasm of the club's early years was undiminished. Nikolich, Forth, and Kramer picked up the editing duties, and were soon joined by Mike Jeziorski and Kevin Mikell.

The 1980s were the club's most active years. The bulletin was in a booklet style, and the members, some of whom were in advertising and publishing, took pride in its design and production. Nor did it lack for substance and imaginative content. There were articles on Latin American DXing, TV DX, sunrise and graveyard medium wave DXing, and foreign medium wave, along with regional stories, station profiles—even a comic strip, DX song lyrics, and some radio fiction. The publishing duties were shared by several people over the years, including Paul Kowalski of Milwaukee, Austin Kelly of Chicago, Joe Farley of Downers Grove, Tom Ross of Lyons, and Ed Stroh of Thornton. In 1987, CADX went online with a DX/SWL special interest group on the BBS Chicago bulletin board.

When support of the bulletin waned in the 1990s, the club's tradition was continued in the form of a website and an e-mailed bulletin of loggings and station news prepared by Christos Rigas of Wood Dale, Illinois, an active shortwave and medium wave DXer who had played many roles in the club over the years.

CADXers made the most of the club's main strength—its local orientation and the ability of its members to get together. Although geography made DXpeditions difficult, there were numerous gatherings in various venues. CADX placed a premium on the social side of its activities, and many of the friendships made years ago continue to this day.

International DXer's Club of San Diego (1978–1983)

It was June 1978 when 75-year-old Larry Brookwell founded the San Diego DXer's Club (renamed San Diego DXer's Club International in February 1980 and International DXer's Club of San Diego in May 1981). Although nominally a club, with the usual entreaties for more member involvement, the IDXC was primarily an extension of Brookwell's irrepressible, and sometimes irreverent, personality.

Although the club bulletin occasionally carried some material from "Sweden Calling DXers" and the BBC Monitoring Service, the focus of IDXC was on receivers and related equipment topics, usually offered in plain, non-technical language. Brookwell had been active in radio since 1924 and he knew his subject well. The club bulletin contained innumerable receiver descriptions, evaluations and commentary. Brookwell understood the difference between receiver specifications and actual receiver performance, and his concentration was always on the latter. In 1981 he published the 179-page *Shortwave Hobby Equipment Review 1981* which consisted of selected material from all the past IDXC bulletins. Supplements were issued in 1982 and 1983.

Most of the bulletin was written by Brookwell. However, for various periods a few columns were written by others. These included "Over Thirty," about VHF-UHF equipment; "The Benelux Conexion," covering the equipment scene in Europe; "The Tokyo Connection," about Japanese equipment; and "The World of RTTY," written by Fred Osterman, who later would become President of Universal Radio, Inc., of Reynoldsburg, Ohio, a leading supplier of shortwave equipment and supplies.

Readers liked Brookwell's friendly, frank and down-to-earth style. His original hope was to attract 20 or 30 local members. By July 1981, club membership was almost 300. By September 1982 it had grown to over 500 members in 15 countries. Nine months later it was almost 600. It became an ANARC associate member in 1980, and a full member in 1982.

While Brookwell usually stuck to his topic, he was not averse to offering his personal observations on the issues of the day. Starting in November 1980, members were treated to "The Odyssey of an Aficionado," a serialization of his "life and loves in and out of the radio business." In 1983 he added some supplementary material and republished "Odyssey" as a separate volume titled *Tail Gunner on a Superheterodyne — The Confessions of a Radio Nut.*

Club members, and the entire DX community, were saddened when Larry's son informed them that Larry had died of a heart attack on December 23, 1983, just months after his wife had passed away. Although most had never met him, many considered him a personal friend. His son sought someone to continue his father's work. A bimonthly *Radio Equipment Review* newsletter edited by Ronald Pokatiloff of Zion, Illinois, was begun in March 1985, in part to fill the void left by the absence of Larry Brookwell and his club.

OZARK MOUNTAIN DX CLUB (1981–1986)

The Ozark Mountain DX Club was founded in 1981. It published a bi-monthly, two to four page newsletter, *The DX Log*, which emphasized the more challenging SWBC DX. Mitch Sams, then of Farmington, Arkansas, was editor-publisher, and was assisted (from November 1985) by Dave Valko of Dunlo, Pennsylvania and Dan Sheedy of Encinitas, California. *The DX Log* featured loggings, bandscans, and occasional features. The club merged with the *Fine Tuning* newsletter in 1986.

GREAT CIRCLE SHORTWAVE SOCIETY (1982–1988)

Although nominally intended to encourage long wave, medium wave and shortwave listening, the real goal of the Great Circle Shortwave Society was to provide a link among those who were active SWLs in the 1950s and 1960s, particularly WPE call sign holders (see p. 358). (There was a flirtation with the idea of issuing "WGC" SWL call signs a la the WPE program, which by then was being run privately by Hank Bennett but had lost much of its cachet.) The club's bulletin, first issued bi-monthly, then quarterly, took the name *WPE Call Letter* in the summer of 1985.

The GCSS was nothing if not ambitious. While not all of the items on its agenda reached fruition — one in particular that would have been nice was the *WPE Callsign Directory*, which was to contain the names and addresses of as many WPE callsign owners as could be found — the GCSS bulletin was substantive, even if editors did change frequently. Although it had nominal content in traditional areas like loggings, card swapping and

QSLing, its main focus was on hobby nostalgia and vacuum-tube receivers. The club had an awards program and a GCSS Ham Radio Net.

The group's original president, Vern A. Weiss of Kankakee, Illinois, was followed in office in 1987 by Harold N. Cones of Newport News, Virginia, an active shortwave hobbyist who later, as one of the two "Radio Professors," would co-author several books on Zenith radios. Also involved in a strong supporting role was club secretary Richard H. Arland. GCSS was ahead of its time, however, interest in hobby nostalgia not yet having firmly taken root. The editorial talent could not compensate for the limited participation beyond a hard core of hollow staters, and the club closed its doors in 1988.

ASSOCIATION OF CLANDESTINE RADIO ENTHUSIASTS (ACE) (1982–2005)

Pirate broadcasting had been on the increase during the 1970s. With the demise of the Free Radio Campaign's newsletter, *The Wavelength,* a new club devoted to tracking the activities of various "unofficial" broadcasters— the Association of Clandestine Radio Enthusiasts— was born. The first president and publisher of ACE (sometimes written "A*C*E") was Darren S. Leno of Moorhead, Minnesota. Although its application for membership caused much debate within the Association of North American Radio Clubs over the propriety of admitting a club whose principal focus was illegal transmissions, ACE was granted associate member status in October 1982 and full membership a year later.[45]

The club's monthly bulletin was called *The Monthly ACE* (*The ACE* from June 1985). The first issue, dated March 1982, was sent free to many DXers, and regular publication began in May, by which time there were 21 members, a number that would grow to 115 in a year and almost twice that number in the club's second year.[46] (The club gained many short-term members through a prominent mention in *Utne Magazine,* which focused on alternative media.) The bulletin covered loggings of North American pirates and European pirates heard in North America, as well as clandestine broadcasters and numbers stations (commonly thought to be "spy" transmissions). It also contained a QSL column, station profiles and interviews with pirate operators, technical information on pirate broadcasting, news of pirate "busts," information on pirate mail drops, reviews of other pirate literature, and additional material supplied by members or obtained from other sources.

Although the focus of the bulletin was on North American shortwave pirates, there was also a monthly report from Europe on the active pirate scene there, as well as reports on the reception of medium wave pirates. Among *The ACE*'s key editors in its early days was John T. Arthur, then of Arcata, California, whose "Veried Response" column contained news about QSLs received, a QSL scoreboard, member buy-sell-trade announcements, pirate-related media observations, and members' comments. Arthur was a long-time pirate fan who had published a similar column in *The Wavelength,* and would be one of ACE's most faithful supporters. Two other key editors were Kirk Baxter of Shawnee Mission, Kansas, whose "DiaLogs" column contained the most extensive pirate loggings available; and Lani Pettit of Sioux City, Iowa, who edited the "Spy Centre" column focusing on numbers stations. Tim Corcoran of Shavertown, Pennsylvania; Keith Hill of Pine City, New York; William J. Martin, Wilmington, Delaware; Scott McClellan, Battle Creek, Michigan; Podney R. Sixe, Camborne, Cornwall, U.K.; and Keith Thibodeaux, Baton Rouge, Louisiana, also edited early ACE columns.

"DiaLogs" contained the full addresses of member-reporters so that pirate stations which lacked postal addresses could QSL members' loggings that were reported in the bulletin. John Arthur also prepared the periodic "Free Radio Directory," a list of pirate addresses. This appeared in *The ACE* twice yearly through December 2003 (from July 2000 it was prepared by Chris Lobdell).

ACE was a leader in the use of computers for DX-related purposes. In November 1984, with over half the membership using computers, the club established a remote Bulletin Board System (BBS) where both listeners and stations could post information (including loggings for *The ACE*). This was one of the earliest shortwave bulletin boards, and it was particularly valuable in the case of pirates, whose operations were often unscheduled. The first ACE member get together was held at the ANARC convention in Milwaukee in July 1985, by which time *The ACE* was the main source of pirate radio information in North America. This and similar gatherings supported a higher level of fraternalism than existed in most other national-level radio clubs.

In keeping with the club's name, in June 1985 *The ACE* added a column devoted exclusively to clandestine, as opposed to pirate, broadcasting. It was edited by Andrew O'Brien of Buffalo, New York, who was assisted by George Zeller, then of Ashland, Ohio. O'Brien handled logs, while Zeller did station profiles and other features until August 1988 when both were consolidated under Zeller. "Clandestine Profile" was one of the bulletin's anchor columns and a prime source of information on clandestine station activity, and Zeller would become a major figure in the pirate listening community.

Despite occasional late or missing bulletins, and ups and downs in member participation, ACE prospered. The booklet-style bulletin was typically 20 to 24 pages, occasionally more, and while it was not as professional looking as those of some of the major clubs, its appearance steadily improved.

The club's first organizational change was in April 1985 when Keith Thibodeaux became publisher. Soon thereafter, in December 1985, Leno was succeeded as president by Kirk Baxter. Baxter would hold the position, and be the club's key man, for 11 years.

Publishing duties passed to William J. Martin in September 1987, but a year later returned to Thibodeaux, who took the operational reins of the club for about a year. The practice of club leaders supplementing club revenues with personal donations was foresworn, and a dues increase contributed to a firmer financial footing. The bulletin was renamed *Free Air*, and an eight-to-twelve page, full-sheet format was adopted with little loss of content thanks to the expanded use of desktop publishing. In October 1988, "Spy Centre" was replaced by "Covert Corner" edited by Harry L. Helms of San Diego, California, an author well-known in the monitoring community. "Covert Corner" covered utility signals that did not fit into established categories, e.g., number stations, unidentified beacons, markers, insurgent communications, government transmissions, etc. ("Covert Corner" would be replaced by another Helms column, "Special Features," in April 1994.)

Baxter resumed publishing in June 1989, re-adopting the bulletin's old name, *The ACE.* "DiaLogs" became a very busy column and would have a number of different editors over the years. In July 1989, another known hobby author, Andrew R. Yoder of Beaver Falls, Pennsylvania, came aboard with the "DX Clip Board." In September 1989, "numbers" stations again got their own column, "Los Numeros," edited by "Havana Moon," a pseudonym of William T. Godbey of Briarcliff Manor, New York. ("Los Numeros" was consolidated into

"Covert Corner" in May 1993, after which there would be little coverage of numbers stations until 1997.) For several years, Don Bishop of Overland Park, Kansas, contributed features about station operations and FCC enforcement activities.

An ACE awards program was established in 1991. It offered several different awards for North American and European AM, FM and shortwave pirate and clandestine stations verified. There were periodic pirate popularity contests. In the mid–1990s, the "freE Mail list," a list of club members' e-mail addresses, appeared occasionally in the bulletin. Around the same time, ACE member (and then "DiaLogs" editor) Kirk Trummel of Springfield, Missouri, maintained two pirate databases on the ACE section of the ANARC BBS: "40 Meter Notes," which showed non-pirate station activity in the 41, 43 and 49 meter bands (thus serving as an aid to pirates seeking open channels), and Trummel's "Little Black Book," showing known addresses of clandestine stations and North American and European pirate stations.

Other columns came and went throughout the 1990s. Two columns, "The Flipside" and "Free Radio Forum," started and ended in the same year, 1995. "Euro-Logs," a column by Rainer Brandt of Germany, appeared during 1994–95 and occasionally thereafter. Of the new columns, only "Micro-'casting" by Bud Stacey of Satsuma, Alabama would have any longevity. It began in March 1994 and continued for nine years. It covered low power FM pirates. The FM pirate community was different from the shortwavers, however, and FM was never a major topic within ACE, whose focus was always on shortwave.

ACE membership was over 300 by early 1992. It had dropped to around 240 in 1995, and it slipped below 200 by 1997, a year that was a busy one for ACE. The club established an ACE section on the Free Radio Network website in 1997. More importantly, a major reorganization took place in February when Baxter was succeeded as club president by Pat Murphy. Murphy appointed a Managing Editor, Steve Rogovich, who had served as "DiaLogs" editor during 1993–94. Both lived in Norfolk, Virginia. In May 1997, the title of the bulletin was again changed, this time to *The ACE Newsletter*, with Murphy and Rogovich jointly serving as publishers. Several new columns were tried. Coverage of numbers stations was resumed in a "Covert Comms" column edited by Tom Sevart of Frontenac, Kansas. For eight months, a technical column, "Tech Topix," also appeared. However, the heart of the bulletin was always in its three main columns—"DiaLogs," John T. Arthur's "Veried Response," and George Zeller's "Clandestine Profile."

In 2000, after three years in charge, Murphy and Rogovich wished to pass the baton. ACE had done an able job servicing a niche element of the SWL universe, but it was struggling against the same declines in membership that the mainstream clubs were facing. These problems notwithstanding, in May 2000 John Arthur, then of Belfast, New York, stepped in and assumed all headquarters duties—"interim" club president, managing editor and publisher. (He also returned the bulletin to its original name, *The Monthly ACE*.) Long time pirateer Chris Lobdell of Tewksbury, Massachusetts took over Arthur's "Veried Response" column and the compilation of the "Free Radio Directory." Helms served as publisher from May 2001 to February 2003, when that duty returned to Arthur, who also resumed editing "Veried Response" in January 2004.

Club membership was dropping, and in the summer of 2000 a persistent deficit required an increase in the membership fee from $21 to $22. Fewer member contributions meant a smaller "DiaLogs" column. The bulletin was down to 12 to 16 pages, and both quality and quantity were suffering. A new column, "Random Transmissions," ran from May 2002 to November 2003. However, keeping columns alive, and editors interested, was a problem. "Micro-'casting" disappeared in March 2003. Chris Smolinski of Westminister, Maryland, took over "Covert Comms" the same month, but it was dropped in February 2004. In April 2003, Zeller, who had continued editing "Clandestine Profile" even after becoming "Outer Limits" editor for *Monitoring Times* in May 1992, closed out his nearly seven year editing career with ACE. A year passed before he was succeeded by Vincent R. Havrilko, whose military obligations interfered with the regular appearance of the column.

Arthur persevered despite these difficulties and the general loss of energy in the club until November 2005, when he informed the members that, after more than 23 years, ACE was ceasing publication. The precipitating event was the departure in September of Lee Reynolds of Lempster, New Hampshire, who had been editing the "DiaLogs" column for almost five years. "DiaLogs" was by now the bulletin's main column, and there was no Reynolds replacement in sight. With membership now around 50 (it had peaked above 400 a few times), there was little choice. The September 2005 bulletin, the club's last, had eight pages. Recent renewals were refunded, and while some thought was given to a possible migration of club activities to the internet, it was decided instead to close down, bringing to an end this valuable meeting place of the pirate radio community.

MICHIGAN AREA RADIO ENTHUSIASTS (MARE) (1984 TO THE PRESENT)

On October 21, 1984, following an informal meeting in May, nine SWLs met in Dearborn, Michigan to create a vehicle for improving contact among area SWLs. The idea was Don Hosmer's, developed following his attendance at the 1983 ANARC convention which had been sponsored by a local DX group in Washington, D.C., the Washington Area DX Association. He thought their model would be a good one to follow. The Michigan Area Radio Enthusiasts was born, and in December Hosmer issued a four-page bulletin, introducing "MARE" to the world.

Members were sought from Michigan and the border areas of Ohio, Indiana and Ontario. The bulletin would be bi-monthly. Meetings were scheduled for twice a year, but soon they were being held more frequently — in members' homes, in libraries, and at the Dearborn Civic Center, usually with 10–20 people in attendance. The club gained publicity on the air, in radio magazines and in other club bulletins, and by August 1986 membership had reached 100. ANARC membership followed.

The club adopted a constitution and bylaws providing for a three-person Board of Directors, and elections were held every year (every other year starting in 1995). The club organized as a non-profit corporation in 1988.

Hosmer became ANARC Executive Secretary on January 1, 1987 and passed the reigns of MARE bulletin editor to Bob Walker of Wixom, Michigan. Soon the bulletin boasted a professionally printed, 16-page booklet format. Regular columns covered shortwave broadcasting, medium wave, scanning and QSLs (limited TV and FM coverage was added in 1989). Kenneth Vito Zichi of Williamston, Michigan provided a wide ranging column which was the bulletin's anchor. Although called "Q & A," soon it ceased being based on members' questions alone. There was also an informative headquarters section, member ads, and a calendar of radio events, plus, at various times, equipment reports, book reviews, new member bios, "The CPRV Page," and other features (including, for several years, a humorous April Fool's page). In 1989, Harold Frodge of Midland, Michigan, assumed the duties of bulletin editor and publisher, and Walker became business manager. (Karl Racenis of Livonia, Michigan, took over as business manager in 1994.) The bulletin was renamed "Great Lakes Monitor" in 1990.

DXpeditions to a site near Brighton, Michigan, were a favorite MARE activity. Six were held in the club's first two years, and by December 1992 the club had held its 25th DXpedition. Reports on DXpedition activities were often included in the bulletin. Other club events included tours of radio stations and radio museums, picnics, and occasional seminars to introduce DXing to the public. Special meetings were held to welcome well-known New Zealand DXer Arthur Cushen (1986), Michael Murray, Secretary General of the EDXC (1986), and Ken MacHarg of HCJB (1989). A computer bulletin board was begun in 1991. A utilities section of the bulletin had not been well-supported when it was tried in 1987, but it reappeared for several years starting in 1993. The following year, Zichi inaugurated a weekly e-mail "Tip-Sheet," with various MARE members serving as editor. The club inaugurated a website in 1998, and utility coverage returned to the bulletin in 2002.

Although as a regional club MARE benefitted from personal contacts among members, it could not escape the widespread membership deflation of the 1990s which all clubs experienced. At the start of 2001, membership was 75, roughly half the 1992 figure of about 140. While in later years the club sponsored several DXpeditions each year, along with spring and fall picnics and occasional other meetings, and continued to receive support from a hard core of members, in 2007 membership was around 45. Although they were determined to keep the club alive, with reduced attendance at club events and fewer contributors to the bulletin, the club faced an uncertain future.

French Canadian Clubs

The earliest of the French-language clubs in North America was Club des DXers du Quebec (CDXQ). It was formed in September 1970, and two months later the first issue of its bulletin, *A l'Ecoute,* appeared. It was a substantial publication which soon grew to approximately 20 pages. It was unique in the North American SWL world because it was bi-lingual, with almost all items written in both French and English. The bulletin covered primarily shortwave broadcast listening, but with some attention to medium wave as well, and included members' loggings, station news, QSLs, heard-verified statistics, member profiles and short feature articles. The club president was Jean-Luc Vachon of Quebec. By mid–1971 the club had 33 members and was growing. In October of that year the bulletin's name was changed to "Hello DXers." (At the same time, Eddie Startz agreed to be an Honorary Director of CDXQ.)

The club withdrew from ANARC in 1972, and thereafter the bulletin went unpublished for some time. In January 1974 a new CDXQ publication emerged called the "Hello DXers News Sheet." It was an all-English bulletin of at most a few pages containing no station information but short features relative to SWLing. It was intended mainly as a vehicle for CDXQ members to keep in touch, and it was published by Frank G. Barratt, the club's vice-president. It appears that the club closed in early 1975.

The most successful of the Canadian French-language clubs was Club d'Ondes Courtes du Quebec, founded in September 1974 by Guy Marcotte and originally known as the Club Ondes Courtes Candiac. Its good looking, 40-page, offset-printed monthly bulletin, *L'Onde,* was entirely in French, and covered shortwave broadcasting, utilities, QSLs, features, and member letters, and also gave limited attention to medium wave and ham activities. Periodically the club also published a list of French-language broadcasts, and it had an awards program that was open to non-members as well as members. Membership grew quickly and the club had 500 members by 1981.

Marcotte continued to lead the club until 1984 when Yvan Paquette took over. Later the baton was passed to Marc Giard, and in 1993 to the team of André Hemlin and Daniel Goyette. Member participation declined over time, however, and the last regular issue of *L'Onde* was published in January 1997. The club is recalled on a commemorative website where station schedules are updated from time to time.[47]

A third French-language club, Club DX Quebecois, Ancienne-Lorett, Quebec, was founded in 1977 and was an ANARC member. Its 40-page bulletin, which was published for several years, was called *Echo-DX* (later *L'Echo des Ondes),* and covered shortwave, medium wave, utilities, receiver reviews, and news of radio activities in French areas overseas. The club was reorganized in 1981 and headquarters was moved to Montreal. It ceased operations in June 1982.

Smaller Clubs After 1960

The 1960s and 1970s were surely the golden age of shortwave listener clubs. While some were intended as local or regional clubs, others evidenced national and even international ambitions. Although a few of the smaller clubs grew to be successful, many never got much beyond the kitchen table.

In the following list, the names of club bulletins are indicated in parentheses where known, and ANARC membership is denoted. Clubs varied greatly in their life span, and club headquarters often changed with a change in club management.

The 1960s saw activity by the following clubs:

Absecon DX Club, Ventnor City, New Jersey
Batesville Shortwave Association, Austin, Texas *(Radiogram), ANARC member*
Black Hawk Radio Club, Asheboro, North Carolina
Bronx DX Club, Bronx, New York *(Bronx DX Digest)*
Central Ontario Radio Monitor Association, Ottawa, Ontario
Delta Echo Radio Club, Parry Sound, Ontario
DX Inter-Nationale, Newton, Massachusetts, *ANARC member*[48]
DXers Radio Club, Souderton, Pennsylvania
DX Organization of Winnipeg, Winnipeg, Manitoba
DX Unlimited Shortwave Club, Seattle, Washington *(DXing Hi-Lites)*[49]
East Nassau DX Radio Club, Jericho, New York
Empire State DX Club, Richmond Hill, New York
Fircrest DX Club, Fircrest, Washington *(DX-Telegramme), ANARC member*

Five Star DX Club, Sunnymead, California *(FSDXC Bulletin)*
Fly-By-Night Communications Club, Bronx, New York
Folcroft Radio Club, Folcroft, Pennsylvania *(Monitor)*, *ANARC member*[50]
Golden Gate Shortwave Listeners League, San Francisco, California
Great Lakes Shortwave DX Club, Kalamazoo, Michigan, *ANARC member*
Intercontinental DX Club, Scarborough, Ontario *(Hi)*
International League of Signal Chasers, Denver, Colorado (SWL-card swapping) *(DX Monitor)*
International SWL Club, Regina, Saskatchewan *(Radio Courier)*
Iowa Shortwave DX Club, Davenport, Iowa
Kentucky DXers Association, Raceland, Kentucky *(Kentucky DX Reports)*, *ANARC member*
Latin Brass Radio Club, Long Beach, California
Macador Radio Club, Northbridge, California
Midwest DX Shortwave Radio Club, Decatur, Illinois *(MWDXSWRCer)*[51]
Missouri DX Association
Missouri Shortwave Club, Columbia, Missouri
Morris DX Society, Morristown, New Jersey
New Brunswick Radio Club, Bathurst, New Brunswick
New England Card Swappers, Meriden, Connecticut
New England DXers Alliance, *ANARC member*
New York State DXers Association, Albany, New York *(The DX Journal)*, *ANARC member*
Northeast Shortwave Listeners Club, Manchester, New Hampshire, *ANARC member*
Northwest Short Wave Listeners Club
Plainview Shortwave Club, Plainview, New York *(Plainview DX Monitor)*
Radio Listeners of America
Radio Passaic International DX Club, Passaic, New Jersey
Santa Maria Valley DXing Association, Santa Maria, California *(Santa Maria Valley DXer)*
South East DX Listeners, Gatlinburg, Tennessee
Shortwave Listeners–Certificate Hunters Club, Detroit, Michigan, *ANARC member*
Teen Shortwave Club, Pittsburgh, Pennsylvania
Texas County Shortwave Association, Houston, Missouri
Texas Shortwave Monitors League, League City, Texas
Toronto–Trans World Short Wave Club, Toronto, Ontario
Trans-National DX League of the Americas, St. John, New Brunswick *(Worldwide Review)*
Under 20 DX Club, Binghamton, New York
United States Shortwave Club, Skokie, Illinois *(Band Spanner)*
United World DXers Club[52]
Wasatch DX Club, Brigham City, Utah
West Coast DX Club, Victoria, British Columbia
Worldwide Monitors Radio Club, Caledonia, New York *(DXer's Digest)*, *ANARC member*

Encouraged by the success of the national clubs during the 1970s, other clubs maintained a high level of activity during the decade. New clubs that became active in the 1970s included:

Association of Illinois DXers, Schaumburg, Illinois *(AIDX Journal)*, *ANARC member*[53]

Brooklyn DX Club, Brooklyn, New York *(English Language Broadcasts*, later *All Language Broadcasts*, and *ASDA Flashsheet)*, *ANARC member*[54]

All Indiana DXers, Richmond, Indiana

Calgary Shortwave Club, Calgary, Alberta

DX Club of Dix Hills, Dix Hills, New York

East Coast DX Alliance, Cinnaminson, New Jersey *(ECDXA Newsletter)*[55]

Great Lakes DX Association, Lansing, Michigan *(NEWSMEMO)*

Happy Station Fan Club, Adamstown, Pennsylvania *(HSFC Magazine)*

Herbert H. Lehman College Radio DX Club, Bronx, New York *(HHLCRC News)*

Idaho DX Association, Boise, Idaho *(IDEX)*, *ANARC member*

Long Island DX Club, Old Bethpage, New York *(DX Digest)*

Middle Tennessee DX Association, Alexandria, Tennessee *(Go-DX)*, *ANARC member*[56]

Midwest DX Club, Morton Grove, Illinois *(MWDXC Journal)*, *ANARC member*

Minnesota DX Club, Minneapolis, Minnesota *(MDXC Newsletter)*, *ANARC member*

New England DX Society, Central Falls, Rhode Island *(NEDS News)*

Radio California U.S.A.,Torrance, California *(Newsletter & SWL Guide)* (later Radio U.S.A., Phoenix, Arizona), *ANARC member*

Riverside Radio Club, Riverside, California[57]

Rocky Mountain DXers Association, Englewood, Colorado, *ANARC member*[58]

Southern California DXers, Long Beach, California (founded in 1974; *Calling All DXers* began publishing in 1984), *ANARC member*

Southern DX Association, Alexandria, Tennessee *(Go-DX*, and the mid-monthly *DXTRA)*[59] (known as the Middle Tennessee DX Association until 1980)

Southern Ontario Radio Club, Prescott, Ontario *(S.O.R.C. News)*

Transworld DX Club, Vancouver, British Colombia *(DX World*, later *DX Data)*, *ANARC member*

Triangle DX Club, Raleigh, North Carolina *(TDX News)*, *ANARC member*

United World Radio DX Club, Elizabethton, Tennessee (bi-monthly *United World DX-Tra* and monthly *DX-O-Gram)*, *ANARC member*

University of Manitoba DX-SWL Club, Winnipeg, Manitoba *(Monitoring Notes)*, *ANARC member*

Virginia DX Association, Richmond, Virginia *(Bandscan)*, *ANARC member*

Voice of the World, Newark, New Jersey

Washington Area DX Association, Silver Spring, Maryland *(WADXA News)*, *ANARC member*

Washington International League of DXers, Washington, D.C.

Wingate SWL Club, Wingate, North Carolina *(WSWLC Magazine)*, *ANARC member*

Woodfield International DX Club, Schaumburg, Illinois (renamed the Association of Illinois DXers in 1977) *(Window on the World*, renamed *AIDX Journal)*, *ANARC member*

World Shortwave Listeners Club, White Plains, New York *(DX Journal*, later *DX Journal & World Broadcasting Review*, supplemented in 1980 by the bi-weekly *Express)*, *ANARC member*

Activity was subdued in the 1980s, but the following clubs appeared on the scene for the first time during those years.

American Youth DX Club, Muskogee, Oklahoma
Association of Manitoba DXers, Winnipeg, Manitoba
Austin Radio Club, Austin, Texas (*Austin Radio Club Newsletter*)
Blue Grass DX Fellowship, Lexington, Kentucky
BC Interior Shortwave Radio Club, Kelowna, British Columbia
Capitol DXers Club, Sacramento, California (*Capitol DXers Newsletter & Radio Log West*)
Cascade Mountain DX Club, Kirkland, Washington (*SW Monitor*)
Central Maryland DXers Association, Baltimore, Maryland (*DX-Gram*)
DX Midamerica, West Allis, Wisconsin (*Hot Tips and Targets*)
DX Northwest, Issaquah, Washington (*Grayline Report*)
El Paso SWL DX Club, El Paso, Texas
Gateway DX Club International, Hazelwood, Missouri (*DX Scene*)
New York City DX Association, New York, New York (*The Urban DXer*)
Northern Lights DX Club, Edina, Minnesota (*DX Forum*)
Old Dominion DX Association (ODDX), Virginia Beach, Virginia (*ODDX Newsletter*)
Pacific Northwest-British Columbia DX Club, Seattle, Washington (*PNBCDXC Newsletter*), ANARC member
Radio Communications Society of the World, Ithaca, New York
Ralston Radio Club, Lawrence, Kansas
St. Louis International DXers (SLIDX), St. Louis, Missouri (*The Gridley Wave*), ANARC member
United Northwest Inland DXers, Spokane, Washington
United States DX Association, Atlanta, Georgia (*The Signal*)[60]
Worldwide DX Club, Menomonee Falls, Wisconsin

Although the 1990s saw a further decline in new clubs, some new names appeared.

British Columbia Shortwave Listening Club (BCDX), Vancouver, British Columbia (*Logjam*)
Central Indiana Shortwave Club, Indianapolis, Indiana (*Shortwave Oddities*)
Chicago Area Radio Monitoring Association (CARMA), Chicago, Illinois (*CARMA Newsletter*)
Cincinnati Area Monitoring Exchange (MONIX), West Chester, Ohio
Ft. Wayne Radio Listeners Club, Fort Wayne, Indiana
Global DX Club, Pinson, Alabama (*Radio Waves*)
Memphis Area Shortwave Hobbyists, Memphis, Tennessee
North Central Texas SWL Club, Grand Prairie, Texas
Northeast Ohio SWL/DXers, Westlake, Ohio
Pitt County SW Listeners Club, Ayden, North Carolina (*The DX Listeners*)
Puna DX Club, Keaau, Hawaii
Sandy River SW Radio DXers Association, Norridgewock, Maine (*The QSL*)
Signal Surfer DX Club, Burns Lake, British Columbia
Toledo Area Radio Enthusiasts, Maumee, Ohio
Triangle Area Scanner/SW Listening Group, Raleigh, North Carolina

Club-Related Groups

ASSOCIATION OF NORTH AMERICAN RADIO CLUBS (1964–2005)

There had been some coordination among DX editors over the years. An International League of Shortwave Editors was founded in 1945 among several editors in the United States, Sweden, Australia, New Zealand, and Cuba, although it appears to have been an informal affair. There was no meaningful coordination among clubs in North America until the formation of the Association of North American Radio Clubs in 1964.[61]

Not a club itself, ANARC was in the nature of a trade association for clubs, formed to help achieve objectives that were beyond the reach of clubs acting individually, to foster the common good of the hobby, to do outreach, and to promote cooperation among clubs. ANARC's founder and first Executive Secretary was Donald N. Jensen of Kenosha, Wisconsin.[62] He and others who had considered the matter had in mind an organization similar to DX-Alliansen, an umbrella organization of Swedish radio clubs.[63] Hank Bennett had editorialized in favor of the concept in both his *Popular Electronics* and NNRC columns.[64]

Full ANARC membership was available to clubs with 50 or more members and at least 12 months of continuous operation. Full member clubs were assessed annual dues based on the number of their members. Clubs that had been in operation for more than six but less than 12 months and had fewer than 50 members could seek non-voting, associate membership. (This two-tier membership structure was dropped in 1983.)

Pursuant to its constitution, ANARC was to be governed by representatives of the member clubs, of which 11, including all the major clubs, were charter members. The club representatives then appointed an Executive Secretary. Starting in 1983 the club representatives elected an Executive Council which appointed the Executive Secretary (later called Chairman).

There were many issues on ANARC's agenda over the years. Among them were the problem of non-verifying stations, providing frequency recommendations to international broadcasters, improving coordination of, and more widely distributing information about, courtesy programs (mainly a BCB topic), improving the QSL practices of Latin American stations, and bringing uniformity to club country lists. Others topics included hobby standards and ethics, better listener liaison with equipment manufacturers, and outreach and public relations. Progress in most of these areas was modest, notwithstanding the considerable energy that was sometimes expended. Some of the simplest activities produced the best results, such as the exchange of bulletins among member clubs and the preparation of a club list which ANARC used to respond to inquiries and which improved the visibility of all clubs.

Soon after its formation, ANARC began publishing a newsletter about the doings of member clubs. Originally it was an internal document available only to member clubs and their representatives. By 1982 the *ANARC Newsletter* had become available to individual DXers on a subscription basis. Readership had reached 750, and the newsletter was producing most of ANARC's revenue. It included a calendar of DX meetings and conventions around the country, reports of ANARC committees, updates on club activities, highlights of member club bulletins, and various features. From April 1979 to November 1988 the newsletter also included "The Marketplace Report." Prepared by the ANARC DX Equipment Information Committee (principally Harold Sellers), it included detailed information on equipment, companies and services of interest to SWLs. It was one of the most informative parts of the newsletter. The newsletter also covered the activities of ANARC's European counterpart, the European DX Council, thus providing a useful window on the European DX club scene. ANARC activities were also covered in on-air reports over HCJB's "DX Partyline" and Radio Canada International's "DX Digest."

Probably ANARC's most important undertaking was the annual ANARC convention, the first of which was held at the Muehlbach Hotel in Kansas City, Missouri on July 29–31, 1966. A dozen or so people were in attendance. ANARC conventions were held in many cities around the United States and Canada until 1990, the year of the last convention. The conventions grew in size, at their height attracting over 200 persons, including station personalities, equipment manufacturers, and DXers from other countries. From 1988 through 2000, ANARC also had a booth and presented the Miami Valley DX Club–originated SWL Monitors Forum at the annual Dayton Hamvention.

Although the feeling in later years was that ANARC's accomplishments had been limited, in fact it had successfully undertaken many projects. In the 1970s, ANARC was responsible for getting one of the best-known national shortwave equipment suppliers, Gilfer Associates of Park Ridge, New Jersey, to send ANARC club publicity sheets with customer orders. It also published multiple editions of the *Directory of DX Publications and Services in North America.* Other projects included the ANARC Ham Net (originally called the East Coast SWL Net) which focused on shortwave listening and operated from 1987 to 1999;[65] the ANARC "DX Code of Ethics," written by John Callarman and issued in 1967; the 32-page ANARC/HAP *Guide to DX Radio Listening,* published in 1980; and the ANARC pamphlet, *Choosing a Receiver,* which was issued in 1981.

Research and preservation activities included Robert Horvitz's "Woodpecker Project," which studied over-the-horizon interference in the shortwave bands (1985); the Committee to Preserve Radio Verifications, born in 1986 and still operating today; the submission to the FCC of written comments on efforts to relax radio interference emission limits on unlicensed electronic devices and to require labeling of receivers capable of tuning frequencies protected by the Electronic Communications Privacy Act (ECPA) of 1986 (1988); and the *ANARC Guide to U.S. Monitoring Laws,* a useful reference to the ECPA (the *Guide* was published in 1989 and updated in 1990).[66] ANARC Executive Secretary Terry Colgan also testified before Congress on the ECPA. ANARC sponsored the North American DX Championships during the years 1993 to 2000. In September 1985 it established the ANARC Bulletin Board System (BBS), which offered loggings, club information, and reference material. The BBS was replaced by an ANARC website in 1996.

Another valuable ANARC function was the establishment in 1967 of a Man of the Year award (later known by other names). Over the years, other awards and certificates of recognition were established as well. These were useful in recognizing those who made major contributions to the DX listening community. They were announced at the annual ANARC convention (later the Winter SWL Fest).[67]

Some ANARC projects outlived their usefulness or became redundant in the changing hobby environment. The ANARC QSL committee, whose mission was to try to get stations to QSL more reliably, communicated with many stations but had little success and

This is to certify that

_____

made a valuable contribution
of time and radio expertise to

The Woodpecker Project

October, 1985

Robert Horvitz, chairman
ANARC Over-the-Horizon Radar Committee

One of many ANARC initiatives was a study of interference caused by over-the-horizon radar. Called the Woodpecker Project for the sound made by the OHR, this card was given to those who participated in the effort.

was disbanded in 1986, along with the frequency recommendation committee, whose goal was to help stations select more suitable frequencies.

The year 1989 was the start of a difficult time in the history of ANARC, whose purpose and direction more and more were being called into question. There were debates over whether ANARC's main goal should be to support the radio club community, or whether it should concentrate on public outreach, a favorite topic among those who wanted to increase the number of DX enthusiasts. How to achieve its goals was also a perennial question. Some clubs were indifferent to ANARC's future and offered little support. Member dues went uncollected for several years, and while then-ANARC Executive Secretary Robert Horvitz had greatly professionalized the *ANARC Newsletter*, it was not breaking even financially, and his calls for greater member and subscriber contributions went unanswered.

Also in 1989 the Ontario DX Association dropped its ANARC membership. ODXA no longer found it necessary to be an ANARC member, and felt that because ANARC's activities were not focused on clubs but on a much wider group of radio listeners as a whole, ANARC should restructure itself in order to be "a representative body for all users of radio receivers and those organizations which aid the user," with various forms of membership open to individuals, clubs, broadcasters, manufacturers and dealers.

The ODXA was the second major club to leave ANARC. The NNRC, an ANARC charter member, had left "reluctantly" in 1973, having concluded that ANARC membership served no useful purpose. According to the NNRC, ANARC had "failed to be the useful, productive group which it was intended to be, and more importantly, there is little probability that it ever can achieve the goals for which it was created."[68] The NNRC returned in 1977, however, taking a more positive view of ANARC's accomplishments (and perhaps seeking inclusion on the list of ANARC member clubs which Gilfer sent to its customers).

During this period there was also consideration of whether ANARC should continue hosting a convention, and, if so, whether it should be held every other year rather than annually. Related convention issues involved whether the convention should continue in its usual "transportable" form or be held in the same place each year, whether it should be held on a regional rather than a national basis, and whether it should affiliate with another compatible meeting of some kind (the 1990 convention was held in conjunction with the combined Tidewater Amateur Radio and Computer Festival and the state ARRL convention in Virginia Beach, Virginia). There was some inclination to require that the convention be sponsored by a club or a broadcasting organization rather than by informal groups of organizers. Clubs had been reluctant to take on ANARC convention sponsorship lest it turn into a financial liability. Also, individual club conventions were on the increase, and some clubs did not want to encourage competition. In addition, some clubs that did not deal principally with shortwave felt left out of the ANARC conventions, which tended to be shortwave oriented.

Time constraints led to Horvitz's resignation as Executive Secretary early in 1990. Sheldon Harvey, President of the Canadian International DX Club, was appointed ANARC Interim Coordinator. Although Harvey set forth some innovative plans for ANARC's restructuring, consensus was elusive and the problems multiplied. The last issue of the *ANARC Newsletter* was published in February–March 1990. The following year, for the first time since the ANARC conventions had begun in 1966, there was no ANARC convention. Malaise and unhappiness continued until the end of 1991, when a new Interim Coordinator, Richard A. D'Angelo of Wyomissing, Pennsylvania, was appointed.

Although polls of individual club members reflected support for ANARC, it now adopted a lower key, lower visibility approach to revitalization, with practical goals that were more in line with resources. Most ANARC successes had been the result of individual efforts, and the number of persons with both the interest and the ability to manage major projects had declined. Organizationally, ANARC's finances were put in order, and a new constitution was adopted by vote of the member clubs. Within ANARC's administrative structure, broader use was made of electronic communication. The annual ANARC convention was dropped, having been effectively supplanted by the Winter SWL Fest, an independent DX gathering that had been run by a group of DXers starting in 1988 (and which continues to the present). The annual ANARC awards continued, and were announced at the Fest.

In 1992 there were 17 ANARC member clubs, with two having more than 1,700 members and four having less than 100. The smallest had 35 members, the largest 2,100. At this time, ANARC member clubs had a total of 7,869 members (including duplicate memberships). ANARC membership would soon undergo some dramatic changes. In 1993, Club d' Ondes Courtes du Quebec, a French-language club, dropped its ANARC membership due principally to the language barrier. The Washington Area DX Association likewise dropped out, as did the Association of DX Reporters and SPEEDX, both of which ceased operations in 1995. The Radio Communications Monitoring Association, a scanner club and the largest of all the ANARC member clubs, closed its doors in June 1996. ANARC picked up several small clubs, but they did not offset the loss of the larger groups.

The ANARC newsletter was resurrected in 1993. Published four times a year, *ANARC Update* focused on club doings and ANARC activities, and included news of upcoming radio group meetings and similar events. It was distributed to ANARC staff and administration, club representatives, and leaders in the international shortwave broadcasting community, the goal being to get ANARC news into the hands of DX media leaders, who were encouraged to use it in their respective publications and programs. In order to avoid the appearance of competing with club bulletins (a recurring complaint about the old newsletter), subscriptions to the newsletter were no longer available to individuals, although individual copies of the newsletter were available to them for $1 and a stamped, self-addressed envelope. This policy was changed in 1994, it having been decided that the content of *ANARC Update* was not in conflict with club bulletins. The last *Update* was issued in December 1996.

The late 1990s and early 2000s were difficult years for ANARC. In 1998 the National Radio Club and the DX Audio Service (both medium wave groups), as well as the American Shortwave Listeners Club (which had closed in April of that year), dropped their ANARC membership. The ANARC Ham Net closed in 1999, and the last of the North American DX Championships, which were held under ANARC auspices, was conducted in 2000. ANARC's principal intramural activity became the periodic election of members to the Executive Council (now called the Executive Board). Its main public activities were the operation of an ANARC website and the annual ANARC awards made at the Winter SWL Fest. Indifference to ANARC's future was widespread.

A persistent theme within ANARC and its member clubs was how to interest more people in DXing. This goal eluded ANARC as it had eluded most clubs since the 1930s. ANARC also suffered from organizational problems. Member clubs varied greatly in size and focus, and ANARC included not just shortwave clubs, but clubs for scanning, medium wave, long wave, and TV and FM enthusiasts. The interests of the clubs varied. Many, particularly the larger clubs, were competitors. All ANARC activities were dependent on volunteer effort and leadership, which varied greatly in quantity and quality. Although most active clubs belonged to ANARC and supported it at least nominally, the level of participation varied, and some harbored continuing suspicions that full support of ANARC would reduce a club's independence. As club membership and club revenues declined, the annual ANARC dues were increasingly seen as bringing little benefit. Some saw ANARC itself, with its attractive newsletter, as competing for members.

As a result of these and other factors, ANARC was disbanded in April 2005, by which time both the size and the intensity of the shortwave listening community were much reduced. Computers had expanded personal communication among both individual hobbyists and clubs, and hobby-wide issues of importance to shortwave listeners, such as the threat from broadcasting over power lines and the need for periodic hobby gatherings, were being addressed in other ways. By the time ANARC was laid to rest, it was serving only 2,324 club members, over half of them in NASWA. This was a fraction of the number it had served in earlier years, half those it had reached as recently as 1997. In connection with ANARC's closing it was decided to continue the website with funding from NASWA and the Winter SWL Fest and open it to all clubs, whether previously ANARC members or not.

A European equivalent of ANARC, the European DX Council, was formed in 1967. It would serve as an umbrella organization for European DX clubs. Within a few years it had over 30 member clubs, a newsletter, and committees working on such things as DX contests, a countries list, and improved reception reporting.[69] The EDXC remains in operation, facing many of the same issues and challenges as ANARC. A similar group, the South Pacific Association of Radio Clubs (SPARC), was formed in 1981 at the initiative of Arthur T. Cushen. Most of the Australian and New Zealand clubs were members at one time or another, and SPARC had a newsletter and held conventions every other year. However, SPARC became inactive as the number of member clubs declined, and it eventually closed.

HANDICAPPED AID PROGRAM[70] (1972–1986)

In March 1972, SPEEDX began the Handicapped Aid Program (HAP). Originally conceived as a "SPEEDXtras" project by SPEEDXtras head Jack White, it became a separate department within the club in 1974. It was chaired by Gene Moser of Coloma, Michigan, who suffered from muscular dystrophy.

The purpose of HAP was to interest handicapped persons in shortwave listening. Only handicapped persons could be HAP "members." They were offered information on the hobby, contact with other SWLs, and a free or half-price subscription to a club bulletin of their choice. Additionally, in cases of genuine interest and limited means, they received the loan of a receiver donated by project supporters. At least one VA hospital was equipped with a HAP-sponsored listening post.

The program became popular and it grew quickly. A HAP newsletter was established, and the project received considerable publicity in club bulletins and DX programs. It became widely known within the shortwave community. Although connected with SPEEDX at first, administratively HAP functioned as an inter-club project. Other clubs were asked to designate HAP representatives, and most did. HAP sister projects were established in England, Canada, India, Australia, and South Africa, and a HAP International Coordinating Committee was formed. The European DX Council established a Handicapped Aid Committee. Harry Van Gelder of Radio Netherlands was designated HAP's 1974 Man of the Year. HAP made other awards in subsequent years.

Funds were raised through the sale of HAP jacket patches, rubber stamps, pens, Radio Canada International interval signal and station identification tapes, a "DXing According to NASWA" cassette and other tapes, auctions at ANARC conventions and elsewhere, and donations. HAP also benefitted from the considerable fund raising efforts of Sam Barto of Bristol, Connecticut who was best known as QSL editor for NASWA and who auctioned pennants and stickers at his annual New England Area DX Outing and via a monthly "HAP Page" auction column that was carried in various club bulletins. Proceeds were donated to HAP. He also conducted auctions at ANARC conventions. Barto's fund raising efforts yielded over $9,000 for HAP groups in the United States and other countries.

In August 1975, other obligations caused Moser to turn over HAP leadership to a new director, Ted F. Poling of Mt. Sterling, Illinois, who had been working closely with Moser. By this time 80 persons, some of them already SWLs, had become HAP members and benefitted from the HAP project. HAP itself had become an associate (later a full) member of ANARC. (Moser died in August 1978.)

The number of HAP new member applications was declining considerably, however. The "HAPpenings" column in SPEEDX, which had appeared monthly during Moser's tenure, became quarterly, and overall interest in the project appeared to have passed its peak. A new director, Wayne Davis of Trenton, New Jersey, took over in April 1977. The "HAPpenings" column was phased out in 1980.

The early 1980s saw a major reorganization of HAP. Record keeping had been limited, and much effort was required to reconstruct the status of HAP's activities, membership lists, and an inventory of loaned equipment. Authority and operations had become fragmented. HAP was established as a non-profit charitable organization, making contributions tax-deductible. In January 1981, the HAP newsletter became a four-page insert of the ANARC newsletter, increasing the visibility of HAP among ANARC clubs and ANARC newsletter subscribers. Around the same time, Davis turned over the leadership of HAP to J. Roland Desrosiers of Chicopee, Massachusetts. He served only for the first three months of 1981, however, at which time Al Schwaneke of Rolla, Missouri took over as President and General Manager. Schwaneke also served as HAP Librarian, and Desrosiers stayed on as Vice-President. John M. Kapinos of Shrewsbury, Massachusetts, became Treasurer, and Sam Barto was put in charge of fund raising. In June 1981, James A. Conrad of Waterloo, Iowa, took over the publicity function from Jeff White of Schaumburg, Illinois, who had performed it for nine months after becoming chairman of ANARC's Public Relations Committee and editor of the ANARC Newsletter in September 1980. Conrad remained publicity chair until the post was assumed by Schwaneke in February 1982.

HAP's difficulties persisted. Effective September 15, 1983, Schwaneke was succeeded by John A. McCann, a blind attorney in Arlington, Virginia, who put forth a thoughtful plan of action for ANARC's improvement. However, McCann soon found himself overwhelmed with a level of administration, both current and catch up, that was more than he expected.

Notwithstanding HAP's problems, there were some significant HAP accomplishments in the early 1980s. Well-known shortwave personality Ian McFarland was heading up the Canadian Handicapped Aid Program and providing visibility to that effort. West coast radio supplier Radio West offered a 25 percent discount to handicapped customers. In conjunction with ANARC, HAP issued the *1981 ANARC-HAP Guide to DX Listening.* By 1982, there were approximately 220 HAP members, over half of whom were blind and 15 percent confined to a wheelchair.[71] Twenty-six receivers were on loan to handicapped DXers. HAP was producing a cassette version of its monthly newsletter for members, and by 1983 was sending out 120 newsletter cassettes a month.

In 1985, HAP was dropped as an ANARC member club for non-payment of dues, although, with some $2,000 on hand, its finances were strong. Later in the year, John Kapinos, by then the project's director, proclaimed HAP in desperate administrative straits. It needed a better operational structure and more people to share in the administration. Kapinos could not do it all himself. He received little response to his calls for volunteer assistance, however, and he closed HAP in 1986. HAP assets were transferred to the Handi-Hams Courage Center, a Minnesota volunteer organization helping handicapped amateur radio operators. They agreed to spread the news about shortwave listening as part of their ham-related programs.[72]

Some special efforts for handicapped SWLs continued. Beginning in 1988, the bulletins of the Canadian International DX Club and the Ontario DX Club became available on cassette for sight-impaired members. The sessions of the ANARC convention of the same

International Year
of Disabled Persons

Année internationale
des personnes handicapées

QSL

Radio Canada
International

The Handicapped Aid Program was aimed at introducing shortwave listening to the handicapped. In 1981, Radio Canada International issued a QSL-card celebrating the International Year of Disabled Persons.

year were also made available on cassette. The National Radio Club's DX Audio Service, presenting broadcast band news on cassette, was established in 1985 and continues in operation. For a time the DX Radio Network of Rochester, New York functioned in a similar capacity.

Socializing

With the exception of a few major clubs such as the Newark News Radio Club and the National Radio Club (a medium wave club), which had held conventions and other social events since the 1930s, it was not until the 1960s that organized gatherings of DXers started occurring on a regular basis. These meetings took many forms—a meeting at someone's home, a pot-luck dinner, a picnic, a DXpedition. Some of the meetings were organized by clubs which also published bulletins for their members. Others were loosely connected with clubs, and some were ad hoc groups of DXers that just wanted to meet to talk about radio.

Among the larger clubs, the Canadians were leaders in the convention business. The Canadian DX Club held its first convention in 1962, followed by the Canadian International DX Club in 1963. The Ontario DX Association began holding conventions in 1985. The only one of the large American clubs to hold a convention was the North American Shortwave Association, which held its one and only convention in 1974.

From 1985 to 1990, Fred Osterman of Universal Shortwave in Reynoldsburg, Ohio, published four editions of the *DXer's Directory*. Its stated purpose was to facilitate contact among shortwave listeners, and to that end it included the names and addresses of listeners who wished to be included. It also included a list of clubs large and small. The final edition listed 1,480 listeners in the United States, 162 in Canada.

NATIONAL CONVENTIONS

The major North American DX gathering was the annual convention of the Association of North American Radio Clubs. ANARC conventions were held annually from 1966 to 1990, and took place on Friday and Saturday, a format that would be followed by other radio conventions. A formal program was presented, with seminars, small group discussions, demonstrations, speakers and the like. The location and the sponsorship of the event changed each year, and conventions were held in many different venues, from Boston to Los Angeles, Montreal to St. Petersburg, Florida. Attendance grew from a dozen or so at the start to well over 200 in the early and mid–1980s. In later years, broadcasters also attended the ANARC conventions, their meetings eventually turning into a broadcasters' forum. The last ANARC convention drew about 100 people.

As the ANARC conventions were aging, another major DX convention started taking form. The first "Winter SWL Fest" was held on February 13 and 14, 1988 at the Fiesta Motor Inn, Willow Grove, Pennsylvania. It was attended by about 60 people.[73] Forming the heart of the group were the members of the Old Dominion DX Association from the Tidewater, Virginia area, some of whom had reunited at the ANARC Convention in Mississauga, Ontario in July 1987 and decided to get together. In 1989 the Fest moved to Kulpsville, Pennsylvania where it has remained ever since. Each year from 1989 through 1999 it was presented as an independent event, although the influence of NASWA was strong, the key Fest organizers—Bob Brown of Lansdale, Pennsylvania, Harold N. Cones of Newport News, Virginia, and Kris Field of Horsham, Pennsylvania—all being part of the NASWA leadership. This team stepped down as convention hosts following the 1999 Fest. Thereafter, NASWA assumed formal sponsorship of the event, and responsibility for Fest organization passed to Richard D. Cuff of Allentown, Pennsylvania and John A. Figliozzi of Clifton Park, New York. Notwithstanding NASWA's role, the Fest remains an event where all radio enthusiasts are welcome and all clubs are invited to participate. Attendance at the Winter SWL Fest exceeded 200 in the mid–1990s, and was about 170 in 2007.

Monitoring Times magazine sponsored an annual listeners convention during the years 1990 to 1996. The first two were held in Knoxville, Tennessee, the balance in Atlanta, Georgia. Attendance was in the 300–400 range. However, the event was not financially successful. *Popular Communications* got into the act as well in 1993 with its "Worldwide SWL Conference" in Virginia Beach, Virginia, held in conjunction with the Virginia Beach Hamfest and Computer Fair. It was a one-time affair.

REGIONAL OR SPECIALIZED EVENTS

Various regional or local DX mini-conventions have taken place over the years. In the 1970s there was the joint ASWLC-SPEEDX "inter-club DX symposium" in Tustin, California (1973); "Radio U.S.A." in Redondo Beach, California (1975); the NORCAL DXers meeting in Millbrae, California (1975); the Ohio Weekend DX Seminar, Dayton, Ohio, sponsored by the Miami Valley DX Club (1975 and later); the Bay Area Radio Freaks "All Bands DX Get Together" in San Francisco, California (1976); and the Bay Area DXtravaganza and Bay Area DXtravaganza II (1977 and 1978 respectively). The first annual *Numero Uno* convention was held in 1973 (see p. 217). It was co-sponsored by *Fine Tuning* from 1982 to 1987, and assumed independent status as the "Reynoldsburg DX Weekend" thereafter. It is still held in Reynoldsburg, Ohio each summer.

An annual New England Area DXers Outing (NEADXO) was sponsored by Sam Barto of Bristol, Connecticut and held at Black Rock State Park, Watertown, Connecticut on the first Sunday after Labor Day from 1976 into the early 1980s. The first gathering drew 32 people. By 1978 over 100 were in attendance, attracted by door prizes, talks, auctions, a pre-outing dinner the night before, and occasionally a radio celebrity.

The 1980s saw two annual two-day conferences of the St. Louis International DXers, St. Louis, Missouri (in 1980 and 1981); the Miami Valley DX Club "All-Wave DX Get Together" in Angola, Indiana (1982) and Dayton, Ohio (1983), the latter being the first organized SWL meeting at the well-known Dayton Hamvention; the Chicago DX Convention, sponsored by the Chicago Area DX Club and held in Gurnee, Illinois in 1986 and Rockton, Illinois in 1989; and the East Coast SWL Conference, presented by the Old Dominion DX Association in Virginia Beach, Virginia (1988).

Interest in these gatherings declined in the 1990s, by which time the Winter SWL Fest was the venue of choice for the gathering of large numbers of the hobby faithful.

LOCAL GATHERINGS

Local shortwave groups have long provided a vehicle for DXers to meet and talk DX. One of the first was the Los Angeles DX Club, established in 1956. Such groups proliferated in later years.

One of the better-known groups was the Boston Area DXers (known as the BAD Guys, or BADX). BADX traces its roots to a 1967 Boston-area publishing committee of the National Radio Club, which had borrowed the acronym from the Bay Area DXers (a San Francisco–Oakland area group which subsequently adopted the name Bay Area Radio Freaks).[74] BADX reinvented itself twice. In the early 1970s an informal 30-member group would meet occasionally at members' homes for food and DX talk. These meetings continued through the early 1980s. From 1975 to 1983, the group functioned loosely as the Boston branch of the New England DX Society, home of the bi-weekly *NEDS News* which was created and edited by Henry J. Michalenka of Central Falls, Rhode Island and published by Alan W. Brooks of Oakdale, Connecticut. *NEDS News* was a friendly, energetic publication, typically two to four pages but often more, where New England DXers shared their SWBC DX tips and radio doings. (Chris Lobdell of Malden, Massachusetts assumed the editorship in October 1979.) BADX was reactivated in 1990 as an outgrowth of a small group of computer-savvy hams and SWLs who were members of the Boston Computer Society, which, its name notwithstanding, was a national-level organization of computer users. BADX continues to meet on a monthly basis.

The Minnesota DX Club dates back to 1971–72 when Minnesota SWLs learned of each other via Keith Glover's mailbag program on Radio Australia. They started exchanging information by phone, and on March 10, 1973 seven people gathered for the club's first meeting. The MDXC was modeled on a predecessor group, the Minnesota DXers' Alliance of the late 1930s. By 1975 the club had 30 members and was meeting monthly, and in 1979 it sponsored a successful ANARC convention in Minneapolis. Membership has been limited to radio enthusiasts living in Minnesota and western Wisconsin. The club had about 40 members in 1985, a number that remains about the same today. The MDXC emphasizes personal contact among members, with monthly meetings held either in members' homes or at the Pavek Museum of Broadcasting in suburban Minneapolis. Since its founding the club has also published a monthly *MDXC Newsletter* which is now issued electronically in .pdf format. In 2000 the club established a Yahoogroup for discussion and for breaking news. The club holds an annual DXpedition in February, a summer picnic, and a holiday dinner in December. Over the years it has broadened its scope to include all forms of listening, including shortwave, medium wave, FM, TV and utilities. It has a website that contains information about the club, lists of shortwave stations, links and feature articles.

One of the groups that has been the most successful at promoting shortwave listening is the Southern California Area DXers (SCADS). Associated with the American Shortwave Listeners Club, SCADS was conceived at the 1973 ANARC convention in San Diego and started quarterly meetings the same year. While some SCADS meetings were informal gatherings in people's homes, others were major events that drew 50–100 people. The agenda typically included guest speakers, panel discussions, equipment displays, demonstrations, radio-related movies, visits of radio personalities, door prizes and a HAP auction. Meetings were usually held in Tustin, Huntington Beach (home of the ASWLC), Long Beach, or Seal Beach. Until 1980 the leader of the group was Don Johnson of Elsinore, California. He was succeeded by Don Schmidt of Long Beach in 1981, and William Fisher of Buena Park circa 1997. (Johnson died in 1992.) A summer picnic has been an annual affair. For a time, starting in 1984, SCADS had a bulletin called *Calling All DXers*. However, the group's focus continued to be on the social side. SCADS took on a lower profile in the 1990s, when smaller meetings were conducted on a monthly basis, a pattern that continues to the present.

A gathering born in 1994 and continuing to the present is known as the Madison-Milwaukee Get-Togethers. Its origins go back to 1993 when two Madison, Wisconsin DXers, Bill Dvorak and Tim Noonan, organized an informal gathering to meet with two DXers who happened to be coming to town at the same time. The annual return of one of them formed the basis for an August meeting each year. The 1994 event was attended by eight hobbyists, a number that has grown to over 40. In recent years the gatherings have alternated between the Madison and Milwaukee areas, and have been held in homes or in park settings. They have become increasingly popular for both regulars and new attendees, with much of the day devoted to conversation and reinvigorating friendships, along with equipment demonstrations and a group dinner.

BADX and SCADS started a movement toward the adoption of informal names by some local groups. There were the CAD Guys (Chicago Area DXers), not to be confused with the CADS (the Colorado Association of DXers, Englewood, Colorado); the LADs (Los Angeles DXers), the ODD Fellows (Old Dominion DX Association, Virginia Beach, Virginia); and the YADS (Youngstown Area DXers, Youngstown, Ohio). Even the French Canadians got in on it with the DUM Guys (Les DXers Unite de Montreal).

Similar groups met in many other locations. These included, in the 1970s (in addition to gatherings by clubs already mentioned): the Bay Area DXers, Pleasanton, California; the Connecticut Shortwave Association (CONN-MEN), Stamford, Connecticut; the Detroit Area DXers, Detroit, Michigan; the Memphis Area DXers, Brighton, Tennessee; the Metropolitan Area DXers (MAD-MEN), New York, New York; the Milwaukee Area DXers, Milwaukee, Wisconsin; the Nebraska DX Association, Lincoln, Nebraska; the Northern California DXers (NORCAL), Castro Valley, California; the Ontario DXers, North Bay, Ontario; the Selma Area DXers, Selma, California; the Suburban Philadelphia Area DXers (SPADES), Swarthmore, Pennsylvania; the Three Mile Island DX Association, Bellefonte, Pennsylvania; and the Vancouver Area DXers Association, Vancouver, British Columbia.

The 1980s saw fewer new groups, but there were some, including the Atlantic States DX Association, Brooklyn, New York; the Austin Area Radio Listeners, Austin, Texas; the Central Kentucky DX Association, Lexington, Kentucky; the Denver Metro Area DXers, Denver, Colorado; the East Central DX Association, Brooklyn, New York; the Hawaii Shortwave Ohana, Honolulu, Hawaii; the International Fellowship of Christian DXers, Lexington, Kentucky; the Rocky Mountain Radio Listeners, Aurora, Colorado; and the Washtenaw Area DXers, Ann Arbor, Michigan.

There was little new activity in the 1990s. The Tidewater SWLs of Virginia Beach, Virginia met regularly from 1991 to 1997. The Blue Grass DaVinci Fellowship, Lexington, Kentucky, was also active, and its members still meet occasionally. Several of the local groups that meet today (Boston, Philadelphia, Washington, D.C., and Ohio) are loosely affiliated with NASWA.

An offshoot of some of these social get togethers is the DXpedition, an event where a number of listeners set up their equipment at a location that affords superior reception, usually on the coast or at a park or similar place that is comparatively free of manmade interference. Most DXpeditions are local affairs, although some DXers have traveled long distances to attend a DXpedition that affords reception that is impossible at home.

Foreign Clubs

Although countless SWL clubs have been formed around the world since World War II, comparatively few have impacted the North American shortwave scene. The foreign clubs are generally beyond the scope of this book, save for some that became well known here.

AUSTRALIA

The first radio club in Australia, the Australian DX Radio Club, was formed in 1933.[75] It was closely affiliated with the weekly program magazine, *Listener-In*. A separate club bulletin, *The Skyrider,* was established in 1938. The club changed its name to the Australian DX Club, and by 1939 it had 300 members. Although both *Listener-In* and the club were principally oriented toward medium wave, they also included some shortwave coverage.

Wartime constraints caused the Australian DX Club to suspend publication of its bulletin in 1942. It reappeared in 1943, but came to an end in 1949, with a core of members in South Australia carrying on as the South Australia DX Club, a group which survived into the early 1970s.

Except for the limited DX coverage of *Listener-In* and *Radio & Hobbies* magazine, the main information vehicles for Australian SWLs in the 1950s and early 1960s were the New Zealand DX Radio Association and the New Zealand Radio DX League. In 1965, the Australian Radio DX Club was formed out of the Sydney branch of the NZRDXL, and the branch's publication, *Australian DX News* (ADXN), became the monthly publication of the ARDXC. There were 23 members at the time. In 1967, club headquarters moved from Sydney to Melbourne, where the members of an active branch of the New Zealand DX Radio Association assumed the ARDXC's leadership. At that time the club had 100 members. Starting in 1968, the club adopted a restricted system whereby membership was lost if a member did not contribute. This requirement was dropped in 1970, however. The bulletin went from stencil to offset in 1975, and by 1978 the club had 300 members.

The ARDXC billing as "Australia's premier DXing organization" was well deserved. The bulletin, *Australian DX News,* covered shortwave broadcasting, medium wave, utilities, amateur, FM and TV. The chief editor from January 1967 to October 1974, and the key man in the club for some 20 years, was Executive Committee Chairman (later club President) Robert J. Padula. Padula was a prodigious worker and one of the world's most accomplished DXers. For his service to shortwave, he was awarded the Medal of the Order of Australia (O.A.M.) in 1981. He was also a prolific writer, and produced many special club publications on such topics as DXing Indonesia, DXing the U.S.S.R., DXing China, medium wave DXing, etc. The club grew, particularly during the 1975–1985 period when the bulletin was typically 30-plus legal size pages. Robert Wagner and Robert Hanner served as chief editors at various times during the years until February 1979 when that duty fell to Peter Bunn, who served during most of the 1980s. The 1980s also saw many social activities among the club's Australian branches.

Membership peaked at 680 in November 1983. Ultimately, however, the decline of shortwave listening had the same negative impact on the ARDXC as it did on North American clubs. Membership fell sharply. By December 1, 1995, it was at 115. The bulletin was much reduced in size and content, though still of good quality, considering the times. A new club president, Richard Jary, took over in January 1996, followed by Robert Fitzgerald the following year. A club restructuring let to a firmer footing. Many others served in important club capacities, including John Wright, who is presently club secretary and bears much of the burden of club operations. Today the ARDXC has been re-established, with a professional looking *Australian DX News* published 11 times per year and available electronically at reduced cost, making membership more economical for both Australian and overseas members, who together number about 180.

The Southern Cross DX Club, based in Adelaide, was formed in June 1973. Although intended primarily as a local club, it had members worldwide. Its modest bulletin was called *DX Post,* and in 1984 it also published the *Guide to DXing Utility Stations.* It closed in 1999 due to declining membership.

In July 1982, a group of Melbourne DXers formed a club called DX Australia. It published a lively monthly bulletin, *DXers Calling,* which became quite popular. The emphasis was on the tropical bands and U.S. medium wave DXing, but it also covered utility DXing and (later) FM. The bulletin, which was generally 30–40 pages long, contained information on loggings, QSLs, musings, propagation, features, etc. Many top Australian DXers and editors supported DX Australia, and by 1984 it had about 170 members. At the time there was talk of a merger with the Southern Cross DX Club, and there were several joint bulletins, but the merger did not take place. DX Australia later published several editions of a Pacific AM Log and a Pacific FM Log. Like many clubs, it suffered a decline in membership during the 1990s, and in 1998 membership dropped below 100. A potential merger with another small club, the South Pacific Union of DXers (which would itself soon close down), did not materialize, and DX Australia ceased operations in June 1998.

DENMARK

A club called the Danish Short-Wave Club (Dansk Kortbølgeklub) was formed in 1946 by O. Lund Johansen, editor of *Populaer Radio* (and, in 1947, creator of the *World Radio Handbook*). After about a year, the contents of the club bulletin, *The Short Wave Listener,* were incorporated into *Populaer Radio,* and there is little trace of the club thereafter.[76]

A Danish national shortwave club was resurrected in November 1956 with the formation of the Danish Short Wave Club, or Danmarks Kortbølgeklub.[77] The first edition of its monthly bulletin, *Kortbølge Nyt [Short Wave News],* was published in January of the following year. It had four pages. In 1959 the club began publishing a separate English-language edition. This attracted an international following and soon resulted in a name change to the Danish Shortwave Club International. By 1961 the club had 234 members.

In 1971 it was decided to publish the monthly bulletin entirely in English. In 1974 the club merged with another Danish club, the Cimbrer DX Club, and the name was changed to the Danish Shortwave *Clubs* International. (It was changed back to the singular in 1996.) By 1977, membership was 496.

The club's Chairman was, from 1956 to 1963, Niels Jacob Jensen; from 1963 to 1972 and from 1981 to the present, Anker Petersen; and from 1972 to 1981, Carol Feil. (Jensen died in 2007.) Kaj Bredahl Jørgensen has been *Short Wave News* editor in chief since 1968, save for 1972–80 when Feil was editor. Although all were Danes, over the years there have been many non–Danes on the editorial staff, and nearly 90 percent of the club's membership has been from countries other than Denmark. Thus the DSWCI has been the most international of all DX clubs.

It did not take long for DXers in North America and elsewhere to appreciate the high quality of the DSWCI's bulletin. Typically it was 32 legal-sized pages, with an extensive loggings section and a focus on the tropical bands. Two areas where *Short Wave News* was particularly strong over the years were pirate and clandestine stations, each of which had its own column starting in 1973. The clandestine section was published until March 1995, and was edited at first by Carol Feil, and then by Mathias Kropf of West Germany. Over the years the pirate section has been prepared by Paul Foged of Denmark, John Campbell of the U.K., and Ken Baird of Scotland. Baird continues to edit the section, which now appears on an occasional basis.

Latin American DX was another area of DSWCI excellence. Among the editors of the "Latin American News" column were Erik Køie of Denmark, Gary Lane of the U.K., and

Tor-Henrik Ekblom, then in Sweden. In 1984 the baton was passed to Finn Krone who edited it until October 1997 (except for a short period in 1990–91 when it returned to Ekblom). He expanded the column to cover DX stations from other parts of the world and home service broadcasts generally. The column was renamed twice, becoming "World DX News" in April 1992 and "DX Mirror" in October 1997. Anker Petersen took over the column in November 1997 and edited it until March 2007, when Dmitry Mezin of Kazan, Russia assumed the job.

In 1982, Noel R. Green of the U.K. began editing the "World News" column, covering schedules and general station news, and he continues to do so today. A column devoted to members' letters ran for many years, and medium wave and utility DX continue to receive coverage. There have also been station profiles and features on receivers and trips or DXpeditions by members. For years there was a "U.K. Corner," and for several years starting in June 1977 there was a column called "The Program Listener," which focused on the program content available on shortwave. Also, *Short Wave News* has long had a very comprehensive QSL column.

From 1995 to 1997 the Union of Asian DXers occasionally prepared several pages about Asian stations. From 1995 to 1999, Don Phillips of the U.K. offered "As I Please," a column of opinion and comment about receiving equipment and a variety of other radio-related topics. From 1997 to 1999 it was complemented by a DX-oriented column called "DX As I Please" prepared by Søren Dippel of Vejen, Denmark. While not every column in *Short Wave News* has appeared as scheduled, the bulletin has remained substantive and devoid of filler.

The club became known for activities besides the bulletin. In the mid–1970s it sponsored band monitoring events, compiling the results and making them available to club members. (These were reinstituted for a short time in 1995.) It also gained major recognition within the DX community for its publication of several highly regarded annuals: the *Tropical Bands Survey* (1973–1988), the *Clandestine Stations List* (1985–1998), and the *Domestic Broadcasting Survey* (1999 to the present). (See p. 211.)

The club also had a DX program, "DX-Window," which was carried over Radio Denmark from 1967 to December 1969, and it produced "World DX News" (later "DX Corner") over Adventist World Radio from 1979 to the end of 1994. The "DX-Window" name became a long term DSWCI trademark. It reappeared in 1987 as a two to four page mid-month (later fortnightly) newsletter edited by club member Bernhard Gründl in Germany and made available to approximately 30 active and experienced DSWCI contributors. *DX-Window* editorial duties passed to Wian Stienstra of the Netherlands in 1989, and *DX-Window* eventually became a two to eight page weekly. (Gründl passed away in 1991.) Declining participation led to its closing in September 1994. However, it was reborn as the weekly *DX-Window* e-mail bulletin in February 1996. It ceased publication in mid–1999, but was revived in May 2001 on a bi-weekly basis.

As with its American counterparts, the club's fortunes have risen and fallen. Membership peaked in 1982 at 833 members, but by 1990 it was in the low 500s. Except for an increase of about 100 during the period 1990–93, membership continued to decline until it leveled out in 2002 at around 235, where it remains. In addition, the financial difficulties which nagged the club for many years grew as competition from the internet intensified. To maintain fiscal soundness, starting in June 2000 the bulletin's length was reduced

The Danish Shortwave Club International verifies your report of the "DX-Window" from Radio Denmark on *15765* kHz, at *1015-1040* GMT, on *Dec. 10, 1981*

"A Must You Can't Miss!"

00097

A special DSWCI 25th anniversary program over Radio Denmark in 1981 was verified with a QSL-card memorializing "DX-Window," the club's program that had been presented over the station from 1967 to 1969.

by half in alternate months, and in mid–2001 the number of monthly issues was reduced (since then it has ranged from eight to ten). Even with these changes, several anonymous donations have been credited with keeping the club solvent.

In meeting this fiscal challenge and improving service to members the club made good use of the internet. In October 2000 it rewarded those willing to download the bulletin from the club's website with a reduction in their annual membership fee from (in the U.S.) $37 to $18. The internet version of *Short Wave News* also now offers color graphics. Early in 2001 members were given internet access to the bulletin's loggings section as it was compiled throughout the month so that they could benefit from new information immediately.

NEW ZEALAND[78]

New Zealand was the home of an energetic band of long distance radio enthusiasts, none more active than world famous DXer Arthur T. Cushen. Cushen was born on January 24, 1920, and died on September 20, 1997. He served as National Vice-President of the New Zealand Association of the Blind for many years, and in 1970 the Queen named him a member of the Most Excellent Order of the British Empire (M.B.E.) for his work in community services involving radio broadcasting, journalism, and the blind. He was a monitor for many stations, and a life member of the New Zealand Radio DX League.

The first New Zealand club was the New Zealand DX Club, established under the sponsorship of *Radio Record* magazine in 1929 and becoming independent ten years later.[79] For many years, Arthur Cushen was the shortwave editor of the New Zealand DX Club's publication, *New Zealand DX-Tra*. As one observer noted, "New Zealand appears to be a DX paradise—for in glancing through the pages of *The DX-Tra*, loggings are noted from all over the world on both SW and BCB. American BCB stations of 250 watts appear to be common loggings in New Zealand."[80] The club closed in 1948.

Another active New Zealand radio club, the New Zealand DX Radio Association, was formed in 1933. Its bulletin, called *Tune-In*, was at first similar to the American *RADEX*.[81] Membership was 180 in 1964. This small club and its bulletin continued in existence for many years, covering shortwave broadcast stations, amateurs and utilities. However, rising costs, declining membership and the absence of replacements for departing editors caused it to close in December 2006.

A new DX group, the New Zealand Radio DX League, was established in 1948. It was formed from a predecessor group of South Island DXers that had published the *New Zealand DX Bulletin* since October 1947. Publication of the eight-page NZRDXL bulletin, the *New Zealand DX Times*, started in October 1948 with Arthur Cushen as shortwave editor. In addition to shortwave broadcasting, the *DX Times* covered other fields of DXing as well. It also featured information on the activities of NZRDXL branches around New Zealand. By 1963 the club had 350 members, peaking at over 700 in the 1970s.

Over the years, many individuals served as NZRDXL National President and handled various *DX Times* editorial duties. (Cushen edited his own column for the *DX Times* continuously from its start in 1948 until his death nearly 50 years later.) The bulletin grew larger, notwithstanding the same drop off in membership experienced by other clubs in the 1990s. In July 1998, through the use of computer software, it adopted a more modern, updated appearance, and feature articles appeared more often. In August 2001 an electronic version was offered at a much reduced rate, making it more broadly available to overseas members. The New Zealand Radio DX League remains an active club, with about 200 members.

SRI LANKA

The Union of Asian DXers (UADX), headquartered in Sri Lanka, traces its history to the Ceylonese Short Wave Listeners' Club (CSWLC), which operated from 1967 to 1970. (Sri Lanka was known as Ceylon before 1972.) The founder of the UADX, Victor Goonetilleke,[82] was, with Sarath Amukotuwa, one of the early leaders of the CSWLC. They were still schoolboys, and with only their pocket money managed to issue the first bulletin in February 1968. The club closed in 1970, however, its founders busy with school and many of its members interested more in the larger stations than in DX, Sarath's and Victor's first love.

May 1972 brought a new attempt at organization in the form of the Union of Asian DXers. The UADX bulletin would distinguish itself over the next 19 years as the only third-world DX newsletter with a major following throughout the world. Its reputation was well deserved. Typically the bulletin was four to eight pages in length, sometimes larger, with monthly publication the goal, even if it was not always achieved. The focus was on Asian stations, but member loggings covering all countries were welcome. In addition, there were station profiles, news about clandestine stations, and other features, plus a few special publications, e.g. the several editions of the *UADX Tropical Band Survey* and the *UADX Afro-Asian-Pacific Survey* published in the late 1970s and early 1980s. The bulletin emphasized shortwave coverage, but medium wave news was covered as well. The club had an awards program, and for members in Sri Lanka there were frequent meetings at members' homes, with Victor Goonetilleke's door always open to international visitors. In later years he became an international traveler himself, attending DX meetings in Europe, the United States and elsewhere.

The bulletins were impressive in content. At various times Victor was assisted in the editing by Sarath Amukotuwa, Oliver Goonawardena, and Sarath Weerakoon, as well as Gerhard Werdin, a German telecommunications engineer who was on assignment in Sri Lanka from 1978 to 1980 and continued assisting the club upon his return to Germany. By the time UADX's 100th issue was published in 1982, the group had 130 members in 25 countries.

The many practical difficulties that the club often faced made publication a triumph. Arrival of the bulletin became less predictable after 1982, with bi-monthly or quarterly issues the rule. Although the appearance of the bulletin never matched that of the larger publications, this was no problem for the many loyal UADX members who enjoyed the presentation of hard DX news in the informal, personalized style which the UADX reflected.

Notwithstanding Victor Goonetilleke's herculean commitment to the UADX, the energy of the earlier years declined during the 1980s. By the end of the decade, publication was down to at most a few bulletins a year. The last one covered the period January-July 1992.

Although UADX is now but a warmly remembered part of DX history, Victor Goonetilleke remains active in both DXing and ham radio. In 2005, he and the Radio Society of Sri Lanka, an amateur radio group of which he was president, were recognized for their emergency relief efforts following the Asian tsunami. According to Victor, it was the knowledge about propagation and radio, gained through long years of DXing, that gave him and his friends the inspiration and confidence to meet the challenges of disaster communications when the tsunami struck Sri Lanka.

SWEDEN

Sweden had numerous shortwave clubs. Although some bulletins contained English-language sections, overall these clubs had little impact in the United States owing to the language difference, the cost of membership, and the local nature of many of the clubs. However, Sweden was the DX capital of the world.

The Sveriges Radioklubb (with DX editor Arne Skoog) was formed in 1944 and soon had 1,500 members. Hundreds of people attended meetings in Stockholm, and numerous local clubs soon formed. Sweden had more listeners that took shortwave seriously than any other country. Stations received more reports from Sweden than anywhere else. This applied to small stations as well as large. In 1952, Radio Diamang in Angola reported that 40 percent of their reception reports were from Sweden (20 percent from Britain, 10 percent from Australia). In May 1944, Radio Tokyo broadcast a Swedish-language program dedicated to the Radio Club of Sweden and received 518 reports.[83] Reports from Swedish listeners were usually detailed and accurate. There was a great deal of publicity in Sweden about shortwave, with many popular radio magazines containing sections about shortwave listening, and a monthly, *Kortvags-Lyssnaren*, devoted exclusively to shortwave. The chief engineer of station HOX, The Voice of Central America, in Panama, once remarked that judging from the mail received from Sweden, "one would think that all Swedes speak English, listen to the radio avidly, and are great stamp collectors."[84] During 1961, while Robert Newhart, who was fluent in Swedish, was shortwave editor, the American Shortwave Listeners Club offered a page in Swedish, presumably to appeal to the large number of Swedish DXers.

Swedish DXers liked to organize. Seven smaller clubs federated into the Sveriges DX-Riksklubb (National DX Club of Sweden) in 1964, and within a few years that number more than doubled. The purpose of the group was to provide a better DX publication than the limited resources of individual clubs, working separately, would support. Clubs were expected to discontinue their own bulletins and support the federation publication.[85]

The DX-Alliansen, a national federation of Swedish DX clubs, was formed in 1956 by Arne Skoog and soon included some 30 clubs. It established the pattern that ANARC would follow upon its founding in 1964. Clubs paid a fee based on the size of the club. There was a monthly newsletter and an annual convention, and a radio program carried over Radio Prague and HCJB. The DX-Alliansen also sponsored the annual Swedish DX Championships and participated in other Scandinavian contest activities.[86] A 1971 list showed 48 DX-Alliansen member clubs, although it was noted that not all were active.

In 1958, the Sveriges DX-förbund (Swedish DX-Federation) was formed and a professionally printed monthly magazine, *Eter Aktuellt,* was published. Sweden also boasted an annual DX Parliament, a convention that attracted many listeners and station representatives.

UNITED KINGDOM

The International Short Wave League was formed in England in 1946. Its sponsor was *Short Wave News,* a small radio magazine started by Amalgamated Short Wave Press, Ltd. the same year. International Short Wave League membership could be had for a small additional subscription. By 1949 the ISWL had 3,000 members.

ISWL activities were covered in a separate section of the magazine. Although the focus of League activities, like those of its magazine parent, tended to be on technical topics and amateur band listening, there was shortwave broadcast coverage in both, and both promoted themselves as serving broadcast as well as ham enthusiasts. Amalgamated Short Wave Press, Ltd. also published two editions of *The Shortwave Listeners' Annual,* one in 1947 and one in 1948. In addition to ham information, the *Annual* included, for broadcast listeners, a shortwave station list, a list of Latin American station slogans, a list of station addresses, and information on propagation and on identifying shortwave stations. A successor to the *Annual* called *The Op-Aid* was published in 1950, but it concentrated on ham radio listening, shortwave broadcasting by then being well-covered by the *World Radio Handbook.*

Short Wave News sponsored a number of interesting projects under the ISWL banner. One, established in 1950, was the ISWL Broadcast Station Identification Service, a group of shortwave monitors who double checked new shortwave information by on-air monitoring and also replied to inquiries about the identification of particular stations (not unlike the "Which Station Was That?" service offered years earlier by the BBC's *World-Radio* magazine). A seven-day turnaround was promised. The ISWL also operated a QSL Bureau for outgoing and incoming QSLs, both broadcast and amateur. (The International Short Wave Club did the same.) It also had a Commercial Identification Service to identify the sources of code interference.

In December 1951, the International Short Wave League became an independent organization and began publishing a monthly bulletin called the *Monitor.* By 1962, membership was said to be about 1,600. The numerical identifier assigned to members, e.g. W2-4736, were used as informal "call letters" by some. The club remains in existence today. *Short Wave News* became *Short Wave News and Amateur Radio* in January 1952 and *The Radio Amateur ("incorporating Short Wave News")* three months later, with a new focus on ham operators as well as ham band listeners. The magazine continued to carry several pages of shortwave broadcast information, plus occasional broadcast features. In January 1954, *The Radio Amateur* was itself incorporated into *The Radio Constructor,* another Amalgamated publication that had been started in 1947.

A similar club-magazine arrangement involved the British Short Wave League, founded in 1935. Its monthly publication, *Short Wave Review,* was oriented mainly to ham band listening and only secondarily to shortwave broadcast listening. In 1949, financial instability led the group to reorganize and affiliate with Britain's monthly *Short Wave Listener* magazine. Publication of *Short Wave Review* as a separate publication ceased, and a 12-page *BSWL Review,* "The Official Journal of the British Short Wave League," was inserted into those copies of *Short Wave Listener* going to BSWL members.[87] *Short Wave Listener* became *Short Wave Listener & Television Review* in May 1950. The BSWL closed in March 1953 when the magazine ceased publication with the promise of a more comprehensive quarterly *Short Wave Manual* which would cover "a far wider field."

A British club that is still in existence is the World DX Club. It was born in May 1968, replacing the World Communications Club of Great Britain which traced its roots to the Sudbury World Communications Club, an activity of a group of enthusiastic schoolboys at the Sudbury Grammar School in Sudbury, Suffolk. The Sudbury club had been formed in October 1963. In 1965 it changed its name to the World Communications Club of Great Britain. The Sudbury club's bulletin had been called *CONTACT,* and the name continued to be used by the World Communications Club.

By 1968, *CONTACT* was providing members with 40 pages of news each month. It was published by club president George Partridge, who was known worldwide for his Joystick antennas. Due to limited club resources, Partridge was subsidizing the production of the bulletin. The club became dormant for several months in 1968 during a needed reorganization, and was reconstituted in May as the World DX Club.[88] A 24-page *CONTACT* reappeared covering shortwave, amateur and FM, and containing members' loggings, DX news, QSLs, letters, and so forth, as well as a "North American Bits and Pieces" column written by North American representative Wesley Huggins of Yadkinville, North Carolina.

Within a year the club had nearly 150 members in 20 countries. Its first president, Alan Thompson, who served until December 1969, was a strong presence. "The birth of WDXC could *never* be described as uncomplicated — I would say that the infant was in intensive-care for the first 18 months of its life due to a shortage of life-giving finance and the fact that many hankered for the much less formalized atmosphere of the old 'World Communications Club of G.B.' — our immediate ancestor."[89]

Starting in 1976, for an extra fee a member could receive the Express News Service, a six to eight page mid-month publication with late DX news and more extensive pirate radio information than was presented in *CONTACT.* This service lasted for five years. By 1978, club membership was at 380, 20 percent of whom lived outside the United Kingdom (half in North America). *CONTACT* was by then 36 pages long. It contained "all the usual features and sections that are to be found in most club publications, plus chat and friendly club news that expresses the British way of life and the DX scene as seen through British eyes."[90] As part of their membership, members agreed to be active in club affairs and contribute to *CONTACT.* From 1972 to 1979 the club also produced the weekly DX program, "World DX News," over Adventist World Radio. By 1979, the club had approximately 400 members. Membership peaked at around 550. The last amateur coverage was in October 1980.

CONTACT, which moved to offset booklet format in 1996, continues to be sent monthly to the club's approximately 325 members, a third of whom are outside the U.K. Over the years, many people contributed to the club's success, including Clive Jenkins, club secretary from 1969 to 1979; Alan Roe, bulletin editor for 25 years and an important force in moving the club forward; and Mike Barraclough, who has edited many different *CON-TACT* columns since 1967 and performed numerous other important roles. Probably no one has performed a more essential role than Arthur Ward, who has been the *CONTACT* publisher for over 30 years, plus its executive editor and, after 1979, club secretary.

Another DX club that is still in operation is the British DX Club. It was founded in September 1974 by two teenagers, David Kenny and David Balhatchet, students at the Thames Valley Grammar School. Back then it was called the Twickenham DX Club, and the first issue of its "monthly journal," *Communication,* had three A4-size pages. The title

"British DX Club" was adopted in 1979. From local roots came rapid national and international expansion. Membership approached 100 by 1976. Years later, in 1991, it neared 500.[91]

The club has long prided itself on the friendly and approachable feel of *Communication,* which went to offset printing in 1985 and which has not only been one of the best looking DX journals but has long had a one-week turnaround time from deadline to mailing. Traditionally, *Communication* has covered not just shortwave, but medium wave and FM as well, with much information on the U.K. domestic broadcasting scene. Regular bulletin content has included members' loggings, QSLs, pirate stations, members' letters, and features on many topics, together with periodic coverage of subjects like satellite broadcasting, utility stations, computers, program listening, and DX programs.

One of the BDXC's special publications was its *QSL Survey,* several editions of which were published in the 1970s and 1980s. It summarized, by country, the numbers and types of QSLs reported in *Communication* from each station, how long it took to obtain replies, whether return postage was needed, etc. Among the club's present-day special publications is the biennial *Broadcasts in English,* which now includes information on World Radio Network satellite broadcasts and Digital Radio Mondiale transmissions, in addition to regular shortwave broadcasts. Approximately every 18 months the club publishes *Radio Stations in the United Kingdom,* a detailed directory of U.K. AM and FM stations. And since 1976, the BDXC Tape Circle has produced a monthly audio cassette of DX catches, talk and radio-related recordings to which about 45 members subscribe.

Over the years a number of well-known names in British DXing have served as the club's general editor — Kenny, Mark Lee, Chris Greenway, Tony Rogers, and Chris Brand. Kenny was with BBC Monitoring from 1979 to 2003, first as a monitor and then working on *World Broadcasting Information* and *World Media,* and the database of broadcasting schedules. Greenway joined BBC Monitoring in 1981. He has also served in the BBC Monitoring East Africa unit in Nairobi.

The club has had a number of long-serving, active members who have provided a measure of stability that many clubs would envy. A 1988 feature article about the hobby, headlined "A Life On the Short Wave," in *The Independent,* a U.K. national newspaper, gave the club a boost and resulted in many new members, some of whom are still with the club. Membership peaked at 550 in 1998 and is now around 475, with non–U.K. members making up about 15 percent of the total. The social side of the club has been tended to as well. DX meetings have been held in Reading every few months for over 30 years, and are still popular.

OTHER FOREIGN CLUBS

There were numerous other shortwave clubs around the world. Among the better-known European clubs in the 1950s and 1960s were the DX-Listeners' Club of Norway, established in 1955; the Finlands DX-Club and Suomen DX-Kuuntelijat, formed in 1954 and 1958 respectively; the Benelux DX Club (Holland), established in 1961; and the German ADDX (Assoziation Deutschsprachiger DX-er und DX-Clubs), created in 1967. From Japan there was the Japan Short Wave Club, which in the 1950s published a regular English-language bulletin called *Shortwave DX Guide*, and which is still in operation.

Closer to home, a promising club was the West Indian DXers Association (WIDXA), founded in Trinidad in 1965. Despite editorial talent like the well-known Victor Jaar of Haiti and the late César Óbjio of the Dominican Republic, and much encouragement from American clubs, the WIDXA closed within a few years. Later, there was the Suriname DX Club International, which operated from 1985 to the early 1990s. Its English-language bulletin, *SURICALL,* was well respected among DXers.

4

Literature

The short-wave DX fan has had to rely, up to recently, on just happening to pick up short-wave, long-distance broadcast reception as he twists his dials. Due to the new and exclusive service that Radio News *is now offering in the DX Corner, listing the complete time schedule of short-wave broadcast stations throughout the world, any listener, young or old, can determine at a glance just what stations are on the air in foreign countries or in any part of the world and on just what wavelength he has to listen for them.... No short-wave listener can afford to do without such a service as supplied exclusively by* Radio News.*

Radio News, *October 1933,
the month it became* Radio News and The Short-Wave[1]

In addition to *Radio News* and other magazines, an important source of DX news prior to the 1940s were the monthly, seasonal or annual publications called "logs." Typically available on the newsstand or by subscription, they included information on the call letters, location, frequency (wavelength in the early years), and power of the stations, arranging the listings in multiple ways for ease of use. They sometimes contained space for a listener to write in his dial settings. Logs became popular in the 1920s when radio was new, and at first covered only standard broadcast stations. Some added shortwave as that medium came into use.

Among the best-known logs with shortwave information was *White's Radio Log*, which was published at various intervals (usually quarterly) over its life. It had covered standard broadcasting stations since its inception in 1924. *White's* included information on U.S. shortwave stations starting in 1930, and worldwide shortwave in 1934. It continued to include shortwave after it had been incorporated into *Radio-TV Experimenter* magazine in 1958. *White's* is described in more detail on p. 169.

Similar to *White's* was *Keller's Radio Call Book and Log*, also begun in 1924 and focusing on standard broadcast stations. By 1928, *Keller's* was also including a brief listing of shortwave stations, and some information on shortwave broadcast hours as well. Soon it was limiting itself to American and Canadian shortwave stations, however, save for a few shortwave entries included among its more ample lists of foreign medium wave stations. But in 1934 it began including a longer list of shortwave stations, "Best Bets on the Short Waves," which included information on times as well as frequencies. "Riding the High Frequencies" was another *Keller's* column devoted to the discussion of shortwave stations and verifications. It does not appear that *Keller's* was published beyond the 1930s, however.

The annual *Stevenson's Radio Bulletin* was widely available. It began including limited shortwave information in 1929, and expanded the listings over the years. *Stevenson's* dropped its shortwave coverage after 1947 (and ceased publication altogether in the early 1950s).

It is worth noting that the logs usually contained only lists of stations and frequencies, not information on power output, addresses, broadcast hours (except for *Keller's*), etc., and so were of limited practical use to shortwave listeners.

Information about shortwave broadcasts was available from a variety of other sources as well. Some of the larger stations issued their own detailed program guides, e.g., the weekly "London Calling," the BBC overseas journal; "Radio EAQ," published by the Spanish national shortwave station; the monthly NHK program guide from Japan; and the monthly English-German guide for North America from Hitler's "Deutscher Kurzwellensender." However, to obtain these you had to be on the station's mailing list, and the information pertained only to that particular station.

From the early 1920s, as domestic broadcasting grew, in addition to logs like *White's*, *Keller's* and *Stevenson's,* there was much station information available in the form of stand-alone radio logs of various kinds, which were distributed by businesses as promotional items. They generally adopted a format similar to the regularly published logs—lists of stations, schedules of "chain" (network) programs, maps, articles about improving reception, blanks for dial settings, and advertisements. The coverage was usually United States and Canadian standard broadcasting stations only. Occasionally there was treatment of foreign stations, but these were always medium wave stations, shortwave not yet having come into regular use.

Information on shortwave stations started appearing in these publications in 1929. It was limited, usually consisting of a list of cities, call letters and wavelengths, and sometimes broadcasting hours. It was typically a short list, for few countries were active on shortwave, and information on their activities was scarce. Among the countries that were known to be operating on shortwave were Australia, Canada, Costa Rica, Denmark, England, France, Germany, Italy, Japan, "Java" (Indonesia), Kenya, the Netherlands, South Africa, Sweden, Switzerland, the United States, the U.S.S.R. and a few others.

During the next few years the coverage of shortwave in radio publications expanded. Station lists grew longer and more detailed, and explanatory articles about what could be heard on shortwave, and the equipment and listening techniques needed to hear it, became more frequent. There were pamphlets and monographs devoted exclusively to shortwave, and more detailed station information such as ID texts, descriptions of tuning signals, and feature articles on particular stations began appearing. In 1932, the *Listeners Official Radio Log* announced "Regular and Short-Wave" on its cover, and featured a five-page list of stations. Broadcast stations and radiotelephone stations (today's "utility" stations) were sometimes lumped together. By 1934, shortwave information was a common, if secondary, component of many station lists, and in the following years many "how to" publications appeared.

Official Sources

WORLD SHORT-WAVE RADIOPHONE TRANSMITTERS

Although much information was being published on shortwave broadcasting stations, what was missing was a comprehensive treatment of the subject. The closest anyone came to that in shortwave's early days was *World Short-Wave Radiophone Transmitters,* a 1934 publication of the Electrical Division, Bureau of Foreign and Domestic Commerce, United States Department of Commerce. The stated purpose of this 96-page book was to meet the increasing need for shortwave station information resulting from the expanded sale of all-wave receivers to the radio-buying public. The book contained information on all the world's known broadcast and utility stations above 1500 kHz., showing frequency and wavelength, call letters, power, and location. It contained some schedule information as well, but at heart it was another version of the station lists that appeared with regularity in radio magazines. Although it was compiled from both official and unofficial listings, as the authors noted in the Foreword, "The variety of sources from which this material has been taken creates a logical supposition that some of the entries may be subject to correction, and that certain stations may have been unintentionally omitted, although the list is as comprehensive as we have been able to make it." A second edition of the book was published in 1935.

FOREIGN BROADCAST INFORMATION SERVICE AND *BROADCASTING STATIONS OF THE WORLD*

A later, similar work was *Broadcasting Stations of the World,* published by the U.S. Foreign Broadcast Intelligence Service (later called the Foreign Broadcast Information Service). The FBIS was the shortwave radio monitoring division of the FCC. It was established in 1941. By 1945 it was monitoring foreign broadcasts from a variety of domestic and foreign locations. The standard FBIS setup was a bank of Hallicrafters SX-28 receivers connected to an array of rhombic antennas. Programs were recorded on wax cylinders (discs if a permanent recording was desired) and translated by FBIS translators. FBIS personnel could handle 34 languages and 30 other dialects.[2] The FBIS monitoring function was transferred several times after the war, and eventually wound up in the CIA.

It appears that the first edition of *Broadcasting Stations of the World* was published in 1945. The year before, FBIS had been looking for Recording Monitors to work in Washington, D.C. The pay was $2,188 per year for a 48 hour work week. The duties were a DXer's dream — the tuning and recording of shortwave broadcasts and, as one FBIS DXer-employee described it, "cruising for new stations and schedule changes."[3]

Broadcasting Stations of the World covered shortwave only, and listed the stations by frequency. By the 1950s it included both shortwave and medium wave stations, listing them three ways— by call letters or station name, by geographic location, and by frequency— each in a separate volume. For a time during the 1950s a fourth volume, covering TV and FM stations, was also published. In the 1960s the publication was reduced to two volumes, one by location and one by frequency, and it was published annually, or nearly so. The last edition, the 26th, was published in 1974. Its information was not always current, however,

Broadcasting Stations of the World was published for nearly 30 years by the Foreign Broadcast Intelligence (later "Information") Service, which monitored foreign radio transmissions for the U.S. government.

and it often listed stations that had long ceased operating. Nonetheless, *Broadcasting Stations of the World* was a widely used reference.

The FBIS also published the semi-monthly "Short Wave Schedule and Reception Notes." It consisted of loggings from some two dozen DXers, and was compiled like the loggings section of a club bulletin. There was some concern among the clubs that, while the FBIS bulletin warranted the support of DXers in wartime, postwar competition from the FBIS would not be welcome.[4] The FBIS noted that it was "primarily and solely a war agency" and "limited by Congressional appropriation to the war period and 60 days after the termination of hostilities."[5] The last issue of the FBIS bulletin was published on September 15, 1946.

Although the FBIS dropped from view among shortwave enthusiasts once publication of *Broadcasting Stations of the World* ceased, it remained in existence within the CIA. In 1967 the scope of its activities was increased to include all foreign mass media, both broadcast and print, and over the years it continued to be "first" in identifying important news events, just as it had been in its early years. The first word of the Soviet decision to withdraw its missiles from Cuba came in an FBIS account of a Radio Moscow broadcast, even before the official text of Khruschev's message to President Kennedy had been delivered. And in August 1991, first news of the attempted coup against President Gorbachev came by way of a TASS report monitored by FBIS.

Since 1974, the main public output of FBIS has been made available through the Commerce Department's National Technical Information Service. *FBIS Daily Reports* has been issued in various forms since 1941. In 1996 it was incorporated into the FBIS *World News Connection,* an on-line news service offering English translations of thousands of foreign media sources, including newspaper articles, conference proceedings, radio and TV broadcasts, periodicals, non-classified technical reports, and other open source materials. In 2005, the FBIS became the CIA's Open Source Center.

BBC MONITORING SERVICE

Although not of much significance to shortwave enthusiasts at the time, a British equivalent of FBIS, the BBC Monitoring Service (later "BBC Monitoring," or "BBCM"), was established in 1939, pre-dating FBIS. Like its American counterpart, BBCM translates foreign media material into English and analyzes it. Its traditional reliance on shortwave monitoring has today been largely replaced by reliance on local AM and FM radio, TV, press, internet, satellite, and news agency sources. BBCM emphasizes information from countries for which other news sources are not readily available. For many years, FBIS and BBCM have partnered in their news gathering duties, dividing responsibilities and sharing information. Together they cover the entire world, excluding the United Kingdom and North America.

BBCM was based at Wood Norton, near Evesham, England, prior to moving to Caversham in 1943.[6] Today it has a number of overseas units, as well as a network of independent contractors. In Wood Norton:

> "It was long work, depressing. But it was a nice atmosphere. It was like being on a ship. It was a completely international crowd" [quoting one of the monitors, the latter-day art historian Sir Ernst Gombrich]. In every way the Anglo-foreign community at Wood Norton was exceptional: happy, co-operative, gifted and committed. "We were all working for a common end. There was

great excitement, the whole place buzzed," recalls Betty Knott, an English supervisor now in her eighties. "The point of the Monitoring Service was that we were trying desperately to do the right thing to win the war. We wanted the war to end. Everyone understood our contribution. We had a teleprinter to Downing Street."[7]

The content side of BBCM's work was the daily, multi-part "Summary of World Broadcasts" (SWB). However, a sub-part of BBCM's wartime organization was the Special Listening Section (SLS) whose task was to monitor the airwaves and note changes in times and frequencies, identify newcomers, assess the political affiliation of clandestine broadcasters, and follow the impact of military events on stations' operations.[8] This was essentially the "DX" side of BBCM. The users of SLS information at the time were limited to the BBC and the British and American governments. In order to get the data to them, SLS published a daily summary of events known as "Reception Notes." This became weekly in 1946. Station schedule information came to be included in supplements to "Reception Notes," as was information on clandestine broadcasting and the structure and volume of output of selected broadcasters, especially Radio Moscow. "Reception Notes" became an authoritative report on broadcasting activities around the world.

"Reception Notes" became available to non-official users in 1972, the same year that a group of DXers was permitted to visit Caversham for the first time. They reported that the main antenna base, containing a multitude of Beverages, rhombics and sloping "V" aerials, was actually at Crowsley Park, about four miles away. There, in a small brick building, signals were received in an "Interception Room" equipped with about 60 receivers of Racal, Eddystone, and Plessey manufacture. The signals were then fed by cable to Caversham, where they went to either the "R.T. Section," which handled teletype signals, or the "Listening Room," for broadcast signals. There were some 50 monitoring positions in the "Listening Room," each equipped with an Eddystone EB36A or a Drake SPR-4, plus a telephone and a belt recorder. The signal was maximized at Crowsley Park, the "Listening Room" receivers being used principally just to relay the signal to the monitor. Reported the visitors: "It was quite a revelation ... to hear the really astonishing quality of the signal being fed to the individual monitor; perhaps this, more than anything else, brings home to one the amazing efficiency of that aerial farm and the other technical facilities at the disposal of the Monitoring service!"[9]

"Reception Notes" was renamed "World Broadcasting Information" (WBI) in 1973. WBI was a weekly publication containing information on schedules, new stations, and other developments in radio and TV broadcasting, and thus the BBCM publication of greatest interest to DXers. There were some format changes to WBI over the years, and in 1993 it was incorporated into the "Summary of World Broadcasts." WBI content was split into two parts, "SWB Media–World Broadcasting Information" and "SWB Media–Broadcasting Schedules." In 1996, BBCM's output about the media was re-launched as the weekly "World Media," again divided into "Broadcasting News" and "Broadcasting Schedules." "World Media" eventually became today's "Media Industry News." It focused on a wide range of media topics, including the classic BBCM areas of new stations and clandestine broadcasters, plus censorship, media-related government activities, media regulation, and media trends. (SWB was replaced by "BBC International Reports" in 2001.)

Today, most BBCM services are available on-line by subscription. Although the high cost of BBCM publications has made it uneconomic for individual hobbyists, some clubs and magazines have subscribed to BBCM publications, and thus BBCM items of interest to the shortwave community usually find their way into the hobby press.

The Berne Lists

Named for the locus of their parent organization, the International Telecommunication Union in Berne, Switzerland, were the so-called Berne lists. The ITU was formed in 1932 out of the International Telegraph Union, which traced its roots to 1865. The ITU became a specialized agency within the United Nations system in 1947.

The ITU's main publication was the massive, multi-volume, multi-lingual *List of Frequencies* (later the *International Frequency List* and the *Master International Frequency Register*) which contained information on stations of all kinds (except amateurs) and on all frequencies. This was *the* "Berne List." It was organized by frequency, and it showed country, station name and call letters, the date the station entered service, the geographical coordinates of its transmitter, power, hours of operation where applicable, area of intended reception, and other details. The *List of Frequencies* covered both broadcast and utility stations, and was of greatest use to utility DXers, for whom reliable information was hard to find. Volumes could be purchased separately, and so a SWBC DXer could purchase just those books covering frequency ranges of interest to him or her. The *List of Frequencies* was published annually from around 1932 to 1947, then less often but with periodic supplements.

The ITU also published lists addressing specific radio services. While most covered utility stations, the *List of Broadcasting Stations* was devoted to medium wave and shortwave broadcasters (TV and FM were added later). It was published annually from 1932 into the 1960s, then less often but with supplements between issues. Eventually it expanded to multiple volumes. The list was just that — a list. Data was arranged by country, then city and frequency, and contained similar information to the *List of Frequencies.*

ITU lists changed in scope and format over time. The main problem was that they were quickly out of date. Of more interest to SWBC listeners was the *Tentative High Frequency Broadcast Schedule (THFBS),* published by the International Frequency Registration Board (IFRB), a branch of the ITU that had been established in 1947. (As part of a 1992 ITU reorganization, the IFRB became the Radio Regulations Board.) The *THFBS* covered shortwave broadcasting activities above 5955 kHz. Its purpose was to facilitate frequency management by showing the expected frequency usage of broadcasters for the upcoming transmission season, and identify potential conflicts.

The *THFBS* was arranged by frequency in a band-survey style similar to today's *Passport to World Band Radio.* It showed station names, transmission times, frequencies, power, antenna gain, and target areas. By looking at the entries for each frequency, cases of potential interference could be spotted. There was a new *THFBS* for each transmission season, plus a "Weekly Circular" containing changes. Because the *THFBS* was based on information received by the IFRB months in advance of the start of the transmission season, by the time of its operative date it was often inaccurate, especially for domestic stations, where entries for long-inactive broadcasters would be repeated for years.

Even before their prices jumped in the mid–1970s, ITU publications were expensive, and for many years had to be obtained directly from the ITU in Switzerland and paid for by International Money Order, a cumbersome and time-consuming process. In 1971, some

ITU publications became available from the U.N. and from Gilfer Associates. Although they were purchased by those SWBC listeners who wanted at their fingertips all available information, for non-professional users they were generally overkill and of less practical value than the *WRTH*. As a result, even among DXers they were considered a niche publication, and were not widely used.

Electronics Magazines

Shortwave news had largely disappeared from the newsstands even before U.S. involvement in the war became a certainty. This was a result of the continued development of AM and, later, FM domestic broadcasting, and the realization that international shortwave radio, with its reputation in the lay community for low fidelity, unreliable signal quality, and complicated tuning, was not going to be a serious competitor. *Short Wave Radio* magazine had lasted but a year (1933–34). *All-Wave Radio*, with its excellent shortwave broadcast section, was absorbed into *Radio News* in 1938 and shortwave coverage was dropped in 1939. *Short Wave Craft* became *Short Wave & Television* in 1937, and *Radio & Television* in 1938. Its treatment of shortwave also stopped in 1939.

The war years further contracted the availability of shortwave information in commercial publications. *Radio Index (RADEX)*, the venerable DX magazine that began life in 1924, went from ten to six issues per year in January 1940. In October 1941 it announced that as of November it would be going to eight issues, subscription only (no newsstand sales). But *RADEX* published its last issue in February 1942. Although there was some effort to revive it in the form of an annual yearbook of information for DXers, nothing materialized.

Radio & Television News (1944–1955)

Shortwave news returned to the public press in a major way in June 1944 when *Radio News*— a publication created in 1919[10] by Hugo Gernsback and now owned by Ziff-Davis— inaugurated the "International Short-Wave" column edited by Kenneth R. Boord, a short wave enthusiast since the 1930s.[11] "In view of the many requests received from our readers," said the magazine, "RADIO NEWS presents this new department on International Short-Wave, which includes latest information for DX fans, both in this country and abroad. News of happenings in the field of short-wave broadcasting, tips for listeners, latest program information and other interesting material will be presented monthly in this column."

Ken Boord offered advice on receivers and antennas, readers' comments, "Best Bets for Beginners," monthly "Last Minute Tips" and "Press Time Flashes," contributors' loggings, and charts compiling the schedules of particular broadcasters, groups of stations, English news schedules, etc. The column grew quickly. One of the best station lists of the day was Ken Boord's "Worldwide Log of Short-Wave Broadcasting Stations," a by-frequency list of stations with detailed information on the stations' broadcasting hours. A different part of the list, covering a particular segment of the shortwave broadcasting bands, appeared in *Radio News* each month.

Boord's column did not appear from May through September 1945, but it returned in

October and it soon became a major meeting place for shortwave listeners. It had many top-notch reporters. Boord had extensive contacts throughout the shortwave community, here and abroad, and he would often share his information with club editors even before it appeared in *Radio News*. His column was full of interesting information, and he was often the source of shortwave "firsts." In February and March 1948 he published an extensive list of active and inactive shortwave stations, by frequency, that was surely one of the most comprehensive references yet published.

Boord's column continued until April 1955 when, "because short-wave listening, as a hobby, has been somewhat replaced by ever-increasing 'graduation' to Novice amateur activity," it was dropped. To shortwave listeners it was a major loss, akin to the closing of *RADEX*. SWLs were referred to a Ziff-Davis sister publication, *Popular Electronics*, which was more of a hobby-oriented magazine.

POPULAR ELECTRONICS (1954–1970, 1974–1982, AND 1989–1999)

Popular Electronics (PE) had been carrying Ken Boord's "World at a Twirl" column from its first issue in October 1954. The column was in the Boord tradition, but a fraction the size of the *Radio News* column. Boord's tutelage was brief, however, for in May 1955 he was replaced by NNRC shortwave editor Hank Bennett.[12]

The *Popular Electronics* column was retitled "Tuning the Short-Wave Bands" ("Short-

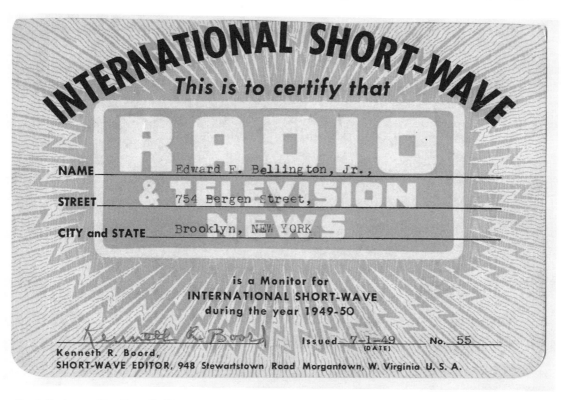

Contributors to Ken Boord's "International Short-Wave" column in *Radio & Television News* received this certificate.

Hank Bennett's column in *Popular Electronics* was must reading for SWLs.

Wave Report" from September 1957), and, while it was modest in size, it was widely read by shortwave listeners of the era. (Within ten years of *PE's* introduction, the magazine announced its paid circulation to be over 400,000.)

> [My dad] presented me with an old military surplus receiver, a BC 312 M. It was big and it was ugly, but it tuned the shortwave broadcasters, and my life has never been the same.... Only those that have ever experienced the thrill could understand the feeling that ran up my spine when I heard the chimes of Big Ben and realized that I was actually listening to London! That evening I left the farm and toured the world from my bedroom closet.... ¶I did not do this all alone. I had a wonderful teacher who came to my house once a month to visit me and explain such things as QSLing and receiver operation. Occasionally my teacher would tell me about a new short-wave station or other people in the hobby. Of great importance to me were the monthly loggings he brought, which allowed me, in my isolation, to learn what was being heard and on what frequencies. My teacher was Hank Bennett ... in *Popular Electronics*.[13]

The heart of Hank Bennett's column were readers' loggings, but each month started off with a station profile, a write-up (with photos) about an individual DXer, or a feature about some aspect of shortwave listening. While shortwave broadcasting was covered mainly in Bennett's column, *Popular Electronics* also carried occasional other articles about shortwave listening. In February 1963 it began carrying the monthly, one-page "Short-Wave Broadcast Predictions" wherein Stanley Leinwoll showed the best frequencies between various parts of the United States and the rest of the world. In March 1964 the column was expanded and renamed "Predicted Radio Receiving Conditions" and thereafter published four times yearly to coincide with the seasonal shortwave schedule changes. The last column was in November 1965.

From June 1962, Bennett's column included a monthly list of English news broadcasts to North America. This appeared under the name Robert Legge (one of Roger Legge's two sons) beginning in November 1965, and was expanded beyond news and renamed "English-Language Broadcasts to North America" in January 1966. Starting in March of that year and lasting until February 1968, most months saw this English-language broadcasts feature supplemented by a separate list covering various groups of stations, usually domestic broadcasts from different parts of the world that could be heard in North America, foreign language broadcasts beamed to North America, or English broadcasts from particular parts of the world. This supplemental list was authored by Bill Legge (Roger's other son), assisted for a time by Bob Hill. Eventually Roger Legge took over both columns, one in December 1967, the other in January 1968.

The shortwave column continued under Hank Bennett's editorship until it was dropped in August 1970 as part of a *PE* makeover that refocused the magazine on the electronics experimenter. Roger Legge's "English-Language Broadcasts to North America" was dropped as well.

It is worth noting that Oliver P. (Perry) Ferrell, the co-owner (with his wife, Jeanne) of Gilfer Associates, a leading supplier of shortwave equipment, books and supplies at the time and for many years thereafter, had served as Associate Editor of *CQ* magazine from 1949, then Managing Editor from 1953 until becoming Managing Editor of *Popular Electronics* in 1955. He served in that position until 1957, later returning to *PE* as Editor from 1960 to 1971.

Popular Electronics resumed regular coverage of shortwave in September 1974 with Glenn Hauser's "DX Listening" column. It included features on various DX subjects, usually shortwave but occasionally TV, utilities and other non–SWBC topics, and was sometimes supplemented with a by-country listing of stations heard. Four times a year, when the international transmission seasons changed, *PE* also carried a list of English-language broadcasts to North America. It was prepared at first by Roger Legge, then by Richard E. Wood, then by Hauser himself (who updated it on off months).

Although before 1980 some issues of *PE* contained neither Hauser's column nor the English-language list, one or the other appeared monthly starting in 1980. A *PE* annual called *Invitation to Electronics, 1982* also contained some limited information for SWLs. In November 1982, *PE* became *Computers & Electronics*. The *Popular Electronics* name would not reappear until 1989.

In 1983, the "DX Listening" column reappeared in a new Gernsback magazine called *Special Projects*. The column was now authored by Don Jensen. *Special Projects* became *Hands-On Electronics* in 1984. Gernsback bought the *Popular Electronics* name from Ziff-Davis, and in February 1989 gave *Hands-On* the "new" name *Popular Electronics*. Through it all, Jensen's column continued to appear, a mix of feature material and DX tips culled from the considerable resources that were at his disposal as a result of his other shortwave activities.

In July 1999, "DX Listening" moved from *Popular Electronics* to *Electronics Now* (formerly *Radio-Electronics*, which itself was *Radio-Craft* until 1948). The last column was in December 1999. In January 2000, *Electronics Now* and *Popular Electronics* were combined into a new magazine, *Poptronics,* which contained no shortwave coverage. It was last published in January 2003.

OTHER ELECTRONICS MAGAZINES

For many years, other popular radio and electronics magazines followed the lead of Radio News and Popular Electronics by including some coverage of shortwave listening. One of these was the bi-monthly Elementary Electronics, owned by Davis Publications. Don Jensen wrote "DX Central Reporting" for Elementary Electronics from 1967 through 1980, and continued with it in Science & Electronics, into which Elementary Electronics was merged in January 1981. (Science & Electronics ceased publishing in 1982 when Davis went out of business.)

Another popular magazine was *Electronics Illustrated,* a Fawcett title which first appeared in May 1958 and went from bi-monthly to monthly publication in October of the same year, back to bi-monthly in 1961. C. M. Stanbury was an early *EI* contributor, and his column, "The Listener," appeared in the magazine starting in 1961. Stanbury also did a "Propagation Forecast" column for *Radio-TV Experimenter* during the 1960s. From September 1964 to May 1970, *Electronics Illustrated* also carried "Notes from EI's DX Club," a one-page summary of "news, comment, tips and trivia," as *EI* called it. Starting in March 1966, *Electronics Illustrated* carried Tom Kneitel's "Uncle Tom's Corner," a communications and electronics question-and-answer column. *EI* was absorbed by its sister publication, *Mechanix Illustrated,* in November 1972.

In addition to their regular columns, these magazines also carried standalone articles on topics relating to shortwave listening. Stanbury also penned radio-related short stories.

Other popular magazines that included some shortwave coverage in the era of the 1960s through the 1980s included *Modern Electronics, Radio-Electronics, Radio-TV Experimenter, Electronics Hobbyist* (small construction projects), and *CB Magazine.* Some amateur radio magazines, such as *CQ* and *73,* also contained occasional articles on shortwave listening and shortwave broadcasting.

Monitoring Magazines

DXING HORIZONS (1960–1961)

Motivated by the many SWLs who missed the comprehensive shortwave coverage in *Radio & Television News,* in 1956 Ken Boord urged shortwave listeners to write to another magazine, *Radio-Electronics,* in order to convince them to start a shortwave column. But the magazine felt that the number of shortwave devotees was insufficient to warrant a short-wave department. Although Boord remained active in DXing, and continued to contribute his own items to various DX bulletins, he disappeared from the editing scene until 1960 when he resurfaced as shortwave editor for a new magazine, *DXing Horizons.*

Befitting its founder, Robert B. Cooper, Jr., who had been the TV-FM-DX Editor for *Radio-Electronics* magazine, *DXing Horizons* at first aimed solely at TV-FM DXers and the secondary TV market — operators of community cable systems, translators, and VHF boost-ers. Ken Boord's column, "The World at a Twirl," appeared in March 1960. It was vintage Ken Boord — full of detailed reports of stations large and small as reported by some of the best DXers around. Shortwave having been given the status of a "department" in the mag-azine, there were other shortwave features as well, including station profiles, biographies of well-known DXers, and an "At Fade-Out" column with last minute DX tips. Nearly half the magazine's 36 pages were devoted to shortwave.

Medium wave coverage was soon added, as was a column on English-language pro-gramming, presented by Al Niblack, and a shortwave propagation column by RFE's Stan-ley Leinwoll. (Leinwoll's column was picked up by *Radio-Electronics* after *DXing Horizons* stopped publishing.) Soon all of the shortwave and medium wave DX material was in a "World of DX" section in Ken Boord's charge. After a while, however, the space devoted to DX within the magazine's usual 30–40 pages was reduced, and the primacy of FM and TV topics was reestablished.

DXing Horizons ceased publication after the April 1961 issue. Taking its place within a few months were three new Cooper magazines: *Television Horizons* (focused on the indus-try rather than viewers), *CB Horizons* (CB, shortwave and medium wave), and *Communi-cation Horizons* (civil defense, VHF and other utility communications). A year later, *VHF Horizons,* covering ham operations above 50 MHz., was added to the group. Thomas S. Kneitel, who had been running the *Popular Electronics* WPE registration program, became an Associate Editor of the new publications group. Boord's and Niblack's columns in *CB Horizons* were soon dropped.

A *DXing Horizons* special project was the World Wide DX League, an effort to bring unity among long distance listeners, improve their public relations, and increase their influ-ence with stations and regulatory bodies. The model was the ARRL and its successful rep-resentation of ham operators. The League also had an awards program based on counties, states, and countries heard and verified. Membership was $2 for full membership if one

MARCH 1960
40 CENTS

DX ing HORIZONS
TV - SHORTWAVE - FM

WELCOME SHORTWAVERS!

VOLUME ONE   NUMBER THREE

"Late February Release...Data on HOT NEW TV Tubes" — Page 13

Ken Boord, Shortwave Editor of *DXing Horizons.*

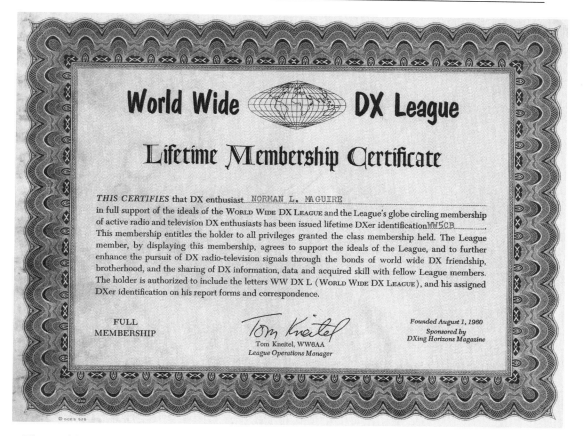

World Wide DX League

Lifetime Membership Certificate

THIS CERTIFIES that DX enthusiast __NORMAN L. MAGUIRE__
in full support of the ideals of the WORLD WIDE DX LEAGUE and the League's globe circling membership
of active radio and television DX enthusiasts has been issued lifetime DXer identification__WW5CB__.
This membership entitles the holder to all privileges granted the class membership held. The League
member, by displaying this membership, agrees to support the ideals of the League, and to further
enhance the pursuit of DX radio-television signals through the bonds of world wide DX friendship,
brotherhood, and the sharing of DX information, data and acquired skill with fellow League members.
The holder is authorized to include the letters WW DX L (WORLD WIDE DX LEAGUE), and his assigned
DXer identification on his report forms and correspondence.

FULL
MEMBERSHIP

Tom Kneitel
Tom Kneitel, WW6AA
League Operations Manager

Founded August 1, 1960
*Sponsored by
DXing Horizons Magazine*

The World Wide DX League was a project of *DXing Horizons.* This member was given the call sign-like identifier WW5CB.

was also a *DXing Horizons* subscriber, otherwise one was an associate member. Full membership also made one "one eligible to participate in League contests for prizes of commercial merchandise value," although none were ever held. An effort to govern the League via a board of directors composed of club representatives was not successful. The World Wide DX League also offered WPE-like identifiers beginning with the "WW" prefix, e.g. WW2AB, WW0CB, etc.

Following the end of his work for Horizons Publications, Ken Boord again asked his readers, whom he called "the greatest private shortwave news-gathering agency in the world," to urge *Radio-Electronics* to institute a shortwave column. This produced the same results as five years earlier,[14] and illness soon forced Boord's departure from the shortwave scene.

Ken Boord's place in the shortwave hobby was captured well by URDXC editor Bob Hill in a dedication of his monthly column to Ken in 1956:

> To the DXer who has done more for other DXers than anyone else I know of.... To the guy who has spent the best years of his life helping us to catch those "rare ones".... To the guy whose unfaltering zeal for, and devotion to, BC SWLing has carried him through over 20 years of QRM/QRN/QSB (undismayed by aches, bumps, bruises and operations).... To— Well, why ramble on? You know who I mean! With sincerest gratitude ... KEN BOORD.[15]

S9 (1962–1982)

Although *S9* was principally a citizens band magazine, it carried some shortwave broadcast information as well. Its editor was Tom Kneitel, a contributor to various publications and editor of the "On the Citizens Band" column of *Popular Electronics* from August 1959 through July 1961, and an Associate Editor of *DXing Horizons*. He published the first issue of *S9* in July 1961.

In November 1964, in response to CBers who wanted to learn more about DXing, Kneitel began writing the "SWL Shack" column covering all phases of listening — shortwave, medium wave, hams and utilities. Later it would be written by Rick Slattery. The focus turned to shortwave broadcast listening when the column became "DX Korner" in September 1968. "DX Korner" was edited by Thomas R. Sundstrom of Willingboro, New Jersey until March 1976 when the editorial baton passed to Don Jensen. Jensen wrote the column through May 1979 when C. M. Stanbury II took over. Stanbury's last column appeared in October 1981.

S9's scope broadened somewhat over the years. From 1979 it was called *S9 Hobby Radio* and it started featuring some SWL-oriented articles (as did its rival, *CB Magazine*). In December 1980, well-known "unofficial radio" author John Santosuosso began a column covering pirate, clandestine and numbers stations. The heart of the magazine was always CB, however, and it was one of the last of many publications devoted to that subject when it ceased publication in March 1982. Reader subscriptions were transferred to *CB Magazine,* and Kneitel went on to found *Popular Communications* in September of the same year.

COMMUNICATIONS HANDBOOK (1963–1979)

The *Communications Handbook* was published annually from 1963 to 1979 by Ziff-Davis Publishing Co., which also published *Popular Electronics* magazine at the time. Indeed, the cover of the soft-cover *Communications Handbook* also bore the name *Popular Electronics*.

The *Handbook's* original purpose was to provide an introduction to four topics: shortwave listening, amateur radio, citizens band, and "business radio" (similar to CB but for commercial use). True to its *PE* ancestry, the shortwave listening section of the *Handbook* was at first written by *PE* shortwave editor Hank Bennett. He covered the usual topics for an introductory shortwave guide — receivers, antennas, propagation, band allocations, QSLs, clubs, monitoring registration (the *PE* WPE system), some "best bet" stations, etc. After a few issues the structure and authorship of the *Handbook* changed. The four basic sections were replaced with over 25 separate articles. Bennett continued to write the basic SWL article, but now there were others as well, and with a broader scope — AM band DXing, QSLing, marine band DXing, aero mobile DXing, SWL-card swapping, etc. Bennett's name was but one of many well-known DX authors in the 1969 edition — Robert LaRose, Jr., Gerry Dexter, Grady Ferguson, Don Jensen, A. R. Niblack, etc.

In 1971 the content was once again consolidated under several major headings, and the section on broadcast station DXing was broadened to include medium wave and shortwave, nearly all of it written by Richard E. Wood, with some contributions by Bennett and

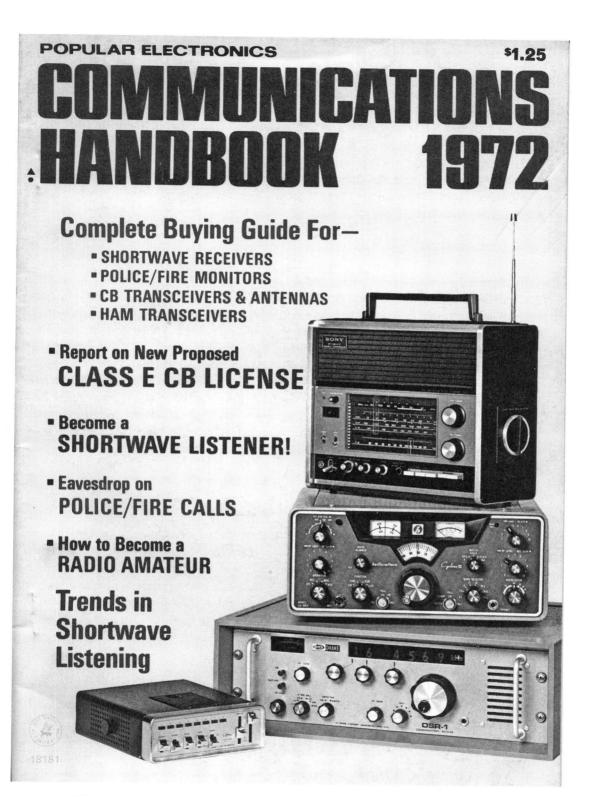

POPULAR ELECTRONICS $1.25

COMMUNICATIONS HANDBOOK 1972

Complete Buying Guide For—

- SHORTWAVE RECEIVERS
- POLICE/FIRE MONITORS
- CB TRANSCEIVERS & ANTENNAS
- HAM TRANSCEIVERS

- Report on New Proposed
CLASS E CB LICENSE

- Become a
SHORTWAVE LISTENER!

- Eavesdrop on
POLICE/FIRE CALLS

- How to Become a
RADIO AMATEUR

Trends in Shortwave Listening

A sister publication of *Popular Electronics,* the annual *Communications Handbook* was published annually from 1963 to 1979. It contained articles of interest to shortwave listeners and other radio hobbyists.

then-*PE* Editor Oliver P. Ferrell. For the first time, *Communications Handbook* also included several extensive "equipment catalogs" of receivers and antennas.

Wood wrote most of the broadcast station content from 1972 through 1977. It was supplemented by material on TV DXing by Glenn Hauser, medium wave DXing by Tom Sundstrom, public safety monitoring by Tom Kneitel, and so forth. In 1978, Wood was succeeded as the main broadcast station author by Harry L. Helms, Jr. Over half the *Handbook* in the years 1971 through 1979 was taken up with product directories, with a corresponding reduction in text. By its last edition the *Handbook* had become primarily an advertising vehicle for equipment manufacturers.

SWL News (1966)

SWL News was published briefly starting in January 1966 by James J. Howard of Kansas City, Missouri, the then-recently retired Executive Editor of the American Shortwave Listeners Club who was doing business as SWL Publications. It was produced in a 20-page, 8½ × 11" magazine format, and in content and overall style it was similar to a club bulletin, specifically an ASWLC bulletin. It contained features, book reviews, editorials, letters from readers, Bob LaRose's "Hello North America" column with information about English broadcasts to North America, and the "International DX Column" by Richard E. Wood. Howard was willing to forego a profit as long as he came out even, for *SWL News* gave him a vehicle for advertising two of his other products, logging forms and reception report forms. Originally intended to be a monthly, a quarterly publication schedule was soon announced. However, despite ambitious plans, only a few issues of *SWL News* were ever published.

Radio Today (1970–1971)

Radio Today was parented by 73, Inc., the publishers of the amateur radio magazine, *73*. The first issue was published in May 1970. Its main focus was citizens band and mobile radio, but it carried material on other bands as well, including a monthly shortwave broadcast column by Hank Bennett titled "SWL DXplorations." As the publisher explained, "We want to appeal to the fellow who is interested in radio but does not know much about it. We want to let him have fun. We want to make him want to have fun and to learn more."[16] Initially published monthly, then bi-monthly, the magazine closed in early 1971 after only a half dozen issues.

Communications World (1971–1981)

Communications World was the successor to *Radio-TV Experimenter* magazine, published by Davis Publications. *CW* was published twice annually from 1971 through 1977, annually thereafter.

The first half of each 100-plus pages of *CW* was a primer on DX listening written principally by the magazine's Communications Editor, Don Jensen, who was also Shortwave Editor of the Davis sister publication, *Elementary Electronics*. The second half of *CW* contained *White's Radio Log*, which was mainly a medium wave log.

02003
Science & Mechanics

THE OFFICIAL SHORTWAVE DXers GUIDE

SPRING-SUMMER 1972 $1.25

Communications WORLD

Formerly Radio-TV **EXPERIMENTER**

INCLUDING THE **COMPLETE**
WHITE'S RADIO LOG
AM-FM-TV-SHORTWAVE

CERTIFIED $4.50 BONUS VALUE

Discover the World of Shortwave Listening

- Lowdown on buying your Dream Receiver!
- Antenna facts on long-wires and dipoles!
- How to succeed at Shortwave DXing!

Plus–how to pull in those BCB toughies!

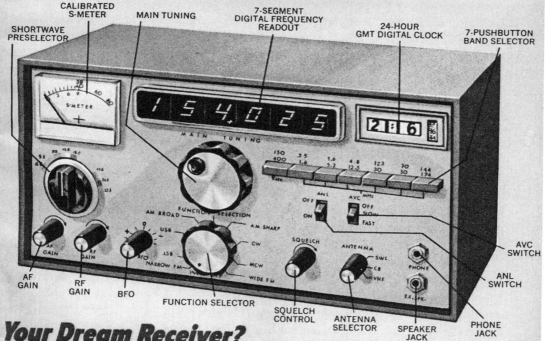

Your Dream Receiver?

Designed by the Editor
with features taken from
quality receivers reviewed in this issue!

The semi-annual *Communications World* was geared mainly to shortwave listeners, but also contained "White's Radio Log," principally a listing of AM stations.

Jensen covered all the topics that would be of interest to the long distance listener — receivers and antennas, propagation, when to tune, languages, world time, reception reports, QSLs, clubs, etc.— sprinkling the presentation with photos of shortwave listeners at their receivers. Each issue presented the material with a slightly different emphasis. Although the overall orientation was toward shortwave broadcast listening, other facets of listening, including utility stations, broadcast band DXing and even scanning and citizens band, were addressed as well. There was news of the stations one could expect to hear, plus coverage of new products, and letters and questions from readers. Starting in 1977, *CW* also contained some standalone articles, usually equipment profiles or SWL-oriented construction projects.

Although *CW* was of greatest value to beginners, experienced hobbyists, always attracted to a newsstand treatment of their seldom-acknowledged pastime, enjoyed it as well, and usually learned something from it.

SHORT WAVE LISTENERS SPECTRUM MAGAZINE (1973)

Short Wave Listeners Spectrum Magazine was a modest magazine-style publication, the first issue of which was published in May 1973, the last in August of the same year. Appearing twice a month, it covered all bands in its 15–27 pages. It had the look of a small club bulletin, but for its modest content relied exclusively on "staff" writers, several of whom were also SPEEDX editors. With a subscriber base of 160 at its highest, the magazine proved to be financially unviable.

FOR SWLs (1974–1976)

This DX "magazine" appeared in May 1974, the creation of Todd Graves of Syracuse, New York who ran an SWL supply business, SWL Guide, from the mid–1960s through the 1970s. Graves' plan was for an SWL publication that would contain material not found in club bulletins (particularly items of value to new SWLs), and that would provide an advertising vehicle for his business. The goal eluded him, however. The publication's start was delayed in part by Graves' ill health and in part by largely unsuccessful attempts to get advance cooperation from stations, equipment manufacturers and listeners. The intention was to publish bi-monthly, with page count dependent on subscriptions and advertising. *For SWLs* turned out to be a modest, mainly home-grown 12-page booklet, only a few issues of which were ever published. The first edition contained a few loggings by an eastern loggings editor (it was planned to have east, west and central logging editors), a couple of features, and an advertisement for some other SWL Guide products.

Early in the short life of *For SWLs* it was decided that the magazine would join with *The Shortwave Reporter,* a four-page newsletter begun in 1973 by one Carl Daugherty of Kirkwood, New York, who announced to his readers that the publications had "merged" and would be "co-published." The limited content of *The Shortwave Reporter* was aimed at program listeners. Even combined, the likelihood of success for these two publications was small. In October 1974, Graves announced that the magazine would be issued quarterly. In August 1975 he had decided to make it semi-annual, to coincide with the slow seasons of his business, and by March 1976 he had dropped the project altogether.

SHORT **50**ᶜ
WAVE
LISTENERS

JUNE 15, 1973

SPECTRUM MAGAZINE

THIS WEEK

IS MUSIC & NEWS
THE ONLY THING
THE VOA TRANSMITTS?

(story page 10)

MESSAGE DATA FLOW

SWL Spectrum was one of a number of ill-fated attempts at magazines about shortwave listening.

INTERNATIONAL LISTENER (1976)

Both begun and ended in 1976, *International Listener* was a 40-page, newsprint stock, monthly magazine. Edited and published by Ed Mayberry of Austin, Texas, its purpose was to promote interest in the programs available on shortwave. As Mayberry explained, there was much published on times and frequencies, but not on the programs themselves. His hope was to "introduce, educate, inform and entertain listeners of varying experience."

Although its life was short — only four issues appeared — *International Listener* was a hopeful attempt to fill a void. It printed the schedules of those broadcasters who would provide them — the unresponsiveness of broadcasters was a constant problem for those wishing to promote shortwave programing — along with feature articles about basic short-wave reception, particular stations and programs, receivers, antennas, propagation, clubs, and other publications. It also contained readers' letters and some shortwave-related advertising. Years later its publisher would host the *International Listener* website.[17]

VOICES (1980–1981)

Voices was unique in both content and appearance. Published in Finland, and the creation of Radio Finland's Patrick Humphreys, it was designed for the English-language short-wave listening market. Well-illustrated and printed on glossy paper, *Voices* was a comparatively high-style product.

Subtitled "The Guide to International Radio," its purpose was just that, to serve as a kind of *TV Guide* for shortwave broadcasting. *Voices* had rightly concluded that the relative difficulty of finding one's way around the shortwave bands meant that the medium had lost its message. In addition, the absence of good audience measures was making it increasingly difficult for broadcasters to justify their existence. *Voices*, with a combination of monitoring and journalistic resources, and a fast turnaround time, viewed itself as the portal into an as-yet undeveloped world of international shortwave broadcast programming, and a vehicle for serving both listeners and broadcasters.

The magazine contained 20–36 pages. It focused on shortwave programs and program personalities, international news broadcasts, and trends and problems in international broadcasting. Articles were interesting, well-researched, and well-written. *Voices* also contained an advance, day-by-day listing of selected programs for the month, together with frequency tables. This was later supplemented with a by-time "Voices Checklist." Advertising came from receiver manufacturers and other shortwave vendors, as well as stations, and Humphreys had a *Voices* segment about shortwave programming on both Radio Canada International's "DX Digest" ("What's On") and Radio Netherlands' "DX Juke Box" ("Voices On the Air").

The success of *Voices* depended on sales, and there the results were disappointing. Although *Voices* appealed to DXers and shortwave listeners alike, it was the latter group, the non-specialists, who were the magazine's principal target. Unlike the DX community, the larger world of shortwave listeners was not well organized, and thus not as easily reachable, nor was it used to paying for shortwave information, especially from a foreign country. While subscriptions, at $16 for 12 issues (later raised to $19), showed a steady increase, after six months the print run was about 7,000, considerably short of the 10,000 goal.

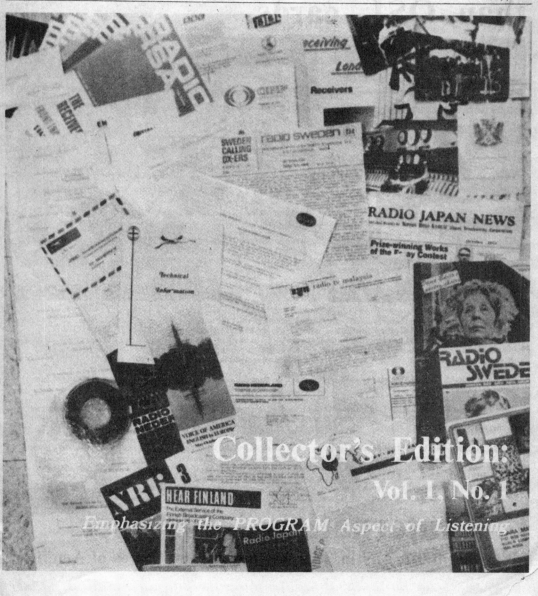

International Listener

Program Guide for English Language Shortwave Broadcasts

Vol. I, No. 1   P. O. Box 3782, Austin, Texas 78764, United States of America   January, 1976

A worthy attempt at an SWL magazine, *International Listener* closed after four issues.

VOICES

SEPTEMBER 1980

The Guide to International Radio

Dream receivers

Aims of propaganda

RNW's Ian de Stains

Australia	AUD 1.80	Japan	JPY 450
Britain	GBP 0.85	Saudi Arabia	SAR 62
Canada	CAD 2.30	South Africa	ZAR 1.80
France	FRF 8.50	Spain	ESP 150
Fed. Germany	DEM 3.60	Sweden	SEK 8.50
Italy	ITL 2000	United States	USD 2.00

VOA's Mary Bitterman on International Broadcasting in the 80's

A promising program-related shortwave magazine, *Voices* proved not financially viable.

Voices also received relatively little support from vendor advertisers in either North America or Western Europe where 60 percent of the magazine's subscribers were located. (Japanese advertisers were slightly more supportive.)[18] And station advertising and informational support were stronger from Soviet bloc states, who seemed to place more value on the magazine than did western countries.

This combination of factors made it impossible for *Voices* to enlarge its size, a precondition to expansion into the more lucrative newsstand market. As a result, the magazine closed in 1981 after only seven issues. Although DXers were not the magazine's main target audience, they were among its most disappointed subscribers.

RADIOVOICE (1982)

Radiovoice was to be a bi-monthly "magazine" on a 90-minute cassette, featuring station schedules, programming notes, technical articles, equipment reviews, features, etc. One of its strengths was its ability to include some off-air station recordings. The creation of Jim Tedford of Marquette, Michigan, it closed the same year it was introduced.

MONITORING TIMES (1982 TO THE PRESENT)

The first issue of *Monitoring Times* was published in January 1982. It was eight pages in a newspaper format and it was labeled and bundled at the kitchen table of Bob and Judy Grove, *MT's* Editor and Advertising Manager respectively, and sent to several hundred people on the Grove mailing list. Although conceived in part as an advertising vehicle for the Grove business— the sale of monitoring equipment and books— the Groves hoped that *MT* would be a serious publication with broad appeal. By the end of the year a 24-page *MT* was being published bi-monthly. It went monthly in January 1984 and soon had a readership of 10,000.

The plan was for *MT* to cover the radio spectrum from the highest to the lowest frequency. As such, non-broadcast transmissions and the technological advances that were so drastically changing the "utility" landscape were its early focus. An *MT* strength in this area was that it published much information not easily found elsewhere. It covered military transmissions, maritime and coastal services, trucking and other specialized transportation communications, police radio, radioteletype, aviation radio, plus government intelligence activities, "spy" transmissions, and space-related radio. Eventually it would also address topics such as communications law, receiver restoration and nostalgia, radio-related computer topics, and ham radio. Over the years there would be frequent equipment reviews, information on new products, articles about antennas and propagation, book reviews, and answers to technical questions. At present *MT* also covers digital communications and internet resources, and satellite communications were covered until 2005 (a separate *Satellite Times* was published from 1994 to 1998). There is hardly a radio-related topic that *MT* has not covered at some time, usually in detail and with original content.

Notwithstanding its utilities orientation, *MT* offered much for the shortwave broadcast listener. From mid–1983, Hank Bennett was an *MT* columnist. His "Monitoring Times Shortwave" was a traditional column of station loggings and related information. Soon, however, it was restyled as "Broadcasting ... Hank Bennett on Shortwave" and it became

M◯NITORING TIMES

$2.00

Volume 5 -Number 2   BRASSTOWN, NORTH CAROLINA 28902   **February, 1986**

Does basking under a palm tree sound good to you this chilly February? Then let MT transport you to the sunny Caribbean, the South Pacific and the Mediterranean (see "High Seas") via the magic of radio!

"What's New?"
THE VERY LATEST IN RECEIVERS ON PAGE 20!

Signals from the Islands

by Ed Noll

Bonaire is one of the six Caribbean islands, including vacation-popular Aruba and Curacao, that make up the Dutch Antilles. About 50 miles north of Venezuela, it is cooled by tradewinds resulting in an average annual temperature of about 80°F.

Bonaire's population of over 9,000 persons live on an island 24 miles long and five miles wide with life centered about the small town of Kralendijk. Other important residents that grace the island are flocks of pink flamingoes.

Remember all of this when next you tune in the Radio Nederland's relay station, located on the higher-elevation northern part of the isle. Two transmitters with a carrier power of 300 kilowatts supply energy to a complex grouping of antennas including omnidirectional and highly directional curtain types.

A curtain aerial is an array of many dipoles suspended between giant towers and guarded by a

☞ *Please turn to page 6*

Radio Happy Isles

by Ken Wood

Radio Happy Isles is the slogan of the Solomon Islands Broadcasting Corporation--one of the few South Pacific Islands which not only maintain a short-wave broadcasting service but find it an invaluable aid.

World War Two and the U.S. Armed Forces brought the first broadcasting to the Solomons. That was in June, 1944, when the Armed Forces Radio Service set up WVUQ on Guadalcanal and a few months later added WVTJ at Munda on New Georgia Island.

Although intended as a news and entertainment service for U.S. military personnel serving in that area of the Pacific theater, the radio caught the fancy of the islanders as well. Soon after the end of the war operators at the government's wireless and telegraph station began to work towards building the first crude broadcasting station which would serve the people of the Solomons.

Those first broadcasts over a small transmitter were on the air for a half hour Sunday program once a week. By 1948 the government of the then British Solomon Islands could count on small license fees from the owners of 57 radios in the islands.

That same year part of the government's budget was set aside for broadcasting. Though only a small amount, it was to be significant that the government already saw the small service as "most useful as a means of keeping residents of the territory in touch with local developments."

In 1952 the Solomon Islands Broadcasting Service was formed and by then had a 400 watt station, VQO at Honiara, operating on 1030 kHz and running a daily schedule of one hour programs consisting of local news, shipping and weather reports, sports news, talks, and local music.

☞ *Please turn to page 5*

Radio Nederland Installation on Bonaire Island.

Presstop from Washington:
LISTENING LAW UPDATE

Report and Commentary
by Bob Grove

At press time there are several important turns of events regarding House of Representatives (Congressional) Bill HR3378 and Senate Bill S1667 concerning monitoring. While the twin bills are still in the initial hearing process and only supportive testimony has been allowed on the House floor, the outrage of the radio users is being felt in Washington.

Two hearings had taken place before the holidays and at least two more were scheduled after the House business resumes January 21st. A witness from the

American Radio Relay League, Perry Williams, representing amateur radio interests, was expected to give testimony in support of an amended bill by early February.

Congressman Robert Kastenmeier, who championed the original bill, is receptive to excluding any amateur radio use of the spectrum including autopatch, a system by which an amateur repeater is tied into a telephone line and broadcasting both sides of the conversation.

More important, Kastenmeier will offer an amendment to his bill allowing inadvertent monitoring of

☞ *Please turn to page 4*

One of two monitoring magazines that have enjoyed long-term success (the other is *Popular Communications*), *Monitoring Times* was originally published in a newspaper style. (Reproduced with the permission of *MT.*)

more of a commentary on shortwave listening. It was published until July 1986. A traditional loggings-and-station information column, "SWL World Watch," was begun in January 1985. Other shortwave broadcast-related columns appearing around the same time were Tom Williamson's "English Language Broadcasts" and John Santosuosso's "Pirate Radio" (also covering clandestine stations). There was also a column about radio club activities.

In July 1986, Larry Miller's *International Radio* (see p. 166) merged with *MT,* and soon Miller assumed the post of *MT* Editor. *MT* changed to a tabloid format and went from 40 to 60 pages. Although some of the old *MT* SWBC columns were dropped, there was a considerable expansion of *MT's* coverage of shortwave broadcasting, which had been the focus of *International Radio.* Up-to-date station news was included in various formats, and there were more SWBC-related features. In addition, the extensive monthly by-time listing of shortwave broadcasts which had been the "Frequency Section" of *International Radio* was preserved in *MT.* (Today called "Shortwave Guide," it remains a major feature of the magazine.)

In January 1988, *MT* adopted a standard magazine format and a glossy cover and went to 80 pages, soon to increase again to over 100. This yielded an increase in subscriptions, advertising, and newsstand sales. In March 1988, Glenn Hauser joined the *MT* editorial staff with a monthly SWBC column called "Shortwave Broadcasting" (from August 1994 called "Global Forum," a name later applied to a department incorporating several *MT* columns). The *MT* "Frequency Section" was supplemented by monthly propagation charts and by the "Program Guide," a by-day-and-time list of many programs that were available on shortwave. (For a short time, programs editor Kannon Shanmugam actually "rated" specific programs.) Soon "Program Guide" was supplemented with "News Guide," which covered the news broadcasts available on shortwave throughout the day. Other monthly columns for SWBC enthusiasts around this time included a loggings column and a "QSL Report" (both later placed within "Global Forum"); "Magne Tests," containing reports on shortwave receivers and accessories by *Passport to World Band Radio* Editor in Chief Larry Magne; W. Clem Small's "Antenna Topics"; and John Santosuosso's pirate radio column, now called "Outer Limits."

Rachel Baughn became *MT* Editor in 1991. (Larry Miller continued writing for the magazine until 1997.) The "Frequency Section" became "Shortwave Guide." Folded into it were the propagation charts, "Newsline" (formerly "News Guide") and "Selected Programs" (formerly "Program Guide"), as well as a quarterly list of DX programs.[19] A new column, "Computers & Radio," was begun in January 1992. George Zeller became "Outer Limits" editor in May 1992. In 1996 the "Newsline" section of the "Shortwave Guide" was dropped, and the propagation charts were replaced with a one-page column and graphic display on propagation forecasting by Jacques d'Avignon.

In 2000, "Magne Tests" was replaced with "Shortwave Equipment," an equipment-review and discussion column. It was renamed "Radio Equipment" and broadened to cover non-shortwave equipment topics in December 2001, then became "On the Bench" in January 2003. Tests of various kinds of monitoring equipment also appeared in the "MT Review" column. The last propagation column appeared in December 2000. Thereafter the magazine carried periodic seasonal propagation outlooks.

Among *MT's* greatest attractions for the SWBC listener were its feature articles, usually highlighting particular stations or aspects of shortwave broadcasting that were not widely covered elsewhere, and frequently authored by persons with professional shortwave broadcasting experience. From 1990 to 1996, *Monitoring Times* also sponsored a series of annual listener conventions.

An interesting aspect of *MT* has been its work in the area of shortwave programming. From 1992 to 1994, *MT* issued as a separate publication Kannon Shanmugam's *Guide to Shortwave Programs*, basically a listing of English-language programs from approximately 100 shortwave stations (see p. 202). Starting in 1994, *MT* tried a quarterly "Program Spotlight" column about shortwave program listening and program content. It lasted only two years, but it returned as a monthly feature in February 1998 under well-known shortwave programming author John Figliozzi. It was relocated to the "Global Forum" section in 2001. When the magazine's monthly "Selected Programs" listing was dropped in 2005, Figliozzi's column was expanded to include a monthly "themed list" of shortwave broadcasts. In September 2006, Figliozzi was succeeded by Fred Waterer, editor of the "Programming Matters" column in the Ontario DX Association bulletin, *Listening In.*

In 1999, as an alternative to the traditional print version, Grove began offering *Monitoring Times* in .pdf format via the internet (it was called "MT Express"). While probably not as convenient for most readers as paper, for a reduced price the reader received the magazine more quickly, with more color and with full searchability of each issue. Grove said that the *MT* frequency tables could be converted to text files and imported into computerized frequency databases, and that the .pdf version was compatible with software that could convert it to audio for the sight-impaired. Starting in 1999, Grove also offered the previous year's *MTs* in "anthology" CDs. (Print anthologies, *The Best of Monitoring Times,* had been published from 1982 to 1985.)

At its height, *MT* had a circulation of 35,000. It was about 15,000 in 2005. *MT*'s greatest assets have been Bob Grove's imagination and entrepreneurial spirit, and the magazine's personal touch with readers. Even after transforming itself from a newspaper to a full-fledged magazine in 1988, it continued to stake out a place where both radio professionals and hobbyists could be comfortable. Its content has been distinctive, and while much of it is specialized, it is sufficiently unintimidating to appeal to both technical and non-technical readers.

Grove Enterprises grew into a major producer of radio accessories and supplier of receivers, antennas, publications, and radio-related supplies. While *MT* carries advertising for Grove and many other vendors, the range and objectivity of *MT*'s content long ago eclipsed its role as an advertising vehicle. It is one of only a few commercial publications today that provides an outlet for informed writing about shortwave broadcasting topics. The wide scope of *MT* has lived up to Grove's goal of an "all frequency" publication, and its authoritative place within the listening community is one to which others can aspire.[20]

POPULAR COMMUNICATIONS (1982 TO THE PRESENT)

In September 1982, six months after the close of *S9* magazine, *S9*'s editor, Tom Kneitel, inaugurated a new publication called *Popular Communications*. In general, the scope of *PopComm* was similar to that of *Monitoring Times,* which had appeared in January of the same year. However, unlike *MT,* which at the time was still a newsprint publication,

PopComm was a full-size glossy (if mostly black and white), illustrated magazine. It was less technical than *MT,* and still reflected a bit of the "gee whiz" style of earlier popular writing about radio.

In the inaugural issue, Kneitel explained that there was a "communications explosion" afoot, "[a] spectacular, multicolored smorgasbord of signals there for you to flavor," and that *PopComm*'s aim was to cover it all. The magazine was to be devoted to the listener and the monitor. With a wide range of topics—shortwave broadcasting, AM and FM, radiotele-type, every variety of utility communications, satellites, alternative (pirate) radio, "survivalist" and paramilitary communications, scanning, citizens band, beacons, military radio, police radio, "underground" and "spy" radio, "numbers" stations, telephone technology, computer-related monitoring, governmental regulatory news, even radar detection—it would make good on that promise over the years. *PopComm* also covered receivers, antennas, other equipment and new products, and it included Kneitel editorials, readers' letters, and relevant advertising—just about everything except ham radio (which was added in August 1987).

For the shortwave broadcast enthusiast, Kneitel enlisted well-known DXer Gerry L. Dexter to present the monthly "Listening Post" column (called "Global Information Guide" from June 2001). It contained information about new stations, QSLs, publications, and other goings on in the SWBC world, plus loggings submitted by readers. Dexter also authored numerous feature articles on SWBC topics. And from February 1983 to August 1996 the magazine carried a quarterly listing of English-language shortwave broadcasts, arranged by time.

Although *PopComm* was careful to point out that it did not support the mission of illegal broadcasting, it offered a pirate radio column, "Free Radio Focus," written by Al Muick. Muick had long been active on the pirate scene, and was the head of Free Radio Campaign–U.S.A., publisher of *The Wavelength*, a pirate newsletter. His *PopComm* column featured news about active pirate stations, mainly in the U.S. It became "Pirate's Den" in October 1983 when the editorial duties shifted to Darren Leno, another pirate activist. Leno was succeeded by Gerry Dexter (writing under the pseudonym Edward Teach) in June 1986. The column was renamed, this time "Pirate & Alternative Radio," in June 2001. A column on clandestine broadcasting, "Clandestine Communiqué," also authored by Gerry Dexter, was added in December 1986.

In February 1990, *PopComm* added a two-page pull-out section called "POP'COMM's World Band Tuning Tips." It was a by-frequency listing of shortwave broadcast stations, showing the best times to try to hear them, and the languages they used. These were not necessarily English broadcasts, nor broadcasts beamed to North America. The list was intended to help readers hear more stations.

Although there were fewer SWBC-related features after 1992, *PopComm* sponsored a "Worldwide SWL Conference" On October 2–3, 1993 in conjunction with the Annual Virginia Beach Hamfest and Computer Fair, Virginia Beach, Virginia. In addition to product demonstrations and equipment and dealer exhibits, there were speakers and panels of presenters. And from 1992 to 1996, *PopComm* published the 140-plus page *Communications Guide.* Half the *Communications Guide* was a buyer's guide containing specifications and prices for SWL and CB receivers, amateur transceivers, and accessories, and half consisted of articles on various aspects of radio monitoring—receivers and antennas, setting up a listening post, tuning tips, QSLing, pirate radio, utilities, tips for listening while traveling, etc.— and obtaining a ham license.

A favorite Kneitel topic, and one unique to *PopComm,* was radio history. The magazine offered many features about wartime broadcasting, propaganda, early stations, early radio ephemera, and similar topics. Many were written by Kneitel himself. Starting in 1984, *PopComm* featured a monthly radio history column authored by "Alice Brannigan."[21] The radio history column continued through October 2000, and resumed in October 2002 with "Shannon's Broadcast Classics" by Shannon Huniwell. A propagation column was added the same month.

Kneitel took Senior Editor status in September 1995, retiring as *PopComm*'s editor in chief.[22] He was succeeded by the magazine's scanning editor, Chuck Gysi. In October 1996, Gysi's senior in the organizational structure, Harold Ort, took over. Ort had been the editor of the *PopComm Communications Guide* and *PopComm*'s short-lived *CB Radio* magazine, which was being phased out in a company downsizing.

Since 1996 there have been some changes in *PopComm*'s layout and focus. While there has been an expansion of the number of logs contained in Dexter's "Global Information Guide," both the pirate radio and clandestine radio columns were dropped in April 2004. Ort was succeeded by the magazine's Managing Editor, Edith Lennon, in June 2007.

THE INTERNATIONAL SHORTWAVE LISTENER'S PROGRAM GUIDE, THE SHORTWAVE GUIDE, AND INTERNATIONAL RADIO (1984–1986)

Published by Miller Publishing (Larry Miller), the first issue of *The International Shortwave Listener's Program Guide* appeared in April 1984. Prior to publication it had been promoted as *The International Shortwave Listener's Program Quarterly.* However, when it appeared it was bi-monthly, and reflected the *Program Guide* title. By the third issue it was monthly. Soon the name was shortened to *The Shortwave Guide,* and in January 1986 it became *International Radio.* (All the titles are referred to collectively below as the *Guide.*) It merged with *Monitoring Times* following the June 1986 issue of *International Radio.*

The *Guide* began in a magazine format, eventually adopting an *MT*-like tabloid style. In many respects it was to the shortwave broadcast community what *MT* was to utility aficionados. It carried original articles about stations and the listening scene in general, plus interviews and editorials, usually in greater depth than could be found in club publications. In this respect it complemented the club bulletins. Many regulars within the long-distance listening community were contributors to the *Guide,* among them David E. Crawford, Arthur Cushen, Ken MacHarg, Adrian M. Peterson (Adventist World Radio), Jeff White, and Richard E. Wood. An interesting series of columns by Dr. John Santosuosso, "Politics and Shortwave," looked at the program content of individual stations under the title, "A Political Scientist Listens to...." ("Radio Jamahiriya," "WINB," "the Voice of Free China," etc.). Santosuosso was an associate professor of political science at Florida Southern College in Lakeland, Florida.

What set the *Guide* apart was its attention to shortwave programming. It carried much material, both descriptive and analytical, on shortwave programming, along with major station features. The focus was on material that would be of direct assistance to listeners in locating programs of interest. This included an extensive table, arranged by date and

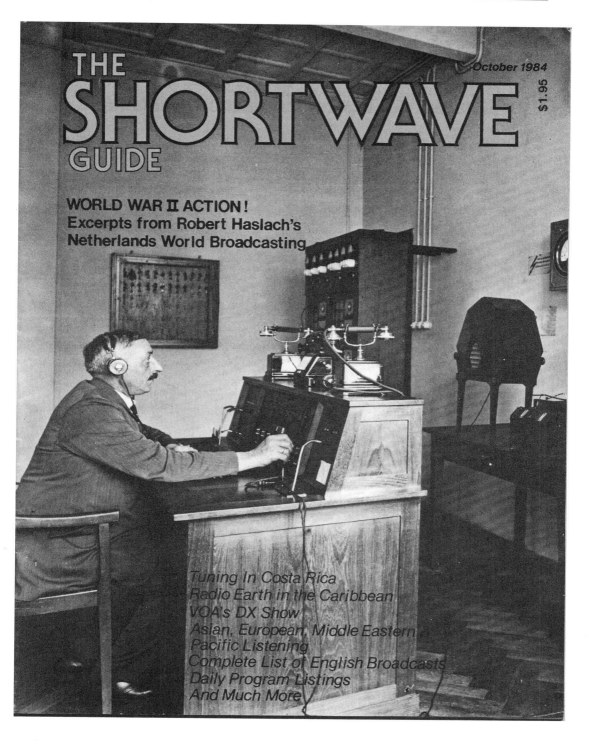

The International Shortwave Listener's Program Guide soon became *The Shortwave Guide.* A well-written, substantive publication, it merged with *Monitoring Times* in 1986 and contributed to the improved coverage of shortwave listening in *MT.* (Reproduced with the permission of *MT.*)

time, identifying many of the English-language programs that would be carried on short-wave during the coming month, and providing brief descriptions—a kind of *TV Guide* of shortwave. As this section was fine tuned it received various titles, including "Programs at a Glance," "[October] on Shortwave," "Day to Day," and finally "Advanced Program Details." It was the heart of the *Guide*. From January 1986 it was supplemented by another section, "Prime Time Shortwave," a by-time listing of daily programs by day of the week rather than date.

To the program material was added Roger Legge's "English Broadcasts to North America," a by-time listing of the frequencies that the stations' used. Legge's material alternated with an expanded version, "English Broadcasts to the World," by Tony Jones. These columns would eventually become the *Guide*'s "Frequency Section." (The Jones column replaced Legge's in 1986.) The *Guide* also carried some non-program news in the areas of equipment and pirate broadcasters, as well as letters from readers and a beginners column written by Clayton Howard of HCJB.

Notwithstanding the obvious service that the *Guide* provided to stations in reaching listeners, Miller received little assistance from the broadcasters, a problem endemic to an industry where most international stations were funded by governments and not audience driven, and the rest were interested mainly in a domestic rather than a worldwide audience.

According to Miller, the merger of *International Radio* with *Monitoring Times* was a product of his wish for relief from the administration of a publication with over 4,000 subscribers. As he put it in the final (June 1986) issue, "[I]n the end, the magazine simply grew beyond our ability to manage it. Slowly but surely, it took its toll on the health and welfare of everyone involved in its production. The magazine simply outgrew itself, making more and more demands, but returning less and less."

By the time of the merger, the *Guide* had become less newsy and the page count was closer to 24 than its earlier 32–40. While the depth of the *Monitoring Times* shortwave broadcast content would never quite reach that of the *Guide,* it improved noticeably, and *Monitoring Times* would strengthen its coverage of shortwave programming in later years.

The *Guide* had been a valuable and enjoyable publication, and its coverage of short-wave programming a first among radio magazines, and still unequaled 20 years later.

WORLD RADIO REPORT (1987)

Another short-lived publication was *World Radio Report. WRR* was inaugurated in connection with the establishment in 1987 of the non-profit Foundation for International Broadcasting. The foundation's goal was to provide a central vehicle to promote short-wave, do audience research, etc.— in short, to be the shortwave listening equivalent of ham radio's American Radio Relay League. The timing seemed right. There was an increased level of private shortwave broadcasting in the United States, and some expected a heightened interest in shortwave as the sunspot count increased and reception improved. Said Michael Poulos of Radio Earth, a Foundation board member: "We're looking for people who are serious. Serious about shortwave and serious about the goals of [the] Foundation. People who want to get in on the ground floor of a hot, new, shortwave organization.... It's time to get shortwave up, on its feet and into the mainstream and everyone can help."[23]

Among the other board members were Larry Miller; Larry Magne, publisher of *Radio Database International* (later *Passport to World Band Radio*); attorney Michael Poulos of Radio Earth; and Ken MacHarg of HCJB.

Larry Miller served as the Foundation's director. Miller Publishing acted as an arm of the Foundation, with Miller in the role of *WRR* editor. Said the advertising about the monthly shortwave-only *World Radio Report*: "It's nothing flashy. It's just the best."

The 44-page magazine carried feature articles about shortwave stations and shortwave personalities; by-country information on loggings, schedules, programs, etc. from many sources; a by-time listing of English broadcasts of stations large and small, including their frequencies; a by-day, *TV Guide*–type program listing; and equipment news. Regrettably, by the end of 1987 only four issues of the promising magazine had appeared, and a limited supply of both energy and funds led to the demise of both it and the Foundation.

HOBBY BROADCASTING (1998–2001)

Created and edited by well-known shortwave pirate devotee Andrew R. Yoder, and first published in the spring of 1998 by his Cabinet Communications of Mont Alto, Pennsylvania, the quarterly *Hobby Broadcasting* magazine was designed to cover all aspects of "personal broadcasting." This included shortwave pirates ("Shortwave Bandscan"), as well as internet radio, FM microcasting, public access radio and TV, pirate radio and TV on all bands, and college and high school broadcasting.

Hobby Broadcasting was a well-written, professional-looking product that was soon 50 pages in length. It contained articles on technical topics, applicable law, governmental and non-governmental activities related to unofficial broadcasting, and radio history, plus thoughtful editorials by Yoder, station profiles, interviews with pirate personalities, reader feedback, book reviews, and even an alternative music column. The magazine was supplemented by a website.

The originality of its content made *Hobby Broadcasting* an interesting read even for those whose listening was limited to shortwave broadcasting. Unfortunately, Yoder's own changing circumstances, plus printer-related delays, limited advertising and a small customer base conspired against the magazine's long-term success. Although the eleventh issue (winter 2000–2001) was its last, *Hobby Broadcasting* had been a commendable effort.

Basic References

WHITE'S RADIO LOG (1924–1985)

Among the best-known logs with shortwave information was *White's Radio Log*, which was published at various intervals over its life span. It covered standard broadcasting stations from its inception in 1924. *White's* started including information on "short wave relay broadcasting stations" located in the United States in 1930, and in 1934 expanded its coverage of shortwave broadcasting to include stations throughout the world. However, *White's* was principally an AM-FM-TV station list — the shortwave list showed no broadcast times, only station names and frequencies, and because the latter were so variable it was never of

White's Radio Log was a separate publication from 1924 to the mid–1950s. From 1958 to 1981 it appeared as part of *Radio-TV Experimenter* and several successor publications.

much practical value to shortwave listeners. *White's* dropped its newsstand sales around 1954. For a short time it still could be obtained by subscription, but soon it disappeared all together.

It made its public reappearance in 1958 when it was incorporated into *Radio-TV Experimenter* magazine, a title of Davis Publications, Inc. The "World-Wide Shortwave Stations" information was upgraded to include the times at which the various stations could be heard. This data was compiled from readers' contributions and from monitoring at *White's* own "DX Central," which the magazine described as "a completely equipped, professionally manned DX monitoring station in New York City and one sub-station near the top of Long Island."[24] The New York City station was actually the listening post of *Radio-TV Experimenter* editor Julian M. Sienkiewicz, and the sub-station that of Tom Kneitel. In the late 1960s, much of the work on *White's* was assumed by Don Jensen.

When *Radio-TV Experimenter* began publishing *White's* in 1958, the magazine was an annual. From 1964, when the magazine became bi-monthly, one third of the list appeared in each issue, and thus the entire list appeared twice annually. From 1969, one sixth of the list appeared in each issue, so it took a year for the entire list to appear. In 1969, *White's* was expanded to include police, fire and other emergency radio stations in various parts of the country.

In January 1970, *White's* moved to *Radio-TV Experimenter's* successor publication, *Science & Electronics,* and then to the new semi-annual Davis title, *Communications World,* which commenced publication in 1971. From 1971 to 1981, each issue of *Communications World* included the full *White's Radio Log,* covering U.S. and Canadian AM, FM and TV stations, and worldwide shortwave. Although *White's* shortwave list continued to show times as well as frequencies, it consisted of only a few pages and was clearly secondary to the list's main focus, which was still AM, FM and TV.

White's disappeared along with *Communications World* in 1981, not to return until 1985 when a final standalone issue of *White's Radio Log* was published by a former Davis employee doing business as Worldwide Publications, Inc. of North Branch, New Jersey.[25] Although there were plans for future editions of *White's,* as well as other shortwave and electronics publications, nothing materialized, thus ending the long run of this venerable, if by then somewhat outmoded, publication.

World Radio TV Handbook (1947 to the Present)

The need for a comprehensive source of information about shortwave broadcasting stations was finally met in November 1947 when the inaugural edition of the *World Radio Handbook* appeared.[26] The *Handbook* came to be known popularly as the "*WRH,*" "*WRTH*" after it changed its name to the *World Radio Television Handbook* in 1961 (although it had included coverage of television since 1954). (It is usually referred to below as the *WRTH.*) It was edited and published by Danish radio enthusiast O. Lund Johansen, who had been writing, editing and publishing in the broadcasting field since 1923.[27]

In the years since 1947 the *WRTH* has grown from 97 pages to almost 700, mainly due to the expansion of AM and FM listings, and while the level of detail has steadily increased, the basic structure has remained largely unchanged. In the first edition, information for each continent was arranged by country.[28] Within each country listing were shown the local

It was the *World Radio Handbook* that first made comprehensive data on shortwave broadcasting available in one place. The annual has been published continuously since 1947.

time, followed by information on each station in varying levels of detail: name and address; names of station officials; schedule information, including times, frequencies, power, languages and programs; ID texts and descriptions of interval signals where available (interval signal melodies often were shown on musical scales); and occasionally a "message to listeners" from the station (usually about reception reports).

In the back of the book were two lists of stations by frequency and wavelength, one for longwave and medium wave stations in Europe, North Africa and the Near East, the other, "Short-Wave Stations of the World," covering the shortwave bands worldwide. The first edition also contained summary information on various broadcasting organizations, endorsements from broadcasting officials, continental maps, a primer on shortwave listening, photos of stations and personalities, and advertising. In the latter department the *WRTH* came of age quickly, for by the third edition the back cover was occupied by a full-page ad from Hallicrafters for its SX-62 receiver.

O. Lund Johansen explained the purpose of the *WRTH* in the foreword to the first edition:

> The increasing importance and rapid growth of international broadcasting justifies the publication of a handbook which supplies practical information concerning the organisation and activities of the different broadcasting corporations. Development has caused international broadcasting to become so complicated and extensive that a World Radio Handbook must be of importance to both listeners and broadcasting companies. Its aim will be to give those interested practical information and instruction regarding what is going on in the ether at all hours of the day and night, and to provide the broadcasting corporations with such particulars of each others activities that the contents of the book will serve to promote the development of international broadcasting.

In 1959, while on a trip to the United States, Johansen talked more about the *WRTH* with a *New York Times* reporter.

> "All things were disturbed [in 1947] ... I thought we could use broadcasting to collect the world again, to give all exact information that would help people get programs." This is done by sending out at least 15,000 pieces of mail to stations, governments and broadcasting agencies each year. The letters ask the authorities for all possible details. The response is generally prompt from both sides of the Iron Curtain, from the young countries as well as the old. Questions are asked in major languages. For instance, a query in Russian to Mongolia (Outer) elicits such details as that Ulan Bator's Radio chairman of the board is Sodnomdarsja and that the station's announcer identifies it by saying "Ankhararai. Ulanbatras jarsj baina." This may not mean much outside of the yurt country, but the student of the handbook learns that it means "Attention. This is Ulan Bator calling."[29]

The name and format of the *WRTH* may have been influenced by the BBC's *World-Radio,* and, to a lesser extent, by the British magazine *Short Wave News*. Titled *The Radio Supplement* at its inception in 1925 and renamed the following year, *World-Radio* was a weekly, newspaper-style publication covering the programs to be heard on foreign stations, mainly European medium wave stations that could be heard in England. Later it included some shortwave information as well, but the emphasis was always on medium wave. It was incorporated into *Radio Times* magazine in 1939. One of the early features of *World-Radio* was the front-page "Station Identification Panel," a bordered text containing information on a particular station—frequency and wavelength, power, interval signal, identification announcement, distance from London, etc. In the early 1930s, the BBC issued several editions of a booklet called *World-Radio Station Identification Panels* in which it republished updated versions of many of the *World-Radio* panels (and added some for a few U.S. medium wave and shortwave stations). The British publication *Short Wave News*, which first appeared in January 1947, picked up the station panel idea and often carried a "Broadcast Station Country Panel" containing a list of the shortwave stations of a particular country, with frequency, power and broadcast times.

The major difference between the *WRTH* and other available information sources was that the *WRTH* was based principally on information received from broadcasters rather than on listener reports. Because the *WRTH* assembled in one place the information on all the broadcasters in a given country, the reader could better understand the overall broadcasting scene there. The full names and addresses of the stations also provided listeners with authoritative location information for individual stations. This was of particular value to those interested in places like Spain, Angola, China, and Latin America, where many small private stations were in operation, or Germany, Japan and Indonesia, where political events had led to a complex shortwave scene. The *WRTH* gave structure to corners of the broadcasting world that had never been fully charted.

The second edition of the *WRTH* appeared in May 1948, and the third in November 1948, reflecting an early intention to publish a summer edition and a winter edition each year. This plan proved too ambitious, however. The fourth edition was not published until October 1949, and thereafter the book was published annually toward the end of each calendar year. At least one German (1953) and Spanish (1960) edition are known to have been issued.[30] The *WRTH* continues to be published every year.[31]

In the U.S., people started learning about the *WRTH* in 1948. At the start it had to be purchased direct from the publisher in Denmark at a cost of 14 International Reply Coupons (which cost 9 cents each at the time). Soon it was available for $1.25 from the *WRTH* U.S. agent, Ben E. Wilbur of East Orange, New Jersey (later Washington, D.C., and still later Livingston, New Jersey). It became available from Gilfer Associates in 1955. The book received rave reviews. Ken Boord called it comprehensive and highly accurate, and one of the best international radio guides he had ever seen. The list price of the *WRTH* increased over the years: $2 in 1955, $8.95 in 1975, $19.50 in 1985, $29.95 in 2005.

Much of the content that is now associated with the *WRTH* was added during the book's first ten years. These include lists of clubs, club representatives and correspondents, standard frequency and time stations, and a list of broadcasts in the international language, Esperanto (all in 1948); a world time table, and the population and number of listeners in each country (1949–50); verification information and stations' future plans (1950–51); a callsign allocation chart and propagation information (1952); FM and television stations (one page each) and late news and corrections (1953); a broadcast frequency allocation chart (1954); brief information on language lessons by radio, a list of receivers by country, and a list of advertisers (1955); and the principal language(s) spoken in each country, and a DX program list (1957).

Medium wave information for areas outside Europe, North Africa and the Near East did not start appearing in the *WRTH* until 1950–51 when the editor felt able to reliably verify their operations. The list of longwave and medium wave stations was expanded to include the world beyond Europe, North Africa and the Near East beginning with the 1967 *WRTH Summer Supplement*.

To keep readers current on summer broadcasting schedule changes that were adopted by many stations, from 1959 to 1971 the *WRTH* published the *WRTH Summer Supplement*

WORLD RADIO
BULLETIN

blisher: O. Lund-Johansen Ltd. - World Publications - Published fortnightly
1, Lindorffsallé, Hellerup, Denmark, Telephone Helrup 3808

Editor: Torsten Ingelsson - All Times GMT - All Frequencies (KHZ)

No 1083 July 15, 1971

(Reproduction either in full or in part without prior
approval of the publisher is not allowed)

<u>LATIN AMERICAN REPORT</u> by Richard E Wood, U.S.A.

<u>ARGENTINA.</u> Radio Nacional, Mendoza, 6180 kHz, noted
with news 1030, local identification and music 1045.
Numerous local identifications, and all own programming, not parallel Radio Nacional, Buenos Aires,
then heard on 6060 kHz. Remarkable strength. Do not
confuse Radio Nacional, Colombia, at present inactive.
WRTVH listed sign-on time of 1600 is wrong.

<u>BOLIVIA.</u> Radio Progreso, La Paz, noted on 6005 kHz
with solid signals 0900-1100. All in Aymará with
rare Spanish words. Emisoras Pio XII, Siglo Veinte,
recommends mailing address Casilla 434, Oruro. Admits frequency drift from nominal 5955 kHz to current 5961.5 kHz. Noted daily 0000-0300 and 0930-
1100. Comments that numerous reception reports have
been received from Sweden.

<u>CHILE.</u> Acc. to a news item first heard over Radio
Victoria, Lima, Peru, 6020 kHz, Radio Presidente
Balmaceda, Santiago, 1300, 5975 and 9590 kHz, was
closed by the Chilean government because of alle-
ged excessive reporting of unrest and strong government activity at the time of the assassination of
former Inetrior Minister Pérez Suchovic. It is,
suggested that DX'ers obtain verifications from all

Above and following two pages: For those requiring the most current information, the *World Radio Handbook* offered a periodic supplement —first the *World Radio Bulletin,* and later the *WRTH Newsletter* and *WRTH Downlink.*

WORLD RADIO TV HANDBOOK

 ## NEWSLETTER

1983
WORLD
COMMUNICATIONS
YEAR

Editor-in-Chief:
J.M. Frost
Søliljevej 44 (P.O. Box 88)
DK-2650 Hvidovre, Denmark

1983
WORLD
COMMUNICATIONS
YEAR

November 1983

Period: 0100 UTC 6 November 1983 to 0100 UTC 4 March 1984

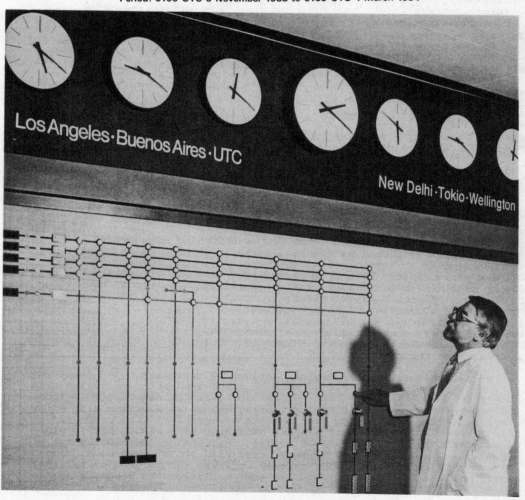

WRTH DOWNLINK

Winter 1990 THE WORLD RADIO TV HANDBOOK NEWSLETTER

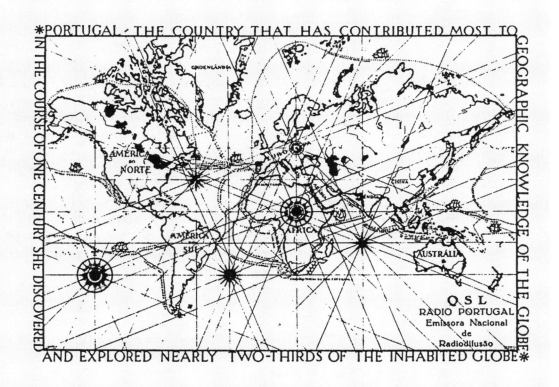

In this issue:

Gulf Crisis Update
Winter Schedules of the International Broadcasters
Broadcasts in Esperanto
Shortwave Table Updater for the Tropical Bands

(called the *Summer Edition* from 1967 to 1971). For those who wanted even more current information, the *World Radio Handbook Bulletin* (later called the *World Radio Bulletin*) was published every two weeks from 1952 to 1976.[32] In 1976 it became the thrice-yearly *WRTH Newsletter,* renamed *WRTH Downlink* in 1988 and published four times in 1988 and 1989, three in 1990 and 1991, the last year of its publication. The *WRTH Latin American Newsletter* was published until 1994. Edited by Tor-Henrik Ekblom, then of Finland, it was provided to DX clubs and *WRTH* monitors on a limited-circulation basis.[33]

A by-time listing of broadcasts in English to various parts of the world was included from 1967 to 1998. (A by-country English-language broadcast list was included for one year only, in 1997.) Some other lists were tried for various periods, including a list of foreign service English news programs (1962 through 1974), a list of language lessons broadcast over the air (1967 and 1968), a by-language listing of stations that broadcast in the various languages of the world (1967 and 1968 summer supplements), and several FM lists (1967 summer supplement). In the May 2006 electronic supplement to the 2006 *WRTH,* by-time lists of international broadcasts in English, French, German, Portuguese and Spanish were included, and these were continued in subsequent *WRTHs.* Save for some European offshore medium wave pirates in the 1960s and 1970s and a few miscellaneous stations, the *WRTH* did not cover pirate broadcasting.

The maps and photographs improved over the years. A world time map was added in 1963. Advertising increased greatly. Station FAX numbers, e-mail addresses and URLs were shown as they came into general use. Responsibility for the annual prediction of shortwave conditions, begun in 1958 by BBC engineer T. W. Bennington and presented in 1963 by RFE's Radio Frequency and Propagation Manager, Stanley Leinwoll, was assumed by George Jacobs, Chief of the VOA's Frequency Division, in 1964, and continues as a regular *WRTH* feature.

Satellite broadcasts were covered during the years 1990 to 1995. (A separate *WRTH Satellite Broadcasting Guide* was published from 1994 to 1996.) To cater to program listeners, an unattributed article, "Listening Tips—Some Tips on Listening to International Radio Programs," was included in 1996. The following year this became a detailed table authored by John Figliozzi under the name "Program Tips—News & Current Affairs on Shortwave 1997." It did not appear in subsequent years, however.

Some items of information were dropped. The individual wavelength equivalents to the frequencies shown in the "Shortwave Stations of the World" list were replaced by a conversion table in 1974; and to the chagrin of many DXers, the musical scale depictions of interval signals were dropped in 1990. DXers were similarly unhappy in 1997 when the international shortwave broadcasters were placed in a separate section from their domestic counterparts, and in 2000 when the domestic part of the book was rearranged solely by country rather than by country within continents. The 1997 change was instructive: the international broadcast country listings occupied 50 pages, while the listings devoted to AM, FM, and the comparatively few domestic shortwave broadcasters, occupied 300.

The *WRTH* had included some articles during its first 20 years of publication. They dealt mainly with propagation, the international broadcasting industry, and the future of radio communications. In the latter part of the 1960s some DX-related articles began appearing. In 1976, following the demise of the separate publication, *How to Listen to the World,* a new features section, "Listen to the World," was added. It appeared (not always under that name) through 1995, although feature articles continued to be a staple in the *WRTH* thereafter. *Listen to the World* contained articles on stations, equipment, propagation, clandestine broadcasting, etc. A popular "Listen to the World" feature were the receiver reviews by Larry Magne which appeared in 1978 and then annually from 1980 to 1986. (Magne created *Radio Database International*, later *Passport to World Band Radio*, in 1984.) Receiver reviews continue to be published in the *WRTH* each year. From 1987 through 1997 they were usually attributed to one or more of Jonathan Marks, Willem Bos, and Thomas R. Sundstrom, and thereafter to *WRTH* Technical Editor John Nelson.

The "Listen to the World" section also contained the announcement of the *WRTH* annual Industry Awards which were made from 1989 to 1998, and which acknowledged industry excellence, mainly in receivers and software.

O. Lund Johansen had been assisted in his editorial duties by journalist Jens M. Frost since 1949. Frost assumed the *WRTH* editorship in 1964 when Johansen retired.[34] From 1978 he was assisted by Andrew G. Sennitt, who had come to the *WRTH* from the BBC Monitoring Service. Although the *WRTH* was bought by Billboard Publications, Inc., around 1967, it continued to operate autonomously, with Frost remaining the publication's face to the shortwave world.

Circulation when Frost took over, 18,000, would more than double by 1977 and grow to 66,000 in 1981.[35] Although more people were reading the *WRTH* in the 1980s — in 1982 it credited a "50 percent increase in readership over the past three years" to a new shortwave market resulting from a "new generation" of shortwave receivers[36] — the process of assembling the book was becoming more difficult, with greater reliance being placed on DXers, who served as contributing editors and assumed responsibility for preparing various parts of the book, and as much reliance on monitoring as on official sources. Frost and Sennitt described the situation in 1984 thus:

> When the WRTH was first conceived in the late 1940s, broadcasters were wildly enthusiastic about it, and it was considered essential to supply the Editor with accurate information about their operations. Precise details of powers and frequencies were willingly supplied, even though such details were not always easy to verify with the equipment then available. ¶In 1984, things have changed. Many of the broadcasters, even the big ones, have long since stopped making the effort to assist us — some can not even be bothered to send printed information which is available in any case! So, we are dependent to an increasing extent on the collaborators and monitors who send us the "missing" information. Even some of the official details in this edition came to us indirectly via a collaborator.[37]

Sennitt became *WRTH* editor upon Frost's retirement in 1987, the same year that editorial offices moved from Denmark to the Billboard building in Amsterdam and the preparation of the *WRTH* entered the computer age. (Frost died on October 18, 1999.[38]) By this time the readership of the *WRTH* was changing. In 1989 Sennitt observed:

> The WRTH owes much of its success to DXers. They are still around, and still making a major contribution to the Handbook through their dedication to accuracy and their constant pursuit of new information. ¶But now we also have another category of reader, whose needs and interests are quite distinct from those of the DXer. This listener is more interested in the message than the medium. He or she may be using a state-of-the-art digital receiver, but is quite happy to listen to news and information in his or her native language. This is certainly not DXing, but is much more akin to the normal everyday process of tuning in to one's local station. The broadcasters are delighted, and have been motivated to serve these listeners more efficiently. Hence the sudden rush of agreements to use transmission facilities of other stations closer to the target audience.[39]

In 1991 he said:

> International broadcasting has entered a new era. In many cases where shortwave used to be the only viable medium, signals of domestic audio quality can now reach the listener through satellite distribution. More and more countries are taking these services and rebroadcasting them locally on mediumwave or FM. ¶Some of those for whom shortwave radio has been a way of life are finding it hard to accept the new reality. Operations which in former times would have used shortwave as a matter of course — such as AFRTS broadcasts to American Forces on active service — no longer use it on a regular basis.[40]

World Radio TV Handbook circulation in 1991, a year in which the Gulf War led to heightened interest in shortwave, is said to have been some 73,000 copies.[41] The *WRTH* continued to grow in size over the years. The decrease in the number of shortwave stations was more than offset by AM and FM stations and the increasing number of international shortwave relay arrangements. There was much information to report. The *WRTH's* coverage of television was expanded as that medium grew throughout the world. The level of detail for each country's listings increased, and the book benefited from more use of color and graphics, and a generally more professional appearance. All of these changes notwithstanding, however, a 1947 user would still be at home with today's *WRTH*.

A major change in the *WRTH* occurred in 1998 when Billboard sold the publication to WRTH Publications Ltd., a U.K. company set up specifically to edit and publish the *WRTH*. (Billboard remained a co-edition partner.) The editorial baton was passed to David Bobbett, a ham operator and British telecommunications journalist who relocated *WRTH* headquarters to Milton Keynes, England. Nicholas Hardyman handled the business side. The transition was somewhat rocky, for while the *WRTH* continued to appear on schedule, regular users quickly spotted deficiencies in the new team's first (1999) edition — omissions and errors of various kinds, misspellings, etc. Bobbett acknowledged the problems and set about to make them right, largely avoiding what could have been a major drop in the *WRTH's* main stock in trade — its credibility. In terms of content, the by-time table of broadcasts in English from international stations returned in 2000 after a one-year hiatus, now including German and Spanish and extended to French in 2001 and Portuguese in 2003. Of greater interest to DXers, the *WRTH* began including information on clandestine broadcasts.

Editorial responsibilities changed again in 2003 when Bobbett was succeeded by *WRTH* Assistant Editor Sean Gilbert and Hardyman became more involved in the editing of the book. Hardyman took on much of the national section and Gilbert handled the international section and worked with contributing editors.[42] A major advancement that year was the availability of free periodic updates to the *WRTH* in the form of downloadable .pdf files. This was a first for the *WRTH*, and a development which DXers welcomed, periodic updates to the *WRTH* not having been available since 1991. Three *WRTH* .pdf updates were made available in 2007, one being 110 pages long. A table of DRM broadcasts was added in 2005, and in 2006 the "WRTH Monitor" was begun wherein *WRTH* posted individual shortwave news update items on their website.

The appearance of the *WRTH* changed shortwave listening in a fundamental way. For the first time, a listener could obtain basic station schedules and other shortwave tuning information easily, without the need to join clubs, follow magazine columns, etc. At the same time, the *WRTH* facilitated growth in the number of people who would move from casual listening to DXing. Although the *WRTH's* present annual circulation of some 30,000 reflects the decline in shortwave listening over the years, even now no dedicated listener would be without a current copy.[43]

Radio Database International and Passport to World Band Radio (1984 to the Present)

Radio Database International (RDI) was the creation of Philadelphia DXer Lawrence E. Magne. Magne, who has a background in computers and finance, had been providing technical assistance in international broadcasting under the name International Broadcasting Services, Ltd. (IBS). Perhaps of more importance to *RDI's* eventual success was the wide respect within which Magne was held within the DX community for his pioneering work in tracking clandestine broadcasters. His 10-page list, *Broadcasting Stations of Exile, Intelligence, Liberation and Revolutionary Organizations,* issued in 1971, was based on the monitoring of a number of experienced DXers and was one of the first compilations of its kind. He was also highly regarded for his shortwave receiver reviews which had been published in the *WRTH* since 1978.

Magne wanted to provide something not previously available to the ordinary listener: a graphic representation of shortwave band usage similar to that used by stations and professional monitoring organizations. In basic design, *Radio Database International* was essentially a bar graph, with time along the X axis, stations and frequencies on the Y axis, and horizontal bars showing the hours when each station used each frequency. Through a combination of notes and formatting, *RDI* was able to display power output, the major languages of the broadcasts, fade-in and fade-out times, less-than-daily operation, channels that suffered from jamming, and similar details.

RDI addressed a major problem in shortwave publishing—the general unavailability of advance schedule information from the stations themselves, and the unreliability of much that was available—by relying on information from a worldwide network of *RDI* monitors. In order to facilitate the preservation and updating of the information, the *RDI* charts were prepared using specially written computer software. This was one of the first major uses of then-new computer technology to manage frequency data and make it available to the shortwave consumer in an attractive and useable form. Resolving the myriad technical issues that arose was no small feat.

RDI was the first major shortwave broadcast reference work to be published since the *WRTH* had appeared nearly 40 years earlier. While *RDI* had been under development there had been much conceptual, business and legal discussion between IBS and potential collaborators and competitors, some of it ending in conflict. Late in 1983, Gilfer Associates announced that it would be the exclusive distributor of *RDI*. And at one point Jens Frost, the *WRTH's* editor, expected to distribute it in Europe. However, with the assistance of his two chief co-architects, Tony Jones of Paraguay and John Campbell of the U.K., the final decision was made that *RDI* would be distributed by IBS itself.

The initial plan was for a Tropical Bands edition (2–5.73 MHz.) to be issued annually, and an International Broadcasting edition (5.73–26.1 MHz.) to be issued every six months. The first Tropical Bands edition was issued in January 1984, followed by a second Tropical Bands edition and an International Broadcasting edition in 1985. The two were

Today's *Passport to World Band Radio* was originally called *Radio Database International.*

combined into a single 1987 volume which was released around September 1986. Thereafter, publication has been in the form of one volume issued annually.

Although the heart of *RDI* was the graphs, or "blue pages" as they came to be known, Magne wanted to utilize *RDI* to facilitate the expansion of the shortwave listening audience, a goal he pursued through both *RDI's* name and its content. In order to de-emphasize the technical side of shortwave listening, wherever possible he replaced "shortwave" with the more user friendly "world band," a term first used by Sony in its 1981 *Directory of World Band Radio*. *Radio Database International* changed its name to *Passport to World Band Radio* in 1988, by which time the graphs were being supplemented with much additional consumer-oriented material, the most important being the "Buyer's Guide to World Band Radio," consisting of receiver reviews that included both descriptive information and performance ratings. How-to articles helped demystify shortwave listening for the novice. Also added were features written by both professionals and hobbyists, a by-country listing of the languages and frequencies utilized by the major shortwave broadcasters (those using at least 50 kw.), a lexicon of terms, and much glossy advertising.

As editor in chief, Magne preserved this basic *Passport* design, improving upon it incrementally over the years. Paper quality, layout and printing improved, and an abundance of photos and other graphics were added. Soon *Passport* was the best looking of the professional shortwave publications. By 1990 it contained 384 pages; the 2008 edition contained 562.

More important than looks, however, was content, and at this Magne was a master. The articles, whose authors were usually recognized experts, improved in depth and number. A "Getting Started" section helped attract readers who were unfamiliar with shortwave, and much attention was paid to the easier-to-hear stations and to shortwave programming—news, sports, music, the "best shows." A detailed, hour-by-hour program review, "What's on Tonight," was added, along with a guide to English broadcasts. In 1992, *Passport* featured "Voices From Home," a parallel guide to foreign-language broadcasts. *Passport* also addressed how shortwave could serve travelers and help listeners understand world trouble spots. In recent years the book has featured many station and area profiles, plus articles on radio history.

Magne knew how to make "world band" attractive to those not technically inclined, a task in which he was aided by the proliferation of better-performing receivers at all levels. The *Passport* receiver reviews expanded in depth and number, and among both the technically adept and the technically challenged, *Passport* became recognized as the best one-stop source for detailed, objective information on shortwave receivers, this notwithstanding the considerable receiver advertising which *Passport* enjoyed. It also adopted a useful multi-level structure for classifying shortwave receivers ("pocket," "compact," "portatop," etc.), and often re-tested receivers to identify improvements in operation and performance. *Passport* has also addressed such specialized receiver topics as radios for emergencies, auto radios, radios with built-in recorders, and kits.

Although Magne was serving a wider audience, DXers were not forgotten. Basic station contact information was added in 1991, including clandestine stations in addition to mainstream broadcasters. "Addresses Plus," as this section was called, became increasingly authoritative. In addition, Magne added antenna reviews and information on collateral products such as atlases, clocks, and recorders, and included articles about the internet and changes in broadcasting technology.

Passport has taken its place alongside the *WRTH* as one of the most valuable short-wave references. A persistent issue, how to deal with seasonal frequency changes in a single annual publication, was highlighted in 2003 when the *WRTH* began making updates to its publication freely available on the internet. *Passport* explained its approach:

> To be as useful as possible over the months to come, *Passport's* schedules consist not just of observed activity, but also that which we have creatively opined will take place during the forthcoming year. This predictive material is based on decades of experience and is original from us. Although inherently not as exact as real-time data, over the years it's been of tangible value to *Passport* readers.[44]

While the details of "creative opining" have not been revealed, the *Passport* data has been surprisingly valid all year round, especially in the case of home service shortwave broadcasters where changes are less frequent. Much of this is thanks to the experience of Magne's two principal collaborators, *Passport* Editor Tony Jones, who is in charge of time and frequency information, and Assistant Editor Craig Tyson of Australia, who is in charge of "Addresses Plus."

There have been a number of other *Passport* projects over the years. In 1988, IBS started publishing "RDI White Papers," detailed laboratory and user evaluations of the specifications, ergonomics, and performance of individual shortwave receivers. There have also been white papers covering antennas and the interpretation of lab tests and measurements, and in 1997 a summer "Addresses Plus" update was released as a white paper. IBS also published *Passport to Web Radio* in 1997 and 1998. The IBS website contains post-publication updates to the receiver reviews and information for newcomers to shortwave and *Passport.*

KLINGENFUSS *SHORTWAVE FREQUENCY GUIDE* AND *SUPER FREQUENCY LIST* (1997 TO THE PRESENT)

Joerg Klingenfuss of Tuebingen, Germany was well-known among utility DXers as the publisher of utility references as far back as 1970. One of his main publications was the annual *Guide to Utility Radio Stations.* A CD version of the utility guide called the *Super Frequency List* became available in 1995, and was expanded in 1997 to include shortwave broadcast stations (international, domestic and clandestine) in addition to its usual utility content. Klingenfuss continued to make the *Guide to Utility Radio Stations* available annually in a print version, and in 1997 added an annual print publication that included both the SWBC material and most of the utility material on the CD. This was called the *Shortwave Frequency Guide.* The first edition contained 11,500 broadcast entries and 13,800 utility entries, and was over 500 pages long.

The *Shortwave Frequency Guide* is divided into utility and SWBC parts. In the SWBC section, the stations are organized in both a by-frequency section and an alphabetical section. Each includes information on station name, country and transmitter site, frequency, time on and time off, target area, language and remarks. Clandestine stations are contained in a small separate section. The *Super Frequency List* is searchable by a variety of parameters, including, for broadcast stations, station name, country, frequency, language, and broadcast start time, as well as by any word in the database.

The Klingenfuss shortwave broadcast publications never gained the same acceptance in the American shortwave community as did the *WRTH* and *Passport.* Unlike those publications, the *Shortwave Frequency Guide* and *Super Frequency List* contain no station contact information, receiver reviews or feature articles. Although not updated during the year, Klingenfuss emphasizes that his publications are based on actual monitoring. However, this is no less true of *Passport* and the *WRTH*.

THE SHORTWAVE GUIDE (2002–2003)

Published by WRTH Publications Ltd., parent organization of the *WRTH*, *The Shortwave Guide*, with a band survey format similar to *Passport's*, appeared to have the potential of a worthy competitor to that publication. Its time-and-frequency tables showed the hours when each station operated, the transmitter power, and the target area. In addition, the language of each broadcast was depicted using an attractive color-coded bar system. *The Shortwave Guide* also contained station contact information, an introductory essay on how and why to listen to shortwave, and a list of clubs. Added in the second edition was information on standard frequency and time stations, internet resources, and DX programs.

Only two issues of *The Shortwave Guide* appeared, however — in June 2002 and June 2003.[45] At $17.95, some may have considered the price of the 200-plus page book high, considering that, for only slightly more, *Passport* offered similar content plus much other material.

Other Station Lists

A list published by Britain's *Wireless World* magazine was the *Guide to Broadcasting Stations*.[46] First appearing in 1946 as *Broadcasting Stations of the World*, the name was changed in the second edition (perhaps to avoid confusion with the publication of the same name issued by the FCC's Foreign Broadcast Intelligence Service in 1945). In its early editions, the *Guide* was a 64-page booklet containing a list of European longwave and medium wave stations and worldwide shortwave stations, showing the frequency, wavelength, and power of each station, with the data arranged both geographically and by frequency. The information was originally supplied by the BBC's Tatsfield receiving station, and was later supplemented by information from other sources. (Tatsfield was a facility separate from the BBC Monitoring Service and was devoted mainly to frequency measurements, identification of interference, and relays of other broadcasters. Its operations were transferred to BBC Monitoring in 1974.)

By the time the *Guide* became known in the United States around 1970 it was a 160-page book. Information on the seasonal usage of each frequency had been added, together with some introductory material about equipment, propagation, station identification, and reception reports. In the 1980s these items were expanded, and supplemented with information about clubs, DX programs and other topics. By the 19th edition in 1987 the *Guide*, then edited by Philip Darrington, was almost 250 pages long. The last edition, the twenty-first, called *Guide to Broadcasting Stations: 21st Century Edition* and edited by Simon Spanswick, was published in 2001.

The absence of broadcast times, languages, station addresses, etc. made the *Guide* of limited practical use, particularly with the *World Radio TV Handbook* readily available, and the *Guide* never developed a significant following in the U.S.

A well-known utility frequency reference was Bob Grove's *Shortwave Directory,* first published by *Monitoring Times* in 1982 and followed by eight subsequent editions (the first edition was called the *Shortwave Frequency Directory*).[47] A major update took place in 1988 when there was added a list of out-of-band feeder frequencies for international broadcasters and a list of the broadcasters' most commonly used in-band frequencies, plus some information on pirate and clandestine stations. This was principally an accommodation to utility listeners, however; *Shortwave Directory* was never intended to be a major resource for SWBC listeners. The last edition, the ninth, published in 2003 (on CD), contained the time-and-frequency guide published monthly in *MT,* and an *MT* article describing the SWBC bands.

Fred Osterman of Universal Radio published the *Shortwave Log* in 1983, with a second edition in 1984.[48] It consisted of Osterman's personal log of stations heard over a ten-year period. Though mainly a utility reference, it included some SWBC material as well. Logs were listed in multiple formats—by frequency, country and city, and mode and time. It also included a "spectrum analysis," illustrating graphically the various types of radio services occupying each channel at various times of the day. The *Shortwave Log* was one of the earliest examples of the power of a computer to aggregate and display data in new ways so as to yield information not otherwise apparent from the individual pieces of data in the database.

Sometimes radio manufacturers packaged their own station lists with their receivers, or sold the lists separately. The best known is Sony's *Wave Handbook,* a well-designed, 100-plus page booklet which was produced in various dated and undated versions during most of the 1980s and 1990s.[49] For the major shortwave stations it typically showed broadcasting times, frequencies, and languages, station address and identification text, all arranged geographically using a band survey format similar to that of *Passport.* Other content varied by version but included such things as lists of North American AM and FM stations, worldwide AM and FM lists, and shortwave tables of frequency and language usage.

At least as ambitious as the *Wave Handbook* was the Philips *Short Wave Handbook,* which was of similar design but limited to shortwave. Among the smaller, pamphlet-style lists of varying worth have been the *RCA Short-Wave Station List* (1950) and *RCA International Short Wave Chart* (1953 and 1954); Telefunken's *Invitation to Better Reception* (1955); Hammarlund's *Where to Find Them* list (reprinted from *Popular Electronics,* 1957); General Electric's *Short Wave Guide* (1950s), *Short Wave and Marine Band Guide* (1950s), *Shortwave Listener's Log & Guide to Shortwave Radio* (1964), and *Around the World Listening Log Book* (reprinted from *Electronics Illustrated,* 1968); the Hallicrafters *Guide to Short Wave Listening* (1961); and Grundig's *Passport to the World of Shortwave Broadcasting Stations* (c. 1989) and *Shortwave Listening Guide* (c. 1991).

In 1991, a company called FB Enterprises in Vancouver, Washington published "Quick-N-Easy Shortwave Listening," a series of laminated cards containing a sampler of various stations and frequencies. Each card covered a different time period. The list had a west coast orientation, and was useful mainly for newcomers.

The *North American Shortwave Frequency Guide* by James D. Pickard appeared in 1992 and was updated in 1994 and 1995.[50] Although this list, a by-frequency mix of utility and shortwave broadcast stations, was of some value to the novice band scanner, it was selective in its contents, omitting nearly all broadcasters in the 60 and 90 meter bands and focusing on the more powerful stations in the higher ranges.

Books and Monographs

Many books and monographs have been written about shortwave listening.

NOTEWORTHY SINGLE- OR MULTIPLE-AUTHOR BOOKS ON SHORTWAVE LISTENING

The World at a Twirl **(1956)**   Although much had been written about shortwave listening in magazines and various standalone publications during the 1930s and 1940s, in the United States there was no "book" about shortwave until 1956 when Ken Boord published *The World at a Twirl*.[51] Boord was between editorial posts at the time; he had left *Radio & Television News* and *Popular Electronics* in 1955, and *DXing Horizons* was still in the future (1960).

"*WT,*" as this first of several Ken Boord "World at a Twirl" shortwave projects came to be known, was 137 pages long, soft cover, and self-published. Although dated Summer 1956, it was actually published late that year. It could be purchased from Boord for $2.50 postpaid anywhere in the world (surface mail). Plans for subsequent editions never came to fruition.

The book, which was dedicated to well-known San Francisco DXer August Balbi, consisted of three parts. In Part I were dozens of photos (and detailed descriptions) of DXers, stations, station personalities, and QSLs. Part I also contained station profiles and "how to" featurettes on such things as clubs, the ham bands, the 24-hour system, hobby ethics, propagation and the like. Part II was a detailed, by-country station log of the times and frequencies of the principal stations in all countries, along with helpful notes on what to listen for and an hour-by-hour table of English news broadcasts. Part III featured more photos, and "Press-Time Flashes."

Although the cover and title page of *WT* showed the book to have been "compiled" by Ken Boord, he was much more than a compiler. *WT* was a strong reflection of Boord's personality and writing style, and his ability to blend information of interest to both newcomers and experienced listeners. *WT* evidenced the passion that Boord brought to long-distance radio listening, and the important clearinghouse function that he performed. In the 1940s and 1950s, there was no one more deeply involved in shortwave listening in the United States than Ken Boord.

Better Shortwave Reception **(1957)**   As one of the first commercial books about shortwave listening, *Better Shortwave Reception* found its way onto many hobbyists' bookshelves when it was first published in 1957, and many years thereafter.[52] The book's author, William I. Orr, was a well-known ham operator (W6SAI), and while the book was not limited to ham radio, it did have more of a technical bent than would interest the typical SWBC listener. In the book's 141 pages the author covered propagation, buying a receiver, antennas, the shortwave bands, QSLing, and the like. Nearly half the book was devoted to the innards of receivers, aligning receivers, and building equipment. Subsequent editions of *Better Shortwave Reception* were co-authored with Stuart D. Cowan, W2LX, and were published into the 1980s. Save for one chapter modestly updating some topics, the original text, and nearly all the photos, remained unchanged for decades.

Short Wave Listening (1966)   Though published in the Netherlands, this volume became fairly well-known in the United States.[53] The author, Jim Vastenhoud, Deputy Director of Engineering at Radio Netherlands in the late 1970s, is best-remembered as the co-host of the "DX Juke Box" program broadcast over the station from approximately 1961 to 1981. He was also the editor of (and a contributor to) the *World DX Guide,* an anthology published in 1978 which is sometimes thought of as the last of the *How to Listen to the World* series.

Most of *Short Wave Listening* treats such semi-technical topics as the principles of shortwave transmission, propagation, interference, the differences among the shortwave bands, receivers, antennas, tape recorders, and frequency measurement. Covered in less depth are reception reports, the Q code, and clubs.

Shortwave Voices of the World (1969)   As indicated in its preface, this book approaches the subject from the perspective of actual listening experience, of which the author, Richard E. Wood, had a great deal, for he had been among the most active DXers on shortwave and medium wave since his start in DXing in 1957. Written in 1969 and nicely illustrated, the book describes the various kinds of stations, the divisions of the frequency spectrum, languages and station identification (Wood was a language professor), jamming, propagation, and QSL collecting.[54] Among its features were the discussion of how broadcasting is organized around the world, and the interesting thumbnail sketches of various stations.

The Complete Shortwave Listener's Handbook (1974)   As its title correctly implies, this first "handbook" of shortwave listening was indeed comprehensive in its almost 300 pages.[55] Although the author, Hank Bennett, was a ham (W2PNA), he was better known as the long-time editor of the shortwave broadcasting columns in the *NNRC Bulletin* (1949–1982) and *Popular Electronics* (1955–1970). However, *The Complete Shortwave Listener's Handbook* covered not only shortwave broadcasting but other aspects of long-distance listening as well, including hams, utility stations, longwave, the standard broadcast band, FM (written by Bruce Elving), TV (written by Glenn Hauser), even citizens band. It covered terminology, receivers, antennas, propagation, the various frequency bands, record keeping, reception reports and QSLs (written in part by John Beaver), clubs, etc. The book's greatest strength was a large section covering the shortwave broadcast stations that a listener could expect to hear from each part of the world. It truly was a handbook, for someone new to DX listening could learn from it the basics of just about every aspect of listening.

The Complete Shortwave Listener's Handbook was modestly updated in 1980 in a second edition co-authored by Bennett and Harry L. Helms.[56] A further update in 1986 by David T. Hardy improved the content on receivers, accessories and antennas, and also updated the station information and added some foreign-language reception reports.[57] The fourth edition, prepared by Andrew R. Yoder in 1994, preserved the structure and organization of the book.[58] It updated the station information to reflect then-current broadcasters and the new geopolitical realities in the eastern bloc, changed most of the graphics, and expanded the treatment of a few topics, such as U.S. broadcasters, pirate radio, publications and clubs.

Yoder also authored a fifth edition, issued in 1997. As with the earlier editions, it covered other types of DXing besides shortwave broadcasting.[59] The organization of the book was substantially revised, and while some of the chapters were unchanged, many were rewritten, updated and expanded, most of the graphics were changed, and several new topics were added, including citizens band, use of computers in DXing, and radio-related collecting. The by-country shortwave station information was improved as well. Although much of the book is now outdated, the fifth edition of *The Complete Shortwave Listener's Handbook* remains overall the best single book on shortwave listening since its publication.

DXing According to NASWA **(1975)**   This 104-page book, written for the North American Short Wave Association, has become a hobby classic.[60] Intended as a club handbook, it was a practical, hands-on guide to SWBC listening written by Edward C. Shaw, one of the club's leaders. The cost was $3, and it was available to both club members and others. Four editions were published between 1975 and 1979, and several thousand copies are believed to have been sold.

DXing According to NASWA covered just about everything: shortwave history and the history of the club; receivers, antennas and accessories; publications; propagation and time; writing reception reports; etc. In a section called "Real-Time DXing," the author discussed the propagational characteristics of the various bands (with an emphasis on "The Marvelous 60 Meter Band"), noise and interference, recognizing languages and identifying stations, keeping records, and tape recording reception. The appendix contained maps, darkness pattern charts and great circle charts, information on awards, lists of hobby vendors, club lists, a world time chart, and other useful tools. It was written in a friendly, informal style that matched the author's well-known interest in encouraging newcomers to shortwave listening. Some of it was based on other Shaw writings which had appeared in the NASWA bulletin. A two-hour condensed version of the book was available on cassette.

The World in My Ears **(1979) and** ***Radio Listeners Guide*** **(1988 and 1990)**   No name has been associated with DXing longer than the late Arthur T. Cushen. He began listening from his native New Zealand in 1935, and over the years his accomplishments were heralded far and wide, not the least because he became blind in 1954 after a lifetime of deteriorating sight. He set the pace in many areas of the hobby.

His first book, *The World in My Ears,* written in 1979, is in two parts.[61] Part I is a personal memoir of Cushen's life at the dials. He begins by describing how he got interested in radio, and what the medium wave bands were like in New Zealand in the 1930s and earlier. He goes on to address World War II propaganda broadcasting, POW monitoring in World War II and the Korean War, the role of shortwave in some major news events, and his life after he lost his sight. Part II is a well-written treatise on shortwave listening covering some of the same topics as *Shortwave Voices of the World:* equipment, frequencies, propagation, languages, reception reports, long-distance medium wave listening, stations and personalities, and so forth.

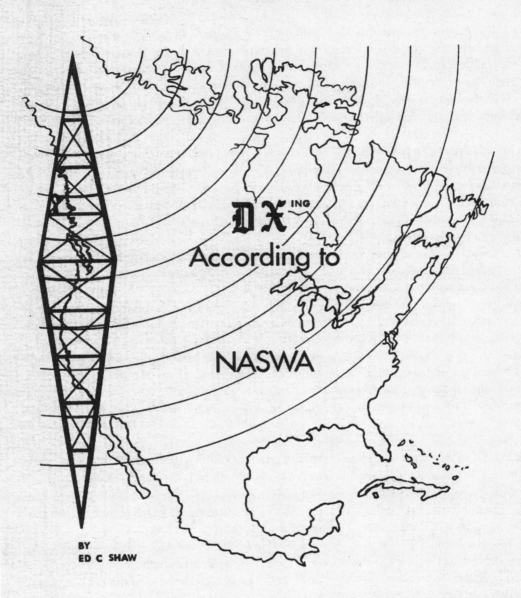

DXing According to NASWA was one of the most successful shortwave club publications ever issued.

The *Radio Listeners Guide,* was first published in 1988 and was based on a series of Cushen articles that had been published in New Zealand's *Electronics Today* magazine.[62] It became better known in the United States when the second edition appeared in 1990. It is a practical exposition of the standard topics of shortwave listening, with many references to the author's personal experiences. Both medium wave and shortwave topics are covered — frequency bands and propagation, international stations, the tropical bands, equipment, languages, world time, programs, clubs, QSLing, and the like. Of special interest in the second edition is an eight-page "supplement," "Secrets of Wartime Listening to Enemy broadcasts," which contains a discussion of World War II propaganda broadcasting relevant to the Pacific area based on the author's monitoring during the years 1941–45.

Arthur Cushen died in 1997. His contributions to the development of long-distance radio listening are memorialized in these books.[63]

Shortwave Listening Handbook (1987) and *Shortwave Listening Guidebook* (1991)

Harry L. Helms is a prolific author of shortwave and electronics books, among them the *Shortwave Listening Handbook*[64] and the *Shortwave Listening Guidebook.*[65]

The *Handbook* was a well-written guide to all aspects of shortwave listening. Although the shortwave coverage focused on shortwave broadcasting, ham radio and utilities were covered as well. The book was strong in the area of receivers, antennas and propagation, and also contained interesting profiles of some of the larger stations, plus sections on domestic shortwave broadcasters, pirates and clandestines. Its depth matched the author's extensive hands-on listening experience.

The *Guidebook* was basically a re-issuance of the *Handbook.* The organization of the *Handbook* was preserved, and while the text was re-edited, some sections were updated and rewritten, and many of the graphics were changed, the *Guidebook* is essentially an updated version of the *Handbook.* The updates in the second edition of the *Guidebook* (1993) were comparatively minor.

OTHER BOOKS ON SHORTWAVE LISTENING

Many introductory books on shortwave appeared over the years. Often these were republished in multiple editions (not all of which are reflected in the following descriptions). Some covered not just shortwave broadcast listening, but medium wave, VHF and UHF, ham radio, utilities, and other related topics.

In the 1960s and 1970s there were Len Buckwalter's *ABC's of Short-Wave Listening* (1962 through 1970),[66] *The Fun of Short-Wave Radio Listening* (1965),[67] and *99 Ways to Improve Your Shortwave Listening* (1977).[68] Charles Woodruff wrote *Short-Wave Listener's Guide* (1964 through at least 1980),[69] and *Questions & Answers on Short-Wave Listening* (1970).[70] Other books were *The Fascinating World of Radio Communications* by Wayne Green (1971),[71] Norman Fallon's *Shortwave Listener's Handbook* (1972 to 1981),[72] Forest H. Belt's *Easy-Guide to Shortwave Listening* (1973),[73] *Introduction to Short-Wave Listening* (1972), published by Radio Shack,[74] *Short Wave Listener's Handbook* (1975)[75] by John Schultz, and *Hear All the Action* (1978).[76] Also in the 1970s, the Deutsche Welle engineering office issued a 43-page, multi-lingual guide to shortwave listening called *The Short-Wave Reception.* (The same guide, with the Grundig name added, was packaged with Grundig shortwave radios.)

In the 1980s there were Sony's *Directory of World Band Radio* (1981 and 1983)[77]; *The Shortwave Listener's Handbook* (1982)[78] by Robert J. Traister; *The World Is Yours on Shortwave Radio* by Sam Alcorn (1984 and later editions through at least 1999)[79]; *The Listener's Handbook* by Bob Grove (1985)[80]; William Barden, Jr.'s *Shortwave Listening Guide* (1987),[81] published by Radio Shack; and two introductory booklets, *Introducing International Radio*[82] (1987) by HCJB broadcasting personality Kenneth D. MacHarg, and *So You Bought a Shortwave Radio!*[83] (1987) by well-known DXer Gerry L. Dexter, whose company, Tiare Publications, was the source of many books on radio and other topics. In 1987, Tiare also published a pamphlet, *How to Log 100 Countries on Shortwave Broadcast,* marketing it alongside the Tiare Century Club which provided a certificate to anyone submitting a log evidencing reception of 100 countries.

The 1990s brought *Bob Grove's Scanner and Shortwave Answer Book* (1990)[84]; *Quick-N-Easy Shortwave Listening* (1993)[85]; *Shortwave Radio Listening for Beginners* (1993)[86] and *The Shortwave Listener's Q & A Book* (1994),[87] both written by Anita Louise McCormick; and *Radio Monitoring— The How-To Guide* (1997) by T. J. "Skip" Arey.[88] A number of stations published introductory pamphlets on shortwave radio, including the 32-page "DX Tips for Beginners" (KNLS, 1992),[89] and "An Introduction to Shortwave Radio," a 24-page pamphlet written by Richard McVicar and published by HCJB in 1995.

Several books addressed shortwave listening for travelers. Intended for immigrants, naturalized citizens, tourists, foreign students, diplomats and businessmen who might not know that their home country could be heard on shortwave, Gerry L. Dexter's *Voices From Home* (1988)[90] provided station schedules and addresses, along with an introduction to shortwave.

A novel approach was taken by the *Traveler's Guide to World Radio* (1990),[91] published by *WRTH* parent Billboard Books. The pocket-size *Traveler's Guide* was a by-city review of the local medium wave and FM stations, and the major English-language international shortwave signals that can be most easily heard in each of 51 cities around the world (BBC, VOA, Deutsche Welle, Radio Australia, etc.). It showed graphically the times and frequencies to check, whether the stations transmitted daily, weekends only, etc. With its survey of portable radios authored by Jonathan Marks of Radio Netherlands, the *Traveler's Guide* was just that — an easily carried primer on "what's on," wherever the user might be. Updated volumes were published from 1993 through 1998.

Although much of Andrew R. Yoder's *Shortwave Listening on the Road — The World Traveler's Guide* (1995)[92] was a by-country review of active stations, it also covered receiver and antenna topics of interest to travelers, plus the computer bulletin board scene at the time.

ANTHOLOGIES

***How to Listen to the World* (1950–1974)**   Early in 1950, *WRTH* editor O. Lund Johansen announced plans to produce a "world listening guide" called *The World Listener* in order to promote listening to foreign broadcasts, especially on shortwave, and give SWLs a chance to share their experiences. Various SWLs would write about their experiences and give practical advice on listening, writing reception reports, etc.[93] The result was *How to Listen to the World,* published later in 1950 at the price of 30 cents.

HTL: The First Round. The "*HTL*" series did not start out as a true anthology. The first five "editions" were basically reprints of the original 32 page pamphlet.[94] These early editions were marked 1st ed., 2d ed., etc., but were undated. However, their advertising indicates that they were published between 1950 and 1953. Half of *HTL* was introductory material about propagation, antennas, and writing reception reports. In the other half, called "World Listeners Write About Their Experiences," ten contributors wrote about such things as tuning, identifying stations, equipment, forming a club, etc.

The sixth edition[95] (1956, 54 pages) preserved the overall approach, but with new contributors and a somewhat harder-core treatment of such topics as receivers, identifying stations, QSLing, medium wave and ham band DXing, listening from New Zealand, language identification, time conversion, and so on.

The seventh edition[96] (1959 and 1960, 70 pages) consisted entirely of listener contributions. The topics were the familiar ones, with some new material on Latin American DXing, TV DXing, and satellites. There was more content than in earlier editions, and *HTL* took on a somewhat more permanent look.

HTL: The Second Round. In the 1962 edition, *HTL* became more professional in content and appearance. This volume, along with the three that followed through 1967, contained no edition numbers.[97] However, the next volume, issued in 1969–70, was denominated the fifth edition, suggesting that the 1962 *HTL* and the three that followed it were the first four in a "second round" of HTL. There were three further numbered editions after the fifth, for a total of eight volumes in this second round.[98] Volumes were usually in the range of 150 to 200 pages.

The topics in *HTL* included propagation, receivers and antennas, organizing a listening post, listening to particular bands or particular parts of the world, identifying languages and learning languages by radio, sending reception reports, listening to pirate and clandestine stations, and so forth. There were also station profiles, descriptions of the listening experiences of various DXers, and information on forming clubs. *HTL* contained some "house" articles that were not attributed to particular authors. However, most of the *HTL* content was authored by well-known listeners, shortwave broadcasting personalities, and professionals. There was also advertising, principally from stations and equipment manufacturers.

Although the overall focus was on shortwave broadcast listening, there was some coverage of amateur, medium wave, utility, VHF, satellite, and TV DXing as well. There were also some technical articles, including a few about building equipment, although DXing had long left its radio construction origins behind. *HTL* content gained depth and variety year by year.

The eighth edition (1974) was produced in association with the BBC and contained many articles on various aspects of BBC operations. Although it was the last edition of *HTL* that was published, the 1976 *WRTH* contained an 86-page section called "Listen to the World" that was a mini–*HTL,* with articles of the same type, authored by some of the *HTL* regulars and a few others.

HTL: World DX Guide. Although not part of the *How to Listen* series, the *World DX Guide,* published in 1978, was an *HTL* surrogate and is sometimes considered the last edition of the *How to Listen* series.[99] It was published by the *WRTH* publisher, Billboard, and was arranged and compiled by well-known Dutch DXer and Radio Netherlands personal-

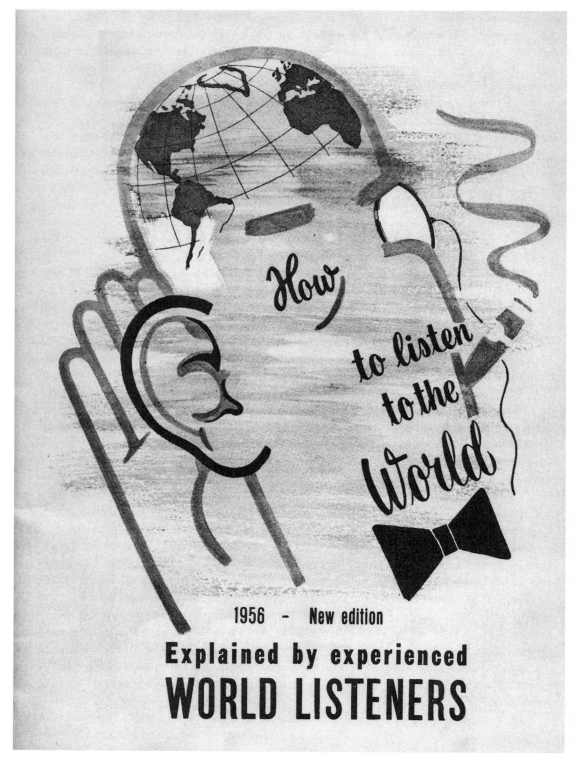

How to Listen to the World began in 1950 as a pamphlet-sized collection of articles about DXing.

HOW TO LISTEN TO THE WORLD

HOW TO HEAR THE WORLD
SHORT AND MEDIUM WAVES
INTER-CONTINENTAL TELEVISION

By 1969–70, *How to Listen to the World* had grown to over 200 pages.

ity Jim Vastenhoud and edited by *HTL* and *WRTH* editor Jens Frost. Although somewhat more technically oriented than *HTL*, it was in the *HTL* genre. Much of the material was written either by Vastenhoud himself or by former *HTL* authors. Although denominated a "1st Edition," this was the only *World DX Guide* that ever appeared.

***The Shortwave Book* (1984)**   Although only 54 pages long, the content of *The Shortwave Book*[100] was broad. It covered basic shortwave listening, choosing a receiver, how a shortwave station operates, jamming, audience research, the new field of computers and radio, and pirate radio. It also offered profiles of several stations. Most of the dozen or so contributors were well-known DXers or shortwave radio professionals.

***Shortwave Radio Listening with the Experts* (1986)**   In the preface to this volume, the editor, Gerry L. Dexter, described how shortwave listening had changed over the years.

> Thirty years ago, Indonesia was an uncharted land from a DX standpoint. Music heard on a Venezuelan station was described as "Latin American," all Chinese stations were "Radio Peking," all Russian stations were "Radio Moscow," and the use of single-sideband reception techniques to clear up a signal transmitted in amplitude modulation was unheard of. "Grayline" might have been thought to be a transportation company, and "krongkong" considered as a monster in some grade-D Japanese horror film. ¶ ...Read a shortwave club bulletin from the 1950s and compare it to one from the 1980s.... [T]he remarkable technical advances made in equipment over the past several years grow pale in comparison to the advances made in the general knowledge of shortwave listening and DXing.[101]

The point was well-taken, for by the mid–1980s the DX knowledge base on many topics—languages, national music, propagation, tuning techniques, and the like—had greatly expanded. To preserve these advances in knowledge, Dexter asked some of the leading figures in the shortwave community to expound on their areas of expertise. The result was *Shortwave Radio Listening with the Experts*, probably the best single compilation of original articles on DXing that had yet been assembled.

The book picked up where most of the "handbooks" and the popular books ended. While many of the topics in its 500+ pages were familiar, here they were treated in greater depth. And there were some topics that had not been the subject of widespread attention before—European pirate radio, DXing the USSR, DXing Indonesia, domestic shortwave broadcasting in the Andes, recognizing languages, and computers and DXing, to name a few. Also addressed were non–SWBC subjects like medium wave and FM listening, the VHF-UHF and FM bands, and utility monitoring. The authors were the DX household names of the day—Magne, Wood, Campbell, Herkimer, Lazarus, Helms, Elving, Perry, Legge, Sparks, and others. Although long out of date and out of print, this book remains one of DXing's standard works.

***Proceedings* (1988–1995)**   The same advances that resulted in *Shortwave Radio Listening with the Experts* were the *raison d'etre* of the six volumes of *Proceedings*.[102] Produced under the "Special Publications" umbrella of the *Fine Tuning (FT)* DX newsletter, each volume was a collection of original articles. The idea grew out of a 1987 survey of *FT* members. The content was oriented mainly toward SWBC listening, and all articles were authored by prominent DXers or others with specialized technical knowledge. The senior editors of the first volume were Professor John H. Bryant and Fr. Fritz Mellberg. Subsequently the editorial group expanded to include, at various times, Guy Atkins, Kevin Atkins, Elton Byington and David M. Clark. *Proceedings* was an effort in which many persons played active roles as authors, reviewers and production staff. Bryant served as executive editor. The preparation of the series reflected a concept common in academic circles: peer review. All articles were vetted by an editorial review panel.

Because *Proceedings* catered to hard-core hobbyists, the content was more advanced than in most other publications. The focus was on propagation, receivers, antennas, other equipment, and the use of computers to aid listening. There were equipment reviews and performance comparisons of new and old receivers and peripheral equipment. *Proceedings* also included features of various kinds on such topics as tuning techniques, DXing particular parts of the world (New Guinea, Brazil and Bolivia), DX reference tools, DXpeditioning, shortwave history, and other subjects. The topics for which *Proceedings* will probably be best remembered are antennas and advanced propagation. In both areas, theory and practice were combined to produce groundbreaking findings.

Special Topics

Equipment   Receivers and antennas have long been staple topics in the shortwave press.

Probably the best modern equipment review collections are those that have appeared annually in *Passport to World Band Radio* since the 1985–86 edition and that are now one of the book's major features. As for full-length book treatment of shortwave receivers, the performance of the various models of the day were well covered in Rainer Lichte's two books, *Radio Receiver — Chance or Choice* (1985) and *More Radio Receiver — Chance or Choice* (1987),[103] as well as the *WRTH Equipment Buyers Guide* by Willem Bos and Jonathan Marks (1993).[104] An early publication on the technical side of shortwave receivers was *How to Improve Short Wave Reception,* published by World Publications around 1961.[105]

The subject of communications receivers was addressed exhaustively in *Communications Receivers, The Vacuum Tube Era: 1932–1981* (1997 and three earlier editions) by Raymond S. Moore[106] and *Shortwave Receivers Past and Present — Communications Receivers 1942–1997* (1998 and two earlier editions) by Fred Osterman.[107] Other books on shortwave receivers include *Inside Your Shortwave Radio* (1992) by Ted Benson,[108] *Buying a Used Shortwave Receiver: A Market Guide to Used Shortwave Receivers* (four editions, 1992–1998) by Fred Osterman,[109] and the *CQ Shortwave Listener Handbook.*[110]

Among the many books on antennas have been those by prolific author Edward M. Noll, including *SWL Antenna Construction Projects* (1970),[111] *25 Simple Tropical and M.W. Band Aerials* (1984),[112] *25 Simple Shortwave Broadcast Band Aerials* (1984),[113] *25 Simple Indoor and Window Aerials* (1984),[114] *Easy-Up Antennas for Radio Listeners and Hams* (1988 and later editions),[115] *Shortwave Listener's Guide for Apartment/Condo Dwellers* (1991),[116] and *73 Dipole and Long-Wire Antennas* (1992 and earlier editions).[117]

Others have included *Antennas for Receiving,*[118] *The SWL Antenna Survey* (1989) (a description of the shortwave antennas available commercially at the time)[119]; Traister's *The Shortwave Listener's Antenna Handbook* (1982)[120]; *The Easy Wire Antenna Handbook* (1992)[121]; *The Antenna Handbook: A Guide to Understanding and Designing Antenna Systems* (1993) by *Monitoring Times* columnist W. Clem Small[122]; *The Antenna Factbook* by Bob Grove

(1995)[123]; three books by Frank P. Hughes—*Limited Space Shortwave Antenna Solutions* (1988),[124] *Easy Shortwave Antennas* (1992)[125] and *Long Wire Antennas* (1994)[126]; and three by Joseph J. Carr —*Practical Antenna Handbook* (1989 and subsequent editions),[127] *Joe Carr's Receiving Antenna Handbook* (1993),[128] and *Joe Carr's Loop Antenna Handbook* (1999).[129] (Carr passed away in 2000.) Another antenna book was Andrew R. Yoder's *Build Your Own Shortwave Antennas* (1994).[130] There were many other antenna books written principally for the technical or the amateur communities, and these were sometimes of use to SWLs.

As local interference sources grew over the years, they were the subject of some attention as well, e.g. William R. Nelson's *Interference Handbook* (1981).[131]

Propagation   Much has been written on this topic, but only a few book-length works became well-known in the listening community.

Stanley Leinwoll was associated with VOA and RFE/RL engineering and frequency management for many years, and his *Shortwave Propagation*,[132] published in 1959, was one of the popular books on the subject at the time. Though certainly not the first book about shortwave propagation, and intended more for broadcasting professionals and hams than the shortwave listening community, it addressed the major topics—the ionosphere, sunspots, sky wave and ground wave, maximum useable frequency, forecasting—in relatively non-technical terms.

Another book that gained some popularity (partly because it was available from Gilfer) was J. A. Ratcliffe's *Sun, Earth and Radio*.[133] This 1970 volume by an English scientist was quite technical, but contained much of value to DXers, including descriptions of the impact of the sun and the ionosphere on radio wave propagation.

The most popular volume on propagation was *The Shortwave Propagation Handbook*,[134] edited by George Jacobs and Theodore J. Cohen. Jacobs had been Chief of the VOA's Frequency Division for 27 years, then Director of Engineering for the U.S. government's Board for International Broadcasting. For decades he also served as *CQ* magazine's propagation editor. Cohen had advanced degrees in physics and geophysics and had published widely on the subject of ionospheric propagation. Both were active hams. The 1979 book was based on the more important contributions to the field which had been published in *CQ* during the preceding 15 years, updated, expanded and supplemented to reflect new material and to focus on propagation forecasting. A second edition of the book was published in 1982, with updated material on sunspot cycle predictions and other topics. Thirteen years later, in 1995, Jacobs and Cohen, joined by ionospheric scientist Robert R. Rose, published *The NEW Shortwave Propagation Handbook*.[135] This was a major rewrite, preserving the organization of the earlier editions but updating the material on solar phenomena and sunspot cycle behavior, and adding new material about computer propagation prediction programs.

Jacques d'Avignon's *Propagation Programs — A Review of Current Forecasting Software* did what it advertised: review the ten or so programs that were available for this purpose in 1993.[136]

Pirates and Clandestines   The principal author of books on North American pirate radio is Andrew R. Yoder, whose writings include *Pirate Radio Stations: Tuning In to Underground Broadcasts* (1990),[137] *Pirate Radio: The Incredible Saga of America's Underground, Illegal Broadcasters* (1995),[138] and *Pirate Radio Stations: Tuning In to Underground Broadcasts in*

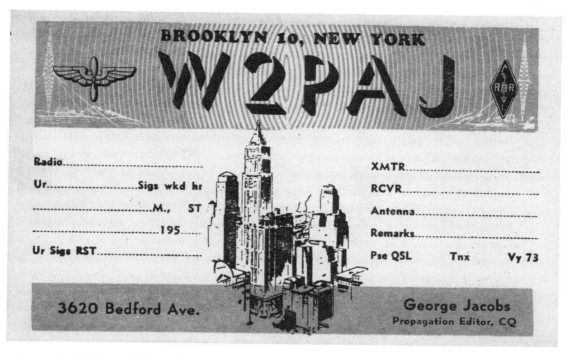

George Jacobs is one of the world's leading authorities on shortwave broadcasting. This was his ham QSL-card in 1952.

the Air and Online (2002).[139] The latter two are essentially expansions of the first, and include detailed accounts of the history and practice of pirate radio in North America. The 1995 edition preserves the organization of the 1990 volume, with a re-edit and a substantial updating of the contents. It also adds sections on FCC enforcement actions, European shortwave pirates, and local and FM pirates. In addition, it includes a CD containing recordings of many stations.

At 367 pages, the 2002 edition is the definitive work on North American pirate radio. It represents a major rewrite of the book, adds a section on computers and pirate radio, and expands the CD to include many more stations. Another Yoder book, *Pirate Radio Operations* (1997),[140] is a "how to" book for pirate station operators rather than listeners. Both *Incredible Saga* (1995) and *Pirate Radio Operations* contain bibliographies of pirate-related books, articles and other resources of the time. *Tuning In* (2002) also discusses where current information on pirate radio can be found.

During the 1990s, pirate radio aficionados benefitted from a series of soft cover volumes called *The Pirate Radio Directory* which appeared annually from 1989 through 1997.[141] Authored by George Zeller from 1989 through 1992, and jointly by Zeller and Yoder thereafter, each volume was a station-by-station review of the pirate radio activity of the previous year, with information about programming, frequencies, contact information, QSLs, etc. Also included was a detailed introduction to pirate listening. A similar book was the *1993 Worldwide Pirate Radio Logbook* by Yoder.[142] It listed all widely reported North American and European shortwave pirate stations in 1992, sorted by time, frequency, and station name, and included a Europirate address list and an overview of the year in pirate radio.

A little known but well written book on pirate radio is *Free Radio–Electronic Civil Disobedience*,[143] authored by Prof. Lawrence C. Soley, better known for his book on clandestine radio (see below). Although focused principally on FM pirating, where the dynamics are different from shortwave, it helps put pirate radio into a broader context. (A good primer on the culture of FM pirating is *Seizing the Airwaves: A Free Radio Handbook*.[144])

European pirate broadcasting, with its implications for the creation of an entirely new industry of private broadcasting in Europe, led to a major cultural transformation there. In North America, European shortwave pirates had little impact save for their DX value, and thus comparatively little that was written on the subject in Europe made its way to North America. However, a few books by Paul Harris became classics. The offshore era (1964–71) was well covered in his three works, *When Pirates Ruled the Waves* (1968 and later editions),[145] *To Be a Pirate King* (1971),[146] and *Broadcasting from the High Seas: The History of Offshore Radio in Europe, 1958–76* (1977).[147] A later book on the subject was *Rebel Radio — The Full Story of British Pirate Radio*.[148]

Clandestine broadcasting has not attracted much in the way of book-length authorship, save for the definitive academic work on the subject by Lawrence C. Soley and John S. Nichols, *Clandestine Radio Broadcasting — A Study of Revolutionary and Counterrevolutionary Electronic Communication* (1987),[149] and the related volume by Soley, *Radio Warfare: OSS and CIA Subversive Propaganda*.[150]

Several publications have appeared from within the listening community itself. *How to Tune the Secret Shortwave Radio Spectrum* (1981) by Harry L. Helms[151] gives some attention to clandestine and pirate broadcasting, but focuses mainly on other kinds of "unofficial" communications. Gerry L. Dexter's well-illustrated *Clandestine Confidential* (1984)[152] presented informative writeups of many of the clandestine broadcasting stations of the time. (Dexter was also editor of the *Clandestine Confidential Newsletter*, see p. 226) *The Clandestine Broadcasting Directory* (1994)[153] by Mathias Kropf consisted of lists of the active clandestine stations of the day in frequency and time order, along with a "database" of information on many stations (start dates, parentage, location, etc.). It also contained the 1993 Clandestine Activity Survey showing the total broadcast hours of clandestine stations by target area. This survey was first compiled in 1986 and is still published annually in various places within the shortwave community. (Kropf has been editor of the "Clandestine" column of the World DX Club (U.K.) bulletin, *CONTACT*, for many years.)

Programming　　Programming was always the poor cousin of shortwave publications, for the most active SWLs were more interested in distance than content.

Within the DX community, probably the first publication that was devoted to programming was the *SWL Program Guide*, created in 1965 by Todd Graves. Although now long forgotten, it established a basic format that would be used by others many years later. It provided information on the English-language broadcasts of some 40 major shortwave stations whose signals were beamed to or could be heard in the United States. Its content was arranged by hour, with between five and 20 listings for each hour. They showed the station, the frequency, the target area, the length of the broadcast, the hour at which the news could be heard, and a brief description of the nature of the program ("DXing," "Drama," "Politics," "Sports," etc.). To this data was added a brief introduction on shortwave reception, and the names and addresses of the stations. The *SWL Program Guide* listings were said to be based on information received directly from the stations rather than on listener reports. It was to be published each December, with a summer bulletin containing frequency changes available at no cost.

Only a few issues of the *SWL Program Guide* were published. Graves conducted an SWL supply business called SWL Guide, and in 1974–76 he attempted at least one other publication called *For SWLs*. Neither made a major impact on the SWL community, however. In the years between them he focused his publishing efforts on the *SWL Guide Newsletter*, which combined short articles with advertising for his business.

Kenneth D. MacHarg's 1983 book, *Tune In the World*,[154] was one of the few publications about shortwave program content. It reviewed the various types of programming available on shortwave, and profiled the origins and programs of some 70 stations. It was also the basis for a series of programs over HCJB in 1984. A few years later, *Muzzled Media*[155] focused in on the scope of news that was largely ignored by the domestic American media but was available on shortwave. In addition to providing an introduction to shortwave radio and a listing of times and frequencies for news from various countries, it contained an interesting comparative analysis of the news broadcast by many different shortwave stations on a single day in June 1986.

Better known was Kraig Krist's *The Shortwave Listener's Program Guide*,[156] an 80–100 page listing of programs for most of the world's English-language broadcasters. First issued in 1989 and published for two years thereafter, it listed programs by name, arranging them by day of the week and time of day. It did not provide advance program details or a summary of the content of the named programs. Krist soon added frequency information in a free supplementary publication, *The Shortwave Listener's Frequency Guide*, in order to make the *Program Guide* more useful. At first *The Shortwave Listener's Program Guide* was published twice a year, in May and November, when the international broadcasters changed their frequencies. This was changed to annual publication and three updates. *The Shortwave Listener's Program Guide* was one of the first program databases that became available for download or for use on line (it was on the ANARC BBS).

Similar to Krist's publication was *The Shortwave Radioguide*, introduced by John A. Figliozzi in May 1990.[157] As with *The Shortwave Listener's Program Guide*, the purpose of *The Shortwave Radioguide* was to provide listeners with an accurate survey of what could be heard in English in North America from almost 50 shortwave stations. The 144-page *Radioguide* consisted of seven sections, one for each day of the week, showing by time the programs that were aired. The *Radioguide* was updated in November when major schedule changes were made by broadcasters. The third and fourth editions were issued in May and November 1991 respectively.[158] They were published jointly by the North American Shortwave Association and the Ontario DX Association, and periodic updates were available for a stamped, self-addressed envelope.

The next edition was published in 1993 (for that year only it was called *The New Shortwave Radioguide*). It was expanded to incorporate English language broadcasts to other parts of the world, as well as some foreign-language broadcasts and some listings by program type. Further editions were published in 1994 and 1995 (in 1995, columns were added classifying programs by type and showing the frequencies for each program). A mid-year supplement was offered in 1995, and the 1995 edition was the first that was available on floppy disk in addition to paper.[159]

In 1996, *Radioguide* became *The Worldwide Shortwave Listening Guide.*[160] Now distributed by Radio Shack, *WSWLG* adopted a new format, with all program data arranged by time, allowing the user to quickly see which programs were on the air at any hour of the day from about 100 stations. Only English-language programs were included. Each time entry showed the program name with a brief description, the general program category, the days of the week on which the program was broadcast, the frequencies, and the target area. Programs to the Americas were in bold and also were included in a second "Menus for the Americas" listing which grouped programs by program type, e.g., news, science and technology, music requests, etc. A second edition of the *WSWLG*, also sold by Radio Shack, was published in 1998. A third, 2000–2001 edition was published in 2000, by which time the *WSWLG* contained over 7,500 entries. (This edition was not sold by Radio Shack.) Part of the data in *WSWLG* has been updated and made available online at the *NASWA WWW Shortwave Listening Guide* website.[161] Figliozzi has authored many articles about shortwave programming for both commercial and club publications. From 1998 to 2006 he was the "Programming Spotlight" columnist for *Monitoring Times* magazine.

In the early 1990s, *Monitoring Times* published Kannon Shanmugam's *Guide to Shortwave Programs.*[162] Issued annually from 1992 to 1994, it contained a by-time listing of English-language programs for each day of the week — basically the same information that was found in *MT's* monthly "Selected Programs" section, which Shanmugam prepared. The *Guide to Shortwave Programs* had two parts, one covering the October–March season, the other covering April–September, and included information on approximately 100 stations. Frequency data was contained in a separate frequency table.

Transmitters   In 1994, the first publication of the Transmitter Documentation Project, *TDP SW-94,* was issued. An undertaking of Ludo Maes in Rijkevorsel, Belgium, the purpose of TDP was to provide historical documentation of shortwave transmitters worldwide. *TDP SW-94* was a 52-page booklet showing for many stations in each country the transmitter site coordinates, the number of transmitters, their power output and their manufacture, as well as the years when they were placed in and taken out of service. Several tables also provided aggregate data on the power of the transmitters. *TDP SW-94* was based on information from a variety of sources within the shortwave community and from the transmitter manufacturers themselves. Annual editions of *TDP* were published from 1994 through 1998. Latter editions included some relevant articles and manufacturers' histories. Thereafter *TDP* information was posted on the TDP website. By then TDP had begun marketing shortwave transmitters and brokering shortwave airtime.

SPECIFIC PARTS OF THE WORLD

Latin America   In addition to the Latin American newsletters mentioned below, several larger works focused on this topic. Perhaps the best known is the six-volume series, *LA DXing,* that was published by the Radio Nuevo Mundo group in Japan between 1980 and 1997.[163] This was a substantial publication, the first edition containing 78 pages, the last 270. The heart of *LA DXing* was the detailed information on individual stations, which typically included station descriptions, program schedules, logos and the texts of station

LA DXing was aimed at DXers specializing in Central and South American stations.

identifications. This information was often obtained through station visits or local moni-toring. Each volume of *LA DXing* also included a list of Latin American shortwave stations, plus feature articles.

Another valuable book on this topic was Henrik Klemetz's *Latin America by Radio.*[164] Readers benefited from Klemetz's near-encyclopedic knowledge of the subject matter. Fol-lowing a discussion of the role of broadcasting in Latin America, Klemetz offered a wealth of practical information on how to identify Latin American stations. This included how to identify and distinguish among different kinds of station announcements, and the various kinds of music, commercials and advertising formats typically heard on the Latin Ameri-can airwaves. QSLing techniques were covered as well.

Two interesting publications about Latin American radio were published by Carl Huf-faker, who lived in Mexico and described himself as "an American displaced by [his] own quest for adventure and a comfortable life." *Carl Huffaker's Latin Notebooks,*[165] published in 1992, was a collection of his columns published in SPEEDX between 1986 and 1991. While their main topic was radio, they often benefited from Huffaker's knowledge of the people, geography, music and culture of the Americas. Earlier, Huffaker had teamed up with John Cereghin to produce for SPEEDX a monograph entitled *The SPEEDX Guide to Latin American DXing.*[166] The *Guide* contained a station list, plus extensive information on understanding the Spanish language and on such radio-specific topics as time announce-ments, call letters, the use of shortwave, harmonics, propagation, Latin American radio net-works, etc. It also contained a country-by-country look at Latin American shortwave broadcasting.

Among the *Fine Tuning* Special Publications was the *DXer's Guide to Latin America,* a by-country and by-frequency listing of stations and their basic operating parameters, e.g. frequency, hours of operation, typical quality of reception, etc. The *DXer's Guide* was based on loggings of Latin American stations reported to various bulletins during the previous year. It was published in 1987, and was followed by a second edition in 1989. The 1989 edi-tion also included several articles.

Australia   Bob Padula has issued many print publications oriented toward Australian DXing. These included the fortnightly, small-circulation *DX Press* from 1986 to 1996 when it went on line as the *Electronic DX Press* (and later *World Broadcasting Magazine*), the *Aus-tralasian Shortwave Guide,* published from 1996 to 2005, and the *Australasian Shortwave Digest,* published from 1999 to 2004.

The *Australasian Shortwave Guide* was originally published four times a year, later twice a year, and covered shortwave broadcasts to the Pacific, South Asia and the Far East. It was arranged by country and by time, and eventually grew to over 40 pages. The *Aus-tralasian Shortwave Digest,* which was originally a monthly, covered international and domestic shortwave broadcasting both to and from the area. It eventually became an annual publication containing articles as well as station listings. In 2005 Padula published a one-time successor to the *Digest,* the *Australasian Shortwave Digest CD Edition 1.* It contained a variety of databases, articles, graphics and weblinks. In 2006 the *Guide* was replaced by a new print publication, *Shortwave Broadcasts to Australia and New Zealand,* covering broadcasts in all languages to the area.

Other Padula print publications have included the many editions of the *High Frequency*

Spectrum Study (begun in 1996), the *Shortwave Guide to South East Asia* (1999), the *Asia-Pacific Shortwave Guide* (2000), and various editions of the *Spectrum Study* which aggregated monitoring results in Australia, usually for the period 0000–0400 UTC on frequencies below 10 MHz. The *Spectrum Study* was issued twice yearly starting in 1970, and was originally in print form, later electronic.

China and the Far East   An important publication about China was *Broadcasting Stations in the People's Republic of China.*[167] Two editions of this 75-page publication were issued, both in 1973. They contained the most comprehensive study of Chinese stations yet published, including a wealth of detail about the home service and foreign service stations, all arranged geographically and accompanied by a frequency list.

DXing China, by Bob Padula, was published by the Australian Radio DX Club in 1980. This 12-page pamphlet covered both mainland China and Taiwan. It contained information on the structure of broadcasting and the transmission times and frequencies of the domestic and external services, plus geographical and historical information. In 2001, Padula also published the *China Shortwave Guide,* a tabulation of all shortwave stations in China and Taiwan. (Available in both paper and electronic formats at first, subsequent editions were electronic only.)

In 1980, the Asian Broadcasting Institute, a Japanese DX group, began publishing monthly reports in English and Japanese about shortwave stations in East Asia. The English reports were discontinued in 1982, but were renewed later in the form of *East Asian DX News.* Eventually the ABI reverted to Japanese for most of its publishing. However, in 1987 it published the *China Radio Handbook,* a 50-page work covering broadcasting in China, past and present, and including detailed schedules of more than 200 stations, frequency lists, and program, ID and QSL information.

Vietnamese Shortwave Stations, 1972–1981, was a nine-page list issued in 1981 by Japanese DXer Isao Ugusa, publisher of *DX Front Line.* In 1985, *DX Front Line* published the more extensive *Broadcasting Stations of Vietnam.* A 114-page book, it contained a thorough history of Vietnamese radio from the 1930s on. It covered over 100 stations, and provided complete schedules and frequency information, detailed maps, etc., much of it previously unpublished. Carefully researched, it was probably the best single-country DX reference ever published, and was followed in 1986 by the 22-page *Frequency List of Radio Stations in Vietnam.*

India   *India BroadBase* was the creation of Manosij Guha, editor of *DX-Grapevine,* a DX bulletin published sporadically by the Universal DX League (UDXL), New Delhi, India. The first two editions of *BroadBase* were published in India in 1990 and 1991, but enjoyed very limited circulation. In 1992, *BroadBase* became a joint project of the UDXL and the Ontario DX Association, with Guha providing the data and ODXA doing the designing, publishing and marketing. This cooperative spirit produced a third edition which appeared on the North American scene in 1993. *India BroadBase* offered the most accurate and comprehensive information available about Indian shortwave stations, along with much interesting data on medium wave, FM, TV and satellite operations. It contained background information about the organization of All India Radio and the domestic network of regional and feeder stations on shortwave, along with detailed operating schedules of all services, addresses, maps and the like. A mid-year updater was published most years. The sixth and last edition of *India BroadBase* was published in 1997.

Indonesia   Indonesia was a favorite DX target. One of the earliest in-depth examinations of Indonesian shortwave broadcasting was the "Special Section on Indonesia" that was published in the bulletin of the North American Shortwave Association in January 1967. It was followed by other articles on the same topic.[168]

An early standalone survey of shortwave broadcasting in the country was the *ARDXC Indonesian Survey*, published by the Australian Radio DX Club in 1974 and updated in 1975. It was principally a station list. Also of interest was the *Indonesian DX Review*, published in the late 1970s by the Indonesian DX Circle of Osaka, Japan, and the *Indonesian Survey* issued in 1979 by Wrightwood, California DXer James Young.

Another important Indonesian reference was the *Survey of SWBC Activity in Indonesia*, the best known edition of which was published in 1992. Its genesis was in the mid–1980s when Mitch Sams of Wichita, Kansas produced a listing of Indonesian stations that had been reported in various (mostly North American) sources the previous year. The *Fine Tuning* Special Publications division, under John Bryant, took over the project in 1986, expanding it to include European and Australian sources, and enlisting the help of several other DXers, including Sams, Kirk Allen, and Jon L. Williams of Indianapolis, Indiana.

The first edition of the 13-page *Survey* was published in 1988. In 1989 the Australian *OZDX* group updated and confirmed the database, whereupon *Fine Tuning* issued a four-page updater. Later that year a second edition of the full *Survey* was published jointly by *FT* and *OZDX*. In 1992 the *Survey* was substantially rewritten by Bryant and David M. Clark. The principal content of its 25 pages was a database of key information on all stations, arranged by frequency and cross-referenced by location. Maps showing station locations were included as well. It was the most complete North American reference yet available on the shortwave broadcasting activity in this much sought-after DX country.

The *DXer's Handbook — Indonesia* was first issued in 1985. Updated editions were published by *Fine Tuning* Special Publications in 1988 and 1991. Written by John Bryant, the *Handbook* was a long-term reference work containing information on the government, history, culture and political subdivisions of the country, a description of the organization of the broadcasting system, maps, time charts, frequency-wavelength conversion tables, and valuable information on identifying and verifying Indonesian stations (including graphics for designing form QSL cards).

USSR   In 1975, the Australian Radio DX Club published Bob Padula's *DXing the USSR*. It explained the broadcasting organization of the USSR, provided detailed schedules and related information on the Soviet regional broadcasters, and discussed various aspects of QSLing Soviet stations. It was revised in 1977.

In 1976, SPEEDX published *A Guide to Soviet Broadcasting*.[169] Written by Robert Butterfield and Doug Snyder, it contained a wealth of information about Soviet shortwave broadcasting. It explained the Soviet system of national and regional broadcasting and the various Soviet radio services. It also included a list of Soviet frequencies and transmitter sites, a map, and a Russian-language reporting form. Nearly ten years later, in 1985, a similar publication, the SPEEDX *Guide to Soviet Radio*, was issued.[170] Authored by Michael

Nowicki, the 32-page booklet covered domestic and international broadcasting, plus feeders, military, maritime and other utilities, jamming, QSLing, etc.

The *USSR DXing Handbook* was published in 1984 by the USSR DXing Circle of Japan, and was based largely on monitoring done in that country during 1983. It consisted of maps, articles, schedules, program and frequency charts, political information, and tips on the Soviet domestic and international broadcasting scene. A second edition was published in 1985.

The dissolution of the Soviet Union in 1991 led to huge changes in the broadcasting scene of the former Soviet Union. In 1992, Alexy Osipov of St. Petersburg, Russia issued a 16-page handbook, *DXing the USSR and Former Soviet Territories,* bringing the post–Soviet broadcasting scene up to date. The same year, to help DXers keep track of changing events, *Fine Tuning* Special Publications issued the *Monitoring Log of the States of the Former USSR.* Designed by John Bryant, it contained a separate page for noting loggings of each of the 87 then-known transmitter sites within the former Soviet Union, together with maps, geographic coordinates of transmitter sites, and sunrise-sunset tables for each location.

MEDIUM WAVE PUBLICATIONS

A number of publications designed mainly for broadcast band DXers became popular with shortwave listeners. From the National Radio Club (NRC) came *The DXpedition Handbook,* written by Shawn Axelrod of Winnipeg, Manitoba, and published in 1994. It covered all the basics of DXpeditioning, and also gave detailed accounts of two Newfoundland medium wave DXpeditions held in 1991 and 1993, as well as "mini–DXpeditioning," a favorite of Mark Connelly of Billerica, Massachusetts, who also contributed to the volume.[171] Other useful NRC publications included the *NRC Antenna Reference Manual,* published in 1975 and followed by two other works in the series, *Volume Two* in 1981 and *Volume Three* in 2004; *Loop Antennas — Design and Theory,* originally published in 1983 and revised and updated in 1986; and *Beverage and Longwire Antennas — Design and Theory,* also originally published in 1983. The *NRC Receiver Reference Manual,* first published in 1975 with a *Volume Two* published in 1982, was also popular.

From the International Radio Club of America came *A DXer's Technical Guide.* It was first published in 1980, with subsequent editions issued in 1983, 1998 and 2004. The first edition was a 98-page compendium of receiver, accessory and antenna reviews, and receiver modifications, all prepared by well-known medium wave DXers. By the time of the fourth edition, the *Guide* had doubled in size and was reminiscent of *Proceedings* in style and content. The scope of the *Guide* remained the same as in previous editions, but the fourth edition was fully updated to reflect the many technological advances in DXing equipment.

Other Publications by Clubs and Individuals

MAJOR SHORTWAVE LISTS

A valuable source of information were the station lists published by clubs or by individual club members. The BCB DXers pioneered this technique, largely because the growth in BCB stations was quickly making the commercially available logs out-of-date. (In October 1946 there were 2,052 licensed stations, mainly AM, but some FM and TV as well, and fully 1,176 others with applications pending.[172])

During the 1940s and 1950s, the Universal Radio DX Club published some excellent shortwave station logs. These were lists of stations, in frequency order, with call letters, station name and power, along with details of the times when the stations could be heard. These logs were compiled by various individuals. The 1946 URDXC list of *Active Short Wave Broadcast Stations* was prepared by William Howe. By 1951–52 it had become the *URDXC Short Wave Log,* and at that time was organized by Weldon H. Wilson. The 1952–53 editions were compiled by Marvin E. Robbins, 1954 by A. R. Niblack, 1955 by Niblack, Jack Fairs and Mike Christie, and so on. These were substantial documents, 40–60 pages in length, somewhat like the modern day *Domestic Broadcasting Survey* published by the Danish Shortwave Club International but covering international as well as domestic shortwave broadcasters. Updates to the log were published in the URDXC bulletin, the *Universalite.* The popularity of these lists declined as the *World Radio Handbook* became widely accepted as the standard reference for shortwave station information.

The NNRC issued a 150-page shortwave log in 1947. Edited by shortwave section editor James J. Hart and M. F. Williams, it listed in Section I all stations by frequency, with call letters, power and known schedules, and in Section II the same material by country. The cost was $1, which included two supplements (issued in 1948).

Although not well known in North America, *The Shortwave Listeners' Annual,* published in the U.K. in 1947 by *Short Wave News* magazine, also carried a good list which had the benefit of being updated monthly in the magazine. *Short Wave News* continued to carry "rolling" station lists after the second (and last) *Shortwave Listeners' Annual* was published in 1948. Each issue covered a different frequency range.

In the 1950s the New Zealand Radio DX League published *World Radio Calls,* a station list covering both medium wave and shortwave stations, and giving names, power, schedule, slogans, etc.

Some clubs benefitted from the special resources of particular members. For example, in the early 1940s, Roger Legge, a long time shortwave and amateur band DXer, was working for the Foreign Broadcast Intelligence Service and was able to compile authoritative lists of times and frequencies of stations in out-of-the-way places like India and the Belgian Congo. In the mid–1950s, a well-known DXer from Johnstown, Pennsylvania, Paul A. Kary, went to work for the U.S. Foreign Broadcast Information Service (as the Foreign Broadcast Intelligence Service had been renamed). From 1954 to 1959 he was based in Kyrenia, Cyprus, where he was able to hear many rare stations. He provided useful information to DXers as the "Middle East Correspondent," and his material appeared in several DX bulletins.

There were numerous lists of various kinds issued by clubs and individual DXers in later years. Usually these were specialized lists that took a close look at a particular class of station or a particular geographic area. During the 1970s, the Australian Radio DX Club published several editions of its *Indonesian Survey,* as well as *DXing the USSR* and other country-specific surveys. Clandestine stations were the subject of a number of publications, including Bernard Chenal's *Les Stations Clandestines* (1971), the Japan Radio Club's *List of the Clandestine Stations* (1971), Larry Magne's *Broadcasting Stations of Exile, Intelli-*

gence, Liberation and Revolutionary Organizations (self-published in 1971, with 1972 supplement) and *Clandestine & Revolutionary Broadcasters of the World Frequency List* (published by NASWA in 1974), and Gerry L. Dexter's *Directory of Clandestine Stations and Programs* (1982). The Worldwide DX Club in Germany issued the *List of Time Signal and Standard Frequency Stations* in 1974. The Union of Asian DXers published the *Tropical Band Survey* (1977) and the UADX *Afro-Asian-Pacific Survey* (1980).

A South American list of particular interest was *Survey of Short Wave Broadcasting Stations Operating in the 60 m. Band, 1957–1974,* published in the latter year by Brazilian DXers Jerzy Sielawa and Jack Perolo. This comprehensive, 81-page English-language publication listed all stations that operated between 4215 and 5290 kHz. during the years 1957 to 1974. It showed frequencies, the year in which each station was last heard on each frequency, and the hours when the station operated or could be heard (in Brazil). Some blank space was left between frequencies for user notes. Also from South America, in 1989 and the years thereafter, came several editions of *Latin American Radio World — Home Service Stations,* issued by well-known Argentine DXers Julian Anderson and Gabriel Ivan Barrera, and containing address and verification signer information in addition to frequency, power and transmission times for each station.

The *Brazilian Medium-Wave, Shortwave and FM Stations List* was published first biennially, then annually, by Brazilian DXer Antonio Ribeiro da Motta, assisted by fellow listener Cláudio Rôtulo de Moraes. It was available from the early 1980s into the 1990s, and it showed the name, call sign, frequency, power, and address of every Brazilian station. Later in the 1990s it was edited by Geraldo Paim da Silva. During the late 1980s, the Suriname DX Club International also issued a number of station lists.

BAND SURVEYS

Different from station lists were band surveys. These were comprehensive examinations of shortwave operations in a particular frequency range. The concept of band surveys appears to have been born in 1950 in an article in the British magazine *Short Wave News.*[173] A member related his experiences in a month of single-band listening on 31 meters and urged it upon others as a means of better understanding reception conditions, receiver and antenna performance, and the program value of the stations on the air. The following year the International Short Wave League formed a Signal Survey Section to compile accurate lists of active broadcast stations. Every other month a different band was examined. The results were compiled and returned to those who participated in the project.

The Universal Radio DX Club promoted a similar idea in 1953, and in May of that year announced the first such "station survey." It was to cover three out-of-band ranges: 4000–4800, 9000–9200 and 15000–15100 kHz. Nothing specific appears to have come of the idea immediately, although in 1955 and 1956 the URDXC's *Universalite* carried several such surveys compiled by individual members.

Band surveys were made more useful when they started appearing in the NNRC bulletin shortwave section in 1956. At first prepared principally by Stewart West, with the participation of Roger Legge and others, a band survey took a single SWBC band and, for each occupied frequency within it, identified which station or stations were operating in each of three time blocks— 0100–1200, 1200–1800 and 1800–0100 EST. As space permitted, collateral information such as language, target area, and sign on and sign off times was also included. The preparation of these surveys was greatly aided by the fact that Legge was doing similar work professionally for the VOA. Over time the surveys were expanded in format. Some were limited to particular groups of countries. They continued in the NNRC bulletin until mid–1961.

In 1969, Gilfer Associates, the SWL retailer in New Jersey, began selling band surveys. The first two that were available covered 17 MHz. ($1.50) and 21 MHz. (12 pages, $1.25). In addition to schedule information, they included station, location, power, beam, and QSL information. The surveys received high marks for accuracy and for reliance on actual monitoring. Additional band surveys, and a subscription service, were contemplated, but they never materialized, in part because of the difficulties in keeping the surveys up-to-date.

The best band surveys were those compiled by Dan Ferguson under the auspices of the North American Shortwave Association. These were generally based on data published in multiple club bulletins, supplemented by his own and others' monitoring. Sixty meter band surveys were published in *FRENDX* in January 1971 and December 1971, and a 49 meter survey was published in September 1972. Several surveys were published as separate NASWA publications: 60 meter and 90 meter surveys in 1972, 31 meters in September 1973, the 1605–4000 kHz. region in 1972 and 1974, and an out-of-band survey in February 1974. A special band survey of clandestine stations, compiled by Larry Magne and Carol Feil of Denmark, was published in February 1974, and a 13 meter survey and two supplements were compiled by Glenn Hauser in 1974–75.

Starting in 1975, NASWA conducted several Coordinated Monitoring Events where members pooled their loggings on particular bands. The results were either published in

Participants in Coordinated Monitoring Events received a card acknowledging their contributions.

FRENDX or sent separately to participants. Years later, in 1981, still on behalf of NASWA, CMEs were conducted for Venezuelan stations and for the 13 and 41 meter bands. A 31 meter band CME took place in 1990.

In 1974 and for several years thereafter, the Danish Shortwave Club International sponsored a dozen or so band-monitoring events wherein members were asked to monitor a particular band for a seven-day period and report their results. With points assigned to each logging, there was a competitive aspect to the activity. A similar event covering Indonesian stations was held in 1982. Results of these surveys were compiled and published in the club bulletin, *Short Wave News,* and served as a detailed guide to the band in question.

In an expanded form, the band survey concept, which was in general use in professional monitoring circles, became the basic design of *Passport to World Band Radio.*

English-Language Broadcast References

In 1977, Dan Ferguson began publishing a list of English-language broadcasts heard in North America. In 1981, and by then maintained on a computer, it became the *Guide to English Shortwave Broadcasts* and it was offered on a four- or six-issue annual subscription basis. Each issue contained four listings: English broadcasts to North America, first by time and then by country, and *all* English broadcasts, also arranged by both time and country. The *Guide to English Shortwave Broadcasts* was issued until 1986.

From the mid–1980s to the mid–1990s, Thomas R. Sundstrom of Vincentown, New Jersey, a prolific writer on SWL, ham and computer topics and an early user of computers in shortwave listening, produced a regularly updated electronic database of English-language transmissions that was available on-line over his popular Pinelands BBS. In 1988 the system (and the viewer which he designed for using it) received the WRTH Industry Award for best computer accessory.

Another computerized list of English broadcasts was "ELBOW," *English Language Broadcasts of the World,* issued by David Snyder, President of the Atlantic States DX Association, in 1981.

The DSWCI Quartet: Tropical Bands, Home Services, Clandestines, Domestic Broadcasting

Starting in 1973, the Danish Shortwave Club International issued a series of valuable surveys. The first of these, and the best known, was the *Tropical Bands Survey (TBS),* published in 1973. It covered all stations in the tropical bands, defined as 2000–5900 kHz. Stations were listed by frequency, with information as to power and operating hours, and a code reflecting how often the station was heard. The *TBS* was updated during the year through supplements printed in the club's bulletin. Because it was based on actual monitoring rather than official information, the *TBS* was among the most authoritative and widely used references.

The *TBS* started out with two special benefits. Its first editor and compiler, Anker Petersen, was an experienced DXer, having started listening in 1957. He knew the bands well and he was highly regarded for his accuracy and thoroughness in reporting. In addition, the DSWCI itself enjoyed a worldwide membership and provided extensive tropical band coverage in its own bulletin, thus providing a good base from which to work.

The *TBS* was published annually, with Petersen editing through 1983. From 1984 to 1994 it was edited by several people: 1984 to 1989, Bernhard Gründl of West Germany (Gründl died in 1991); 1990, Hans van den Boogert of the Netherlands (later Taiwan); and 1991–1994, Julian Anderson of Argentina. Petersen resumed editing in 1995.

In 1984, the DSWCI published *Home Service Stations Outside the Tropical Bands,* compiled by respected Danish DXer Finn Krone. It followed the format of the *TBS,* complementing it by covering those stations operating *above* 5900 kHz. which transmitted home service programs rather than international programs. A second edition was published in 1985 and a third in 1987.

The year 1985 saw the publication of the first annual DSWCI *Clandestine Stations List.* Also edited by Krone, it consisted of two parts. Part I was a by-frequency list of clandestine stations with their operating hours. Part II contained additional information about the stations. This part expanded in scope over the years and eventually contained addresses, organizational parentage, schedules, information on station IDs, verification policies where known, and other information. Later issues of the *Clandestine Stations List* also included descriptions of the political situation in the various countries, which was of great assistance in understanding the mission and provenance of the stations. The first edition of the *Clandestine Stations List* was incorporated into the July and September 1985 issues of *Short Wave News.* Thereafter it was distributed as a separate publication. The last *Clandestine Stations List* was published in 1998.

The *TBS* remained in its original format every year until 1999 when the decline of tropical band broadcasting permitted expansion of the *TBS* to include those stations in the international shortwave bands which broadcast to a domestic audience or relayed such domestic broadcasts abroad, i.e., the stations which had been covered in *Home Service Stations Outside the Tropical Bands.* The resulting publication was named the *Domestic Broadcasting Survey (DBS).* In 2004 the *DBS* was expanded to include a list of clandestine stations (the information that had been covered in Part I of the *Clandestine Stations List*). The *DBS* in this format continues to be published every year, in electronic as well as print form. Since 1995 it has been supplemented by the DSWCI Tropical Bands Monitor, a monthly update published on the DSWCI website which shows those stations in the *DBS* which were logged that month.

For many years during the 1970s and 1980s, the DSWCI also published a *List of Time Signal Stations* prepared by club member Gerd Klawitter of Germany.

TROPICAL BAND LIST

The *Tropical Band List (TBL)* made its first appearance in 1992. Created by German DXer Willi H. Passmann, *TBL* was similar in design to the DSWCI *Tropical Bands Survey,* but it was produced by computer and updated on an ongoing basis. Although more expensive than the *TBS* (especially if one wanted frequent updaters), it was a valuable resource for serious DXers. Early editions covered the frequency range 2.3–6.0 MHz, but the *TBL* was soon expanded to cover home service broadcasts and clandestine broadcasts up to 7 MHz, and later up to 30 MHz.

Part A of the *TBL* was in frequency order, Part B in country order. It included frequency, transmitter power, how often the station was reported heard, sunrise and sunset times at the transmitter (useful for predicting optimal reception periods), and notes covering times heard, languages used, ID texts, etc. The *TBL* also contained a list of inactive stations, a sometimes-helpful tool for identifying unidentified stations. Single copies or subscriptions for four quarterly issues of the *TBL* were available directly from the publisher, or, in North America, from the Ontario DX Association. In 2001 the *TBL* also became available in .pdf format via computer download, making for rapid delivery and reduced cost. It ceased publishing in 2006, however, due to the declining audience for such a labor-intensive publication.

INTERNATIONAL LISTENING GUIDE

Although in North America it never achieved the popularity of some of the other shortwave directories, the *International Listening Guide,* published in West Germany by Bernd Friedewald, was also available. It began publishing in May 1978, and expanded in size, scope and organization over the years. Eventually, *ILG* would be published up to four times per year, corresponding to the standard shortwave broadcast transmission seasons.

ILG consisted of several parts. In May and November, the "ILG Programme Guide" was issued. It included "External and Home Services in English," in time order, and "At A Glance," showing English broadcasts in country order. Also in May and November there was the "ILG Frequency Database." It consisted of "The World Frequency Survey," basically a frequency occupancy chart of all shortwave stations in the 5745–26050 kHz. range. In March and September all subscribers received "Newsflash," which updated both the program guide and the database (the tropical bands were included in the October database update).

One feature that distinguished *ILG* was its early use of computers to update the database on an ongoing basis. Its reputation for accuracy never exceeded that of other publications, however, and the closely packed layout, with its extensive use of symbols and abbreviations, was hard on the eye. However, it did manage to condense a tremendous amount of information — times, frequencies, target areas, languages, programs, days of the week, etc. — in one place.

The *International Listening Guide* stopped publishing in 1990. For some years thereafter quarterly frequency databases were available, first on diskette and then by e-mail. In 1998, Friedewald created an online version of his databases. The public version, *ILGRadio,* was available at first for a small fee, and then at no charge. However, in 2006 it was discontinued. A professional *ILG* database is still maintained.

DX NEWSLETTERS

***DX Journal* (1950)**   *DX Journal* was a small newsletter that appeared briefly in 1950. Most issues were mimeographed, although an eight-page, professionally printed format was tried toward the publication's end. It was edited by Raymond S. Moore, then of Rowley, Massachusetts, who many years later would author several editions of a major work on vacuum tube communications receivers. *DX Journal* covered medium wave, shortwave and amateur DX. It was available by subscription, and was to be published three times per month during the broadcast band season. Although it contained loggings and other contributions from some of the leading DXers of the day, its membership was small and its life did not extend much beyond a half-dozen issues.

World at a Twirl Flash Sheet, DXing Horizons Flash Sheet, and DX Bulletin Board (1950s–1962)   During the late 1950s, following his 11 years of editing at *Radio News* and *Popular Electronics,* Ken Boord remained active in the DX community in part by editing the *World at a Twirl Flash Sheet,* known informally as the Ken Boord circuit letter. It was a DX newsletter consisting of a half-dozen "circuits" with five or six experienced DXers on each. They would send their DX items to Board, who would add his own material, make up multiple carbon copies of the *Flash Sheet* (this was before the advent of photocopiers), and send a carbon copy to the first person on each circuit. That person would take notes and send it on to the next person, and so on. From time to time Board produced other valuable DX aids as well, including his "Touring the World for News in English," a comprehensive schedule, by country and continent, of English news broadcasts available on shortwave.

The *World at a Twirl Flash Sheet* became the *DXing Horizons Flash Sheet* when Boord affiliated with *DXing Horizons* magazine in 1960. When Boord became ill in early 1962 and went on what was hoped would be a temporary hiatus, *Flash Sheet* participant George Cox of New Castle, Delaware, stepped in and continued the *Flash Sheet* concept, albeit on a more limited basis, with the *DX Bulletin Board.* As with the circuit letter, the *DX Bulletin Board* consisted of several circuits, four at the outset with four to six persons in each. In 1962, Cox was succeeded as head of *DX Bulletin Board* by Jack Perolo of Brazil. However, declining participation, and the burdens of travel associated with Perolo's professional responsibilities, led to the closure of *DX Bulletin Board* late the same year.

Shortwave Messenger (1960–1963)   Established in September 1960, *Shortwave Messenger* was a modest, bi-weekly SWBC newsletter, at first available to those who kept a supply of stamped, self-addressed envelopes on hand with its editor and publisher, J. Art Russell of San Diego, California. (A subscription system was adopted in 1961.) In the words of Russell, "It is our purpose to be first with the news, frequency changes, etc."[174] Starting out as a half-page compilation of DX items and other news submitted by subscribers, it expanded to three pages, and soon covered the AM broadcast band as well. The *Shortwave Messenger* came to an end when Russell passed away in February 1963.

Short Wave Newsletter and **Short Wave News Service** (1968–1985)   The *Short Wave Newsletter* was a bi-weekly, small-circulation newsletter established in 1968 by C. M. Stanbury II of Crystal Beach, Ontario. Issues were typically five to ten double-spaced pages and consisted of material submitted by the 15–20 newsletter participants, supplemented with material added by Stanbury.

The *Short Wave News Service,* also begun in 1968, was intended as a service to new clubs and smaller clubs which found it difficult to compete with larger clubs in the area of comprehensive shortwave news. It was a monthly newsletter, about the same size as the

Short Wave Newsletter, and was sent free to clubs that published at least nine times a year and had at least 20 members. In exchange, recipient clubs sent their bulletins to Stanbury. The *Short Wave News Service* consisted mainly of DX items from the *Short Wave Newsletter,* plus commentary by Stanbury on various shortwave-related topics.

These publications provided much of the information for Stanbury's writings in *Elementary Electronics, Electronics Illustrated* and other magazines, as well as his "International Broadcast Memo" column in the Canadian International DX Club bulletin, the *Messenger.* Stanbury was a colorful character. He was known for occasionally using the name "T. X. Thrush" when he wanted to preserve the confidentiality of a source or make a statement that he did not want attributed to himself.

Although the heyday of the *Short Wave Newsletter* and the *Short Wave News Service* was the late 1960s and the 1970s, Stanbury continued to publish them until early 1985. When he closed them he said it was because they were not helping with the fundamental problems of shortwave, which he described as the dullness of most programs and publications, the absence of needed new approaches by clubs, and the attraction of medium wave, which he felt offered a greater DX challenge. Stanbury passed away in 1986.

***Numero Uno,* DXplorer Radio Association and *DXplorer* (1969 to the Present)**   As the name implies, *Numero Uno* was considered by many to be the premier shortwave DX newsletter. It was published from October 1969 until June 2001. *NU's* origins are traceable to a micro organization, the DXplorer Radio Association, founded in 1957 by Don Jensen and Gerry Dexter. The participants of DXRA varied over time, but usually consisted of five to seven active and serious DXers. DXRA operated on a circuit system. Each member contributed a letter with his current shortwave doings—stations heard and stations sought, QSLs received, comments, etc.—in each round. When he received the packet of DXRA letters, he removed his old entry, added a new one and got the package back on its way, hopefully within 48 hours. The goal of making DXRA a source of fresh DX news dictated DXRA's size—content would become stale in a round of mailings to a larger group.

DXRA also undertook what were called DX Service Projects, original research on various aspects of DXing, usually lists of various kinds—corrections to the *WRTH* station addresses, irregular verifiers, Soviet transmitter sites, stations that sent pennants, etc. Much of this work was also published in the NASWA bulletin, *FRENDX.* A major DXRA product was a shortwave broadcast radio countries list that became the basis for the countries list which NASWA adopted in 1967. In addition, DXRA members occasionally sent each other "DXtras," postcards alerting fellow members to some unusual DX opportunity that could not await completion of the circuit.

In 1958, with Jensen in the service and Dexter in college, administration of DXRA was turned over to John Beaver of Pueblo, Colorado. Over time, some members became inactive for various reasons, and DXRA soon ceased operation. It was rejuvenated and reorganized in 1964. However, the delays in completing a circuit encompassing even the small DXRA membership made it an imperfect system for the rapid exchange of DX news. As a result, the focus of DXRA became original research, much of it done by Jensen in the area of clandestine broadcasting. Achievement of DXRA's original purpose would have to await another vehicle: the *Numero Uno* newsletter. The DXRA circuit letter grew into a radio-related social vehicle, facilitating regular mail contact among a small group of DXers who shared a passion for DXing. It continued in operation, with a varying membership, until around 1989.

Founded in 1969, *Numero Uno* was a weekly publication. As with DXRA, the goal was to provide a faster exchange of DX news than was possible via the monthly club bulletins. Also, Jensen's *Elementary Electronics* column and his other authoring projects would benefit from quality news that was fresh enough to withstand the delays of magazine publishing.

NU was in some respects a fraternity. Membership was by invitation, and limited to 25–30 of the most experienced SWBC DXers. In addition to keeping a supply of stamped, self-addressed envelopes on hand for the *NU* mailings, members were expected to be active contributors. The work entailed in compiling and mailing even a two-page newsletter every week was considerable, and so *NU* was a no-frills operation—no membership cards, no contests, no "features" such as might be found in regular club bulletins, and, to keep the content at a high level, no routine DX.[175]

In his expression of hope that *NU* No. 1 would be the first of a lengthy series of newsletters, Jensen was prophetic, for *NU* would have a life of almost 32 years before being reorganized into an e-group called DXplorer, which continues in operation. Although there were many administrative adjustments along the way, the basic *NU* formula was preserved throughout.

Ralph Perry, then also living in Kenosha, Wisconsin, co-edited with Jensen from October 1976 to December 1978, and Dexter edited every third month from March 1979 to June 1985. Otherwise, Jensen was *NU*'s sole editor until 1990. While the goal was to keep *NU* to a page or two by including only the most newsworthy information, by the end of the 1970s it was often four to six pages long, sometimes more.

One of *NU*'s strengths was the membership of several DXers who were in a position to obtain information not readily available elsewhere. Tony Jones in Paraguay provided comprehensive coverage of the huge and constantly changing Bolivian shortwave scene. U.K.–based (and world traveling) John Campbell offered inside knowledge of the European pirate broadcasting scene. Perry Ferrell, co-owner of Gilfer Associates, supplied information on the rapidly changing shortwave equipment scene, as did Larry Magne, who was also well known also for his work on clandestine broadcasting. Several others made similar contributions, including such well-known Latin American specialists as Jack Perolo in Brazil, Pedro F. Arrunátegui in Peru, Henrik Klemetz in Colombia, and Takayuki Inoue Nozaki in Japan. Although much of this information also appeared elsewhere in the DX press, NUers got it first, and all in one place.

NU members also benefitted from the *NU* Index which was prepared by various members at different times. It indexed all *NU* content by country and station, permitting the quick retrieval of DX information long before there were electronic searches, and gave *NU* a unique historic value. Save for some indices of Latin American loggings published by *Fine Tuning* in 1994–95, no other serial shortwave publication was indexed down to the station level.

Part of *NU*'s financial support came from an annual surcharge which was added to the envelope requirement in 1973 and assessed most years. In 1986 this system was replaced by a two-tier membership fee, with active NUers paying $22.50 annually, others twice that amount. A one-tier system was adopted in 1988.

NU was also supported by the proceeds of an auction of DX–related items donated by attendees of the "NUventions." The first of these annual meetings of the *NU* faithful was held in Charleston, West Virginia, in July 1973. It was arranged mainly because the 1972 ANARC convention in Boston had been a great success and the 1973 ANARC convention was on the west coast, too far for most *NU* members, who tended to be located in the eastern and central states. Originally, NUventions started on Friday afternoon and ended on Sunday morning. However, over the years attendees started arriving earlier for more fellowship than the Friday–Sunday program would allow. The close working relationship between *NU* and *Fine Tuning* led to joint annual "*NU-FT*" conventions starting in 1982. In 1988 this gathering, by then always held in Reynoldsburg, Ohio, became the Reynoldsburg DX Weekend, shedding its formal *NU* and *FT* connections and opening attendance to others while preserving the same basic format. It continues to be held annually.

During the years 1972–82, Jensen encouraged friendly competition among NUers through the annual Golden Plunger Award which went to the NUer who did the most "to keep the SWBC news flowing." *NU* membership certificates were issued for a time beginning in 1983, and for the next several years an endorsement seal was awarded to one or more "Outstanding NUers" after every 10 issues.

NU went on hiatus for two months in the summer of 1985, resuming with Jensen promising himself a leaner, more manageable *NU*. Thereafter, *NU* was usually kept to two pages, with an emphasis on the hottest, most interesting DX information. Crisper editing, together with some formatting changes, minimized the loss of information in the reduced page count.

From 1985 to 1990, *NU* had a "Special Projects" division. NUers had always shared the results of their individual research efforts, e.g. special lists of active stations, lists of known verification signers, etc. "Special Projects" formalized this and enlisted additional members in a variety of initiatives coordinated by Dexter. The biggest long term Special Project successes were the Committee to Preserve Radio Verifications, which is still in operation, and the annual North American DX Championships, which was sponsored by *NU* from 1987 to 1992 and by ANARC from 1993 to 2000. A joint *NU-FT* Special Transmissions Committee, whose purpose was to arrange special broadcasts from countries not otherwise on the air, succeeded in arranging such a broadcast from a pirate station in Northern Ireland, a new radio country. During 1985–87, NUers also prepared many impromptu band surveys, as well as such DXing aids as lists of national holidays and station anniversaries, and country-specific station lists, including in particular the "Peruvian Quipu" and several carefully researched Indonesian lists. Other projects — a communications bibliography, a Latin American music course, an *NU* musings section — never took off.

Jensen discouraged publicity that would lead to requests for membership which could not be honored. As a result, to the DX community at large *NU* was a somewhat mysterious, elitist group, often the target of barbs and (usually) good-natured ribbing. In 1978, a humorous *Numero Dos* newsletter was published by Pete Grenier of Azusa, California (he would later serve as ASWLC "Africa" editor). Around 1981, toward the end of its existence, the Three Mile Island DX Association of central Pennsylvania renamed itself Numero Dos and adopted the slogan "We Try Harder." Years later, in 1987–90, Rowland F. Archer of Raleigh, North Carolina, produced a series of articles for NASWA entitled "On to the Middle," so named after Jensen's well-known "On to the Top" series.

Although *NU* was strong in 1989 when it celebrated its twentieth anniversary, in October of that year, to the surprise of most NUers, Jensen announced his retirement. Much had changed over the years. Most of the old DX bulls from *NU*'s early days were gone, and the decline in shortwave broadcasting in the tropical bands, the DXer's favorite haunt, was evident. And *NU* was no longer unique; there were other weeklies that tread the same path (although most NUers would say not quite so well). And after 1,000 issues, Jensen was tired and had developed other interests.

In January 1990, the *NU* leadership baton was passed to John C. Herkimer of Caledonia, New York, a DXer for over 20 years and an *NU* member since 1983. Although, at Herkimer's request, Jensen took the honorary post of Editor Emeritus (and would relieve Herkimer for a month each summer), Herkimer assumed full responsibility for all aspects of *NU*—policy, editing and operations.

Herkimer had a strong sense of *NU*'s history, for he had authored a 20th anniversary history of the group. Called "Looking Back," it was published in year-by-year sections which accompanied *NU* issues during a 20 week period in 1989. ("Looking Back" was updated when *NU* celebrated its 25th anniversary in 1994.) His historical bent notwithstanding, Herkimer lost no time in introducing a major change in *NU*—desktop publishing. This gave *NU* a polished, professional appearance and for the first time permitted the use of graphics, such as QSLs, pennants, etc. Chris Lobdell of Tewksbury, Massachusetts, assisted Herkimer by assuming the job of culling exchange bulletins for material to include in *NU*. The *NU* page count started climbing, and soon *NU*s of four or more pages, often supplemented with copies of *Relámpago DX* and other DX publications, were the rule.

But the 1990s was a period of fundamental change in the DX world. Although many *NU* members had a preference for the good looking, hard copy publication they were now used to, e-mail was coming into wide use, and electronic DX newsletters and on-air DX programs, in which many NU members participated, were able to get the news out faster than paper publications. In the increasingly electronic world, the paper *NU* often served more as a summary of DX activity than the primary source of new information. Some members drifted away. These circumstances, together with Herkimer's own declining DX activity, led to his and Jensen's joint decision to close *NU*. As Jensen put it, "*NU* changed the face of DXing! And now it's time to move on. Let others try to live up to its standards, and good luck to them!"

The bond among *NU*'s core members was strong, however, and with Jensen's and Herkimer's blessing, a proposal for an electronic *NU* gained support. The last paper edition of *NU* covered the week ending December 15, 1995. On December 22, the first electronic issue was distributed. Editorial duties for *NU*, now distributed as a weekly e-mail message, were assumed by Jerome S. Berg of Lexington, Massachusetts, an *NU* member since 1970. He was assisted by Richard A. D'Angelo of Wyomissing, Pennsylvania, who, as *NU* Administrator, handled dues and related matters, and David M. Clark of Thornhill, Ontario, who served as Associate Editor. Most NUers had e-mail. For the rest, D'Angelo printed out the weekly *NU* and sent it to them by postal mail. Eventually only a few members required the paper copy.

Notwithstanding the decline in shortwave listening generally, *NU* prospered. The speed of e-mail served the newsletter's mission well, and many previous members returned. With the physical burden of publishing and mailing a paper newsletter now gone, content could be the exclusive focus. *NU* grew in size; eight to ten page issues were not unusual, and there were frequent mid-week supplements with time-sensitive information. In 1997, a website, NU Sight & Sound, was established for QSLs, audio recordings, web links, the "*NU* Index," and other items. John Herkimer was the webmaster. The "*NU* Indonesia List," a continuously updated compilation of all Indonesia loggings reported in *NU,* was introduced in 1998, and a like "*NU* India List," prepared by John Bryant, was added the following year.

By 2001, both the shortwave bands and DXing had changed greatly. The internet was producing more information on shortwave stations than ever, all of it delivered in rapid if somewhat disjointed fashion. However, the number of challenging new shortwave targets was decreasing as tropical band broadcasting declined and the migration to FM, satellite, internet broadcasting and other means of reaching audiences intensified. And even in its electronic incarnation, *NU* was a jealous mistress. While member participation was strong, and the size of the typical weekly issue still seven closely packed pages, the relative paucity of new DX targets demanded a more streamlined effort. As a result, it was decided to retire *NU.* Over its nearly 32-year history, 119 DXers had been members at one time or another. All their names were reprised in the final issue, which was published on June 24, 2001.

The *NU* spirit lived on, however. *NU* reinvented itself as a Yahoogroup called *DXplorer.* *DXplorer* had begun a year earlier as an *NU* adjunct where members could share informal comments and musings by direct postings. Now it became the prime vehicle for members' loggings and other information. The *NU* website was preserved and renamed DXplorer Sight & Sound. Berg managed the group, supplementing members' messages with a weekly "Review of the DX Press" (RDXP) e-mail message containing selected member loggings and items of interest from other DX publications. RDXP could be produced relatively quickly, and served as a central repository for most of the hard DX news that was scattered among various internet resources. It, together with the member messages, yielded basically the same quantity and quality of information as *NU.* DXplorer continues to operate in this way today.

Bandspread **(1970–1974)**   *Bandspread* was the bi-weekly newsletter of the British Association of DXers (BADX), which was founded in the U.K. in October 1970 by Alan Thompson of Neath, Glamorgan. Thompson, who had been the moving force behind the establishment of the U.K.'s World DX Club in 1968, and had served as Secretary General of the European DX Council in 1970–71, was a widely respected DXer.

BADX was an invitation-only group of approximately 50 highly skilled monitors around the world. The heart of *Bandspread* was the "Bandscan" column, which contained members' loggings. In this respect it was a British version of *NU,* and in fact the two groups had close contacts. *Bandspread* also contained occasional features, such as receiver reviews, station profiles, discussions of propagation, etc. BADX also issued a few special publications, including the *Guide to Turkish S.W. Low Power Stations* (1972), and *BBC Monitoring Service — A Layman Looks at Caversham Park* (1972), a detailed look at the operation of the BBC Monitoring Service at a time when the DX community knew little about its inner workings.

What set *Bandspread* apart was the inclusion of letters from members discussing all manner of DX topics. These were often followed by reactions from Thompson or other members, and the give and take provided an interesting supplement to the DX news. Although *Bandspread* started out as a five-page bulletin, it was soon eight to 16 legal-sized pages, with the last page, called the "DX News Flash Sheet," reserved for material received at press time.

Until May 1972, Thompson had some assistance in his editorial duties. After that he became the sole editor, which contributed greatly to the burden of producing *Bandspread*. In early 1973 the bulletin was reduced in length to four to eight pages. Publication became monthly for a while, but, save for a few longer gaps, resumed a bi-weekly schedule in 1974. Production of *Bandspread* had become too much for one person, however, and publication ceased in October 1974, exactly four years from *Bandspread*'s inception.

In July 1980 another British newsletter appeared, the *Weekly DX Newsletter* of the U.K. DX Association which had been formed by Gordon Darling, Chris Gibbs and Michael Barraclough. The high-quality newsletter was usually one or two pages in length, its membership principally U.K. DXers, about 50 in number at the group's peak. It ceased publication in February 1984.

Tropical DX Newsletter **(1974–1979)**   The *Tropical DX Newsletter (TDXN)* began publication in August 1974. Edited by Fred Heutte, Jr., of Washington, D.C., it was a bi-weekly compilation of loggings and other news and comments reported by its approximately 50 subscribers. As the name implied, the focus was on shortwave broadcasting in the tropical regions of the world, usually on 60 meters and below. However, news of other interesting stations was covered as well, and in fact *TDXN* wound up covering domestic shortwave broadcasting in all parts of the world and on all bands. The content of *TDXN* was high in quality, with issues varying between three and eight pages in length. Toward the end of 1978, a series of professional and personal conflicts burdened the editor and caused publication to become irregular. The last *TDXN* was issued in February 1979.

"Down Under" DX Survey **(1975–1983)**   Although the "Down Under" DX Circle published the *"Down Under" DX Survey (DUDXS)* only two to four times a year, this Australian publication took its DX seriously. It was the most authoritative source of DX information about Asia, which was its focus, and in particular Indonesia. Its purpose was to provide current news on the Asian broadcasting scene by surveying particular frequency ranges. *DUDXS* concentrated on the lower frequencies, preparing monitored schedules of stations about which published information was hard to obtain, and articles about broadcasting in Asian countries.

Issues usually consisted of 12 legal-sized pages, and provided a quantity and quality of information that could not be obtained elsewhere. Membership was free to contributors, but individual issues could be obtained by non-contributors as well. The *DUDXS* was edited principally by "convenor" David Foster of Burwood, Victoria, assisted from 1975 to 1980 by Mike Willis of Nedlands, Western Australia, and from 1977 to 1983 by Geoff Cosier of Melbourne, Victoria.

The *DUDXS* was a bit counter-intuitive by Australian standards. Its targets—China, Russia, Indonesia and other Asian countries—were areas that were generally relegated to third place among Australian DXers, after Africa and Latin America. In addition, many "down under" DXers were accustomed to building their totals by hearing and seeking QSLs from stations on as many different frequencies as possible — a "quantity" approach. The *DUDXS* concentrated on smaller stations, many of which were less likely to QSL.

The source of most *DUDXS* material was a cadre of serious listeners in Australia and other countries. What distinguished the *DUDXS* was the degree to which these listeners followed the DXing scene on a continuous basis and compiled their findings into comprehensive surveys rather than simply lists of loggings. This greatly increased the value of the information, in effect making the DUDXC as much a research organization as a DX group. Some of the surveys covered Asian stations in one or more specific frequency bands, while others covered particular countries, such as Indonesia, China or Vietnam.

In addition, *DUDXS* carried miscellaneous country-specific information ("Echoes From Asia"), data on QSLs, and features about topics such as receivers, music, members' trips through Indonesia, and the like. For a time, Mike Willis edited special South East Asian and South Asian sections. Although the *DUDXS* concentrated on shortwave, medium wave news received attention as well.

The *"Down Under" DX Survey*'s treatment of its subject was exhaustive and represented the high point in specialized DX publications.

USSR High Frequency Broadcast Newsletter and *International Broadcasting Journal* (1976–1989)

Begun in November 1976 by Roger Legge, then of McLean, Virginia, "*UHN,*" as the *USSR High Frequency Broadcast Newsletter* was popularly known, dealt mainly with foreign service transmissions from the former Soviet Union and its constituent republics. Its focus was not simply on the schedules of the various transmissions, although extensive time and frequency data was included, but rather on the transmitter sites in use. The newsletter was useful because of the frequent changes in Soviet frequencies and the Soviet policy of showing transmitter sites on QSLs when requested.

Some of the data in *UHN* was gleaned from the IFRB's seasonal Tentative High Frequency Broadcasting Schedules, while other information was developed by technically adept hobbyists who focused on transmitter locations based on the propagational characteristics of the signal, the examination of charts, maps and satellite photos showing tower locations and heights, and, in one instance, even on-site investigation. Another source of information was the transmitter site information entered by Radio Moscow on its QSLs. Although this QSL data was widely acknowledged to be inexact, the indicated sites were believed to be at least in the general area of the actual sites. Usually four pages in length, *UHN* was issued from five to eight times a year at the start. Publication eventually became less frequent, however, and ceased altogether around 1989.

For a brief time, Legge published a similarly designed, four-page newsletter called *International Broadcasting Journal*. The purpose of *International Broadcasting Journal* was to assist users of the ITU's Tentative High Frequency Broadcasting Schedules by correcting out-of-date and otherwise inaccurate data that was typically reflected in the ITU listings (mainly on domestic rather than international shortwave stations). The first issue of *International Broadcasting Journal* was published in July 1979, and it is believed that publication ceased after a few issues.

Review of International Broadcasting (1977–1999)

Review of International Broadcasting, or "*RIB,*" was born in February 1977, the product of a conflict between its founder,

Glenn Hauser, and the leadership of the North American Shortwave Association over editorial guidelines which Hauser felt infringed on his editorial independence as "Listener's Notebook" editor. Hauser was an experienced DX editor.[176]

Hauser continued with his NASWA column, but moved some of the material to this new, independent publication (the first issue of which was called simply *Glenn Hauser's Independent Publication*). *RIB's* purpose was a broad one: "To provide a forum for discussion of programming in international broadcasting, primarily, but not limited to, shortwave. Also, for the discussion of any topic related to the DX listening hobby. Some timely 'DX news' and schedules may also be included."[177] As Hauser put it later in comparing *RIB* with another of his publications, *DX Listening Digest*, *RIB* emphasized "words over [DXLD's] numbers."

RIB's strength was twofold. First, it was the most substantive publication yet to focus on shortwave programming. Although there had been a few small attempts at this previously, shortwave fans were principally interested in DX. *RIB's* emphasis would be elsewhere. *RIB's* second asset was Hauser himself. Always marching to the beat of his own drum, and sometimes at odds with others in the DX establishment, he had been listening seriously to shortwave since he was 12 years old, and to medium wave and TV even before that. He was a top notch DXer, and unfailingly comprehensive and accurate in his writing, editing and reporting.

On the negative side, *RIB* was, as advertised, a forum for discussion, not an effort to evaluate shortwave programming in an organized way or to apply common standards to what was heard over the air. It was, in Hauser's words, a "grassroots journal of opinion," some of which was thoughtful, much of it idle. Usually presented in a "Listener Insights on Programming" section, there was much point-counterpoint, with no topic too small for comment, particularly by a few regulars who seem never to have had an unexpressed thought. Still, Hauser made the most of it, adding occasional personal postscripts and switching early on from a "by contributor" to a "by country" organization to permit of some continuity of content from issue to issue.

In addition to programming, *RIB* contained discussion about music, equipment, other publications, and technical topics. These too were based mainly on subscriber comments. *RIB* also contained DX news in the form of summaries of Hauser's weekly presentation over Radio Canada International's "DX Digest" ("SWL Digest" from September 1981), along with comprehensive tables of broadcasts in English, DX programs, and "Shows We Like," a listing by time and day of the week of programs which Hauser and *RIB* readers felt worthwhile. A modest medium wave section covering programming on U.S. broadcast band stations was begun in August 1979. Advertising on a small scale began appearing in 1981, and over time, greater use was made of direct reproductions of newspaper and magazine articles.

In 1982, a major DX news column, "DX Listening Digest," was added to *RIB*. By this time, David Newkirk of Seattle, Washington was editing a regular "Radio Equipment Forum" column in *RIB*. (From mid–1985 it would be edited by Loren Cox, Jr., of Lexington, Kentucky, who also began a "Satellite Watch" column at the end of the year.) And in July 1983, *RIB* began carrying detailed DX reports on shortwave stations in a prime DX target area, Peru, from Juan Carlos Codina, a DXer in that country. Notwithstanding the liberal reproduction of BBC material about BBC World Service programming, the DX and technical news had by this time made *RIB* a magazine of broader shortwave content than just programing.[178] However, it regained much of its "programming" roots in October 1985 when *DX Listening Digest* was spun off as a separate magazine. At that time, according to Hauser, *RIB* had about 1,000 subscribers.

Soon *RIB* and *DXLD* were being published on an alternate basis, with one or the other appearing approximately monthly. Subscriptions of both declined in the late 1980s. Although still highly substantive, *RIB* contained increasing photo-reproduction of material from newspapers and other print media. Moreover, by the early 1990s, Hauser's reliance on material written by others provided relief from much of the editorial workload. The Cox equipment and satellite columns, together with another Cox column, "The Media Mind," plus the BBC World Service material, "Broadcasts In English" and related material from John Norfolk, and Alan Roe's periodic "Transmissions In English," reprinted from the World DX Club, took up most of *RIB*. There were also fewer reader contributions. By this time *RIB* was being published four times per year, reduced to twice a year circa 1995. The last issue was No. 154, published in October 1997.

On September 12, 1997, about the same time that *RIB* ceased publishing, Hauser inaugurated *Review of International Broadcasting Online Data (RIBOLD)* on an experimental basis. It was free at first, by subscription starting in July 1999. *RIBOLD* usually appeared daily, sometimes twice daily, and the content generally paralleled the traditional content of *RIB*: international broadcasting news, program news, and selective DX news, along with coverage of some U.S. domestic broadcasting, radio-related internet activities, satellite broadcasting, equipment, propagation, and some amateur-related news. The experiment ended on October 31, 1999, and was followed by the introduction of the on-line version of *DX Listening Digest* on January 1, 2000.

Fine Tuning (1977–1997)   *Fine Tuning* was founded in 1977 by Dan Ferguson, then of South Charleston, West Virginia, and had its roots in earlier Ferguson endeavors, including a usually-weekly DX sheet called *WEEDX* which he published in the early 1970s, and a weekly tip sheet which he distributed among a dozen or so Florida DXers when he lived in Miami in the late 1960s and early 1970s. Issue No. 1 of *Fine Tuning (FT)* was published on June 4, 1977. In his letter of invitation to prospective members, Ferguson described the goal of *FT* as the publication of a newsletter for active and serious SWBC DXers, at least bi-weekly, more often when the information warranted it. Initially, participation was, as with *NU,* by invitation. However, up to 100 participants would be accommodated, compared with *NU'*s 30. Thus *FT'*s membership requirements were more relaxed. A member's only obligation was to supply stamped, self-addressed envelopes for the mailings, and to contribute material. (A subscription fee replaced the envelope system in 1982.)

Ferguson edited the loggings section, "Fine Tuning Dial," which became the heart of *FT*. Two other specialty columns would appear monthly—"Informativos Ahora," which included features on Latin American DXing from John Moritz, Jr., of Youngstown, Ohio, and "The Sked Shed," edited by Bob Hill, then of San Francisco, California. A third column, "The QSL Mail Box," edited by Bill Sparks, also of San Francisco, appeared occasionally starting in October, and by April 1979 was the only one of the three specialty columns remaining.

FT was a success from the start, and publication was soon weekly except in the summer or when the press of other business dictated less frequent publication. The quality of the content was high and there was a lot of it, with the loggings alone often occupying four to six pages. *FT* was similar in concept to *NU* but for a wider audience. Ferguson was an *NU* member, and other *NU* members participated in *FT* as well. These pre-existing personal relationships led to a good deal of *NU-FT* cooperation over the years. The presence of *FT* also relieved some of the pressure on *NU* to take on additional members. By January 1978, *FT* membership was over 50. Eventually the invitation requirement was dropped and membership became available to all on a subscription basis.

Ferguson was one of the first to use a computer in DX newsletter publishing, in 1979 printing the loggings section of *FT* from a database using a Tandy Radio Shack TRS-80 computer. Despite this, preparation of the *FT* loggings was a considerable burden, and but for limited spells of editing by Moritz and Stu Klein of North Brunswick, New Jersey, Ferguson did it all himself until October 1984 when Larry Yamron of Pittsburgh, Pennsylvania, assumed the editorial responsibility.

Ferguson went to work for the VOA in 1985. In November 1986, *FT* and *The DX Log*, the publication of the Ozark Mountain DX Club, merged and the administrative structure of *FT* changed.[179] Dave Valko of Dunlo, Pennsylvania, Dan Sheedy of Encinitas, California, and Kirk Allen of Newkirk, Oklahoma, would handle the editorial duties on a rotating basis. *DX Log* editor/publisher Mitch Sams, then of Farmington, Arkansas, became Managing Editor of *FT* and backup editor, while Larry Yamron served as publisher, QSL editor and backup editor. While it was planned that *The DX Log* would retain some of its own identity by appearing as an occasional feature supplement to *FT*, this never occurred, although *FT* did carry special features from time to time. A "Special Projects" division was set up under John Bryant of Stillwater, Oklahoma. The total number of members at the time of the merger was approximately 80.

The post-merger plan was for *FT* to be issued bi-weekly. In fact, it was issued weekly during the DX season, bi-weekly at other times. Issues were typically four densely packed pages (sometimes more, rarely less), and often included photos of *FT* members. The editorial rotation eventually included, at various times, Kevin Atkins of Pinson, Alabama, Alan Laves of Dallas, Texas, Mike Nikolich of Arlington Heights, Illinois, and John Wilkins of Wheat Ridge, Colorado, in addition to Valko, Sheedy and Allen. By 1989, *FT* membership was over 150. In 1992, Yamron began accepting last minute information by telephone or e-mail, adding it to *FT* just before mailing.

Over the years, both before and after the merger, *Fine Tuning* also served as a publication vehicle for a number of special projects, including a list of English language broadcasts from international broadcasters (1979), sunrise-sunset tables (1979), band surveys (1982 and 1983), the *DXer's Guide to Latin America* (1987 and 1989), and John Tuchscherer's *Latin American DX Helper* (1986 and 1987). Other *Fine Tuning* publications included multiple editions of the *Survey of SWBC Activity in Indonesia*, and several items authored principally by John Bryant: the *DXer's Handbook — Indonesia*, the *Monitoring Log of the States of the Former USSR*, *The Spectrum Log — Tropical Band Edition* and *The Spectrum Log — International Broadcast Band Edition* (for tracking activities in a particular band over a range of hours), *The System* (a set of camera-ready forms for various aspects of SWL record-keeping),[180] and *The Propagation Predictor* (for tracking the 27-day sunspot cycle). Some of the *FT* publications were sold by NASWA. The highly respected *Proceedings* anthologies were the work of many individuals, working largely under Bryant's tutelage.

In the Fall of 1997, personal and professional obligations of Yamron and Sams made the regular production of *FT* more difficult, and publication ended in December. *FT* had made an important mark in the shortwave community for 20 years, and was greatly missed.

***DX South Florida* and the *Mosquito Coast News* (1981–1996)**   To American DXers interested in hearing Latin American stations, *DX South Florida* was a rich source of information. Begun in October 1981 by Robert L. Wilkner of Ft. Lauderdale, Florida, and his son Larry, and soon co-edited by Steven H. Reinstein of Miramar, Florida (with Wilkner as publisher), the small-circulation *DXSF* was a high energy, high quality, pleasantly irreverent bi-weekly newsletter whose purpose was to provide a DX information vehicle for active south Florida DXers. Though *DXSF* was not limited to Latin American DX (nor to DXers living in south Florida), the "LA" stations were the first love of *DXSF*'s "DX Commandos," as the editors liked to call contributors, with Indonesia and Asia a clear second. Terry L. Krueger of Clearwater, Florida, joined the publishing team in 1989. (Reinstein left DXing in 1990.)

While the heart of *DXSF*'s usual two to four pages were members' reports of shortwave stations heard and QSLed (there was limited coverage of medium wave and harmonics as well, plus some occasional FM and utility material), one of the main attractions of *DXSF* was in the variety of "supplementary" material it contained. Band surveys, lists of Peruvian departamentos and provincias, user reports on receivers and antennas, articles about broadcasting and listening in other countries—all were common *DXSF* fare. *DXSF* also covered the cultural side of Latin American radio and the Spanish-speaking world. It often reproduced newspaper clippings on radio topics or radio-related political subjects. These were sometimes contained in a supplement, the *Mosquito Coast DX News*, which Wilkner began in December 1988.

Although the most active *DXSFers* were the Floridians and some other North American DXers, *DXSF* had members in other countries as well, including a number in South America. A nice *DXSF* feature was the reproduction of some of the personal DX sheets that were issued by some DXers in Latin America and elsewhere but not widely available, e.g. *BOLPEBRA DX News* (information about Bolivia, Peruvian and Brazilian stations) and the *Bolivia Newsletter*, both from Rogildo F. Aragão of Bolivia; *El Chasqui DX* (Pedro F. Arrunátegui, Peru); *Radio Panorama* (Daniel Camporini, Argentina); *Dateline Bogotá* (Henrik Klemetz, Colombia); *Relámpago DX* (Takayuki Inoue Nozaki, Japan). Often *DXSF* reproduced all or parts of Latin American–related newsletters and bulletins, such as *Onda Corta, Pampas DXing, Latinoamérica DX*, the *WRTH Latin America Newsletter, Radio Nuevo Mundo, SURICALL* (Suriname DX Club International), etc. The Central American clandestine broadcasting scene, which was very active at the time, was also well covered, there being in the United States no better listening post for both the stations and the related politics than south Florida.

By 1995, most of the original Florida members who were still active were communicating via e-mail, and for them the paper *DXSF* bulletin was becoming superfluous. In October, *DXSF* was replaced with a new paper publication, the *Mosquito Coast News*, with Wilkner as editor and publisher. He handled postal mail contributions, while Krueger summarized the best of the e-mail material. Postal contributors were encouraged to adopt a standard format that could be photocopied directly into *Mosquito Coast News.*

With a decline in the level of DX activity among the Floridians, and a corresponding increase in participation by non–Floridians, both the personality and the *raison d'etre* of *DXSF/Mosquito Coast News* had changed. With the DX needs of its membership being met satisfactorily by other publications, the *Mosquito Coast News* ceased publishing in September 1996. In the sameness of the internet world, the free-spirited enthusiasm of *DX South Florida* and *Mosquito Coast News* were missed.

Clandestine Confidential Newsletter (1984–1993)

Gerry L. Dexter's *Clandestine Confidential Newsletter* was a spin off of his book, *Clandestine Confidential,* published in 1984. The four-page newsletter was issued six times a year beginning in 1984 and contained clandestine station news arranged by country. *Clandestine Confidential* also published separate lists of clandestine stations and their addresses. The newsletter closed in December 1993, by which time contributions from a declining readership had become limited and other Dexter publishing projects had gained priority. A plan to begin producing an annual directory of clandestine broadcasting similar to the *Pirate Radio Directory* did not materialize, perhaps because of the appearance of the annual *Clandestine Stations List* published by the Danish Shortwave Club International starting in 1985.

DX Listening Digest (1985 to the Present) and *Glenn Hauser's Shortwave/DX Report* (1997–1999)

Glenn Hauser's second major publication was *DX Listening Digest (DXLD).* Once a section in *Review of International Broadcasting, DXLD* was inaugurated as a separate publication in October 1985. Its purpose was to facilitate the separation of DX news from the programming information and commentary which were *RIB*'s main focus.

DXLD was a 40-page compendium of shortwave DX information, mainly detailed time and frequency information and news on station doings, planned startups, etc., plus occasional features submitted by subscribers or picked up from the large number of radio publications, DX programs and other sources that Hauser monitored. Issued in booklet format, most of the material was presented alphabetically by country, supplemented by information on recent publications and other general topics of interest to the shortwave community. *DXLD* also carried limited information on developments in U.S. AM and FM broadcasting.

DXLD was comprehensive — always a strength of Glenn Hauser productions — and represented a huge amount of editorial and production effort. There was little DX information of value that escaped *DXLD*'s attention. In the first edition, Hauser called *DXLD* "the start of something great," and it was. A year after it started, Hauser reported having about 500 subscribers.

In most issues through July 1988, *DXLD* also contained a 12–16 page section called "Enjoying Radio." Prepared by David Newkirk of Newington, Connecticut, "Enjoying Radio" was a smorgasbord of material (including occasional poetry) on softer subjects such as receivers, antennas and other technical issues, and to a lesser extent topics like clubs, publications, shortwave programming, etc. It contained material reprinted from other publications or original material from Newkirk and others. Newkirk described "Enjoying Radio" as "a running open forum concentrating on our personal enjoyment of and participation in radio and its effect on and place in our lives."[181]

Over time, Hauser had to give priority to his deadline-driven shortwave projects, including a weekly "World of Radio" program and a monthly *Monitoring Times* column. More material was cut-and-pasted into *DXLD* from other sources, and issues of both *DXLD* and *RIB* were published less frequently. *DXLD* had never been intended as a monthly publication. However, there had been eight issues in 1986, its first full calendar year of operation; there were only three in 1994, and publication became less frequent thereafter. In October 1990, Hauser began receiving assistance in the preparation of *DXLD* from Bruce MacGibbon of Gresham, Oregon. This arrangement lasted for only a few issues, however. The last *DXLD* appears to have been No. 49, issued on May 1, 1998, nearly a year after the next previous edition.

Prior to the closure of *DXLD*, Hauser started another shortwave news service, *Glenn Hauser's Shortwave/DX Report (GHSWDXR)*, which was distributed on-line. While it was principally for the use of other DX editors in their publications, soon it was archived on the internet for general use. (It also formed the basis for columns of Hauser-generated material which commenced in the World DX Club [U.K.] and Canadian International DX Club bulletins in December 1997 and January 1998 respectively.) *GHSWDXR* was issued approximately weekly at first, more often later. Soon after the final issue, which was published on December 31, 1999, it was replaced by a new on-line publication, adopting again the title *DX Listening Digest*. The new *DXLD* would be, as Hauser described it, "the primary outlet for our personal observations, for material contributed directly to us, and important timely or relevant reports from elsewhere." Issues of the on-line *DXLD* continue to be published every few days. In April 2004 it was supplemented by a *DXLD* Yahoogroup where users can post messages with current DX news, some of which is used in the on-line *DXLD*.

DXers everywhere have benefitted greatly from the republication of Glenn Hauser's material by many shortwave clubs. Over the years Hauser has also authored shortwave columns for *Popular Electronics* and *Monitoring Times* and edited columns for the Canadian DX Club, the Worldwide TV-FM DX Association, and the North American Shortwave Association. He has also contributed information to many DX programs, and is the creator of "World of Radio," the longest-running DX program still on the air.

OZDX (1986–1998)   *OZDX* was a high quality Australian DX newsletter. Although its official start was in May 1986, its origins date to 1984 when, following the demise of the "Down Under" DX Circle, Geoff Cosier produced seven untitled and informal DX newsletters, adopting the name *OZDX* in number eight. Two more informal editions were published in January 1985, followed by a long break and, in December, a final number in the informal series. The first of the "official" *OZDX* issues was published in May 1986 and edited by well-known Australian DXer Peter Bunn of Highett, Victoria.

Open to all DXers, *OZDX* followed some of the same principles as the *DUDXS* but it used a more conventional, non-survey format. Varying in length from four to 14 pages in length, *OZDX* consisted of shortwave and medium wave loggings, by-country station information, and QSL data, all contributed by its membership of serious DXers located in both Australia and elsewhere. The shortwave focus was the tropical bands, but interesting stations on the higher frequencies were covered as well. In 1992, the editorship passed to another top notch Australian DXer, David Martin of Mansfield, Victoria, and four years later membership peaked around 65. Official publication ceased in June 1997, mainly because, by that time, internet resources were able to provide DX news faster than paper publications. A few additional informal numbers were issued during the following year.

The DX Spread (1987–1990)　　*The DX Spread* was a two to six page monthly SWBC DX newsletter issued by Bruce MacGibbon of Gresham, Oregon. It was begun as a replacement of the *SPEEDXgram* which MacGibbon had edited and which had been dropped by SPEEDX. The first *DX Spread* was in April 1987. Soon it went approximately bi-weekly, sometimes with two issues combined in one. It merged into Glenn Hauser's *DX Listening Digest* in September 1990.

Radio News Bulletin (1993–1997)　　From 1993 to 1997, Dr. Adrian M. Peterson of Indianapolis, Indiana, Coordinator for International Relations of Adventist World Radio, produced an informative newsletter called *Radio News Bulletin*. Issued monthly at first, less often later, its distribution included approximately 250 DX publications and programs, leaders in the DX community, and AWR listeners. A well-known DXer, radio personality and shortwave historian, Peterson knew what would interest DXers, and his professional role in broadcasting gave him access to information not otherwise available. Among the topics covered in *Radio News Bulletin* were radio history, information on the doings at various stations, club activities, and highlights from other publications and DX programs, plus updates on AWR schedules, news and information about AWR activities around the world, and occasional pieces about his own travels and DX experiences. *Radio News Bulletin* also included photos of many interesting QSLs from Peterson's extensive collection. (Peterson would often arrange for the issuance of specially endorsed QSLs on special AWR occasions.)

Jihad DX and ***Cumbre DX*** (1994 to the Present)　　*Jihad DX,* which commenced publication in September 1994, was the first high-quality American DX newsletter to adopt e-mail as its sole method of delivery. It was founded by Hans Johnson, then of Columbia, Maryland, who had just completed a year editing NASWA's "Listener's Notebook." Johnson, an Arabic speaker, explained that there was more to jihad than the "holy war" connotation usually ascribed to it by westerners—that a jihad was a "supreme effort," which is what he promised in *Jihad DX.* This did not satisfy some, and in December 1995 the name was changed to *Cumbre DX,* cumbre meaning "top" or "summit" in Spanish.

　　Cumbre DX was a weekly and was intended to emphasize rare DX and get the information around faster than weekly paper bulletins. It covered SWBC only, excluding pirates (until 2006). Contributions were made by e-mail, and special between-weeks editions were issued when the news warranted it.

　　Anticipating a problem encountered by virtually all contributor-reliant DX publications—recipients who did not contribute—*Cumbre* had no memberships or subscriptions. Rather, one who contributed to a particular issue received that issue and that issue only. Although soon this was broadened to accommodate requests for individual issues to which one had not contributed, even this was found too restrictive, and eventually the present rule, which seeks monthly contributions, was established. Johnson edited the newsletter himself at the outset, but soon enlisted other DXers for that role, and at one time nearly a dozen shared the editing on a rotating basis.

Cumbre attracted many well-qualified DXers. In 2002 the weekly electronic newsletter was replaced by a list service which automatically distributes contributors' messages to all list members, and *Cumbre* continues to operate on that basis. Late in 1995 a weekly half-hour "DXing With Cumbre" DX program was begun over shortwave station WHRI.

The *Cumbre* website contains an archive of the early weekly newsletters as well as some station audio files (mainly from 1997 to 2001), but otherwise serves mainly as an informational vehicle for those interested in joining or in hearing recent "DXing With Cumbre" programs over the internet. Special reports on various DX topics were occasionally posted as well, and CumbreLite, a database, updated weekly, of digests of *Cumbre* loggings, was available on request during much of *Cumbre's* life as a weekly.

Cumbre has also been active in urging contributors to send elder copies of the *World Radio TV Handbook* and *Passport to World Band Radio* to DXers in developing countries and has provided contact information for those wishing to do so.

OTHER NEWSLETTERS

There were a number of other newsletters that deserve mention, most of them serving specialized audiences.

Asia An English-language newsletter from Japan was the *Far Eastern DX Review,* published on a monthly basis starting in 1976 by the Japanese Association of DXers in the person of Isao Ugusa and Hiroyuki Nagase. Later it became the *New Far Eastern DX Review,* then *DX Front Line.* It was highly regarded, and focused on East and Southeast Asia, covering in detail the shortwave transmissions from Vietnam, Laos, Kampuchea, Korea, etc.

Latin America A number of newsletters specialized in Latin American DXing. One was *Radio Nuevo Mundo (RNM),* a small-circulation newsletter published by the Radio Nuevo Mundo group in Japan starting in 1978. *RNM* covered Latin American stations exclusively, and included reports from members— several of whom traveled to Latin America specifically to visit radio stations—along with loggings, QSL information, and material from other sources. The group declined in size over the years, and the paper edition was dropped at the end of 2004. An electronic edition has been published occasionally thereafter. The club was best known for its series of books called *LA DXing.* Periodically it also issued *RNM* pennants, in keeping with a widespread practice among Latin American stations.

Well-known Japanese DXer Takayuki Inoue Nozaki published a newsletter called *Relampago DX* beginning in 1986. *Relampago DX* contained detailed reports of his loggings, often with the texts of station identifications. In addition, "TIN," as he was called, traveled to Latin America to visit stations, and *Relampago DX* often contained interesting accounts of his trips.

Latin American shortwave enthusiasts could also enjoy the English-Language *Latin American DX Report* published by Emilio Pedro Povrzenic of Argentina from 1982 to 1985 when it was merged into Povrzenic's Spanish-language publication, *Latinoamérica DX,* which began publication in 1978 and continued into the mid–1990s. Another English-language Latin American newsletter was *Pampas DXing*, published by Julian Anderson of Buenos Aires, Argentina. It was issued on roughly a monthly basis from 1987 to 1992 when it ceased publication. It reappeared briefly in 1994, by which time Anderson was also editing the annual *Tropical Bands Survey* for the Danish Shortwave Club International. *Pampas DXing* included loggings by Latin American DXers, as well as feature articles on the technical and historical aspects of various stations and the culture and geography of the area. Another English-language bulletin was *The Radio News,* published from 1988 to the mid–1990s by Jario Salazar Rojas and Manuel A. Rodriguez Lanza of Venezuela.

Onda Corta was a well-known Spanish-language newsletter published monthly from 1987 to 1994 by Gabriel Ivan Barrera of Buenos Aires, Argentina. Although its scope was worldwide, it was strongest on Latin American news.

Africa   *QTH Africa,* edited by Miki Vcelar of Pretoria, South Africa, was published from 1979 to 1982. A two-page newsletter issued every 10 days, it covered shortwave broadcasters in Africa and occasionally elsewhere. By virtue of his location, Vcelar often benefitted from on-air information that would not ordinarily be available to the DX community, and *QTH Africa* was the best source of information on the changing broadcasting scene in Mozambique and the countries of southern Africa. Subscribers received updated schedules of the external services of African stations four times a year, and a survey of clandestine African stations twice a year. Vcelar also produced monthly African DX reports for Radio RSA's "DX Corner" program, AWR-Asia's "Radio Monitors International," and TWR's "DX Listener's Log" program over KTWR, as well as cassettes containing recordings of African stations.

USSR   *Orbita* was the newsletter of the USSR DXing Circle in Japan. It covered shortwave activities in the USSR and was published for roughly five years starting in 1982. The USSR DXing Circle also published the *USSR DXing Handbook.*

Anatoly S. Klepov began publishing the monthly English-language *Soviet Radio Today* in 1991. It contained news on new stations in Russia and other parts of the former Soviet Union. Eventually Klepov formed the Russian DX League and *Soviet Radio Today* became *DX Moscow,* then *ASK-DX,* and finally *RUS-DX,* which is still published on the internet.

Pirate Radio   *The Wavelength* was the earliest pirate radio newsletter in the U.S. It was the voice of the "Free Radio Campaign–U.S.A.," and was published monthly by Al Muick of Wescosville, Pennsylvania. The first issue was in March 1979.[182] *The Wavelength* contained members' loggings, QSLs, and news of domestic and European pirates submitted by members and by the station operators themselves. Its P.O. Box 542, Wescosville, address soon became a popular pirate maildrop. Operation of *The Wavelength* briefly passed to Darren Leno of Moorehead, Minnesota, in July 1979 when Muick entered the army, but Muick was back at the helm in October. Another name that was involved with Free Radio Campaign–U.S.A. was John T. Arthur, then of Potter Valley, California, who was QSL editor. *The Wavelength* went bi-monthly in April 1981, but Muick stopped publishing soon thereafter when continuation of *The Wavelength* became incompatible with his military service.

A number of pirate newsletters made their appearances in the early 1980s. Based in

DX FRONT LINE

Issue nr. 38   1 Nov 1979

Published monthly.

World's first and only DX magazine that specializes in Indo-China.

This month's front page: "Entering the Monkhood", painted by a child in Thailand.

PROPER CREDIT: Other publications may use material from DX FRONT LINE provided that credit is given both to the original reporter and DX FRONT LINE (abbreviated DXFL). BBCMS material is copyright and should not be reproduced without their permission.

CONTRIBUTIONS to DXFL are highly appreciated. Reports should be written in any of the following languages: English, Japanese, French, Bahasa Indonesia, Bahasa Malaysia, Thai, Lao, Kampuchean, Vietnamese, Burmese or Korean. When reporting, be sure to write clearly.

GENERAL CORRESPONDENCES on this publication should be sent to:

Isao Ugusa, 1-3-10 Mikage-yamate, Higashinada, Kobe, 658 JAPAN/JAPON.

DX Front Line was one of several publications that focused on DXing stations in particular parts of the world (in this case, East and Southeast Asia).

Seattle, Washington, *The Unofficial Radio Bulletin* was issued by the "International Association of Free Radio" during the years 1980–82. The names of the principals were said to be James T. Jameson, publisher, and Mickey Anderson, editor.

Another pirate publication was *Selectivity,* the bulletin of the "Society for Hearing Illegal Transmissions." Published from May 1981 to September 1983, it was edited by Gregg Bares of Watertown, Connecticut. Its off-color parentage notwithstanding, *Selectivity* contained a good deal of interesting shortwave pirate material in its five or so pages. A successor *Selectivity*, edited by Scott McClellan of Battle Creek, Michigan, was published from October 1983 to August 1985 when it was renamed *Monitoring Crimes* and published until January 1987. Both Bares and McClellan had also been connected with *The Wavelength.* (From January 1983 to February 1986, Bares, then living in Washington, D.C., also published the periodic *BBI Newsletter [Bootleg Broadcasters International]* for pirate station operators.)

A well-known pirate newsletter was the bi-weekly *Pirate Pages*, which was edited by Andrew R. Yoder, then of Beaver Falls, Pennsylvania. Yoder started publishing *PiPa*, as it became known, in April 1989 in order to provide a substitute for the ACE bulletin, which was in a temporary publishing hiatus. *PiPa* was available by subscription. Typically it contained two to four pages of contributors' loggings, QSLs, and other pirate news. The promptness of publication made its content current and thus of high value to active listeners. Although the ACE bulletin returned in June, *PiPa* continued publishing, and remained in operation until late 1998 when Yoder turned his attention to another project, *Hobby Broadcasting* magazine. *PiPa* was published in e-mail as well as paper form during approximately its last two years.

Free Radio Weekly (FRW) is an e-mail pirate newsletter. It was begun in October 1995 and is still being published. It was founded by Chris Lobdell. In 1999 he was succeeded as editor by Neil Wolfish of Toronto, Ontario. A rotating system of co-editors was instituted the same year, and Wolfish served as editor until early 2006. *Free Radio Weekly* was established principally in response to a decision of *Cumbre DX* not to cover pirate loggings, and while the *FRW* distribution policy was flexible, *FRW* was intended mainly for contributors. It is now posted on the internet.

A variety of pamphlets on pirate radio, mostly self-published, have circulated within the pirate community at various times. Keith Thibodeaux of Baton Rouge, Louisiana published *Pirate Activity for 1983,* the *Guide to North American Pirate Activity* (1984), and the *Pirate Radio Report* (1986). These were collections of loggings from the relevant years, along with some pirate radio–related articles. The *Free Radio Handbook*, a 20-page booklet first issued around 1984, contained basic information about pirate radio— terms, transmitters and antennas, audio, propagation, operating tips, and sources of help. A second, 28-page edition edited by John T. Arthur was published in the early to mid–1990s and covered similar topics but with different content.[183] It was probably 1988 when "Zeke Teflon" wrote *The Complete Manual of Pirate Radio,* a 20-page pamphlet covering mainly the technical side of pirating. A second, slightly expanded (48-page) edition was published in 1993 and became something of a cult classic.[184]

In 1994, following his issuance of several databases of pirate and clandestine DX information, Kirk Trummel produced his *Little Black Book,* a 29-page listing of hundreds of pirate and clandestine station addresses. Trummel also maintained bandscans of the shortwave bands most used by pirate stations. These were available on paper or via download from the ACE section of the ANARC BBS, and updated via ACE. And in 1996, Kirk Trummel and John Cruzan published *The Pirate Radio Survival Guide*, a primer for pirate radio operators.[185] (Trummel died in 2001.)

Equipment   For vacuum tube radio enthusiasts, the *R390 Users Group Newsletter* began publishing in the spring of 1983. It was edited by T. J. "Skip" Arey, then of Edgewater Park, New Jersey, and it contained extensive information on the venerable R390 receiver. Soon it was renamed the *Hollow State Newsletter* and it expanded its coverage to include vacuum tube receivers in general. Editorial duties shifted to Dallas S. Lankford of Ruston, Louisiana in 1985, Reid C. Wheeler of Olympia, Washington, in 1993, and Barry Hauser, East Meadow, New York, in 2000. Fifty-three issues were published before the *Hollow State Newsletter* closed down in 2002.[186]

Another publication covering equipment was *Radio Equipment Review*. It was edited by Ronald Pokatiloff of Zion, Illinois, who hoped to fill the void in the area of equipment reviews left by the closure of the International DXer's Club of San Diego following the death of its president, Larry Brookwell. As with the IDXC, *Radio Equipment Review* featured reviews of new shortwave receivers and accessories, user comments, Pokatiloff's views on the best and worst receivers, etc., plus coverage of some radio-related computer topics. It was published bi-monthly beginning in March 1985, and issues were typically around 15 pages in length. The subscriber base was small, however, and it appears that *Radio Equipment Review* ceased publishing in early 1987 after 12 issues.

Women   One newsletter catered to women, of whom relatively few have been attracted to shortwave listening. The newsletter was *Wavelengths,* published in Cambridge, Massachusetts, from 1994 to 1996. It covered readers' letters, reviews of programs and books, and the like.

Education   One of the interesting sidelights of shortwave has been its use in the classroom. In 1989, educator Myles Mustoe wrote *Shortwave Goes to School,* a practical guide to the use of shortwave in the classroom. It explained how to set up a listening post in the classroom, and included model forms and student activity assignments.[187]

The subject received additional attention in 1994 when Neil Carleton of Almonte, Ontario, a museum administrator turned elementary school teacher, began *The Shortwave Classroom* newsletter. Published for four years, the newsletter's goal was to help establish a network of teachers who use shortwave listening and amateur radio in the classroom to teach languages, as well as subjects such as global perspectives, media studies, world geography, and social studies. It benefitted from the experiences of contributors worldwide, as well as Carleton's many projects, which included the establishment of a shortwave listening club in his school, and making ham radio contact between his students and the astronauts aboard the International Space Station. In 1996 and 2001, the students in Carleton's Shortwave Listening Club class prepared a special program that was broadcast over HCJB and issued special QSLs for the program. Carleton's work has been widely acknowledged in the press and on the air, and within both the shortwave and the educational communities.[188]

Recordings

Audio recording provided a means of introducing people to shortwave and sharing recordings of stations that were audible in one part of the world but not another. One of the best-known examples of an introduction to shortwave was a 45 rpm record produced by Hallicrafters in 1959. Narrated by "Man on the Go" Alex Drier, *The Amazing World of Short Wave Listening* was a promotional item. It opened to the sound of Morse code, with Drier saying, "This is the sound of adventure, and I am Alex Drier, your 'Man on the Go,' here to escort you on a most unusual trip around the fabulous, fantastic world of shortwave radio." It contained stories and recordings about learning languages, time station

A world of excitement was out there waiting to be heard, according to the Hallicrafters 45 rpm record, "The Amazing World of Short Wave Listening."

WWV, air and marine radio, ham radio, broadcasts of emerging events, etc. As with many other commercial shortwave promotions, it promised more than shortwave could actually deliver.

Reel-to-reel tape recorders were widely available to consumers by the 1950s. In 1964, the *WRTH's* World Publications offered the *Broadcast Identification Tape*, a dual track, 3¾ ips reel-to-reel tape containing an hour of studio-quality recordings of identifications and interval signals of some 65 stations. In 1967 they offered a second tape, *Latin America ... In 45 Minutes,* this one containing recordings of 40 Central and South American stations.

Also in 1964, SWL Records in Culver City, Colorado, a company founded by ASWLC member Win Klontz of Los Angeles, California, produced a 12-inch, 33 rpm record containing 40 minutes of station identifications and interval signals. Among the 18 stations included were large broadcasters like Radio Berlin International, Radio Finland, and the Voice of Free China, plus smaller ones such as Radio Clube de Moçambique, Radio La Cruz del Sur (Bolivia), and the Solomon Islands Broadcasting Service. A second volume followed in 1967. Called "The Rarest DX in the World," it presented recordings from such stations as Radio Andorra, Radio Nepal, the Fiji Broadcasting Commission, Radio Wewak (Papua New Guinea), and Radio Cook Islands. In 1970, Edward C. Shaw of Norfolk, Virginia, an active NASWA member, made available a reel-to-reel tape of almost 100 different interval signals, most recorded off-air.

In 1974–75, the Radio Canada Shortwave Club program produced two features which were then made available on tape to DXers. Of particular note were the *Identifications and Interval Signal* series consisting of several tapes, each covering a different part of the world,[189] and the *Foreign Language Recognition Course,* a 90-minute tape with spoken examples of 55 languages and comments by well-known DXer and linguist Richard E. Wood. These were sold in both cassette and reel formats through the Handicapped Aid Program even after the Radio Canada Shortwave Club closed in 1975. They were re-issued on CD in 2006 and 2007 respectively by Ian McFarland as part of a charitable fund-raising effort.

In 1978, *Long Live Short-Wave* was published by Trans-Island Productions, Ltd., on the Isle of Man, U.K. Available either in record or cassette format, and accompanied by a booklet, the 52-minute recording included an introduction to the shortwave hobby, identifications of over 30 stations, and discussion about various kinds of transmissions audible in the shortwave spectrum, plus information on receivers, antennas, QSLs, satellites, propagation, how to identify slow scan TV, radioteletype, etc. Henry Hatch of the BBC World Radio Club was one of the narrators. It was available in the U.S.

With the coming of cassettes, the 1980s saw the increased availability of station recordings. In 1980, Bob Grove, who would go on to create *Monitoring Times* in 1982, offered *Sounds of Shortwave,* a professionally produced, 60-minute cassette containing examples of jamming, "spy" transmissions, slow scan TV, radioteletype, etc., along with answers to questions about receivers, antennas, improving reception, etc.

Also in 1980, the Free Radio Campaign–U.S.A. offered a series of cassette recordings of American and European pirate stations, with the proceeds to benefit the Handicapped Aid Program. In the mid–1980s, the Canadian and U.K. chapters of HAP offered a variety of tapes for sale. These included the Radio Canada International tapes mentioned above, along with the *HAP Unofficial Radio Series* (six tapes covering clandestines, pirates, hoaxes, etc.), the *HAP DXer's Course* (four tapes on various topics), and the *Hitchhikers Guide to*

DXing, a three-tape series parodying the famous radio and TV series (and international broadcasting itself), originally broadcast on Radio Netherlands' "Media Network." Some of these tapes were also available from other sources. HAP-U.S.A. also sold tapes of ANARC convention proceedings.

Several other tapes that were sometimes found in the United States in the 1970s and the 1980s were produced by the Finnish DX Association. The recordings on these tapes were made from the vicinity of the station and thus were of very good quality. One, produced in 1976, was *World on the Air,* a 60-minute tape of identifications and interval signals from 90 stations around the world. It was similar to the group's *U.S.A. and Canada on the Air,* available in LP or cassette form and offering airchecks of North American stations. *World on the Air* was followed in 1980 by *America Latina en el Aire,* a 90 minute cassette tape of announcements, jingles and program excerpts from some 58 stations of Central and South America. It was accompanied by a booklet containing partial texts of the recordings and some useful information about Latin American broadcasting practices.

Station recordings had become less of a novelty by the 1990s, but production was easier and a number of tapes were in circulation. In 1992, Harold B. ("Hausie") Hausenfluck of Richmond, Virginia, made available a set of cassette tapes of over 100 interval signals. A tape from the pirate radio community was *The Pirate Zone,* a 1993 cassette of 28 pirate IDs (many weak and distorted). Tapes of pirate air checks were also available from Panaxis Productions, Paradise, California. In 1998, pirate expert Kirk Trummel also offered a CD of many pirate stations. And in 1994, Radio Shack, borrowing the name of Bob Grove's 1980 recording, issued *Sounds of Shortwave,* a book-tape combination produced by Ken Winters. Mainly for beginners, it was an introduction to shortwave, and included various sounds that are encountered on the shortwave bands—jammers, data transmissions, etc.

More recently, collections of station recordings have been made available on CD. These have included, in 2000, *Chuck's DX Audio Archive,* a collection of 50 station recordings together with images of QSLs, receivers and DXpeditions prepared by the late Chuck Mitchell of Indianapolis, Indiana; in 2001, *Latin American Radio Stations, 1975–2001,* a 371-recording CD produced by Max Van Arnhem of the Netherlands; and several CDs containing recordings of pirate stations.

DX Humor

The seriousness with which some hobbyists took their shortwave resulted inevitably in a few well-deserved attempts to poke fun at the DX culture. One of the earliest was an occasional column that appeared during the years 1962–1964 in the Canadian DX Club bulletin, *CADX.* At first called "The Eavesdropper," then "D.O.T. Images," it purported to report on the conversations of roving reporter VE7ZZZ with various hobby persons on radio-relevant topics, e.g. SWL-card swapping vs. listening, the doings of local chapters, adding a "tube multiplier" for contesters using small receivers, club convention activities, etc.

Another early attempt at humor was the *Monitor,* a publication of the "Won-Mor DX Club." To the best of the memory of its principal author, Edward C. Shaw (who was assisted in its production by Alvin V. Sizer), there were four annual issues which were passed out at various DX conventions from 1967 to 1970. By popular demand, a final edition of the *Monitor* appeared in 1973 for distribution to those in attendance at the 1973 South Charleston, West Virginia *Numero Uno* convention. The *Monitor* parodied many of the issues and the personalities on the DX scene at the time.

More widely known were the Charlie Loudenboomer columns which appeared monthly from January 1966 to January 1977 in *FRENDX*, the bulletin of the North American Short-wave Association. Charlie was always happy to recount the activities of his favorite club, the International World Wide Continental DX Association. Many topics were fair game, including clubs, contests, QSLs, receivers, radio conventions, DXpeditions, super DXers, SWL-card swappers, etc.

So popular was Charlie that in 1973 NASWA published *The Best of Loudenboomer — The Collected Works of Charlie Loudenboomer, Vol. 1.* (There was no Vol. 2.) However, Charlie was eventually dropped from the bulletin because of a satirical column about a club member who took offense. The secrecy surrounding the real identity of CL continues to this day.

Other clubs had some occasional humor as well. In the early 1970s, the ASWLC "Potpourri" column carried some hobby satire under the name "Hi Lights." Beginning in March 1978, a CL wannabe, the "Clara Listensprechen Report," appeared in *SWL*, the bulletin of the American Shortwave Listeners Club. Clara appeared through 1979 and a few times in 1980 and 1981, but the humor was well-below Charlie's standards. In 1985 and a few years thereafter, the Canadian International DX Club had its annual "April Fools" article where Dr. C. Q. Gigglehurtz (alias Shawn Axelrod of Winnipeg, Manitoba) addressed the impact on one's physical and mental well being of Rigatis Mortis, Sun Spotus Absentius, Non ID'us Transmittus, and other DX ailments. NASWA returned to the humor circuit with Harold N. Cones' "Dr. DX" series that appeared in *FRENDX* most months from November 1987 to April 1989. And when the NASWA "Technical Topics" column was edited by Joe Buch (1993–2006), he typically celebrated April 1 with a story spoofing some technical aspect of shortwave listening. (Some of the spoofs were not immediately recognized as such by casual readers.)

The best known modern-day DX humor is *BLANDX*, a hobby parody bulletin in a format similar to *FRENDX*. The first edition was published in 1981 by the Three Mile Island DX Association, a central Pennsylvania radio group. Material was submitted by various members, with most of the editing being done by Brett Saylor of Altoona. The original plan was to submit the material to NASWA for publication in *FRENDX*. Instead, it was decided to produce *BLANDX* with a private, limited distribution. Soon, however, *BLANDX* was a hit throughout the hobby. The authors became busy with other pursuits, and *BLANDX* did not reappear until six years later when it was resurrected by one of the original Three Mile Islanders, Don Moore, then of Glendale Heights, Illinois, who proceeded to edit and publish seven issues during the years 1987 to 1994. (Brett Saylor also published an issue in 1991.)

Unlike some attempts at hobby humor, *BLANDX* was genuinely funny. Moore's sense of humor was well-honed, and he, along with other hobbyists who contributed to *BLANDX* over the years, were hobby insiders whose experience led them easily to topics and personalities deserving of attention.[190] This was often expressed through some favorite *BLANDX* names, like Ray Framus and Sven Gonzalez, and "SASWA" columns like "Listeners' Trash-bag" and "DXers' Kindergarten" (takeoffs on NASWA's "Listeners' Notebook" and "DXers' Classroom").

Among *BLANDX*'s imaginative pieces: a movie review of *The Terminator* ("This film neglects the subject of propagation entirely! Unbelievable!"); an ad for the *International DX Inquirer* newspaper ("Yet *another* sex scandal at Grønlands Radio! ..."); the *BLANDX* Awards ("'THE' DXer, for verification of 50 stations using 'The' in their name, slogan, or ID"); and a review of Kenya's Bleene JDG-1001 receiver ("At 25 pounds it's the first Bleene that can easily be carried by just one person. It can be powered by either AC or 96 AA batteries. That's right — no messy gasoline!"). Also from *BLANDX*:

> Are *You* a DX-aholic?—"I first realized I had a problem when I woke up after Memorial Day Weekend, 1989, and realized I hadn't been outside the house in over 62 hours, and had been wearing headphones continuously for three days."

> *Mistreatings 1990* [a parody of *Proceedings*]—Highlights In This Year's Edition: "Rod Curtain explains: 'Filter Modifications for the Cramosonic BRX-41 Receiver Using Just Sand Paper and a Monkey Wrench,'" "Bill Dernoff's superb 'The File Cabinet as a DX Accessory,'" "Ryan Finelli on 'Waterproofing Your QSLs,'" and "Active Antennas— Why Are They So Hyper?"

> Van Helder's DX Therapy Clinic—"Learn to be one with your receiver through transcendental meditation."

> The DXer's Revenge—"Can't hear the Marshall Islands due to all the RFI QRM in your neighborhood? Use the new DXer's Revenge from the Union of Exiled Ugandan DX Clubs. It's a washing machine sized cyclotron which can be used to generate all-band QRM for your entire neighborhood (they won't even be able to watch cable TV!). It's a great stress relieving device; you just fire it up on weekends and leave town to DX."

And from the *BLANDX* QSL columns:

> FINLAND: Radio Finland 15400 Rcv'd form ltr saying they don't QSL because of the expense involved. Also rcv'd a 20 lb package of tourist info.

> BELGIUM: BRT F/d cd of Renoir's famous painting, "Four Men Examining A Bent Fork." I'm glad some stations have good taste. I use these to teach art appreciation to my kids.

> SINGAPORE: Radio Singapore 9530 F/d "Greetings from Singapore — A Country That Really Knows How to Raise Cane" cd in 52 days.

> GREAT BRITAIN: BBC 7205 n/d pic of truck w/caption "It's a lorry, you dumb yanks."

> COSTA RICA: Casino Radio 5945v "Queen of Hearts" cd, #24 of 52 in the Stack the Deck series. 21 days for uncashed Florida Lotto ticket. V/s Huberto Cantillo, who says he corrects stamps as a hobby. In fact, the ones on the envelope had the old values crossed out in red ink and new ones written in.

If you knew anything about shortwave, you liked *BLANDX*. (Don Moore reintroduced *BLANDX* as a website in March 2007.)

5

Listener Programs

Every Wednesday night [in 1936] at 2 A.M., Edward Startz installs himself in the Philips Laboratory with a quantity of gramophone records, a pile of letters received from listeners, coffee and cigarettes, and converses through his microphone, alternating with music, for at least three and sometimes four hours a night. The broadcast, heard in New York at 7 P.M. Wednesday, begins with the Wilhelmus (the national anthem), which is directly followed by a few Sousa marches. And then, scarcely is the last sound faded than we hear that well-known voice, that audible contact with the homeland: "Good evening, ladies and gentlemen! This is the PCJ transmitter of Philips in Eindhoven. Good evening, Surinam, Curacao, Aruba and the Antilles. Good evening, Netherlanders in North, South and Central America. Here we are again![1]

Many stations carried mailbag programs, where listeners' letters were acknowledged. Beginning in the mid–1940s, a new kind of program, the "DX program," appeared. Its stated purpose was to provide listeners with current information on stations and frequencies by what was then the fastest method — on the air. More subtle was the value of the programs in building a listener base and generating letters that served to validate a country's shortwave service. Many such programs came into existence. Some were of marginal value, while others were sources of DX information not easily obtainable elsewhere.[2]

All the programs discussed below were broadcast in English. There were similar programs on other languages.

The Top DX Programs

"DXERS CALLING" (AUSTRALIA)

"Australian DXers Calling" appears to have been the first DX program. Sponsored by the Australian Radio DX Club, it took to the air on Saturdays in July 1946. Later it moved to Sunday. The program's first DX editor was Ernest H. Suffolk, one of a group of Australian DX club members who developed the idea and sold it to Radio Australia. Some years later Suffolk was succeeded by Australian master DXer Graham Hutchins. Hutchins passed away in 1964, after which the DX news was read by the Director of Radio Australia, Peter Homfray. Bob Padula wrote the scripts from 1966, sharing the honors with several others after 1972.[3] By 1971 the 10-minute program, by then called "DXers Calling," had a wide following. Throughout the North American shortwave community SWLs would start their Sundays by tuning in to the DX news from Radio Australia.

The last "DXers Calling" was on October 30, 1977. (A Japanese version continued for some time thereafter.) In later years, Radio Australia made several attempts at DX–related programming. Some DX items were included in "Club Forum," a 15-minute weekly program hosted by Keith Glover and aired from 1972 to 1981 in connection with the Radio Australia Listeners' Club. In July 1981, Radio Australia introduced a monthly program (later fortnightly) called "Spectrum." It was hosted by Dick Speekman, formerly of Radio Netherlands. It was produced in cooperation with the ARDXC and the Southern Cross DX Club, and its content was wide-ranging. Although it never developed the following of "DXers Calling," the information was some of the most timely available. The program was broadcast until 1983, and soon was succeeded by "Talkback," which also included some DX news. "Talkback" came to an end in 1985. "Communicator" followed and was on the air until 1993, but was devoted more to electronics and general communications than DX news. Padula has continued to contribute regular segments (now known as the "Australian DX Report") to various international DX programs over the years.

"Sweden Calling DXers"

Though not the first DX program, Radio Sweden's weekly 10-minute "Sweden Calling DXers" was one of the most popular, an important albeit brief part of a North American DXer's Tuesday night listening.

The first edition of "Sweden Calling DXers" was broadcast on February 28, 1948 over the station's modest 12 kw. shortwave transmitters. Arne Skoog, one of Sweden's foremost

Photos of the Australian contributors to "Talkback" are shown on this Radio Australia QSL-card.

Radio Sweden's Arne Skoog was one of the foremost authorities on shortwave broadcasting.

authorities on shortwave radio, was the DX Editor. He produced the scripts, and the program was voiced by a professional announcer. It consisted of hard DX news compiled from listeners' contributions. Beginning in 1948, "SCDX" began airmailing to DX clubs and contributors a two-page stenciled "script" summarizing the items presented on the program. Although the wide range of international contributors, together with the sometimes-relaxed editing, made the program susceptible to honest mistakes and occasional hoaxes, SCDX became a valuable, and very popular, source of DX information.

The program was canceled for budget reasons in 1954. However, popular demand soon brought it back. Notwithstanding occasional pruning of the mailing list, the initial circulation of 50–100 copies of the scripts grew to 400 by 1960, 700 by 1963, and over 1,200 by 1969, when the mailing was stopped. So popular had the scripts become, however, that the mailings resumed six months later, albeit on a bi-weekly basis.

Some 40 percent of all the mail to Radio Sweden was for SCDX. The mailing list reached 1,400 by 1973. From time to time SCDX would also issue other helpful listening publications, the first being the 1949 publication, *The World in Your Loudspeaker.* Others included periodic club lists, and, much later (in the 1980s), *The Beginner's Guide to DXing* and *The DXer's Guide to Computing.* For a time the program would also include a tape recording of a station, supplied by a DXer. However, this feature was cancelled in 1964 when the program was shortened by several minutes. In 1978 a new feature was added. "Tune In" made suggestions for interesting programs to listen to from the international broadcasting stations. It did not last long, however, for SCDX listeners were interested in DX, not program content.

Eventually there were English, French, German, Spanish, Portuguese and Russian editions of SCDX, and in preparing the program Skoog would have the assistance of several fellow DXers. The 1,000th program was aired on January 7, 1969, and a special QSL-card issued. SCDX celebrated its 25th anniversary in 1973 with a week-long contest where participants were asked to log the most DX programs from other stations.

The last program prepared by Skoog was on October 4, 1977.[4] Thereafter George Wood produced most of the programs and scripts. Skoog retired in 1978. (He died in 1999.[5]) The SCDX scripts were mailed monthly rather than bi-weekly starting in 1980. Over time the program took on more of a "media" flavor, with emphasis on Nordic activities and satellite and computer topics rather than shortwave. Later Wood authored a widely read pamphlet that contained detailed information for satellite radio enthusiasts. Called *Communications in Space: The DXer's Guide to the Galaxy,* it went through several editions during the early 1990s.

In 1990, SCDX was cut to twice a month and electronic delivery of the bulletin replaced postal delivery. In late 1991, SCDX was renamed "MediaScan," and the on-line scripts were replaced with a *MediaScan/Sweden Calling DXers* on-line newsletter. Eventually the program lost its shortwave following, and the last "MediaScan" program was broadcast on July 17, 2001. *MediaScan/Sweden Calling DXers* continued on-line and by e-mail, always maintaining the chronology of the original SCDX numbering system. The last newsletter, No. 2490, was issued in March 2005. "MediaScan" was transformed into a blog and an RSS feed, and maintains a presence on the Radio Sweden website.

"DX JUKE BOX" AND "MEDIA NETWORK" (NETHERLANDS)

The first offering of a DX program from Radio Netherlands was "DX Juke Box." It began circa 1961, having grown from a five-minute DX report that first came to air in 1958. Its host, Harry van Gelder, had been a tax consultant for 15 years prior to joining the station's staff in 1955. He assembled the program in cooperation with the Benelux DX Club and the Finnish DX Club, and presented it with the help of Jim Vastenhoud of the station's engineering staff and others, including, toward the end of van Gelder's tenure, well-known American DXer Serge C. P. Neumann, who was then working at the station. As would be the case with the program's successor, "Media Network," "DX Juke Box" contained informative features, interviews, DX news, equipment updates, propagation and technical information, answers to letters, and some music. The music, there to attract music lovers to DXing (hence the name of the program), soon became incidental.

Van Gelder retired due to ill health in 1974[6] and was succeeded by 30-year-old Dick Speekman who had migrated to Australia at age 17 but been born in the Netherlands. Vastenhoud stayed on. Speekman returned to Australia in 1979 and passed the reins to Frits Greveling. Greveling returned to his native South Africa in 1980 and joined the Dutch service of Radio RSA. Harry van Gelder died in 2003.

The program took a major turn when Jonathan Marks took over from Greveling and became the show's producer-presenter. Marks, who was from England, had come to Radio

Netherlands by way of Radio Austria International and the BBC Monitoring Service. "DX Juke Box" became "Media Network" in May of 1981. The music part of "DX Juke Box" was dropped. The 30-minute program, now promoted as a "communications magazine" for both hobbyists and professionals, became broader in scope. DXers were far from forgotten, however, and, with Marks' likeable style, "Media Network" soon became the premier program of its kind.[7]

Starting in 1967, "DX Juke Box" had included a monthly DX report by Glenn Hauser. Over the years similar reports were presented regularly by other DXers, including, in 1974, a monthly Asian DX report by Canadian diplomat Gregg A. Calkin, then working in Pakistan. By 1977, four reporters were presenting five minutes of DX tips on a rotating basis: Arthur Cushen of New Zealand, Victor Goonetilleke of Sri Lanka, Jan Tunér of Sweden, and Hauser. Such reports were de-emphasized by Marks, although Goonetilleke was heard often, as were John Campbell of the U.K. on pirate and clandestine news and Richard Ginbey of South Africa on stations of that continent. Hauser's reports were dropped in 1982.

What set "Media Network" apart were its well-produced, polished format and its original content. As a working journalist, Marks was interested in real news—new stations, new receivers, interviews with correspondents and radio personalities, field reports from stations and from world trouble spots, major radio-related documentaries, etc. Often he would pick up the phone and interview key people in the middle of a story's development. In addition, "Media Network" was unusually timely because it was recorded close to air

Jonathan Marks of "Media Network," shown here with his co-presenter, Diana Janssen. This QSL-card was issued in 1996 on the occasion of the program's 750th show.

time. When a European radio ship began testing on a Thursday morning in 1984, the news was on the program within an hour. Also, editions of the program broadcast to various parts of the world were updated on a continuous basis as developments dictated, and so a listener might hear different material in different versions. Marks ceased being a one-man team when Diana Janssen, the station's head of New Media and Strategy, joined him as co-presenter in 1994.

Despite its broadened scope, "Media Network" was a specialty program, and maintaining the high quality for which it had become known required resources that were needed elsewhere. In addition, Janssen was leaving the station, and Marks had taken on additional responsibilities, having left radio and TV production to head the English language service from 1991 to 1994, then serving as Director of Programmes for all of Radio Netherlands. As a result, "Media Network" closed at its height on October 26, 2000. Since then it has maintained its name and presence through an expanded website and a daily media news service maintained by regular "Media Network" contributor and former *WRTH* editor Andy Sennitt. Marks left the station in 2003 for a new career in writing, producing, consulting and research.

Even before "DX Juke Box," Radio Netherlands offered its listeners a wide variety of homegrown publications for SWLs. This developed into the Radio Netherlands "DX Information Catalog" (later the "Listeners Service Catalog") of information sheets and booklets on such topics as antenna construction, receiver selectivity improvement, BFOs, crystal calibrators, antenna tuners and other accessories, foreign-language reception reporting, etc. During the 1960s and 1970s, the information service offered on-air courses that were presented as part of "DX Juke Box." The courses included printed texts sent free on request. There was a radio servicing course (1965), a shortwave propagation course (1967), an "All Round DXers Course" (1968), an audio course (1971), and a communications system course (1977), plus courses on receivers, antennas, transistor theory, and other topics.

For some 20 years beginning in 1980, the station offered many editions of two popular monographs. The "Receiver Shopping List" started out as a simple product list and developed into a booklet of prices, specifications and comments on many of the shortwave receivers on the market. In later years it was available on the Radio Netherlands "Media Network" website until its final iteration was transferred to the website of the U.K. Medium Wave Circle in 2006. "The Booklist" was a pamphlet describing a wide variety of print resources for DXers. It eventually grew to some 50 pages. The station also made available booklets on shortwave-related software, as well as monographs such as "DXing Indonesia," "Latin American DXing," "Writing Useful Reception Reports," and the "Weather Satellite Fact Sheet." In 1994 it published "InfoDutch," a booklet about computer bulletin boards, the internet, e-mail, receiver control software, etc., and "Antenna Advice, v. 1.0," replacing an earlier brochure called "Give Your Antenna Some Air."

Radio Netherlands knew what DXers liked, and produced it well.

"DX Partyline" (HCJB, Ecuador)

On Mondays in 1960, HCJB had a program called "Party Line" which gave missionaries serving in Ecuador a chance to send greetings to families and friends back home. Each missionary was given a regular spot on a monthly basis. However, in months with five

Mondays the extra Monday was left unscheduled and was usually devoted to transmitter maintenance. When this was no longer necessary, it occurred to HCJB staffer Hardy Hayes that the station had never had a DX program. "DX Partyline" was born, and the first broadcast went out over the airwaves on May 29, 1961. At first offered only on the first Monday of five-week months, it was soon scheduled on a regular monthly basis, then twice monthly, and, after a few years, weekly. By 1973 the program was airing three different 30-minute versions each week, to North America, Europe, and the South Pacific. At its height in later years, "DXPL" would be 50 minutes long.

Hayes was the host for just a few months, and was followed by Clayton Howard, assisted by his wife Ruth. Clayton had arrived at the station in 1941 as a recent physics graduate, and Ruth soon joined him. They were married on the air. He worked in the station's engineering department and over the years played many roles in the station's technical development.

Clayton Howard's easy-going if somewhat deliberate style was a welcome counterpoint to some of shortwave's more strident political voices, and the program became strongly associated with Howard during the 24 years he served as host. It offered DX news, technical and other information about HCJB, interviews with staff and visitors, contests, news of DX events, and many other features, usually with an American flavor. DX tips were obtained from club bulletins amd other third-party sources, and from direct contributors. DXPL also hosted recorded contributions from DX groups and others, and over the years there were regular segments from SPEEDX, ASWLC, ODXA, ANARC, the South Pacific Association of Radio Clubs, the Association of Illinois DXers, the Miami Valley DX Club, the European DX Council, the Benelux DX Club, the Handicapped Aid Program, etc., plus Neil Carleton's "Radio Stamps" and DX reports from Arthur Cushen, Peter Bunn, Paul Ormandy, Henrik Klemetz, and others. In 1983, DXPL started carrying a radio version of Ken MacHarg's book, *Tune In the World,* which featured station profiles and program information.

The Howards retired in 1984,[8] and the last of their DXPLs was broadcast on June 18 of that year. John Beck hosted the program for the next four years. Brent Allred took over for several years after that, followed by Rich McVicar from 1991 to 1996, Ken MacHarg for the next two years, and Allen Graham from 1998 to the present. Since HCJB was one of the first stations a new SWL would run across, DXPL was always orientated toward beginners. Over time, however, especially during the post–Howard years, the DX content increased. McVicar introduced a mid-week DXPL updater called "The Latest Catch." In 1995 DXPL sponsored a special DX test of long distance reception of the HCJB medium wave transmitter which a number of North American listeners were able to hear.[9]

True to its evangelical mission, DXPL also included "Tips For Real Living" and other brief inspirational messages. But even those hobbyists whose sole interest was DX enjoyed the program. Unfortunately, it was a victim of a restructuring which eliminated most of HCJB's English-language programming in 2003. It remained on the air via a once-weekly transmission from Quito to North America and in several airings from the less well-heard HCJB station in Kunnunura, Australia, as well as from U.S. stations WWCR and WRMI and on the internet. However, the last DXPL broadcast from Quito was on May 6, 2006, when the station eliminated most of its remaining Quito-based English-language programming. Reduced now to 15 minutes, it is still presented over the other outlets, but it lacks the major on-air presence of earlier years.

"Radio Canada Shortwave Club" and "SWL Digest" (Canada)

Canada's DX program was an outgrowth of the program of the Radio Canada Short-wave Club. The bi-weekly Saturday program of the same name started in 1962, offering answers to listeners' questions and presenting information on various communications topics. The club president was S. Basil (Pip) Duke, a Londoner (U.K.) who had an extensive military, engineering, and broadcasting background before joining the CBC in 1954 and becoming Supervisor of Engineering Services for the CBC's shortwave broadcasts.[10] From 1967 to 1970 the club had a six-page monthly bulletin with news about the CBC International Service, member profiles and lists of members seeking pen pals, member loggings, shortwave-related features and the like. It went to 10 pages, bi-monthly, in July 1970, but lasted only until January 1971, just as the club reached 10,000 members. After that, some club news appeared in the station's program schedule. As a result of the need to cut costs and concentrate on CBC program content, the club was disbanded and left the air in 1975 (along with the RCI mailbag program, "Listeners Corner").

Radio Canada International reintroduced a DX program in 1977 in the form of the weekly "DX Digest" hosted by Ian McFarland. (It became "SWL Digest" in September 1981.) The program, which was soon a half-hour in length, made McFarland a well-known short-

The Radio Canada Shortwave Club was in operation from 1962 to 1975.

wave personality.[11] A Montreal Institute of Technology graduate, he had worked in the CBC's London, England, bureau before joining the CBC's International Service, where he was soon supervising the production of the Radio Canada Shortwave Club. His businesslike but friendly style appealed to both beginners and experienced hobbyists, and his frequent appearances at SWL gatherings made him a familiar presence in shortwave circles. The program contributed as well to the fame of Glenn Hauser, whose weekly five-minute "DX News Report" began airing in 1977 and continued throughout the program's nearly 14 year life.

Other "SWL Digest" features included ANARC and EDXC news, equipment reports from Larry Magne, and news of the Handicapped Aid Program, of which McFarland was a supporter. A series of "Radio Database International" reports by top DXers Don Jensen (U.S.) and Noel Green (U.K.) were presented in 1984–85. This was a joint project of "SWL Digest" and Switzerland's "Swiss Shortwave Merry-Go-Round," done in cooperation with *Radio Database International*. Each station carried a different edition of the program. Jensen continued with his reports after the series ended, and was then joined by Tony Jones of Paraguay. In the mid to late 1980s, "SWL Digest" also carried "Don Jensen's Journal," which covered topics of modern DX history.

Unfortunately, "SWL Digest" left the air in March 1991 as part of a major RCI downsizing. A nominal replacement, called "Media File," lasted only a few months.

"WORLD OF RADIO"

One of the longest running and most informative DX programs, Glenn Hauser's half-hour "World of Radio" has been heard over many shortwave stations, mainly in the U.S., for almost a quarter century. Originally called "Shortwave Review," the program came to air on the FM band circa 1979 when Hauser worked at WUOT in Knoxville, Tennessee, and used shortwave news to fill the time. It became "World of Radio" in October 1979, and made its shortwave debut on WRNO, New Orleans, Louisiana, in 1982, soon after the station commenced operation. Hauser's program could make DX news available even faster than the weekly DX newsletters. Over the years many other shortwave stations have made air time available at no charge for "World of Radio," including WWCR, WHRI, WJIE, WBCQ, WGTG, WRMI, KCBI, KAIJ, and Radio For Peace International, Costa Rica. "World of Radio" has also aired on a few BCB and FM stations, plus World Radio Network (satellite) and the internet.

Hauser was experienced in providing DX news over the air. From 1967 to 1982 he had done a monthly DX report on Radio Netherlands' "DX Juke Box" which had attracted a large following, and in 1977 he began holding forth weekly on RCI's "DX Digest." Around 1981 he had also provided DX news for Austria's "Shortwave Panorama," and from 1976 until the program was canceled in 2007 he presented a monthly DX report on Radio Netherlands' Spanish media program, "Radio Enlace" (originally "Espacio DXista").

"World of Radio" has always been broad in content, reflecting Hauser's interests in general broadcasting, shortwave programming, and DX news. The focus has been on shortwave, but with frequent references to other media as well. The program continues to the present, and is also repackaged into a monthly Spanish-language program, "Mundo Radial," which is carried on WRMI, WWCR and the internet (and was on RFPI when they were on the air). Another Hauser program, the monthly "Continent of Media," was also carried by RFPI and continues to be available on the internet. It covers mainly domestic broadcasting news from the Americas.

February 1993 saw Hauser experiment for five weeks with a daily DX program. The 15-minute "DX-Daily" was aired every weeknight at 0400 UTC and contained up-to-the-minute DX news. It was produced by Hauser with the assistance of Jeff White and Radio Miami International, sponsored by Grove Enterprises, Universal Shortwave and C. Crane Co., and transmitted over WRNO. Although the program resumed for a brief period in April, Hauser felt that it did not receive enough editorial support from the DX community and ended "DX Daily."[12] A year later he experimented briefly with a "DX-Daily" information line where, for 95 cents a minute, listeners could call a 900 number and receive four minutes of updated shortwave information.

DX Programs of Note

"DX WINDOW" (DENMARK)

Radio Denmark began a weekly 10-minute DX news bulletin, "Denmark Calling World Listeners," in 1950. It was compiled by *WRTH* editor O. Lund Johansen, and it continued until replaced by "DX-Window" in November 1967. The 20-minute "DX-Window" was prepared by the Danish Short Wave Club International. It had a modern, energetic style that gave it a unique sound among DX programs, and it was widely listened to until the end of 1969 when it was dropped in conjunction with Radio Denmark's elimination of English-language broadcasts.

"THIS RADIO AGE" AND "ARTHUR CUSHEN'S DX WORLD" (NEW ZEALAND)

In 1951, Radio New Zealand started carrying a monthly DX program produced by Arthur Cushen. It was called "This Radio Age," the title borrowed from that of an electronics magazine of the day. It became the 15-minute "Arthur Cushen's DX World" in 1964 and it focused on Pacific stations or stations heard by Arthur from his New Zealand location. It was on the air until 1982.

"DX CORNER" AND "SWISS SHORTWAVE MERRY-GO-ROUND" (SWITZERLAND)

The Swiss Shortwave Service began a monthly DX program in 1951. It became the weekly "DX Corner" in 1959 and it featured both broadcast and amateur news, plus acknowledgments of listener letters and other features. In 1963 it was succeeded by the 40-minute weekly "Swiss Shortwave Merry-Go-Round." One of the contributors was Bob Thomann, who became host of the program in 1966, to be joined five years later by fellow ham Bob Zanotti. The program focused more on listeners' technical questions than DX, and "The Two Bobs" presented it in an informal, impromptu style. It was eventually broadcast twice a month, but reduced to 20 minutes, then ten, and left the air in June 1994.

"Around the World," "SweDX," "Finn DX" and "World of Radio" (Finland)

Although they were not widely followed in North America, Finland had a number of DX programs over the years. In 1957, the Finlands DX Club inaugurated a monthly 20-minute DX program over Radio Finland called "Around the World." It was on the air twice a month by 1960, and featured DX news and special reports from other Nordic countries. By 1965, Radio Finland was carrying a bi-monthly English-language DX program produced by Sweden's DX-Alliansen called "SweDX." Also in the 1960s there was a weekly, hour-long program called "DX Mailbag." In 1972 a program called "Finn DX," produced by the Friendly DX Club, was begun, and over time grew in length from five minutes to 15 minutes. And in 1978, Radio Finland inaugurated a new fortnightly program called "World of Radio," the original source of the name which Glenn Hauser adopted in 1979 after Radio Finland dropped the program.

"DX Corner" and "Media Roundup" (Japan)

Radio Japan included a monthly "DX News" segment in its "Listeners' Corner" program beginning in 1958. It was prepared by the Japan Short Wave Club. The program improved over time, and in 1982, renamed "DX Corner," it occupied a ten-minute monthly segment of the "Tokyo Calling" program. It expanded to 25 minutes in 1990 and became "Media Roundup" in 1992. It covered both DX and media news, with features on communications, technology, domestic broadcasting, etc. There were also occasional appearances by Ian McFarland when he worked at the station (1991–93).

In 1997 the program was replaced with a new program, once again called "DX Corner" and now occupying a monthly five-minute slot on the "Hello From Tokyo" program. "Hello From Tokyo" was dropped in April 2006, and the DX program was renamed "DX Express" and became a twice monthly segment of Radio Japan's "World Interactive" program. However, it was eliminated from the Radio Japan program lineup in 2007 when "World Interactive" was shortened.

"Shortwave Listeners' Corner," "BBC World Radio Club" and "Waveguide" (U.K.)

The BBC's "Shortwave Listeners' Corner," begun circa 1961, consisted mainly of technical and non-technical features for SWLs presented by BBC personnel. The 15-minute program was on the air weekly in the General Overseas Service and was hosted by Dorothy Logan.[13] It was succeeded in 1967 by the 15-minute "BBC World Radio Club," a program with a similar mission, namely, to give listeners advice on reception and treat communications topics of broader interest as well, all in layman's language. The program also offered occasional DX news, sometimes gathered through the resources of the BBC Monitoring Service. Reg Kennedy was the program's producer, and the distinctive voice of Henry Hatch was heard regularly.[14] The program remained on the air until the end of 1980.

The BBC's "Waveguide" ran as a weekly program from 1982 to March 1996, returning in September as an eight-part series on basic listening and then a monthly program

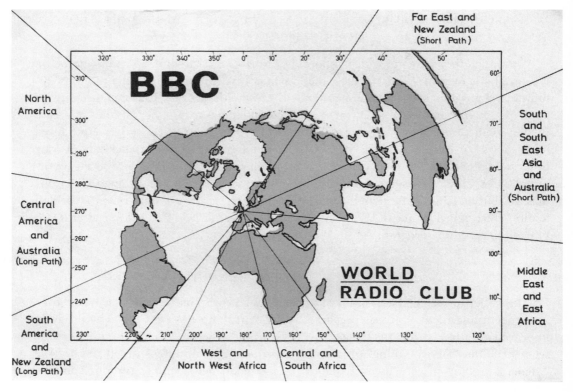

The BBC World Radio Club was both a program and a club. Not a DX program in the usual sense, occasionally it offered valuable DX tips from the BBC Monitoring Service.

until leaving the air for good in 2001. The brief program —five minutes at first, then ten — was hosted by Simon Spanswick, now Chief Executive of the Association for International Broadcasting. The focus was usually on the newcomer to shortwave, and on BBC–related topics. Listeners' letters were read, and some informational pamphlets were made available to listeners. "Waveguide" also carried some limited DX information developed by the BBC Monitoring Service.

"DX CORNER" (SOUTH AFRICA) AND "RADIO SAFARI" (PORTUGAL AND SOUTH AFRICA)

Radio RSA had a DX program as early as 1966. It became known as "DX Corner" and there were two different editions each week. For most of its life it was hosted by Gerry E. Wood, widely acknowledged as one of the best DX program editors. The six-minute program was a mix of club news, technical items, information on African stations, and thoughtful comments, plus an annual contest. Wood was succeeded by Pieter M. Martins in 1978. Martins was well-versed in shortwave by virtue of his position as engineer in charge of frequency planning at Radio RSA's monitoring facility. The program was on the air until 1990 when Radio RSA canceled all broadcasting to points outside Africa.

From 1966 to 1970, well-known South African DXer Richard Ginbey had a monthly English-language DX program over Radio Portugal ("Voice of the West") called "Radio

Safari." The program resumed for a time over Radio RSA in the early 1970s when Ginbey lived in South Africa. (Ginbey died in 1991.)

"World DX News" (U.K. and Denmark)

England's World DX Club produced a program called "World DX News" beginning in November 1972. It was broadcast Sunday mornings as part of the Adventist World Radio programming over the Deutsche Welle relay station in Sines, Portugal (and later the DW station on Malta). In 1979, production of the program shifted to the Danish Shortwave Club International, which had been out of the DX program business since its "DX Window" had ended in 1969. In November 1994, after a sojourn on the small AWR shortwave station in Forli, Italy, the program, by now called "DX Corner," was moved to the 250 kw. transmitter at Rimavska Sobota, Slovakia, which was used by AWR.

"DX Corner" was about 12 minutes long, with half devoted to DX tips and half to a feature. DSWCI stalwart Finn Krone handled the shortwave news, and medium wave was addressed first by Barry Davies, then Gordon Bennett, both of the U.K. For a time, a summary of the program's content was also carried in the DSWCI monthly bulletin. The program was high in quality, but broadcast at a time that made reception in the U.S. difficult. In January 1995, it was supplanted by the new AWR DX program, "Wavescan" (see below).

"DX Panorama" and "Radio Monitors International" (Sri Lanka)

A twice-monthly DX program called "DX Panorama" was produced by the Ceylonese Shortwave Listeners Club and broadcast over Radio Ceylon beginning in 1968. The scripts were prepared by Victor Goonetilleke and Sarath Amukotuwa. It was on the air for three years and was cancelled because there was no announcer available to tackle the technical terms and exotic place names.[15]

"Radio Monitors International" was broadcast over the Sri Lanka Broadcasting Corp. from 1975 to 1985, with re-runs heard occasionally after that. It was produced and presented by well-known DXer and then–Director of AWR-Asia, Adrian M. Peterson, at first (for about six months) from Columbo, Sri Lanka, and then from Poona, India, to which Peterson had relocated. The program was originally a ten-minute segment of a variety program called "Radio Journal." It became a separate, 15-minute program in 1978, and was extended to 30 minutes the following year. It usually began with a station profile, followed by DX news and "Window on the World," which consisted of material received from various DX clubs and DX programs which did not typically have a large Asian following.

While the Sunday morning program was not always easily heard in the United States, Peterson's prodigious knowledge of the medium gained through his personal DXing and his extensive travels along the world's DX trails made "Radio Monitors International" a unique listening experience. During 1985–86, a North American edition of the program was broadcast on Radio Earth.

"Worldwide Shortwave Spectrum" and "Communications World" (VOA)

The VOA started airing "Worldwide Shortwave Spectrum" as a segment of the weekly "VOA Magazine" show in 1984. The feature was hosted by Gene Reich, and focused on the telecommunications world generally rather than shortwave and DX. It became the longer (20 minute) standalone program, "Communications World," in 1987. The program grew to a half hour in 1995, the year Reich was succeeded by Kim Andrew Elliott of the IBB audience research office, who was a well-known figure in the DX community. Although the program remained broad in scope, Elliott was more attuned to the DX world, and the program often included DX tips and items of interest to the DX community. From 1998 it aired in three separate segments, which made continuity difficult even for devoted listeners (a combined version was available on WWCR and via satellite and the internet). It returned to a single program format in February 2000. However, the program ended in February 2002.

"DX Asiawaves" and "Wavescan" (AWR)

In addition to AWR's "World DX News" (see above), another branch of AWR, AWR-Asia, also inaugurated a DX program. It was called "DX Asiawaves" and it was broadcast over KSDA, Guam, starting in 1985. Produced by Greg Scott, then KSDA program director, in cooperation with the ARDXC, it was on the air until 1994.

The AWR DX program best known to today's DXers is "Wavescan," which began in January 1995 as "Worldscan" and was broadcast over various AWR shortwave outlets, plus WRMI and Radio Copán International (Honduras). Until the program went on hiatus in December 2004 it was researched and written in Indianapolis, Indiana by Adrian Peterson, and recorded in various AWR facilities. Peterson's extensive contacts throughout the radio world always resulted in interesting items of shortwave information, and his radio monographs on numerous long-forgotten topics of radio history were unique. "Wavescan" returned to shortwave in January 2006 as an Asian-oriented program, consistent with an AWR reorganization which decentralized English-language operations. It is now researched and produced in the AWR studios in Singapore, and is on the air over various AWR channels and on WRMI.

"DXing With Cumbre" (WHRI)

"DXing With Cumbre" was born in late 1995 as a weekly half-hour program over shortwave station WHRI. Its voice has been Marie Lamb of Syracuse, New York. In its early years, "DXing With Cumbre" focused mainly on material from the newsletter. Over time, however, the DX items were supplemented with more general information. Since 1997 the program has also included a weekly pirate segment by Chris Lobdell of Tewksbury, Massachusetts, and since 1999 a monthly report from Bob Padula. "DXing With Cumbre" has also had segments about QSLs and shortwave programming. It is now also available as a podcast.

Other DX Programs

THE 1950S AND 1960S

There were countless other DX programs over the years, many of them sharing the name "DX Corner." Most made little attempt at originality and consisted mainly of DX tips, often out of date, taken from other sources, or talks on technical topics or general communication subjects of limited interest. Others were worthy efforts, but had a limited following in North America because they were not as well heard or not as well known as those mentioned above — or not as interesting.

One of the oldest DX programs, dating back to at least 1949, was "OTC Calling DXers" from OTC, Radiodiffusion Nationale Belge, Leopoldville, the Belgian Congo (now Zaire). It broadcast DX tips on Wednesday nights (in North America). It was continued from Brussels when the Congo station assumed relay status in 1952, but was cancelled when Brussels ceased all foreign-language programming in 1955.

Germany's Deutsche Welle inaugurated a monthly DX program in 1954. At one point DW also published a DX news bulletin, *Deutsche Welle DX*, which was available to contributors. The DW DX editor for many years was Deutsche Welle chief engineer Gustav-Georg Thiele. (Thiele died in 1995.) The program catered to hams as well as SWLs and went through various formats. A new monthly program, "Deutsche Welle World DX Meeting," was begun in 1991 and continued until canceled in early 2007. It took the place of the

"OTC Calling DXers" contributed to the popularity of the famous station in the Belgian Congo.

Radio Berlin International DX program that closed when Germany unified. The program host was the same as from the RBI days, now on the DW staff. Ten years later it had become a ham-oriented DX segment on DW's monthly mailbag program. Because the DW DX programs were brief and on the air only monthly, they had very low visibility in North America.

Beginning in 1956, 4VEH in Haiti had a weekly DX program, "DXers Corner," which was hosted by Miriam Stockton, who was also an NNRC member. The program presented station news and information on various DX activities gleaned from club bulletins and direct contributors. It was on the air until approximately 1969.

WRUL began a weekly DX program in 1960. "Calling SWLs" (renamed "DXing World-wide" in 1964) was produced for a time by the ASWLC. By 1966, WRUL had become WNYW and the program was being sponsored by R. L. Drake Co. It was not the best source of in-depth DX news, but it was lively and topical, and covered new products, technical tips, etc. It came to an end in 1970.

Several other programs are worth noting. The twice-monthly "Radio Portugal DX Club" was begun in 1964. It focused on stations in the Portuguese world, at the time a rich repository of DX targets. Trans World Radio's weekly "DX Special" first appeared in 1966. It was produced in Monaco and was heard with a strong signal in North America via TWR's Bonaire station. (It was also transmitted from Monaco, and for a time from Swaziland.) While material tended to be dated, and the program suffered from low listener support, it remained on the air until the late 1970s. In the 1960s and 1970s, FEBC in the Philippines had a weekly program called "Manila Calling DXers." And the Korean Broadcasting System had a DX program as far back as the mid–1960s. A later incarnation, "Media Roundup," became "Multiwave Feedback" in 1997.

THE 1970S AND LATER

Achievement of forgettable status came easily to many of the DX programs which were inaugurated in this period. Few had any large following among hard core DXers.

Belgium's twice-monthly "DX Corner Belgium" commenced in 1970. The casually presented program hosted by Frans Vossen contained letters and DX tips from listeners. By 1980 it was weekly and focused mainly on features rather than DX information. Later it was called "Radio World" and was heard until Radio Vlaanderen International ended its foreign-language broadcasts in 2005. For about five years starting in 1971, Radio Norway offered a monthly program called "DX Radio Norway." And Austria's weekly English-language media magazine, "Shortwave Panorama," began in 1975 and continued for 18 years before closing early in 1993. The 15-minute program focused on general broadcasting developments and radio-related feature stories, adding DX tips from Glenn Hauser in the 1980s.

"DX Corner," a three or four minute segment of Israel's Sunday "Calling All Listeners" mailbag program, started in 1974. Produced by ham operator Ben Dalfen, who liked to focus on technical issues, it was broadcast for some 20 years. In the Seychelles, FEBA's twice monthly "DX Postbag"— more a mailbag show than a DX program — also began in 1974. It was discontinued in 1976. (The station would offer the weekly "World of Short-wave" program in the 1990s.) Also in the 1970s, Radio Veritas in the Philippines had a program called "Listeners International." Although it was seldom heard beyond the west coast, it carried DX loggings, features on equipment, answers to listener letters, etc. SPEEDX assisted in the preparation of the program, which was on the air until approximately 1983.

The Voice of Turkey's 20-minute "DX Corner," begun around 1978, contained a bit of DX news, along with music and acknowledgment of reception reports. It was reduced from three weekly editions to one in 1981, and later became bi-weekly. It is still on the air. In 1979, Radio Exterior de España inaugurated a brief DX program called "CQ, CQ" which catered to both SWLs and hams. It ran for almost 20 years. Radio Exterior de España still has a weekly DX program called "Radio Waves."

Several programs came on the air in the early 1980s. From Guam, KTWR's "DX Listeners' Log" began in 1980. Produced by Gary Whitmore, it featured some DX tips, the naming of a DXer of the week, and examples of different languages and tuning signals that could be heard on shortwave. Later in the same decade, KTWR produced "Pacific DX Magazine," a program useful mainly for beginners. It ended when program producer Bill Damick moved to TWR Swaziland. Another program from KTWR, "Pacific DX Report," was presented from 1995 to 2000. Unfortunately, reception from Guam was not the most reliable and so these programs were not widely followed in North America.

All India Radio began a "DX Corner" program on an experimental basis in October 1980. It was broadcast twice a month and was prepared by DXer Alok Das Gupta. It contained some DX tips, along with various DX–related topics and acknowledgment of listeners' reception reports, and was on the air for many years. Also in the early 1980s, TWR had a monthly program called "Callsign" over its station in Monaco.

As the decade wore on, fewer new DX programs appeared. Two that bucked the trend were Trans World Radio's beginners program, "Bonaire Wavelengths," which was on the air from 1989 to 1993, and Radio Havana Cuba's twice-weekly "DXers Unlimited" which also appeared around 1989. "DXers Unlimited" was, and continues to be, hosted by well-known Cuban amateur and broadcasting professional Arnie Coro, whose friendly voice has covered many different aspects of telecommunications over the years, with a focus on technical topics, propagation and ham radio.

In the U.S., soon after KCBI, Dallas, Texas came on the air in 1985 it began airing a DX program called "The DX Connection." The program asked DXers to send in recordings to play over the air and had a hotline on which listeners could call in DX information. Quality control was weak, however, and the program was gone by the end of the next year.

Several American DX programs, none very successful, were initiated in the 1990s. "Signals" was an hour-long program carried over WWCR from 1991 to 1993. Hosted by Christopher King and "Kristin Kaye" (Christine Klauberg Paustian), it was billed as "a weekly magazine of communications and technology." It covered current events in the world of communications generally, with a focus on clandestine and pirate broadcasting, satellite radio, AM and FM, and other topics. Although its eclectic content seldom included hard DX news, the well-produced program was interesting nonetheless. From 1993 to 1998, WWCR carried "Spectrum." KNLS started a new program, "Radio Today," in 1993. And a few years later, WHRI had its "DX Radio Show," and WVHA its "DXtra."

There were some other DX programs on the air in the 1990s. Early in the decade Spain had its English "DX Spot." FEBC had multiple DX programs—a weekly 15-minute program called "Communication," plus "DX Dial," "DX Report," "DX Update" and "DX Spot"

"Signals" was on the listening calendars of many SWLs during 1991–93.

(for beginners). In 1994 Finland had a five-minute program on Fridays called "Media Roundup." It covered shortwave and satellite news. Portugal had a DX program around the same time. It was on the air until 1997. By the 1990s, however, several of the major programs, such as "Media Network," "World of Radio" and "DX Partyline," had occupied the DX field, and other DX programs were largely ignored by most listeners.

By the end of 2001, among the major DX programs only "Communications World," "DX Partyline" and "World of Radio" remained. "Communications World" closed in 2002, and the "DX Partyline" broadcast from Quito was dropped in 2006, after which the program retained a much lower visibility via HCJB in Kununurra, Australia, several U.S. stations and the internet. Other DX programs of varying content, quality and frequency — some with little hard DX news and more properly categorized as "media" programs— were still carried by Radio Australia, Radio Vlaanderen International, Radio Havana Cuba, Radio Budapest, All India Radio, Radio Korea International, Radio New Zealand International, the Voice of Turkey, and a few others.[16] "Wavescan," now produced mainly in Singapore, returned to the air in 2006 but with a reduced presence. "World of Radio" and "DXing With Cumbre" were still on. For the most part, however, the era of the DX program was over.

Other Listener Programs

A corollary of the DX programs for SWLs were the programs for radio amateurs. From 1949 to 1965, VOA had an amateur radio program hosted by Bill Leonard, W2SKE. It was called "Radio Amateur's Notebook." Leonard was known among New York radio listeners as the host of the WCBS program "This Is New York." Propagation analysis was offered by George Jacobs, who would become chief of the VOA Frequency Division and the chief engineering consultant for many of the country's private shortwave stations. The program had its own distinctive verification card. Leonard went on to become president of CBS News.

In the early 1960s, many of the Soviet and Eastern European shortwave stations inaugurated what they called DX programs. Most of these focussed on ham radio rather than shortwave listening, and usually gravitated toward technical topics, propagation news and the like. The style was usually drab and the content not very newsworthy. Stations with such programs included Radio Budapest (its program commenced in 1957), Radio Bucharest, Radio Prague, Radio Sofia (also begun in 1957), Radio Warsaw, Radio Berlin International, Radio Kiev, Radio Tashkent, and Radio Moscow itself. Many of these programs were on the air for many years, and a few are still on.

Among other amateur radio programs have been HCJB's "Ham Radio Today," which was on the air for many years until HCJB canceled its English programs to North America in 2003, and "Amateur Radio Spectrum" from Radio RSA, South Africa, which was on until 1990. SENTECH, South Africa's national telecommunications provider, still broadcasts a weekly hour-long program for the South African Radio League. "Radio Techniques," which came on the air in 1992 over WWCR but closed the following year, also emphasized ham radio. WBCQ, which started broadcasting in 1998, soon had a ham radio program, and still does. Radio Romania International dropped its "Special Program for Radio Amateurs" in 2001.

Although very much a specialty, and under the radar of most listeners, in the 1970s there were a number of programs about stamp collecting. Some of them had begun many years earlier, and at least one, Radio Sweden's monthly "Philately Corner," was still on the air in the late 1980s.[17] In the 1950s the VOA had a VOA Stamp Club with over 100,000 members.

Of the numerous mailbag programs, Radio Australia's was the most popular. From its inception in the late 1940s, the station's weekly North American "Mailbag" had a wide following. From 1956 until the program's last airing on December 28, 1980 it was hosted by the man with arguably the best-known voice on shortwave, Keith Glover. Some letters were included in a new program called "You Asked for It," but this could not substitute for Glover's hearty, friendly approach. He had been with the ABC since 1947, and received a number of awards for his role in international broadcasting. He was well known on home service radio and TV as well. Later in his career he moved into ABC management. He retired in 1985 and passed away in 2006.

Another popular mailbag program was the Voice of Denmark's "Saturday Night Club." A light program of listener letters hosted by one of the station's English language announcers, Marianne Linard, its informality and spontaneity made it a favorite of shortwave hobbyists of the 1950s.

Also to be remembered is "Listeners' Corner" over the CBC International Service. It was hosted by popular shortwave personality Earle Fisher until he retired in 1972. The program closed in 1975 after almost 30 years on the air. (A mailbag program was resumed two years later.) Fisher died in 2005.

Surely the best known shortwave program of all time was the Radio Netherlands Sunday show, "Happy Station," hosted by Eddie Startz. Experimental shortwave broadcasts

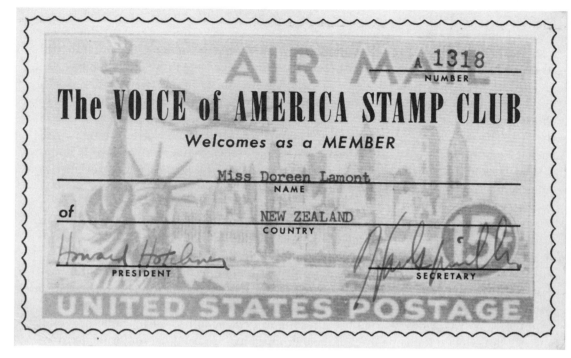

A number of stations had programs about stamp collecting. This is the membership card of the "VOA Stamp Club."

from Holland had begun in 1927, and Startz appeared soon thereafter. Although no tangible proof has been found, Startz claimed that he first used the phrase "Happy Station" on the air on November 19, 1928 (*"This Is the Happy Station of a Happy Nation"*). As the station's listenership increased, the multi-lingual Startz became an international star, a shortwave ambassador, the personification of what was hoped would be a new era of goodwill and understanding among nations that would be facilitated by international broadcasting.[18]

The war years silenced the station, but it returned in 1947 and with it came Eddie Startz and the "Happy Station." A consummate showman, over the years Startz would invent numerous trademarks for the show. The station's callsign, PCJ, was "Peace, Cheer and Joy." The program would always open with a Sousa

Earle Fisher hosted "Listeners' Corner" on the CBC International Service until 1972.

Save for the war years, Radio Netherlands' DX showman extraordinaire, Eddie Startz, was on the air from 1928 to 1969.

march. There was the ever-present "Nice Cup of Tea"; the references to WLHF, "World's Largest Happy Family," and the "University of Light Learning"; the birthday greetings, where he called listeners by name; the Happy Station animal gang (horses Happy, Pappy, Bright and Breezy, Pasha the dog and Ella the cow); the "radio cruises" to distant parts of the world; and the reminder, at the end of each program, to "keep in touch with the Dutch."

Startz retired at the end of 1969 at the age of 71. In 1972 he traveled to the United

Startz had innumerable ways to give meaning to the station's early call letters, PCJ—"Peace, Cheer and Joy."

States where he was the guest of honor and keynote speaker at the convention of the Association of North American Radio Clubs held near Boston, Massachusetts, in August. He died in 1976, and soon thereafter NNRC shortwave editor Hank Bennett recalled his meeting with Eddie Startz:

> Amelia and I did have the opportunity several years ago to meet Eddie, to have the pleasure of taking him out to luncheon at a small French restaurant above Times Square. Our knowledge of French being most limited, to put it mildly, it was up to Eddie to order from the menu. With his French-accented double–Dutch and a waiter who obviously was neither French nor Dutch, we did manage to get what we ordered. A wild and woolly taxi ride through New York City in which Eddie offered his opinion on a number of subjects, including wild and woolly taxi rides, ensued, and before the day was over we felt that our lives had been tremendously enriched by being able to meet this most gracious person.[19]

Finding a replacement seemed an impossibility. However, Tom Meyer, an announcer in the Dutch section for six years who was about to seek other professional pursuits, accepted the challenge. Although 40 years Startz's junior, Meyer spoke six languages and he did a laudable job, bringing the program up to date but preserving its tone and spirit (by this time the program was 80 minutes long). He also hosted the show's Spanish version, "La Estación de la Alegría." "Happy Station" was receiving 19,000 letters a year.

Tom Meyer left Radio Netherlands in 1992, after which the program was hosted by Pete Myers. Pete Myers had been at the station for 16 years following an 11-year career at the BBC World Service, and had extensive experience in many fields of broadcasting. Some of the program's old trappings were jettisoned, more time was given to listener comments, and in general the program was repackaged into a more modern format. Pete Meyers was succeeded by Jonathan Groubert, who had the unhappy job of retiring the program in September 1995.

Call-in Programs

By the mid–1980s there were a number of shortwave programs that took calls from listeners. These included the BBC's weekly "It's Your World," and the VOA's "Talk to America," which was monthly until 1994 when it went daily, Monday through Friday. ("Talk to America" was replaced by a web chat in 2007.) The first call-in program devoted to hearing from distant listeners just for the sake of contact, rather than to obtain their views on substantive issues, appears to have been a series of New Year's programs begun on December 31, 1977 by Radio RSA, South Africa. These programs lasted several hours each, and while they were a novel attempt at listener contact, they were notoriously dull. Nonetheless, they continued throughout the 1980s,

HCJB had its first call-in show in 1984. Thirty-nine people from 10 countries called in. This developed into a monthly program, "Open Line," which covered many topics. The next year the station had a series of call-ins connected with "DX Partyline," along with call-ins for the "Happiness Is" program. Additional call-ins were held in later years, including a joint "Ham Radio Today"–"DX Partyline" call-in in 1991. The HCJB announcers were a more lively bunch and were able to avoid the stiffness of their South African primogenitors.

Perhaps the most memorable shortwave call-ins were those aired over Radio St. Helena

HCJB issued this QSL-card for a call-in program in 1993.

during the ten special, several-hour transmissions for DX listeners that were conducted between 1990 and 2007. Whatever was lacking in the conversation was more than made up for by the novelty of the event.

The best-known call-in program touching on DXing was not on shortwave. From 1984 to 1994, late night ABC talk show host Ray Briem conducted an annual five-hour "DX Special" about DXing. Originating in Los Angeles, it was carried nationwide. Briem was a DXer himself,[20] and he invited many well known DX personages to take part in the discussion, either in person or by phone. Over the years the program was graced by such DX luminaries as Joe Adamov, Arthur Cushen, Bruce Elving (FM), George Jacobs, Tom Kneitel, Stewart MacKenzie, Larry Magne, Tom Meyer, Paul Swearingen (BCB), and others. Listeners could call in and join the conversation. Briem retired at the end of 1994, and the Southern California Area DXers (SCADS) honored him by holding a Ray Briem Appreciation Day banquet at Knot's Berry Farm.

Listener Clubs

During the early and mid–1960s, many stations started "listener clubs" which were intended to win loyal listener following and maximize listener contact. Some were connected with a "club" program. Typically there was a requirement for the submission of a certain number of reception reports on a continuing basis, for which the listener would receive a certificate and a club number, along with other trinkets and benefits. In all cases the listener would be rewarded with a place on the station's mailing list.

Listener clubs were a favorite of the Eastern European stations (and the source of much consternation on the part of the International Short Wave Club which saw them as

This certificate was issued to members of the Radio Prague Monitor Club.

an attempt to circumvent the ISWC anti-jamming campaign). The Radio Prague Monitor Club was typical. An SWL had to submit seven reports a year (15 if you were in Europe), each for a different day, for which a numbered certificate would be issued. Membership could be renewed yearly for seven new reports, whereupon a renewal endorsement sticker would be provided. The club also had a DX award, the Radio Prague DX Diploma, for members who submitted a list of stations heard and verified in 15 ITU zones.

Other Eastern European clubs were a variation on this theme. The Radio Budapest Shortwave Club, formed in 1965, required four reports monthly, for which a member would receive a certificate, a station pennant, stickers, a monthly newsletter, and a copy of the quarterly *Radio Budapest Antenna* magazine. Radio Kiev required reports of three broadcasts, for which the listener received a souvenir each month. The rules of the Radio Berlin International DX Club required two reports per month. Members received a membership card and a certificate, a bi-monthly DX bulletin, and the monthly *RBI Journal.*

The Eastern European stations were not alone in using this strategy to seek listener solidarity. In 1965 the Radio New York Worldwide SWL Club was formed. For an annual fee of $1 a member received a card, a certificate, a pennant, a monthly bulletin containing DX tips, technical information and feature articles, plus the ability to purchase a rubber stamp, a club button, reception report forms, etc. Soon the club had almost 600 members in 35 countries. By 1968 it had grown to 3,500 members in 85 countries. However, by 1970 the club was no longer in operation (and by 1973, neither was the station).

Mention has been made of two program-connected clubs, the Radio Canada Shortwave Club and the BBC World Radio Club. The Radio Canada Shortwave Club required five reports at the outset followed by one every two weeks thereafter. Members received, in addition to a bi-lingual certificate and annual endorsement, a pennant, a decal, a stick pin, an antenna booklet and a membership list. Formed in 1962, by 1968 the club had 5,200 members in 78 countries. By 1972 the membership had grown to nearly 12,000.

In the BBC club, membership was by request, with no requirement for reception reports. The World Radio Club was founded in 1967 and within two years nearly 7,000 listeners had signed up, almost 12,000 within five years. Because it was BBC practice to respond to reception reports with only a general acknowledgment card rather than the true verification that most DXers wanted, one of the most welcome benefits of BBC World Radio Club membership was the eligibility for the full-data QSL certificates which were offered to members for reception reports on various BBC transmitter sites during "DX Award Competition" periods. The club held contests of other sorts as well, sometimes offering a difficult-to-get BBC pennant for correct answers to contest questions.

Another program-connected club was the Radio Portugal DX Club. Formed in 1964, it required five reports on transmissions beamed to the listener's area, plus one report every two weeks. The station provided logsheets for use in reporting, and it sent IRCs to regular reporters. Two pages of the station's quarterly listener magazine were reserved for the Radio Portugal DX Club. It focused on Portuguese DX stations, and had a three-level awards program based on QSLs received from a list of stations in the Portuguese world. The bronze level required ten stations QSLed, silver 20, gold 50. (Due to the difficulty of hearing the Portuguese African stations, North American listeners were allowed to substitute Brazilian QSLs in satisfaction of half the African station requirement.)

In 1967, Radio RSA began a Monitoring Panel which required monthly reports for

Radio Berlin International issued this QSL-card on the RBI DX Club's 10th anniversary.

BBC LONDON

This is to confirm with thanks that

John D. Tuchscherer
....

participated in the World Service Reception Survey

in   November 1972
....

EXTERNAL BROADCASTING   **WORLD SERVICE**
ENGINEERING   **WORLD RADIO CLUB**

The BBC World Radio Club acknowledged listener participation in reception surveys conducted from time to time.

which the participant received a quarterly newsletter and eligibility for various awards. In the early 1960s even Radio Nigeria had a Radio Nigeria Listeners Club, with a weekly broadcast that sought contact with DX clubs around the world. The Radio Australia Listeners' Club operated from the late 1960s to 1981. In its early days it required two reception reports per month for six months, for which a member would receive a club badge and certificate (a new round of reports each year produced a new certificate).

The popularity of listener clubs declined over the years, and many had faded from the scene by the 1990s. However, the Radio Vlaanderen International Listeners' Club operated until 1997, and the Radio Bulgaria Monitoring Club is still in operation.

Perhaps the most famous, and the best, listeners' club was that of HCJB. It was formed in 1974 and was associated with the station's "DX Partyline" program. The idea of a listeners' club had been considered for many years. In 1973, "DX Partyline" hosts Clayton and Helen Howard conducted a contest for a club name and ANDEX, short for "Andes DXers International," was chosen.

The first issue of the club's monthly bulletin, *ANDEX International,* was published in January 1974. It was an entertaining and highly informative six-to-eight page presentation of information about HCJB personalities, programs and facilities, the country of Ecuador, other Ecuadorian radio stations, propagation, antennas and other technical topics, a "DXer of the Month," favorite QSLs, station pennants, and many other topics of interest to DXers. Membership certificates were distributed, and although there was a small fee, by May there

ANDEX

INTERNATIONAL

Vol. 1 No. 1   January 1974

HAPPY NEW YEAR

As we begin the New Year of 1974, Andes DX'ers International is happy to welcome all our members and to wish each of you a Happy New Year. We are glad you decided to join our Club! Not only is this a New Year, but we are a new Club. We start out together with real expectation.

Our purpose is to serve each of our members in every way possible. Each month you will receive your copy of ANDEX which will bring you much information concerning the Club, HCJB, and the country of Ecuador. We do not plan to include DX Tips and information on radio stations in other countries of the world. For this type of information, listen to the DX Party Line and join one of the many Clubs that publish such material. We do plan to feature one shortwave station in Ecuador each month with as complete information as possible.

Enclosed with this January, 1974 ANDEX is your membership certificate for 1974. The border was drawn by a gifted Ecuadorian artist, Sr. Salguero, and represents Ecuador and HCJB. At the center top you will find the Equator Monument which divides the Northern and Southern Hemispheres. This monument is located just a short distance from Quito and is surrounded by the Andes mountains with many beautiful snow peaks. The sides of the border represent HCJB with our hydroelectric plant at the top and our transmitter building and antenna towers at the bottom.

As the year progresses we will have added features. We would encourage you to write to us letting us know your reactions as well as sending along your suggestions. Also, let your friends know about the Club and encourage them to join. They will receive the same cordial welcome we give to each of you.

DX'ER OF THE MONTH

Cornelius Webb

Selected to receive the first "DXer of the Month" award is Cornelius Webb of Dallas, Texas. Corky is shown above at his listening post in his home. He is employed as a salesman, has been interested in shortwave DXing for about 4 years, and finds time to do an average of about 4 hours of listening each evening.

The excellent equipment owned by Corky includes two Drake receivers, R-4B and SPR-4. A Mosley SWL-7 antenna is supplemented by a dipole and a Joystick with Joymatch. Reception is improved with a Gilfer preselector. Two cassette recorders and a reel-to-reel recorder complete the installation. With such facilities it is little wonder that Corky is such a successful DXer.

Other interests which occupy some of Corky's time include hunting, fishing and photography.

Our congratulations go to Cornelius Webb for having been chosen as the first "DXer of the Month." An attractive certificate will be sent to Corky in recognition of this honor.

Opposite: The Radio Portugal DX Club had an awards program that required hearing and verifying stations throughout the Portuguese-speaking world. *Above:* The first issue of *ANDEX International*, bulletin of HCJB's "Andes DXers International" club (ANDEX), appeared in January 1974. The club remained in operation for 22 years.

The Voice of South Africa

radio rsa

Die Stem van Suid Afrika

Monitoring Panel
Honorary Certificate

This is to certify that

DAVID WALCUTT

an Official Monitor of Radio RSA, is hereby declared an
Honorary Life Member of the Radio RSA Monitoring Panel, having
faithfully provided regular reports on reception since 1978

Signed at Johannesburg this   5th   day of   December   19 83

Head External Services   Technical Manager Shortwave Services

Long-time members of South Africa's Radio RSA Monitoring Panel received a certificate for their efforts.

were 500 members, a number that grew to over 1,200 by the club's first anniversary. Clayton Howard headed the club until 1981 and was followed by Ruth Stanley, Doris Hastings, Brent Allred, Rich McVicar, and Ken MacHarg. The bulletin became bi-monthly in 1981.

ANDEX International was a source of interesting reading if not necessarily cutting edge DX news. However, while almost 10,000 listeners had been members of ANDEX at one time or another, by 1996 the rate of new memberships had dropped markedly. With information about the station, about shortwave generally, and about Ecuador becoming increasingly accessible over the internet, there seemed to be no pressing need to continue, and the club closed in June. The demographics had changed from the early 1970s. The teenagers who had been so enthusiastic in getting the club off the ground had become adults, and modern youth wasn't as interested in shortwave radio.

6

Receivers

In 1944–45 at 16, I bluffed my way into a job as "short wave tuner-iner" at the old Hallicrafters show room on Michigan Avenue in Chicago. My sole job was to tune in foreign broadcasts for the dubious public on a battery of SX-28's on display in the show room. I did this after school every day, and would have gladly paid Hallicrafters for the privilege! Naturally (this is what I told the bosses) I needed an SX-28 at home so that I could brush up on my "tuning ability." They obliged by loaning me a receiver (you couldn't buy them in 1944 because of priorities). I'll never forget the day they said "O.K., take it home." I was so delirious with excitement that I hailed a cab to take me and the SX-28 to my home, for $4.50, which represented 25 percent of my weekly salary.[1]

It is in receivers that the impact of technology on shortwave listening is best illustrated. The improvements in receiver design since the 1940s have been revolutionary, paralleling the same kinds of technological advances that have reduced the reliance on shortwave for international broadcasting. Even excluding ham band–only equipment, since the war there have been countless receivers to choose from, in all price ranges.

Prices indicated below are usually suggested retail prices or advertised prices at the time of a model's introduction. Comparing prices can be problematic because a manufacturer's price sometimes excluded speakers or cabinets. Rectifiers and voltage regulators are included among a receiver's tube count. The manufacturers and receivers mentioned do not include all companies or all sets. Excluded are most ham band–only receivers, auto receivers, foreign receivers that had a limited exposure in the United States, transceivers, antennas, and most accessories. DRM receivers are covered (briefly) in *Broadcasting on the Short Waves, 1945 to Today*.

By the 1940s, receiver design had long since left the regenerative, TRF (tuned radio frequency) and neutrodyne phases behind, and the superheterodyne was the universal standard for new receivers. The popularity of the all wave set — basically a home radio with a shortwave band — was also a thing of the past. They had been omnipresent during the 1930s, when the future of international shortwave broadcasting was thought to be bright, but their limited capabilities contributed to their disuse in later years. Save for the big Grundigs of old and a few like sets, the era of the high-end, "high fidelity" shortwave receiver — Scott, Lincoln, McMurdo-Silver — was over as well, as was the notion of home-construction of shortwave receivers by laymen (except for kits).

There was very limited production of civilian radios of any kind during the war, and what there was was subject to distribution on a priority basis. As the war ended and renewed production was anticipated, there was much discussion about the features that the new receivers should possess. Opined one *Radio News* reader in 1945:

> [R]eceivers built after 1939 were, as a whole, inferior to earlier models. I believe this was partly due to both cheap parts and poor design, but probably more to the latter. The majority of receivers, even in the higher priced brackets, were limited to one i.f. stage of doubtful gain, and a preselector on the broadcast band, leaving much to be desired in the way of short-wave performance, a.v.c. action, and signal-to-noise ratio. The audio amplifiers and loudspeakers in these receivers were also often sadly neglected....[2]

At the War's End

There was great pent up demand for receivers after the war. The ham market, which had dried up during the war because hams had not been allowed to operate, was coming back. Most listeners were still using their prewar receivers. Although some manufacturers continued to advertise during the war as if their products were still in production, listeners seeking to make a purchase were generally limited to whatever prewar equipment was still on the shelf. Many companies were represented in this array of aging equipment, including Hallicrafters, Hammarlund, Howard, National, Patterson, Philco, Scott, RCA and others. By the end of the war, many of these names would disappear, leaving five major survivors: Collins,[3] Hallicrafters,[4] Hammarlund,[5] National,[6] and RME (Radio Manufacturing Engineers Co.).[7]

In the 1950s, listeners often received their introduction to shortwave through a Grundig, Telefunken or other European radio, many of which graced parlors in those years. Although their ability to handle difficult reception conditions was small, the warm, room-filling audio emanating from their polished-wood cabinets was impressive, and they were adequate for getting at least some of "London," "Paris," or other locations marked on their dials.

However, a serious listener pondering the purchase of a new, good-quality shortwave receiver after the war would more likely opt for a communications receiver. These sets sought to address many of the difficulties of shortwave reception. While lacking the stylish cabinetry of the Grundigs et al., they were designed to get the most out of a shortwave signal. They had been born in the 1930s to meet the demands for better shortwave reception of both amateur radio operators and an engaged shortwave listening public.

Communications receivers were usually of the "general coverage" variety, that is, they covered the standard broadcast (AM) band plus shortwave up to 30 MHz. Common features, depending on price, included calibrated dials; bandspread dials with a 0–100 logging scale (and often calibrated bandspread on the ham bands); in the better models, multiple RF and IF stages; a beat frequency oscillator (BFO), sometimes with separate pitch control; automatic volume control (AVC); interference-reduction devices such as a multiple-position crystal bandwidth filter (sometimes with a phasing control to help mitigate heterodynes), mechanical filters, a Q-multiplier, or a notch filter; an "S" meter to measure signal strength; and a crystal calibrator. These sets came in all sizes and prices. Except for the entry-level models, early communications receivers generally required an external speaker, and sometimes a separate power supply. The manufacturer usually offered these as accessories.[8]

Communications receivers could be purchased from the many radio and electronics stores then in business in most large cities, such as Hatry Electronics in Hartford, Connecticut, Henry Radio Stores in Los Angeles, California, and Radio Shack in Boston, among numerous others. The large mail order houses, such as Allied Radio Corp.[9] and Lafayette Radio Corp.,[10] both in Chicago, Burstein-Applebee Co. in Kansas City, Missouri, and World Radio Laboratories in Council Bluffs, Iowa, also carried shortwave receivers.

At war's end, many of the mid- and high-performance legacy receivers of the prewar years were still providing good service.

At the mid-level there was the National **NC-45**, a fairly conventional eight-tube, four-band receiver that came on the market in 1941. It was similar to the **NC-44** of the late 1930s, but with a noise limiter and some minor circuitry changes. It had the virtue of selling for only $57.50.

RME receivers were not as popular as Nationals, but knowledgable shortwavers regarded them highly for their design and construction. In 1941–45, RME offered the **RME-43**, a nine-tube, six band set selling for $110. It was a well-featured receiver, with bandspread and two tuning speeds, along with a BFO, noise limiter, and tone switch. It differed from its predecessor **RME-41** mainly in the addition of a crystal filter and an S-meter. (Early RME receivers had an "R" meter, for "readability," marked in units of 1 to 5, rather than what would become the conventional nine-plus unit "S" meter.)

The National **HRO** is not typically thought of as a "mid-level" radio, but when it was introduced in 1934 it had only nine tubes. (The last pre-solid state HRO, the **HRO-60**, went into production in 1952 and had 18 tubes.) The original HRO sold for $167.50. It used a plug-in coil arrangement that was anachronistic even by 1940s standards, but that remained the HRO hallmark until the set went solid state in 1964. Each of the four standard coil modules covered one segment of the shortwave range up to 30 MHz. and one bandspread amateur range (coil modules for other frequency ranges were also available). The HRO also featured the "PW" micrometer dial from the National **NC-100** that had debuted in 1936. To operate the HRO, you inserted the correct coil module in the front panel and then located your approximate frequency by checking the markings on the PW dial with a calibration curve on the face of the module. Although on most bands you could estimate your frequency only within about 25 kHz, by noting the dial settings you could easily return to the same frequency later. The dial was backlash free, and the 20-to-1 ratio gearbox provided a very long dial spread for each frequency range.

In the higher-end market one often encountered the 11-tube, five band National **NC-120**, one of the National "sliding coil" receivers (see discussion under NC-2-40C et al. below). Like the HRO, it had two RF stages, mainly to reduce oscillations so as to meet Navy specifications. Often this receiver found its way to listeners via war surplus, most having been made for military purposes under the designation "RAO."

Another high-ender was the Hallicrafters **SX-28 "Super Skyrider."** It had gone into production in 1940 and seen extensive government and military service. The top of the Hallicrafters line, to most listeners it was the ultimate receiver. With 15-tubes and six-bands, it was loaded with features: six selectivity positions, S-meter, tone control, noise limiter, antenna trimmer, bandspread (calibrated on four ham bands), and something that most

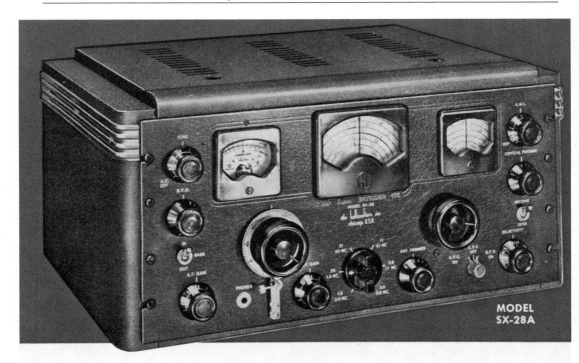

MODEL
SX-28A

WITH the greatly increased interest in short wave reception resulting from the war has come an ever-growing demand for receivers specifically designed for high frequency performance. To meet this demand, Hallicrafters presents the new Super Sky-Rider. Combining superb broadcast reception with optimum performance as a communications receiver, the SX-28A will satisfy the requirements of the most critical user.

This new Hallicrafters Model SX-28A is a further refinement of the famous SX-28 that achieved such popularity with amateur and professional operators prior to Pearl Harbor. Embodying circuit refinements and constructional modifications necessitated by the arduous con-

ditions of military service, the new SX-28A offers the maximum in communications receiver performance to the discriminating buyer.

The traditional sensitivity and selectivity of the SX-28 have been further improved in this new Super Sky-Rider by the use of "micro-set" permeability-tuned inductances in the r.f. section. The inductances, trimmer capacitors and associated components for each r.f. stage are mounted on small individual sub-chasses and may be removed for easy servicing.

Thousands of these fine receivers have seen service with the armed forces in all parts of the world and have maintained and enhanced Hallicrafters reputation for outstanding quality and performance under the most difficult conditions.

The Hallicrafters SX-28A would rank high on any list of classic shortwave receivers. It saw extensive government use during the war.

receivers lacked — good audio. In 1944, Hallicrafters released the slightly modified SX-28A. It is believed that over 50,000 SX-28/28A receivers were manufactured. The cost of the SX-28 was $159.50 when released. Although there would be other contenders, the SX-28/28A is the classic example of a high quality receiver of the era, and a dream for many DXers.[11]

Hallicrafters

In our opinion Hallicrafters is to the radio communications field what Ford was to the prewar auto industry. Hallicrafters offers almost a dozen models in all price ranges, with almost every feature desirable. They also appear to be issuing new models and revising old ones with little advance notice instead of sticking to a set model as the other firms do. Morton D. Meehan, *NNRC Bulletin*, August 1, 1946, p. 2.

The Hammarlund "**Super-Pro**" receivers had been introduced in 1936 and had gone through multiple versions before reaching the **SP-200** series which came on the market in 1940. Hammarlund advertised the SP-200 as the last word in communications receiver engineering. It was an 18-tube set, available in one variation that covered 540 kHz. to 20 MHz. and another that covered 1.25 MHz. to 40 MHz. Among its unique features was, in addition to a five-position crystal filter and phasing control, a bandwidth control that was fully variable from 3 to 16 kHz. Said Hammarlund: "Amateur and commercial operators alike will find reception of weak, distant stations more reliable because selectivity can be adjusted to suit receiving conditions."[12] Although prone to the warmup drift of virtually all tube sets, the receiver produced 14 watts of superb audio, especially at the higher bandwidth settings. The S-meter was adjustable, "to conform with the operator's custom of reporting signal strength." The receiver was available with two external speaker sizes, 10" (the SP-210) and 12" (SP-220), and required an external power supply. The list price, with power supply, 10" speaker and speaker cabinet, was $465, but it was soon being sold for $279.

Mention should be made as well of the pre-war Hammarlund **HQ-120X**, the six-band ancestor of the Hammarlund HQ series that would be among the best DX receivers of the 1950s and 1960s. The HQ-120X was a 12-tube receiver costing $129 when it was released in 1938, at which time it was widely viewed as an advanced receiver. Although of limited interest to shortwave broadcast listeners, its 300 degree calibrated electrical bandspread on four ham bands was an industry first. After the war the HQ-120X would be succeeded by the HQ-129X (see below).

Although not in as wide use as most other shortwave receivers, RCA's entry in the communications receiver market, the **AR-88**, is worth noting. It was a heavy duty, 14-tube, six-band, general coverage model with high sensitivity, high gain, and good selectivity. It weighed 100 pounds. RCA had planned to begin shipping the AR-88 in 1939. However, the war caused the entire production to be diverted to military purposes and shipped overseas. The AR-88 was never sold as new on the commercial market; most found their way to England, Canada, and Russia. The AR-88 was also in wide use among RCA radiotelephone facilities. It is believed that the AR-88 was manufactured only until the war's end. Relatively few U.S. listeners had them, but those that did liked them. Because of its wartime presence overseas, these units were often found in the shacks of foreign hams or SWLs.[13]

One of the "Super-Pro" series, the Hammarlund SP-200 had the behemoth look of the best DX receivers of its time.

1945–49: Communications Receivers in the Postwar Environment

As the end of the embargo on civilian radio production came into sight, rosy advertisements whetted appetites and urged early purchases. In a 1945 advertising tie-in with Allied Radio, Hallicrafters suggested that listeners wishing to purchase communications receivers reserve them with Allied (the 10 percent deposit was returnable on cancellation). RME hailed the features of the forthcoming RME-45, including "advances made while producing for the armed forces," and promised that it was "definitely well worth waiting for." In fact, many of the postwar receivers were old radios in new packages, with a few modest new features. Most real advances in receiver design came well after the war.

In receiver surveys conducted by DX clubs in the 1940s and 1950s, Hallicrafters was always the brand in widest use by members. In a 1957 NNRC survey covering 187 receivers owned by 127 members, 91 radios were Hallicrafters. The next most popular was National (34 receivers), and then Hammarlund (24).[14] Larger surveys showed similar results. In 1960, of almost 13,000 *Popular Electronics* "WPE" registrants, 41.2 percent owned more than one receiver, and Hallicrafters and National were in first and second place respectively.[15]

Entry Level Receivers (Five and Six Tubes)

There were few good entry-level receivers on the new-set market in 1945. True to its policy of producing receivers at all price ranges, Hallicrafters took the lead with the **S-41** "**Skyrider Junior**," a very basic six-tube, $33.50 radio that was identical to one that Hallicrafters had managed to manufacture during the war under the Echophone name.[16] More important than the S-41 itself was its early progeny, the **S-38** line, which was in production from 1946 to 1961 and is one of the best-known shortwave receivers of all time.[17] As was the case with many Hallicrafters receivers, reception was not the absolute *nes plus ultra*, but, at $39.95 to $59.95, depending on the model, the S-38 offered good value.

The old-fashioned "airplane dial" appearance of the S-38 and its -A, -B and -C offspring was replaced in 1954 by the large slide rule design of the -D and -E models. The various versions of this five-tube radio differed slightly, but all covered the standard broadcast band and shortwave up to 32 MHz. in four bands, and offered bandspread, a BFO, and a built-in speaker (the original S-38 also had an automatic noise limiter and a BFO pitch control).

The S-38 introduced many thousands of listeners to shortwave. (Almost 10 percent of the WPE registrants owned an S-38.) It would be the first receiver of the postwar period to become a long-term favorite. Its main competitor was the similarly featured National **NC-33**, a six-tube unit that sold for $65.95 when it appeared in 1948. However, the NC-33 was on the market for only a few years.

Intermediate Level Receivers (Eight to Ten Tubes)

In 1946, Hallicrafters released the **S-40**, a $79.50 receiver which was basically a repackaged version of the popular S-20R "Sky Champion" which had been on the market for seven years. It had all the features of the S-38 but with nine tubes and more controls. The nearly identical **S-40A** and **S-40B** were on the market through 1955. In a 1954 survey of members

NEW HALLICRAFTERS MODEL S-38

The Hallicrafters S-38 went through multiple versions during its 15-year life span, and was the first receiver of countless SWLs.

of the Universal Radio DX Club, the S-40B was voted the "most wanted" receiver in the $100–200 class.[18] In 1957, Hallicrafters introduced the S-40's successor, the eight-tube S-52, selling for $99.50. Another eight-tube model offered that year, the S-53 ($79.50), included the 48–55 MHz. 6 meter amateur band.

National had two receivers in the intermediate category, the **NC-46**, first offered in 1946, and the **NC-57**. The NC-46 was a ten-tube, four-band receiver, somewhat similar to the Hallicrafters S-40 but without the built-in speaker. It sold for $97.50. The lack of an RF section in the NC-46 meant limited image rejection, however.[19] The nine-tube NC-57 came on the market in 1947. Priced at $89.50, it had continuous frequency coverage up to 54 MHz. in five bands, bandspread, and a built-in speaker, and the gear-driven main tuning dial was fairly accurate.[20] Although selectivity was not its strong point, it was a worthy competitor to the Hallicrafters S-40. (As with the S-40, an external S-meter was available as an accessory.)

Also in the intermediate group were three RME receivers. The **RME-84** was marketed during the 1946–49 period. It was an eight-tube unit with a built-in speaker, and at $98.70 was competitive with the S-40 and NC-57, although less popular than those models.[21] During the same period, RME produced the **RME-45**, a nine-tuber weighing 45 pounds and costing $198.70. In reality it was an updated version of the prewar **RME-43**. It had two-speed tuning and a mechanical calibrated bandspread for the ham bands. In a mid–1990s

The ten-tube National NC-46 was one of the company's many mid-level receivers.

comparison, the RME-45 was found not to be quite up to the performance of the HQ-129X and NC-173.[22]

High-end Receivers (More Than Ten Tubes)

In the high-performance category, the main player was National, with its NC and HRO lines.

The 12-tube National **NC-2-40C** and **NC-2-40D** were made in the 1943–48 period and were the offspring of a well-regarded 1940 model, the 12-tube **NC-200**. All were of the "coil catacomb" variety which traced its lineage to the NC-100. These receivers were unique at the time because they accomplished band switching by means of National's "Movable Coil Tuning Unit," a gear driven, sliding tray of coils which, when combined with the receiver's pointer dial, yielded good frequency "repeatability" (the ability to return reliably to a given frequency).[23] The NC-200 was the first National receiver to accomplish both band switching and tuning with one dial. You pulled it out to select one of six general coverage bands or four amateur bandspread ranges, then pushed it in to dial the frequency. The NC-2-40C and NC-2-40D operated the same way. The NC-200 sold for $147.50. The NC-2-40C sold for $225 and was similar to the NC-200 but with only six general coverage bands (the NC-2-40D, like the NC-200, also had dedicated ham band ranges). They all typified the electro-mechanical precision that would distinguish many National receivers.

Two National receivers that were popular with serious SWLs were the 13-tube **NC-**

173[24] and the 16-tube **NC-183**. Both receivers came on the market in 1947, first the NC-173 at $180, then the NC-183 at $255. These were the first truly new postwar receivers by National, the HROs, NC-46 and NC-2-40 being updated versions of prewar sets. The NC-173 and NC-183 used conventional switched coils in lieu of the more complex coil arrangements on some other National receivers. They had many features in common — frequency coverage (both had 6 meter coverage), bandspread, crystal filter, noise limiter, S-meter, etc. The main differences were that the NC-183 was of better construction, it had two RF stages to the NC-173's one, and it had better audio, producing eight watts of audio to the 3.5 watts of the NC-173. In 1952, the NC-183 was succeeded by the very popular NC-183D (see below).

National's HRO series went through countless variations over the years. For the most part these involved relatively minor changes in circuitry or design — the location of controls, the availability of new special-order coil assemblies covering particular ranges, etc.[25] There were many military models, the HRO being widely used by both American and British forces during the war years. In 1944, the **HRO-5** was introduced. An 11-tube model, it incorporated some of the features of the military HROs, but it was fundamentally the same receiver as the basic HRO save for metal tubes and some cosmetic changes. It was followed in 1946 by the **HRO-5A** ($274.35), and in 1947 by the **HRO-7** (by then 12 tubes, $299.36). The HRO-7 had a more modern-looking cabinet. The frequency graph on the front of the coil module was of a slide rule design rather than a curve, and a new **Select-O-Ject** audio filter was introduced as an accessory (discussed below).[26] All of these receivers were of the same fundamental design, however, and budget-conscious purchasers sacrificed little by picking up a wartime military model of the HRO.[27]

The 14-tube **HRO-50** ($335) arrived in 1949 and was a major upgrade of earlier models. It was the first HRO with an internal power supply. More importantly, although the classic HRO coil arrangement was retained, the frequency could now be read directly through the combination of the standard HRO micrometer dial and drum-mounted frequency scales corresponding to the coil sets in use. This eliminated the need to refer to printed curves and graphs, which now disappeared from the front of the coil unit.

Listeners will debate whether the legendary reputation of the HRO was myth or reality. It was sensitive, quiet, ruggedly constructed and well-shielded, but, until the HRO-50, tuning was complicated.

Hammarlund had two entries in the high-end receiver group in these years. The five-band, 18-tube Hamarlund **SP-400** was introduced in 1946 and was basically an upgrade of the SP-200. In the same year, the 11-tube **HQ-129X**, successor to the prewar HQ-120X, appeared. It became a popular receiver.[28] Different from the many receivers which used common, three-gang tuning capacitors, the HQ-129X, like the Hammarlund line generally, had a six-section main tuning capacitor and a nine-section bandspread tuning capacitor. This helped spread the various bands wide over the dial and minimized frequency drift. The receiver had a five-position crystal filter, and, with three IF amplifiers rather than the two of most other competitive receivers, skirt selectivity was sharp. On the negative side, audio quality left something to be desired, and the receiver produced some images.[29] The price of the HQ-129X was intentionally kept low for marketing purposes — $129.

Hallicrafters offered several high-end receivers. In a sense these receivers were a "modern" return to the high fidelity shortwave sets of the 1930s. The replacement for the SX-

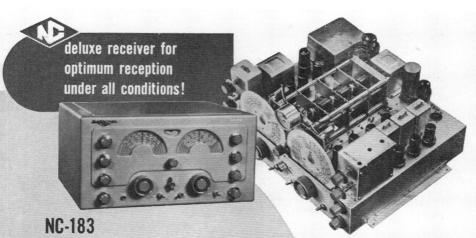

deluxe receiver for optimum reception under all conditions!

NC-183

The flawless design and superb construction of this professional communication receiver make possible amazing performance even under the worst operating conditions. If it's possible to receive a signal, the NC-183 will bring it in!

Continuous tuning from 540 kcs to 31 mcs plus the 48 to 56 mcs band for 6-meter reception. Two tuned R.F. stages provide extremely high sensitivity and image rejection. Voltage regulated oscillator and BFO assure minimum drift on phone and CW. Separate main tuning and bandspread dials calibrated for tuning ease. Main dial covers range in five bands. Bandspread dial calibrated for amateur 80, 40, 20, 11-10 and 6-meter bands. Bandspread usable over entire range. Six-position crystal filter provides any selectivity required from very broad to extremely sharp for cutting through adjacent channel interference. New-type noise limiter effectively minimizes electrical interference. High fidelity push-pull audio output with phono input and front-of-panel RADIO-PHONO switch. Accessory socket for NFM adaptor or other unit, such as crystal calibrator. Uses 2-6SG7 R.F.; 16SA7 1st det.; 1-6J5 osc.; 2-6SG7 I. F.; 1-6H6 2nd det.; 1-6SJ7 B.F.O.; 1-6AC7 A.V.C.; 1-6H6 noise limiter; 1-6SJ7 A.F.; 1-6J5 phase inv.; 2-6V6GT aud. out.; 1-VR-150 volt. reg.; 1-5V4G rect. Accessory socket for Select-o-Ject (see page 4).

$268 net*
(Less speaker)

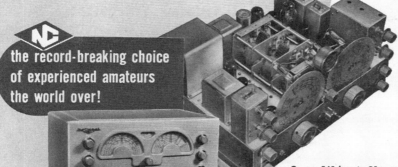

the record-breaking choice of experienced amateurs the world over!

NC-173

The *only* moderate-priced receiver built to National's world-famous standards of sound construction and truly professional performance! Thousands of these sets now in operation attest its popularity and performance.

Covers 540 kcs. to 31 mcs. plus 48 to 56 mcs. for amateur 6-meter band with average sensitivity of 3 microvolts. Separate bandspread dial calibrated for 80, 40, 20, 10 and 6 meter bands. New double-diode noise limiter with variable threshold effective on both phone and CW. Separate AVC usable on phone and CW. New wide-range, 6-position crystal filter, S-meter, antenna trimmer for maximum performance with any antenna, phono input. 1-6SG7 tuned R.F.; 1-6SA7 1st det.; 1-6J5 osc.; 2-6SG7 I.F.; 1-6H6 2nd det. — AVC; 1-6AC7, AVC; 1-6SJ7 BFO; 1-6H6 noise limiter; 1-6SJ7 audio; 1-6v6 output; 1-VR150 volt. reg.; 5Y3GT/G rect.

$199.50 net*
(less speaker)

Slightly higher west of the Rockies.

It was 1947 when National introduced the NC-173 and NC-183 receivers, the company's first postwar receivers that were more than updated prewar models.

the finest amateur receiver National has ever built!

THE NEW *DIRECT READING* **HRO-50**

Now, National presents a great new HRO receiver after more than three years of designing, development and testing. Retaining all the world-famous, performance-proved HRO features, this superb receiver — the finest National has ever made — now incorporates no less than 14 advanced-design innovations. Exhaustive comparative tests indicate the new HRO-50, by far the most modern and versatile in its field, will set an entirely new standard of performance for communication receivers.

14 ALL NEW FEATURES

1. Direct frequency reading linear scale with a single range in view at a time. 2. Provisions for using 100/1000 kcs. crystal calibrator unit, switched from panel. 3. Variable front-of-panel antenna trimmer. 4. Built-in power supply with heat resistant barrier. 5. Front-of-panel oscillator compensation control. 6. B.F.O. switch separated from B.F.O. frequency control. 7. Provision for incorporation of NFM adapter inside receiver, switched from front panel. 8. Dimmer control for dial and meter illumination. 9. Miniature tubes in front end and high frequency oscillator. 10. Speaker matching transformer built into receiver with 8 and 500 -ohm output terminals. 11. High frequency and beat frequency oscillator circuits not disabled when receiver in "send" position. 12. High-fidelity push-pull audio amplifier, 8 watts undistorted output. 13. Tip jack for phono input. 14. Accessory socket for Select-o-Ject (see page 4).

The National HRO-50 offered easier tuning than earlier HROs, on which determining one's frequency was cumbersome.

28A, the **SX-42**, was released in 1947 and was the company's first genuinely new postwar receiver.[30] It had 15 tubes and sold for $275. A major feature was its FM capability. The radio covered the broadcast band, shortwave, and the higher bands up to 110 MHz., which included the 6 meter amateur band and the FM band. It was designed to produce high quality audio, and it had a phono input so that the audio section could be used with other audio devices. The 11-tube **SX-43** (1947) was a trimmed down version of the SX-42 (still

AM, SW and FM capable). It sold for the more competitive price of $179.50. The 15-tube **S-47**, which was also released in 1947, covered shortwave up to 18 MHz. and was sold mainly as a high fidelity AM-FM radio. This $200 unit was marketed under other brand names as well. Finally, in 1948 Hallicrafters released the 16-tube **SX-62**, an SWL-style receiver which was basically the SX-42 in a modernized cabinet and sans bandspread or S-meter. The SX-62 had a slide rule tuning dial; the band you were tuning lighted up as you switched to it. It had a 500 kHz. crystal calibrator and an adjustable dial pointer that could be re-set without detuning the receiver. Its price was $269.50, and, as one observer noted, "it packs a lot of wallop and is easy to operate."[31] (The SX-62 was advertised on the back cover of the *World Radio TV Handbook* during the years 1949 through 1953.)

Collins was at the high end of the high-end receivers. The company had produced amateur equipment since 1931. The **51J** series, introduced in 1949, was basically a general coverage spinoff of the successful Collins **75A** series of amateur band-only receivers that had come to market in 1947.

It is important to note that, prior to the 51J, all shortwave receivers, even high-end models, had been of the single conversion variety; that is, while they may have had two or even three IF stages, they converted the signal to a single IF frequency (usually 455 kHz.). Double conversion—converting the signal to two or more IF frequencies—yielded better selectivity and reduced the receiver's tendency to produce images (signals appearing on other than their actual frequencies). It would become emblematic of the very best of the high-end shortwave receivers, and it was appropriate that it should first be adopted for shortwave in the Collins 51J line.

At $875, the **51J1** was for those who demanded, and could afford, the very best. It was a 16-tube receiver with most of the features common to communications receivers, plus a unique dial mechanism consisting of a slide rule megacycle drum dial containing 30 one-megahertz segments and a circular dial through which the kHz. reading was obtained. The combination of crystal controlled oscillators and the Collins PTO (permeability tuned oscillator) produced rock-solid stability and equally spaced (linear) tuning on all bands. Collins claimed frequency readout accurate to 300 cycles, unsurpassed at the time. The important thing about the 51J1, of which only about 120 units were made, is that it set the stage for much larger productions of the successor (and more expensive) models, the **51J2** (16 tubes, 1950) and **51J3** (18 tubes, 1952), with a combined production of almost 3,000 units, and, in 1957, the last 51J, the 19-tube, mechanical filter-equipped **51J4** (9,000 units).[32]

Listeners lucky enough to have a 51J receiver often obtained them via the military version of the 51J3, known as the **R-388**. There were many war surplus receivers on the market after the war. Prices were good, but much of the war surplus equipment was snapped up by Signal Corps personnel before it made its way to the general market. Also, much of it was battery operated. Over time, high quality military versions of National (HRO), Collins and Hallicrafters receivers appeared on the surplus market.

ACCESSORIES

A war surplus device that was a valuable accessory for serious listeners was a frequency meter, notably the **BC-221**. It cost around $40. Not a meter in the traditional sense, it was a small receiver-size box with various controls and a fold-down cover containing a book of charts (which was important to obtain when purchasing the unit). Without such a device, the common way to determine a receiver's frequency with relative precision was to interpolate between two known points that were usually established with a crystal calibrator or by reference to known stations. The BC-221 was a different concept. As one hobbyist observed: "I marveled at its accuracy and ease of operation. You simply plug the unit into an A.C. outlet and attach a wire that runs from the BC-221 to the antenna post of the receiver. Then you zero beat the unit against an incoming signal, take your reading and look at a graph furnished with the monitor [the charts]. You then get a frequency reading accurate to within one kilocycle."[33] For those who insisted on precise frequency measurement, the BC-221 would remain the standard until the digital era.

Another accessory that was popular with experienced DXers in the 1940s and well into the 1950s was the three-tube RME preselector. There were four incarnations after its 1936 introduction — the **DB20, DB20A, DB22A** and **DB23**. The unit was basically a three-tube tuned RF amplifier which RME said would provide a minimum of 20 db of gain on all shortwave bands. Well-known DXer Charles S. Sutton of Ohio felt that it boosted most signals by 3 to 4 "S" units.[34] As with modern preselectors, it was connected between the receiver and the antenna. The first three models provided continuous coverage and were large in size — roughly a foot square — with an on-off switch, a large main tuning dial and a peaking control on the front. The DB23 was smaller and covered only the 80 to 10 meter ham bands, and was not quite up to the performance of the earlier models. While the DB20/20A/22A were designed to match the appearance of RME receivers, they were useable with other radios as well.[35]

"Panoramic adaptors" (spectrum analyzers) also became available after the war, the most popular one being the 10-tube Hallicrafters **SP-44 "Skyrider Panoramic."** When connected to a receiver with a 450–470 kHz. IF (most receivers), it showed visually, on a two-inch cathode ray tube, the signals present within 100 kHz. on either side of the received frequency. Among its advertised purposes were spotting "frequency modulation or parasitics" on an AM signal, measuring modulation and signal strength, and checking other frequencies. Around 1947 it sold for $49.50.[36]

The 1950s: Improvements on the Basics

The 1950s saw some improvements in receivers, and the introduction of many different models. However, the standard shortwave superhet remained the basic design. The 1950s were also a time when shortwave kits started gaining popularity.

It was during these years that the first specialist SWL retailer appeared. Gilfer Associates of Park Ridge, New Jersey was formed by Oliver P. (Perry) and Jeanne Ferrell in 1953, principally as a vehicle to distribute the *World Radio TV Handbook.* (The Gilfer name derives from the combination of the Ferrell name with Jeanne's maiden name, Gillespie.) Perry had been an editor of various radio magazines, including *CQ* and *Popular Electronics,* and knew the radio industry well. He wanted to establish a one-stop shopping place for SWLs, who, until then, had to rely on stores catering mainly to ham radio. He also wanted to make shortwave listening more attractive and easier. Gilfer quickly expanded

Jeanne and Perry Ferrell, proprietors of Gilfer Shortwave, were well-known in the SWL community.

into other publications and supplies, then into receivers, receiver modifications, and accessories, and soon was a leading source of SWL-related merchandise. It was well-known for a utilities guide called the *Confidential Frequency List* which contained much information not easily obtainable elsewhere. It became Gilfer Shortwave in 1979. Jeanne carried on after Perry's death in 1984. She sold the company ten years later. It closed in 1997.[37]

Another well-known SWL supplier was Universal Service of Columbus, Ohio, later known as Universal Amateur Radio, Inc., Universal Shortwave Radio, and finally (in 1990) Universal Radio, Inc. Universal had served the hobby radio market since 1942, carrying all of the major receiver brands and offering them for demonstration in a well-stocked showroom. It became better known to SWLs as Gilfer took a lower profile in the 1980s. Now located in Reynoldsburg, it is the leading supplier of shortwave receivers, accessories and books, and its owners since 1982, Fred and Barbara Osterman, are well-known within the shortwave community.

Other suppliers and accessory manufacturers of more recent decades have included Grove Enterprises, Inc. of Brasstown, North Carolina; C. Crane Co., Fortuna, California (portables); MFJ Enterprises, Inc., Mississippi State, Mississippi; Kiwa Electronics, Kasson, Minnesota; Sherwood Engineering, Inc., Denver, Colorado; Radio West of Escondido, California (it closed in 1991); Electronic Equipment Bank of Vienna, Virginia (closed in 1998); and EGE, Inc. of Woodbridge, Virginia (also closed). Some suppliers also performed receiver modifications.

Entry Level Receivers (Four and Five Tubes)

While better known today for its intermediate and high-end receivers, National was a leader in the low-end market as well, with two popular entries to compete with the S-38. First came the National **SW-54**, a modernistic-looking, five-tube, four-band set introduced in 1950 for $49.95. With only 5 kHz. selectivity this was no DX machine, but it had bandspread and a built-in speaker, and it is an example of another radio on which many old timers cut their teeth. In 1958 the SW-54 was redesigned, dubbed the **NC-60 "Special,"** and sold for $59.95. It was on the market through the early 1960s.

Although Hallicrafters continued to serve new shortwave listeners mainly through new models of the S-38, during the 1950s it offered a number of other four- and five-tube receivers. These were very basic kitchen radios, some with just a single shortwave band in addition to AM, others with multiple bands but little more than a bandspread control to

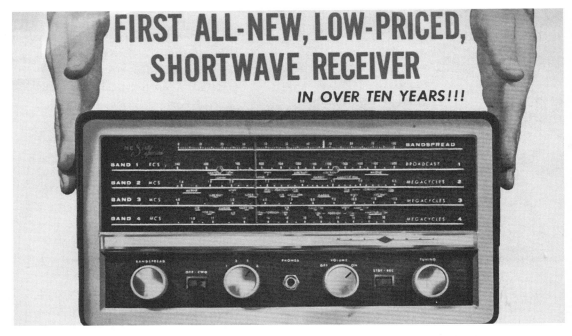

A starter receiver, the NC-60 was brought out in 1958 as a follow-on to National's popular SW-54.

identify them as shortwave sets. These were the **S-103 "Continental,"** and the "R" series receivers—the **5R10, 5R30, 5R35, 5R40,** and **5R50.** Some were portable, and one, the **S-80 "Defender,"** was a battery operated desktop set intended for use in non-electrified areas such as farms. The shortwave range of these receivers was usually 6–18 MHz, and the price was circa $30–45.

INTERMEDIATE LEVEL RECEIVERS (EIGHT TO TEN TUBES)

In the early and mid–1950s, Hallicrafters had several eight-tube models, all of which differed in features and price. One was the most advanced of the "R" series, the **8R40,** a four-band set that had somewhat more of a communications receiver look than the home radio motif of its "R" sisters. It covered the AM band, and up to 44 MHz. in three short-wave bands, and featured bandspread and a variable pitch BFO. It sold for $89.95.

The S-53A, produced in 1951, was a replacement for the S-53 which had been on the market for several years. In selectivity and resistance to images, however, its performance was below that of its predecessor.[38] The S-53A sold for $79.95, and in a 1954 survey of members of the Universal Radio DX Club it was voted the most wanted receiver under $100.[39] Four years later, Hallicrafters introduced the S-85 for $119.95. It was a handsome set in the genre of the more expensive SX-88, SX-99 and SX-100 receivers. It had four bands and calibrated bandspread, but no selectivity control. A better eight tuber was the **SX-99,** which sold for $149.95. It was introduced around the same time as the S-85, but it had more controls and an S-meter.

In 1958, Hallicrafters released three more eight-tube receivers. They were similar in appearance but differed in features. The five-band **S-107** was basically a replacement for the S-53A. It cost $94.95 and had 6 meter coverage as well as BCB and shortwave. The **S-108** was a fuller-featured radio at a higher price—$129.95. The top of the group was the four-band **SX-110** at $159.95. It had variable selectivity, an antenna trimmer and an S-meter. Unlike the S-107 and S-108, it did not have a built-in speaker.

In its intermediate-level radios, National favored nine-tube designs rather than Hallicrafters' eight. In 1953, National introduced the nine-tube **NC-88 "World Master,"** successor to the NC-57. It cost $129.95, and it had four bands, ham-band bandspread, frequency coverage up to 40 MHz., and an internal speaker A year later came the **NC-98.** Its price was a bit higher—$149.95—but it boasted three-position selectivity and an S-meter (and required an external speaker). Both the NC-88 and NC-98 offered an "SW" version which had calibrated bandspread for the 16, 19, 25, 31 and 49 meter international shortwave bands, an unusual feature at the time.

National's 1957 nine-tube offering was the highly regarded **NC-188.** It also had coverage up to 40 MHz. in four bands, plus ham-band bandspread, but, unlike its big brother, the NC-109 (see below), it lacked variable selectivity. It sold for $159.95.

Hammarlund's entry in the intermediate category was the **HQ-100,** the first of the Hammarlunds to sport the "rounded edge" appearance that would characterize most of the "100" series through the HQ-180. The HQ-100 was a four-band, ten-tube receiver, with bandspread, S-meter, antenna trimmer and many other features. Its was also the first of the Hammarlunds to offer an electric clock as an option. The clock incorporated a timer that would turn the receiver on in advance of actual use in order to warm it up and minimize drift during usage. The HQ-100 had a Q-multiplier in place of the usual multi-position crystal filter. It had the large, comfortable controls that would typify the HQ line, and it produced warm, room filling audio through its 9" external speaker.

The HQ-100 came on the market in 1956 and cost $189. Although it was the budget entry of the "100" series that would earn Hammarlund top marks among serious shortwave aficionados, it was highly regarded. One NNRC member observed.

> When comparing the HQ-100 with its predecessors in this shack, I'm impressed with vast technical improvements and at the same time, vaguely conscious of something lost. When a listener shifts from what might be called broad band tuning to pin-point selectivity, the whole approach to DX is modified. Not unlike a pilot flying blind, you grope your way across a once familiar 20 meter "bandscape," seemingly lost without the usual heterodyne chorus, SWBC image milestones and stray CW birdies. The time-honored ability to tune three stations at once is gone, except for those which are zero beat on the same channel. All this is replaced by razor-sharp precision, knocking Radio Moscow for a loop and allowing 50 watt CE4XX of unknown provinces to pin your S meter needle with a 5 by 9 reception.[40]

In 1959, a new company entered the communications receiver market. Lafayette Radio Co. operated a chain of radio stores and a large mail order business and would become well known among shortwave listeners in the 1960s and 1970s. Their first offering was the well-featured, four-band **HE-10,** which looked like a grown up Hallicrafters S-38 with nearly twice the tubes (nine). At $79.95 it was more expensive than the S-38. It was the first popular shortwave receiver on the U.S. market that was made in Japan.

HIGH END RECEIVERS (MORE THAN TEN TUBES)

By the early 1950s, the distinction between single-conversion, and double- and even triple-conversion sets was being clearly drawn. Although single-conversion sets were still being made, many of the new receivers were in the multiple-conversion category (and at higher prices).

Single Conversion   National had several high performers in the single-conversion category, including the **HRO-50-1** which replaced the HRO-50 in 1951. It had a much improved IF section, providing better selectivity, but at a premium price — $383.50.

Of greater interest to the general shortwave listening community were National's NC-125 and NC-109 receivers. The **NC-125** appeared in 1950. For $149.95 this 11-tube unit offered general coverage in four bands on a slide rule dial. While it had no selectivity options, it did incorporate the National **Select-O-Ject** audio filter (available as an accessory on other National receivers for $25).[41] Even in those days, the finickiness of audio filters was apparent, as one observer was quoted in a National advertisement: "[I]t has been found that a certain amount of training and practice is necessary to get the excellent results that the device is capable of. After the first half-hearted try, the operator may be tempted to alter the name to 'Select-O-Junk,' but after mastering the operation of the device, he may change this again to 'Select-O-Ject' as more descriptive of its high-powered performance."[42]

The **NC-109**, also an 11-tube model, appeared in 1957. It was basically an NC-188 with a six-position crystal filter and a product detector for SSB reception (which was gaining broader acceptance within the ham community at that time). It sold for $199.95.

Two single-conversion Hammarlund receivers were of interest to shortwave listeners, the HQ-140X and the HQ-150. Introduced in 1953, the 11-tube **HQ-140X** was the successor to the HQ-129X and was similar to it in many ways, including its classic appearance. However, at $264.50 it was double the price. The 13-tube **HQ-150** succeeded the HQ-140 in 1956 and was similar in features and appearance but was a much better performer. With a Q-multiplier in addition to the crystal filter and crystal phasing control, plus a crystal calibrator, it offered greatly improved signal handling capabilities. The price of the HQ-150 was $294.

RME was still making shortwave receivers in the 1950s. Two of interest to SWLs were the **RME-50** (1952) and the **RME-79** (1953). Both were 12-tube sets of like appearance and features, and were similar to the RME-45B of 1947 save for two-speed vernier tuning (called "Cal-O-Matic" on the RME-50, "Dial-O-Matic" on the RME-79). Later RME receivers were ham band–only in coverage. The company merged with Electro-Voice in 1953 and the RME name disappeared about ten years later.

Double and Triple Conversion   The lowest priced dual-conversion receivers made in the 1950s were the 13-tube Hallicrafters **SX-71** (1950, $179.50) (dual conversion only above 4.7 MHz.),[43] and the 11-tube Hallicrafters **S-76** (1951, $169.50). Next in price was the **SX-96**, a 12-tuber that appeared in 1954 for $249.95. After that, prices were $300 or more, sometimes much more.

The massive, 20-tube Hallicrafters **SX-73** came on the market in 1952. Originally made for the military, it was roughly the equivalent of the SP-600, and one of the most sophisticated receivers ever built by Hallicrafters. It was single conversion to 7 MHz, double after that. At $975, however, few found their way into listeners' DX dens.[44]

At a somewhat lower but still premium price was the **SX-88**, released in 1953. Also a 20-tube set, its original price was $499.95, but soon jumped to $595 and then $675. The

SX-88 was at the very least the best Hallicrafters receiver since the SX-28, perhaps the best Hallicrafters ever made. It was dual conversion on all six bands, and sensitivity and selectivity were rated high. Probably fewer than 1,000 units were made, a factor that has contributed to its distinction.[45] The **SX-100** came along a year later. It was a simplified SX-88, with 14 tubes and, at $295, a much lower price tag. One of its major attractions was a "T" notch filter of the type that would become better known in some of the Hammarlund receivers.

In the National line, 1952 saw the birth of the last of the tube-type HROs, the 18-tube **HRO-60**, which remained in production until the arrival of the solid-state HRO-500 in 1964. The main difference between the HRO-60 and the HRO-50 was the use of dual conversion. Like the Hallicrafters SX-73, however, the HRO-60 was still single conversion below 7 MHz. And true to its heritage, the plug-in coils remained. The HRO-60 was priced at $533.50, and in the URDXC survey it was the most wanted receiver in the over $500 class.

For those who wanted a double-conversion National receiver for less money there was the **NC-183D**, which also appeared in 1952. It had 16 tubes, two RF stages and three IF stages, 6 meter amateur band coverage (47–55 MHz.), six-position crystal selectivity with phasing, and dual conversion above 4 MHz. Nominally an upgrade of the NC-183, it boasted many improvements, including better circuitry, a new front end, good stability, improved selectivity, and easier dial reading. Although latter-day tests showed some significant shortcomings, including lower-than-expected sensitivity and dynamic range, and poor performance on 6 meters,[46] a hands-on review at the time rated the receiver "very high on the important counts of selectivity, useable sensitivity, image rejection and stability."[47] Its price was $383.50, and it was the most-wanted receiver in the $300–400 class in the URDXC survey.[48]

Although the HRO-60 was generally considered National's top-of-the-line receiver, the **NC-400**, introduced in 1959, was in many respects a more advanced set and had several interesting options, including a plug-in mechanical filter to replace the crystal filter, five crystal sockets for fixed-frequency operation, and a diversity modification kit that permitted diversity reception (essentially combining the reception from two receivers, each with its own antenna, into a single audio signal for reduced fading).[49] The NC-400 had 18 tubes, was dual conversion above 7 MHz., and got high marks in the sensitivity and selectivity departments.[50] It sold for a hefty $895. Although, as with the HRO-60, it remained in production until 1964, only 500–1,000 are estimated to have been made, and many were purchased by law enforcement and government agencies.

Also in the super receiver category was the last of the Hammarlund "**Super-Pro**" series, the **SP-600**.[51] Although virtually all were built for the military or other government use during the period 1951–1972, many found there way into DXers' hands, where they rightfully took their place among the classics of the vacuum tube era. The SP-600 was an 85-pound, six-band, 20-tube receiver that tuned to 54 MHz. It was dual conversion above 7.4 MHz., and while retro in appearance even by 1950s standards, its performance was outstanding, as well it should have been for a "new" price of $985.

The SP-600 had an interesting dial mechanism. The bandspread dial was a simple 0–100 scale. However, on the main tuning dial a number appeared under the main dial pointer indicating the number of revolutions of the bandspread dial. If not direct readout, this at least made for easy return to a given frequency. If the main dial read "4" and the bandspread dial "87.6," a return to position "4-87.6" would bring you back to the desired frequency. The SP-600 had six selectivity positions.

The SP-600 was followed in 1955 by the art deco-styled, drum-tuning Hammarlund **PRO-310**, a 14-tube receiver that was manufactured only for a year. In performance it probably belonged somewhere between the HQ line and the Collins 51J. It was a full-featured receiver, with good audio, good selectivity, and double conversion above 2 MHz., but it made little impact on the shortwave market. It sold for $495.

More popular among serious SWLs than the expensive SP-600 were the multiple-conversion Hammarlund "HQs" of the 1950s—the HQ-145, -160 and -180, all of which were introduced circa 1959. The **HQ-145** was an 11-tube receiver; the **HQ-160** 13 tubes. Both were dual conversion above 10 MHz., and both were good performers The main operational differences between them were that the HQ-160 came equipped with a crystal calibrator (optional on the HQ-145), and it had a Q-multiplier in place of the HQ-145's six-position crystal filter. The HQ-145 had the Hammarlund clock-timer, the HQ-160 did not. Most importantly, both were equipped with an IF notch ("slot") filter that was very effective in nulling out narrow ranges within the receiver's bandpass. The Hammarlund notch could often reduce or eliminate unwanted heterodynes and improve the readability of the received signal. The HQ-145 sold new for $269, the HQ-160 $379. While these were popular receivers, they did not reach the level of acceptance of the HQ-129X.

If one were to poll shortwave listeners on the best performing of the affordable vacuum tube receivers, the Hammarlund **HQ-180** (1959), and its slightly upgraded successor,

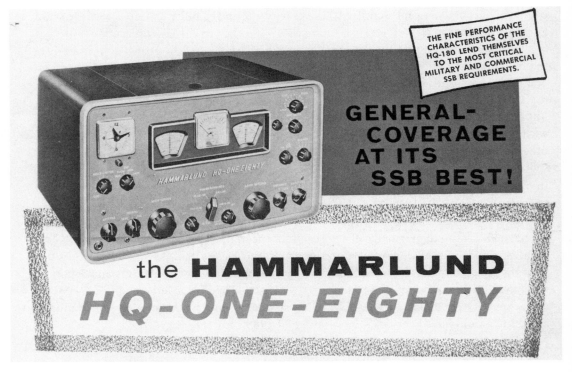

THE FINE PERFORMANCE CHARACTERISTICS OF THE HQ-180 LEND THEMSELVES TO THE MOST CRITICAL MILITARY AND COMMERCIAL SSB REQUIREMENTS.

GENERAL-COVERAGE AT ITS SSB BEST!

the **HAMMARLUND** *HQ-ONE-EIGHTY*

In the opinion of many users, when it came to pure DXing capabilities, the Hammarlund HQ-180 and HQ-180A were the finest vacuum tube shortwave receivers ever made.

the **HQ-180A** (released in 1963), would be the winners.[52] The HQ-180 had 18 tubes, the HQ-180A had 17 (and two silicon rectifiers). They were general coverage versions of the Hammarlund HQ-170 and HQ-170A ham band–only receivers that were released at the same time. They were double conversion up to 7 MHz., triple conversion above. The typical configuration included the optional Hammarlund clock-timer. They sold for $429.

The strength of the HQ-180 and HQ-180A was in their selectivity. They had four positions of IF selectivity. In addition, either sideband could be selected without, as in most receivers, simultaneously activating the BFO. Thus either sideband was available in AM mode. By selecting the sideband opposite the interfering station, the operator could greatly reduce or even eliminate the interference. This was a major advantage, and, when combined with the receiver's notch filter, provided unusual flexibility under difficult reception conditions. Although a bit susceptible to overload in some situations, the selectivity options, combined with room-filling audio and solid, flywheel tuning, made these receivers for the ages, probably not materially improved upon in overall signal-handling ability until digital readout and passband tuning came along. The best shortwave DXers used the 180/180A. And even though the bandspread dial was inoperative on the standard broadcast band (which was split into two bands), the receiver was the universal standard among serious medium wave DXers.

The Best

"[I]f I were to evaluate the best receiver I've ever had, it would have to be the Hammarlund HQ-180A. It was sensitive. Its selectivity was superior. Its tremendous bandspread made finding frequencies and stations easy, in spite of the analog readout. I could really eliminate hets with its notch filter. It gave out plenty of audio. All in all, for digging out the weak DX, I've never used any receiver that approached the HQ-180A." Bill Flynn, "Old Times vs. New," *WPE Call Letter* (Great Circle Shortwave Society), March 1988, p. 13.

As for Collins, in addition to the 51J series of super receivers discussed earlier, there was, in 1959, the equally super **51S1.** At $1,500, it was triple conversion up to 7 MHz., double conversion above. In design it looked like the then-new Collins "S" series line of amateur equipment. The 51S1 had 17 tubes plus semiconductors, and it was equipped with a Q-multiplier.[53]

Better known among shortwave listeners and Collins aficionados was a 26-tube military receiver, the **R-390A,** which was manufactured from 1954 to 1985.[54] Although often associated with Collins, which did the research and development for the receiver, other companies manufactured many more R-390A's than did Collins. The receiver was triple conversion below 8 MHz., double above. It is difficult to imagine a more ruggedly built receiver, inside and out, than the R-390A. Servicing required some know how, however, for the direct frequency readout for which it was justly famous was accomplished through a complex system of cams and gears that physically moved cores through various coils and transformers as the receiver was tuned (an arrangement similar to that of the 51J).

From a DXer's perspective the three main features of the R-390A were the sharp selectivity of its mechanical filters, its near total resistance to overload, and its direct frequency readout. Direct readout was accomplished via a two-knob mechanical odometer arrangement. You selected the megahertz band with one knob, and with the other you dialed to your chosen channel, reading the frequency on a rotating odometer, the last barrel of which was marked in tenths of a kilohertz. There were no shortcuts to this system, however, and tuning between, and within, bands required a lot of knob turning. But to most DXers it was worth the trouble.

Although at a "new" price of over $2,000 the R-390A was beyond the reach of nearly all hobbyists, it appears that over 50,000 units were made, and over time a great many of these made their way to the hobby market where for many years the receiver was fairly plentiful at less than half the original price (in later years much less), and a favorite among shortwave connoisseurs.

Finally, mention should be made of the Racal **RA17**, a 23-tube, triple-conversion set that came to market in 1957.[55] The radio's $1,200 price tag and its British manufacture meant that there would be few of these sets in use in the states in those days (even though there was an American-manufactured version, the **RA71**). More would appear later in the used market, however, and over time the several versions of the RA17 gained legendary status, largely because of its reputation as the receiver of choice of the BBC Monitoring Service and other professional monitors. Although the dial arrangement was different, tuning was similar to the R-390A in that you selected megacycles with one knob and kilocycles with the other, and then enjoyed 1 kHz. tuning accuracy. However, the RA17 was an electronic novelty because it was the first receiver to use the Wadley Loop, a circuit that permitted accurate frequency dialing without the need for multiple crystals. It was invented during the war by Dr. Trevor Wadley, a South African engineer who later worked for Racal. On the RA17, six inches of dial length equaled 100 kHz. The receiver gave better than 1 kHz. dialing accuracy.

Some felt that the strength of the RA17 was more in its audio quality than its DX capabilities, which was not surprising given that deciphering the spoken word rather than DX worthiness was the principal concern of professional monitors. In later years, Racal manufactured many other high quality, high technology shortwave receivers, both in Britain and the United States, mainly for the professional, commercial and military markets, and at prices far beyond the range of most shortwave listeners.

1960–1975: Years of Change

The next 15 years were a period of great improvement in communications receiver design. Solid-state technology replaced vacuum tubes, and the first receivers designed especially for the shortwave *broadcast* listener were introduced, along with general coverage receivers that made SWBC tuning easier. It was also a period of radical change in the industry. Hallicrafters, Hammarlund, National and Collins all introduced their last receivers. Now there were new names—Drake,[56] Kenwood, house brands from Allied Radio and Lafayette, and others—and receivers from Japan, at first manufactured for western companies and then marketed under Japanese brand names.

RECEIVERS DESIGNED WITH THE SHORTWAVE BROADCAST LISTENER IN MIND

The bane of the shortwave listener was trying to easily identify or return to a frequency with precision, i.e., 5 kHz. or better. Typically this required interpolation between known points, a process that discouraged casual listening and reinforced shortwave's reputation as a medium suited mainly for the technically inclined. The bandspread dials of the better receivers eased the process, but usually they were calibrated only for the amateur bands so as to make the receivers more attractive to ham operators.

In the 1960s, receiver manufacturers saw in the SWBC listener a distinct market. This translated into receivers specifically designed for shortwave broadcast listening, and, in other receivers, innovations that facilitated SWBC listening.

Shortwave Broadcast Receivers   Hallicrafters, Drake, Allied and Lafayette all introduced receivers that covered only the shortwave broadcast bands. Save for the advanced Squires-Sanders SS-IBS receiver (below), Hallicrafters was first off the mark with the **S-200 "Legionnaire,"** a four-tube set introduced in 1965 for $59.95. It was followed in 1966 by the **S-210**, a six-tuber selling for $89.95, and the 1967 **S-214**, a receiver which sold for the same price. (The S-214 was made in Japan.) What all these sets had in common was that they covered the standard broadcast band plus four specific shortwave broadcasting ranges— 49, 31, 25 and 19 meters. (The S-210 and S-214 also had FM coverage.) While they were modest performers, their limited frequency range permitted the bands to be spread out for more accurate dialing.

In 1967, Drake released the better-performing but still-average **SW-4A**, a tube-and-transistor receiver that tuned the standard broadcast band and the 49, 41, 31, 25, 19, 16, 13 and 11 meter bands. (A low-production predecessor, the SW-4, had appeared a year earlier.) With the purchase of additional crystals, other ranges could be substituted. The SW-4A's strength was its ease of use. The desired MHz. band was selected and the main tuning dial used to reach the exact frequency, which Drake claimed could be read to an accuracy of 2 kHz. Said Drake: "No more guesswork in identifying stations. Set the dial at a station's frequency, and if conditions permit, you'll hear it, every time!" On the negative side,

One of the early SWBC–only receivers, the no-frills Drake SW-4A offered simplicity of operation.

consistent with the receiver's primary use as a shortwave broadcast rather than a ham receiver, there was no BFO, and only one selectivity position — 5 kHz. The receiver sold for $289.

The SW-4A's simplicity was no accident. The initiative for its design came from short-wave station WNYW, which wanted an easy-to-use receiver to market in connection with promotion of the station and shortwave listening generally. A label that appeared on the front of the SW-4 and SW-4A read, "Designed especially for Radio New York Worldwide."[57]

Two years later Drake introduced the all-solid state **SPR-4** which covered the same bands as the SW-4A (except 11 meters) with even greater dial accuracy. Each band was 500 kHz. wide, and the receiver could accommodate 13 extra crystals of any 500 kHz. range. Adding the number on the main tuning dial to the number of the selected MHz. band yielded the frequency to 1 kHz. A built-in IF notch filter was a valuable tuning aid; how-ever, selectivity was tied to reception mode (4.8 kHz. in AM), and correct setting of a touchy preselector control was necessary to avoid spurious signals. The price was $449, and the receiver became very popular, in part because it accommodated user modifications well. Both the SW-4A and the SPR-4 were dual-conversion receivers.

Worth noting as well was Drake's "R-4" line of ham receivers introduced during the years 1964–73 — the **R-4, R-4A, R-4B** and **R-4C**. These were high performance, mixed tube-and-semiconductor receivers that cost between $380 and $500 when they were introduced. Although they came equipped for coverage of the ham bands only, there was provision for adding additional crystals for virtually any 500 kHz. shortwave segment (except, due to the receiver's IF frequency, 5.0–6.0 MHz.). This flexibility, plus 1 kHz. dial accuracy on all bands, made them of interest to shortwave listeners. The R-4B was preferred because, even though it was double conversion (the others were triple), the receiver's several selectivity positions and its passband tuning — it was one of the early receivers with this feature — were avail-able in all modes, including the standard AM mode normally used for broadcast listening.[58]

Drake receivers were popular. In a 1975 survey of the members of the American Short-wave Listeners Club, Drake and Radio Shack were tied for the two most popular lines. In the case of Drake, the SPR-4 was the most-owned receiver. Among Radio Shack aficiona-dos, it was the DX-150 and the DX-160 that were the most popular (see below).[59]

Both Allied and Lafayette entered the SWBC–only market with several single-conver-sion receivers made in Japan. In 1970, Allied brought out its **A2508** at $119.95, followed in 1971 by the **A2509** at $79.95. These were tube sets that covered the broadcast band plus 49, 31, 25 and 19 meters and the VHF public service band (the A2508 also covered FM). The Lafayette receivers were the "**Explor-Air Mark V**" (tubes), released in 1968 at $49.95, and the "**Explor-Air Mark VI**," a solid-state model that appeared in 1970 at a price of $79.95. Both covered the same shortwave bands as the Allied receivers, plus the standard broad-cast band (and, in the Mark VI, FM).

The two best-performing shortwave broadcast receivers were probably the Allied SX-190 and the **Squires-Sanders SS-IBS**. The SS-IBS was the earliest of the SWBC–only receivers. It had been introduced in 1963. This was a premium receiver — 13 tubes plus semi-conductors. It became famous not because of its widespread use — it was intended mainly for professionals, and, at $1,225, was beyond the price range of most SWLs — but because of the novelty of its tuning mechanism. It had three positions of selectivity, and it covered all the shortwave broadcast bands from 49 to 11 meters, plus the combined amateur–SWBC

75 meter band. It also had crystal sockets for two additional bands of the user's choice. Frequency readout was accurate to 1 kHz. The rotating slide rule dial took you to the nearest 100 kHz., and a 0–99 odometer did the rest. The radio is best remembered for the speedy frequency shifting that was accomplished by up and down buttons that controlled an electric motor connected to the main tuning dial.[60]

Made in Japan, the outstanding performance, reasonable price ($249.95) and wide availability through local Allied Radio stores of the solid-state **SX-190** made it a favorite among SWLs soon after its introduction in 1971. It covered 75, 49, 41, 31, 25, 19 and 16 meters, plus three ham bands and the CB band, all in 500 kHz. swaths and all with crystal calibrator-assisted dial accuracy of 1 kHz. or better. It also had crystal sockets for two additional shortwave bands. It had but one selectivity position (4 kHz.), but it featured a built-in Q-multiplier. The SX-190 is widely viewed as a classic receiver, competing favorably with the SPR-4 at a much lower price.

A surprising aspect of the SWBC–only receivers was their lack of tropical band coverage. None came equipped to tune the 60 or 90 meter bands, or (save for the SS-1BS and SX-190) the 75 meter ham–SWBC band. It is clear that these receivers were intended for the international broadcast monitor rather than the DXer who might favor the more challenging domestic targets found in the tropical bands.

Although intended mainly for amateur work, the double-conversion, solid-state Yaesu **FR-101** was also useable on the SWBC bands. Released in 1974, the FR-101S had a linear display with dial accuracy to 1 kHz., while the FR-101SD had digital readout. The receiver tuned in crystal-controlled ranges of 500 kHz., and the bandswitch was marked for the 13, 16, 19, 25, 31 and 60 meter shortwave broadcast bands, and the CB band, in addition to the ham bands. The FR-101 also accommodated crystals for four other user-designated bands. However, it shipped with ham band crystals only, requiring extra expense to make the $500-plus unit into a SWBC receiver. This, plus the absence of other common tuning controls, contributed to its low use in the SWBC community.

Shortwave Broadcast Bandspread   Mention has already been made of the special shortwave broadcast bandspread dial available on the NC-88 and NC-98 general coverage receivers introduced in the 1950s. This innovation was found on a few more receivers from the early 1960s through the mid–1970s. It permitted the same accurate tuning of the SWBC bands as was available for the ham bands with the typical ham band–calibrated bandspread dial. National re-introduced this feature in 1961 in its **NC-190,** a 10-tube double-conversion receiver that sold for $199.50. The NC-190 bandspread dial was calibrated for both the ham bands and the SWBC bands. A like system was featured on the eight-tube, double-conversion **NC-140** which was introduced in 1963 for $189.95.

Also in 1963, National introduced two other general coverage receivers with special provisions for tuning the SWBC bands. These were the five-tube **NC-77X** and the six-tube **NC-121,** both entry-level, slide rule dial, single-conversion receivers. Unlike most shortwave radios, their bandspread dials were not calibrated for the ham bands. However, they had a 0–100 scale (found on many bandspread dials), and the instruction manual contained diagrams that correlated the 0–100 markings with the frequencies of the international shortwave bands as well as the ham bands. If the main tuning dial was set at a designated point at the end of the band, reference to the diagram permitted matching the digits of the 0–100 scale with the frequencies of the SWBC or ham range chosen. The same approach to SWBC bandspreading was used in some other receivers, including the popular Realistic DX-160 which was released in 1975. While not as convenient as an approach as that of the NC-190 and NC-140, the prices of the NC-77X and NC-121 were much lower — $69.95 and $129.95 respectively.

The Hallicrafters **SX-133,** introduced in 1967, followed the same approach as the NC-190 and NC-140. It featured a slide rule bandspread for five ham bands and the 49, 31, 25 and 19 meter SWBC bands. Although only a seven-tube, single-conversion model, at $249.50 it offered a good range of features— adjustable BFO, antenna trimmer, noise limiter, crystal filter, three-position selectivity, and S-meter. The SWBC bandspread distinguished this model from the **SX-130,** introduced two years earlier.

In 1969, Heathkit offered builders of its **GR-78** shortwave receiver the choice of either a ham-calibrated or SWBC-calibrated bandspread dial. Trio-Kenwood, a Japanese firm, introduced the Kenwood **QR-666,** and its successor, the more advanced **R-300,** in 1974 and 1976 respectively. Double-conversion, solid-state receivers, they also featured optional bandspread dials covering the SWBC rather than the ham ranges. The QR-666 was priced at $129, the R-300 at $239. The bandspread of Radio Shack's **DX-200** (1981) was also calibrated for both the ham bands and the SWBC bands.

Direct Frequency Readout   For those willing to pay the price, a number of general coverage receivers capable of 1 kHz. or better tuning resolution were available. Earliest among

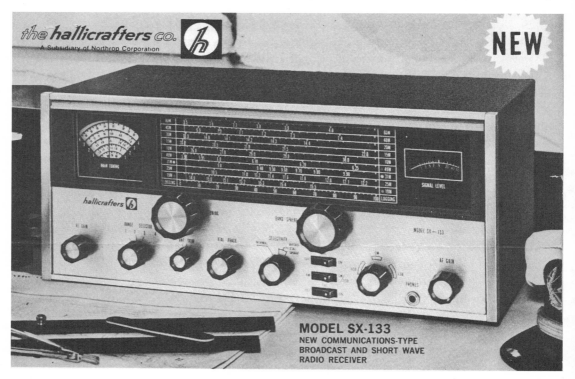

The inclusion of four SWBC bands on the bandspread dial made the Hallicrafters SX-133 of interest to SWLs.

them was the venerable **HRO-500.** Thirty-two pounds at birth in 1964 and selling for $1,295, it was the replacement for the HRO-60 and the first solid-state receiver in the HRO line. Triple conversion below 4 MHz., double above, it was intended mainly for the commercial rather than the SWL or ham market. The HRO-500 preserved the basic HRO PW dial design, but replaced the standard HRO coil mechanism with a frequency synthesizer. As with other HROs, frequency readout took some getting used to, but it was worth the trouble — one dial spin represented either 10 or 50 kHz., at the operator's choice. The receiver had four positions of selectivity, a tunable IF notch filter, preselector, passband tuning, and selectable sideband. While not amenable to owner repair, and a bit prone to synthesizer problems, it was the "dream receiver" of its time.[61]

The HRO-500 is important because it was the first receiver to boast a phase locked loop ("PLL") synthesizer. PLL technology was different from the Wadley Loop but produced like results — extremely accurate frequency readouts, and stable, drift-free reception. Available at first only on advanced receivers, it soon became the industry standard for all but the simplest shortwave receivers, desktop and portable, and inaugurated a whole new generation of receiver design and simplified shortwave radio reception.

In 1970 the HRO-500 was succeeded by the **HRO-600,** selling for $2,900. Among many other improvements, the classic PW tuning dial was scrapped in favor of a more conventional, Nixie-tube frequency display, and the preselector was motorized. The HRO-600 was intended mainly for the commercial market (reportedly only a few hundred were made).

More reasonably priced than the HROs was the solid-state **Galaxy R-530,** introduced in 1967. At $695 it was the first relatively affordable receiver to use the Wadley Loop design theretofore available only on the Racal RA17. You selected the megahertz. band with one dial, and tuned to your exact frequency with the other, a linear control with 0–1000 kHz. markings. While the three-position selectivity switch could have used more choices suitable for voice reception than the two that were available (2.1 and, as an option, 6.0 kHz.), the receiver was a top performer.

Although not popular with shortwave listeners, the line of McKay-Dymek receivers that appeared in 1975 and a few years thereafter is worthy of note. The several consumer and commercial variations of these triple-conversion, general coverage receivers — the **DR-22, -33, -44, -55, -101** — looked like stereo components. Tuning involved setting a series of knobs, each controlling one digit of the frequency. For "9635" you tuned four knobs — one to "9," one to "6," etc., plus a fine tuning knob if necessary. Depending on the model, the chosen frequency appeared in red LEDs or was reflected on the labels of the control knobs. As one observer put it, "To operate just turn it on, set the Mode, tune the readout to the desired frequency, and 'thar she blows.'" Even without feeding the output into an external hi-fi system, the audio quality of these receivers was superior. They were high performers on many other levels as well, including tropical band reception. However, since they were useless for band scanning they were of little interest to even those SWLs who could afford the $1,000 price tag. Drake offered a similar tuning design in several $2,000-plus commercial models sold during the 1970s.

Other Receivers An abundance of entry-level receivers came to market from 1960 to 1975. Hallicrafters introduced two in 1961: the **S-119 "Skybuddy II"** and the **S-120.** The S-119

The Superb GALAXY R-530 Communications Receiver

Featuring Wadley Loop circuitry and modular construction, the high performance Galaxy R-530 had most of what DXers could want in 1967, including visual frequency accuracy to 1 kHz.

was a no-frills, three-tube set that tuned up to 16 MHz. It sold for $49.95. Of greater interest to the new SWL was the S-120, which sold for $69.95. While still a modest, four-tube set, it tuned to 30 MHz. and had bandspread, a BFO, and a telescopic rod antenna for shortwave reception. It replaced the venerable S-38 line and was soon in wide use. In 1962, a year which began with 37 Hallicrafters products in production, the five-tube **S-118** appeared. It cost $99.95. From 1961 to 1965, Hallicrafters also released three S-120 clones in wooden cabinets, along with a wooden-sheathed lookalike for the S-118 and two five-tube receivers of similar design. This was the "**WR**" series, priced from $69.95 to $99.95.

Hallicrafters brought its last two original designs to the entry-level market in 1967. Both were made in Japan. The **S-120A "Star Quest"** was solid state, and different from the S-120 vacuum tube receiver whose number it shares. It sold for $59.95. At a slightly higher level of performance was the solid-state **S-240,** a dual-conversion receiver selling for $109.95. In 1970, Hallicrafters released the **S-125 "Star Quest II."** It was virtually identical to the S-120A, and it was the last of the Hallicrafters S/SX models.

National, Lafayette, and Allied Radio Shack, under its "Realistic" brand, were also offering entry-level receivers during this period. National's **NC-105,** a six-tube set, was introduced in 1961 for $119.95. Lafayette's three-tube **HE-60** (1962, $39.95), four-tube **HE-40** (1961, $54.50), and seven tube **HA-63** (1963, $64.50) were all made in Japan. In 1966, Lafayette entered the market again with the three-tube **HA-226** (a repackaged HE-60) and the six-tube **HA-700** for $49.95 and $89.95 respectively. Allied brought out the four-tube Realistic **DX-75** in 1965, and the **DX-120 "Star Patrol"** in 1970. Both sold for $69.95. Save for the HE-60 and HA-226, which appear to have been very basic sets, these receivers were surprisingly well-featured for the money.

Many single-conversion models were available at the intermediate level. Among the most popular was the Realistic **DX-150,** introduced in 1967. At $119.95, this solid-state general coverage receiver compared favorably with more expensive sets and was considered the best of the Japanese imports at the time. It covered the broadcast band and shortwave to 30 MHz., and although it lacked selectivity options it had a full range of other controls, including a BFO with pitch control, antenna trimmer, fast and slow AVC, noise limiter, S-meter, and bandspread (but calibrated only for the ham bands and without a 0–100 scale). It was quiet, sensitive, and well-built. The popularity of the DX-150 led to the release of the **DX-150A** (1969), **DX-150B** (1972), and **DX-160** (1974). The DX-150A and DX-150B offered mainly circuitry changes that improved selectivity and reduced the intrusion of spurious signals to which the DX-150 was prone. The price increased to $159.95 with the DX-160, which added longwave and, more importantly, a 0–100 logging scale on the bandspread dial. It also included in the operating manual a series of calibration charts that matched the markings of the logging scale to the frequencies of the international broadcasting bands.

Hammarlund's entries in the intermediate, single-conversion market during these years were upgrades of the HQ-100. The **HQ-100A,** released in 1961, incorporated only minor changes to its unlettered namesake. It sold for $189. The HQ-100 was further upgraded in 1968 in the form of the **HQ-200,** which traded two of the HQ-100's tubes for semiconductors and was slightly more modern in appearance than the earlier "HQs." Its price was $249.95.

Two single-conversion receivers by Hallicrafters were the **S-129** and **SX-130,** both introduced in 1965. The receivers had seven tubes and a silicon diode rectifier, and were identical except that the SX-130 had two selectivity choices and an S-meter. With a noise limiter and an adjustable BFO, the SX-130 gained some popularity among shortwave listeners. The receivers sold for $154.95 and $169.95 respectively.

Lafayette had several low-priced entries in the single-conversion intermediate market. These were the nine-tube **HE-30,** released in 1961 and selling for $99.95; the **HA-230,** an eight-tube set that appeared in 1964 and sold for $79.95; and the solid-state **HA-600,** which sold for $99.95 when introduced in 1968. The HA-600 gained considerable popularity. It was readily available at Lafayette stores, and while not a stellar performer, it offered good features—including a long bandspread, albeit calibrated only for the ham bands—at low cost. It was similar to Allied's **A2515,** introduced in 1967 at the same price.

Better single-conversion performers from Lafayette were the **HE-80** and **HA-225,** 14-tube models that were introduced in 1963 and 1964 respectively and were priced similarly (HE-80, $139.50; HA-225, $129.95). All the Allied and Lafayette radios were made in Japan.

Other intermediates were the Hammarlund **HQ-145A** and the Drake **SSR-1,** both double-conversion receivers. The HQ-145A came to market in 1966 and was basically a repackaged HQ-145. It sold for $299. At the same price but appearing almost ten years later (1975) was the solid-state SSR-1. Although the Wadley Loop circuitry and 10 kHz. dial gradations permitted tuning accuracy within 5 kHz., the design, construction and performance of the SSR-1, the only Drake unit made in Japan, was considered below that of other Drake equipment, and the receiver enjoyed only mediocre reviews.

One of the best double-conversion receivers of the era was the Hallicrafters **SX-122,** which sold for $295 when it was introduced in 1964, $100 more when the almost identical **SX-122A** appeared in 1967. This was an 11-tube receiver, mechanically a new design but similar to the SX-100 electronically. Although it lacked a 0–100 scale on the bandspread, and, unlike the SX-100, it had no notch filter, it received generally high marks, and demonstrated good stability and image rejection.[62]

Post-1975: Major Advancements

The year 1976 ushered in a new era in shortwave receiver design. Solid-state circuitry replaced tubes for good. Performance was superior on many fronts, and features theretofore found only on dream receivers now became widely available at the consumer level. Foremost among these was direct frequency readout. With phase locked loop technology, which soon became the standard, the guesswork was gone — you simply read your frequency on the LED display, often to one or two decimal places. (Outboard digital displays that could be connected to tube receivers soon came to market as well.) Frequency drift was eliminated, and the most basic receivers became rock stable, permitting clear single sideband reception and the use of other specialized reception techniques. The miniaturization of components led to smaller and lighter-weight receivers, including full-featured portables that operated like desktops. *Passport to World Band Radio* coined the term "portatop" in recognition of the increasing number of receivers that had characteristics of both and functioned equally well in both environments. Practically all receivers were now made in Japan, home of a shortwave listening youth craze which incentivized manufacturers during the 1970s, and nearly all were double conversion. Receiver capabilities increased, and so did prices.

Shortwave listeners, long used to operating in the shadow of the hams, became a market all their own, and welcomed the abundance of riches. Receiver manufacturers had a quandary, however. While DXers were interested in high performance features that maximized a receiver's signal-handling ability, SWLs who were more interested in ease of tuning and in hearing the programs of the stronger stations with good fidelity comprised the larger market. Most receivers were a compromise, with individual designs that tilted in one direction or the other, but few that could satisfy all comers And for many manufacturers it was still the much larger amateur radio market that provided the research and development which led to better general coverage receivers.

A few analog receivers lacking digital readout were released after 1975. For a year starting in 1978, Lafayette sold the **BCR-101,** a moderately priced ($249.95), adequately performing dual-conversion receiver whose two-step analog tuning dial could be read to about 5 kHz accuracy. In 1981, Radio Shack brought out both the introductory **DX-100** ($99.95) and the slightly better-performing **DX-200** ($229.50). The latter set was not very sensitive, but it had a bandspread which, while crowded, was calibrated for both the ham bands and the international SWBC bands, *including* part of the 60 meter tropical band (down to 4850 kHz.). The DX-200 was discontinued in 1982, the DX-100 in 1983. These were the last receivers of the "old" solid-state generation.

NEW-AGE DESKTOPS, 1976–85

The Good Performers Although lacking digital readout, the **FRG-7** ("Frog 7") by Yaesu, a company well-known in the ham market, is usually considered the bridge to the new generation of receivers. As with virtually all the communications receivers made from 1976 on, it had full coverage of shortwave to 30 MHz. Wadley Loop technology had appeared at the consumer level in the United States circa 1972 on a specialty portable, the South African–made Barlow Wadley XCR-30 (see below), and in 1975 on the underperforming Drake SSR-1. The FRG-7 was the first to offer it on a full-featured desktop that would receive wide acclaim. You set the megahertz dial until a glowing LED went out, indicating "lock," and you then used the kilohertz knob to dial the desired frequency to an accuracy of 5 kHz. Although it had only one selectivity position, sensitivity was high and the audio was good, and many third-party modifications, including digital displays and improved filters, became available. Considered a good all around receiver, it was an example of how far the entry-level receiver market had advanced. In a 1979 study of 400 members of SPEEDX and NASWA, 23 percent owned an FRG-7.[63]

The FRG-7, which was introduced at a price of $299.50 in 1976, remained a favorite for years (it was discontinued in 1982). In 1977 the **FRG-7000** added a digital display and a second bandwidth. The FRG line was improved further in 1981 by the **FRG-7700,** which abandoned the Wadley Loop for a more conventional phase locked loop circuit. Although its selectivity was still subject to criticism, an optional memory unit for the FRG-7700

Introduced in 1976, the Yaesu FRG-7 offered Wadley Loop performance for $299.50. It became one of the most popular receivers of its day.

($150) permitted the storage of frequencies (12) for the first time. In 1985, Yaesu introduced the microprocessor-controlled **FRG-8800** for $599. It added keypad tuning and frequency scanning. However, it was no match for the then-popular ICOM R71A, with which it was designed to compete, and the shortwave community generally gave it only fair marks.

Although the post FRG-7 Yaesu receivers offered good value, they were never a favorite of hard core DXers. Neither was another 1976 offering, the **RF-4800,** the desktop member of Panasonic's "Command Series," which was better known for its portables, the RF-2200 et al. (see p. 317). With its impressively large, black, military appearance, the RF-4800 was a beautiful set to look at. A solid-state analog receiver with a digital display, it had two-speed tuning (main dial "in" for fast tuning, "out" for slow), good sensitivity and good audio (useful on the FM band, which it also covered), and it was easy to use. However, selectivity was mediocre, and it produced many spurious signals. It did not live up to expectations for a $449.95 receiver, nor did the slightly improved **RF-4900** which appeared two years later. The basic **RF-3100** and the more advanced **RF-B600** portatops, introduced between 1982 and 1984, displayed similar limitations and never developed much of a following.

Sony would become best known in 1980–85 for its ground breaking ICF-2001 and ICF-2010 portables. It introduced a number of well-featured receivers during the years 1976 to 1981, including the portatop **CRF-320** and **CRF-330** and portable **CRF-1.** However, at $1,500 and up they were uncompetitive, and little known within the shortwave community. More familiar was the Sony **ICF-6700W,** released in 1978. A somewhat retro, boxy design for the time, and costing an often-discounted list price of $439.95, this semi-portable PLL receiver was only an average performer. Common complaints were the set's single-speed tuning, its propensity to overload, and selectivity options that seemed to suit neither program listeners nor DXers. But it had a digital display in addition to a traditional analog dial, and good audio, which was becoming standard on shortwave receivers by the late 1970s. There were electrical and design improvements in the more expensive ($650) **ICF-6800W,** released in 1980, and additional changes (including improved selectivity) in the orange-lettered version ("6800 orange") introduced one year later. The ICF-6800W was quite popular in Europe, but, surprisingly, it was not widely distributed in the United States.

The closest thing to a lemon in shortwave receivers was Radio Shack's Realistic **DX-300.** Introduced in 1979 at $379.95, there were high expectations for this Wadley Loop receiver. Unfortunately, a defective design caused the receiver to go into free running oscillation, resulting in untold returns, repairs and replacements. The news traveled quickly, and the DX-300 was soon receiving low marks in the SWL community. Once operating properly, the receiver still suffered from images and overloading, and — a common problem in the receivers of the day — the one-position, 6 kHz. selectivity was adequate for program listening in undemanding situations, but too wide for DXing. However, some SWLs felt that, once the oscillation problem was fixed, it was a decent receiver for the money, especially at $249.00, the price at which it was discounted when its successor, the **DX-302,** was released in 1980 at the DX 300's original price. Nearly identical to the DX-300 in appearance, the DX-302 replaced its predecessor's audio filter with an IF filter, but tuning was similarly touchy, and overall performance was only fair. The last of the consumer-grade Wadley Loop receivers sold in the United States, it was phased out in 1982.

It was expensive at $650, but the Sony ICF-6800W PLL receiver had good looks.

The Kenwood **R-1000,** introduced in 1979, was a small, nicely designed unit, with a combination analog-and-digital display and two bandwidth choices. At $499.95 it was a good seller, despite some shortcomings. It was easy to operate, and performed best as an SWL receiver rather than a DX machine.

In 1982, Kenwood brought out the **R-600.** It was an introductory receiver, simple to operate. It shared some of the basic features of the soon-to-be-released **R-2000**, and it offered good value for $399.95. The R-2000 came along in 1983. At $549.95 it was a fuller-featured set than the R-1000 and reflected the technological advances of the previous four years. However, like the R-1000, its reputation was primarily that of an SWL's radio. It had a variety of useful functions—three speed (albeit too-coarse) tuning, digital readout displayed to 100 Hz., ten memories, and fairly good selectivity—but it lacked others, such as passband tuning and a notch filter. And it was prone to overloading. A businesslike receiver that was easy to operate, it was overshadowed by the technically superior ICOM R70 about which the shortwave community was abuzz at the time. However, like the R70's better-known successor, the R71A, it was available for some ten years.

The High Performers   The year 1977 saw the introduction of the first of the post–1975 high performers: the Japan Radio Company **NRD-505**. JRC traces its history to the early 1900s. It became well-known for its postwar commercial and maritime receivers, and, later, ham receivers. The NRD-505 was its first offering in the United States, and even though it was JRC's least expensive receiver, at $2,000-plus the four-memory, dual-conversion set was far outside the range of most hams and SWLs. It had an indirect impact on the short-wave community, however, for many of its features found their way into the NRD-515, brought to market in 1980 at the more acceptable (to some) list price of $1,399.

The solid, semi-military look and excellent workmanship of the **NRD-515** was a delight to users who missed the physicality of the old vacuum tube receivers. Tuning was via the by then-standard twin megahertz-kilohertz dials unless you bought the optional outboard keypad. A fine tuning control operated over a 2 kHz. range, and frequency was displayed to 100 Hz. Two bandwidth filters in AM (6.0 and 2.4 kHz.), and an auxiliary position for adding a third, made for good selectivity. However, the 515 lacked a notch filter, and pass-band tuning did not function in AM mode. It worked in SSB, however, and added appreciably to the 515's use in "exalted-carrier selectable sideband" reception (ECSS), a technique of listening to AM signals in sideband mode in order to clarify weak signals and reduce interference. Unlike the 505 (the first receiver to offer frequency memories), the 515 lacked built-in memories, but a 24- (later 96-) memory unit available as an option made the 515 one of the first receivers capable of storing frequencies in large numbers. The receiver was quiet, and fun to use, and developed a wide following among serious SWLs.[64]

In a class similar to the NRD-515 (many would say a step above) was the general-coverage Drake **R7**, which started out as a ham band–only receiver. When introduced in 1979 it lacked a digital display, which Drake soon added to the analog dial (the display read to 100 Hz.). The digital display, general coverage version sold for $1,295. Drake further upgraded the receiver in 1981 by adding a noise blanker (previously an option) and improving the bandwidth configuration. This was the **R7A**, which sold for $1,649.

The R7A had many features that were important to serious DXers: two bandwidths and provision for three more, full electronic passband tuning that operated in AM mode, an incremental tuning control, and a tunable IF notch filter. An early tendency to spurious signal generation was corrected in later runs. Although the receiver's three-step tuning process made the R7A a bit cumbersome to use, the receiver combined just about all that a DXer with deep pockets could wish for, albeit sometimes only with extra-cost options available from Drake or third-party suppliers. The R7A was probably the best receiver available at the time.[65]

Also in the high-performance category was the ICOM **IC-R71A**. The R71A was the progeny of the **IC-R70**, a quirky but technologically advanced receiver introduced in 1982 for a list price of $749, far below the original price of the NRD-515 and the Drake R7.[66] (ICOM receivers were often available for $100 or more below list price.) With technology derived from ICOM's ham transceivers, the operation of the R70 was in some ways more

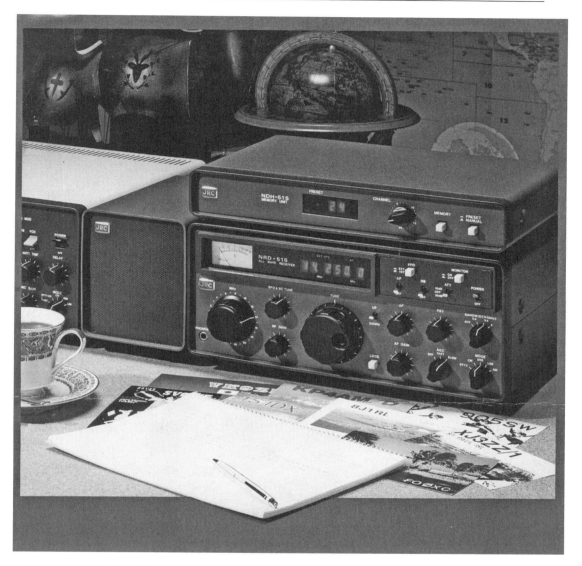

The Japan Radio Co. NRD-515 set a new standard in receiver performance when it was released in 1980, but at a high price.

suitable to the amateur radio world. However, SWLs wanting the best recognized the value of the R70, and by 1984 ICOM had refined the design, eliminating some of the idiosyncracies. The resulting R71A was introduced at a list price of $799.[67]

The R71A was quadruple conversion, and featured two VFOs, each with variable-rate tuning. Although only two selectivity positions (6.0 and 2.8 kHz.) were available in AM mode, others were available in SSB, and the receiver was strong in the area of ECSS reception. In addition, the passband tuning feature operated more like a bandwidth control, and further narrowed the bandwidth selected (albeit not as cleanly as DXers might have hoped). And the radio had a notch filter. True to its microprocessor design, many parameters of the receiver's performance could be operator-adjusted. A design quirk was that the receiver's

A bit more expensive than the NRD-515, the Drake R7A boasted passband tuning that operated in AM mode, a valuable feature for SWBC DXing.

The ICOM IC-R71A really was "The World Class World Receiver" when it was introduced in 1984. It gained a wide following in the United States.

operating system was maintained by an internal battery having a 10–15 year life, and for most users replacing it would require returning the radio to the factory.

The receiver had 32 tunable memories. The front-panel speaker of the R70 was moved to the top of the unit on the R71A, and the extra space dedicated to a keypad. The R71A was largely free of spurious signals, and the synthesizer — a sometimes-noisy element of PLL receivers — was quiet. Audio was decent, although not up to that of some other receivers. The front panel was crowded and the controls small, and the receiver was still complicated to use. However, it accommodated modifications well.[68] Some of the most popular improved the receiver's selectivity and made the notch filter available in AM mode.

Desktops After 1984

Most of the companies selling communications receivers introduced new models in the post–1984 years, and several new firms entered the marketplace.

Established Brands Japan Radio Company's NRD-515, which had been introduced in 1980, had three successors. The **NRD-525** appeared in 1986 at a price of $1,179. While it had lost some of the heft of the 515, it now sported a built-in keypad, 200 memories, an IF notch that worked in AM mode, and a computer interface port. The AM bandwidth filters were somewhat less capable than the 515's, and although the 525 received high marks from many, complaints about the fidelity of its audio (and a persistent hiss) were frequent, detracting from the receiver's overall strong reputation.[69]

The **NRD-535** was introduced in 1991. JRC made some early circuitry modifications, and the improved version of the receiver, known as the **NRD-535D,** was widely hailed as a wonder in shortwave receiver design. The dual-conversion NRD was now triple conversion and boasted all of the features and controls that the serious DXer would want: two well-selected stock filters suitable for AM listening, variable bandwidth control, passband tuning, an IF notch, 200 memories, and synchronous detection, an ECSS-like feature that locked an internally generated carrier on to the received signal, reducing fading and improving audio.[70] The 535 was not perfect — the synchronous detector tended to lose lock, and audio, even with the 525's hiss now gone, was still thought subpar by some. But it was as near perfect as any other receiver that the shortwave community had seen, as well it should have been at a price of $1,699.[71]

The NRD-535D represented the apex of JRC's shortwave receiver line. In 1997, the company brought out a stripped-down version of the 535. At $799, the **NRD-345** was a fairly good receiver by the heightened standards of the day, but gone were some of the features that DXers liked most about the 535 — passband tuning, IF notch, and variable bandwidth. The company returned to the 535 level in 1999 with the last of the NRD line, the **NRD-545.** It reflected most of the features of the NRD-535 with a few improvements (such as 1,000 memories instead of 200). However, digital signal processing, the then-new source of many benefits in the design of the 545, showed some encumbrances as well in the form of niggling design and operational deficiencies that might have been forgiven on a less expensive set but not one selling for $1,799. JRC left these unattended, the result being that the 545's reputation never matched that of its eight-year-old predecessor.

The Kenwood **R-5000,** introduced in 1987 at $899, was basically the receiver portion of Kenwood's popular TS-440 transceiver. It was well-featured, with 100 memories, dual VFOs, analog and keypad tuning, digital display to 10 Hz., two selectivity positions (with more available as options), an audio notch filter, and passband tuning. Although it had some deficiencies—the memories were not tunable, the stock wide filter was too wide, the keypad was of a distracting, non-standard layout, and passband tuning was not available in AM—it was a very good, very sensitive receiver. In performance it was roughly similar to the R71A and the NRD-525, but with better audio. It never achieved the popularity of those receivers, however, and it was the last of Kenwood's communications receivers.[72]

Not so with the Drake **R8.** Drake had stopped production of ham and SWL receivers in 1984, shifting its focus to making home satellite receivers. Its re-entry into the communications receiver market seven years later caused great excitement in the shortwave community.[73] The black-front, microprocessor-controlled R8 was in the Drake tradition: well-built, and designed with the user in mind. Among its many features were two VFOs, digital display reading to 10 Hz., keypad and conventional tuning, 100 memories, five bandwidths (available in all modes), passband tuning in all modes, audio notch filter, and synchronous detection. Operation of the R8 was straightforward, and many felt that its audio was the best of any receiver around.

The receiver was not entirely without critics. The synchronous detector lost lock too easily, and most DXers would have preferred an IF notch to the R8's touchy audio notch. But Drake's strong reputation in the ham and SWL community, and the receiver's "Made in America" label, gave it the benefit of the doubt. Soon countless DXers were taking advantage of the Drake no-risk sales policy: order the R8 direct from the Ohio factory for $999 and if you were unhappy with it for any reason, return it within 15 days for a full refund. There were few returns, and soon the R8 was the affordable receiver of choice among American SWLs.[74]

The **R8A** (1995, $1,099) increased the number of memories to 440 and improved many aspects of the receiver's performance and ergonomics. The **R8B** ($1997, $1,199) increased the memories to 1,000, and improved the unit's synchronous detection, making it sideband selectable.

The strong sales of the R8 and its progeny led Drake to introduce a series of lower-priced receivers. While some DX features like passband tuning and a notch were foregone in these units, they all had digital display, keypad and analog tuning, memories, and excellent audio. The **SW8** came to market in 1994 for $599. It had synchronous detection, sideband capability, three bandwidths, and a built-in swivel antenna, and it covered FM and VHF. While not performing at the level of the R8, it offered good value as a basic DX receiver. The **SW1** and **SW2** were introduced in 1996 and 1997 respectively. They were intended mainly for program listening. The SW1 was a no-frills receiver priced accordingly at $299. The SW2 was a step up, adding sideband reception capability, a second bandwidth (albeit too narrow for AM listening), and sideband selectable synchronous detection, for $499.

The 1991–97 Drake receivers, particularly the R8 series, had been highly successful, but by 2005 orders were no longer strong enough to justify continued production. As a result, Drake stopped producing them in 2005, although, true to a long Drake tradition, they continued to service those already in use.[75]

Yaesu had introduced its last general coverage receiver, the FRG-8800, in 1985. Nearly

There was joy in DX shacks everywhere when the R8 was introduced in 1991. Drake was back in the receiver market with a U.S.–made product that was a winner. The R8A, shown in the illustration, came along in 1995, and the R8B in 1997.

a decade later, in 1993, it re-entered the market with the compact **FRG-100.** Although the FRG-100 had 50 memories, and its mediocre bandwidths were improved in a subsequent "B" model, it lacked built-in keypad tuning, a common feature by that time, as well as DX-critical features like passband tuning, notch filter, and synchronous detection, at least some of which one would have expected, even in a receiver at the comparatively low price of $599. As a result, although it was a competent SWL receiver, the FRG-100 never achieved much popularity with DXers. It was thought discontinued in 1999, but new units continued to appear for five more years.

In 1995, Radio Shack brought out the **DX-394.** Like the Realistic DX-100 and DX-200, the company's last communications receivers which had been introduced almost 15 years earlier, the DX-394 was a modest performer. Designed principally for the international listener seeking a low-cost option, it had few DX capabilities, and some initial design flaws were corrected in "A" and "B" versions that followed. But the price was attractive: $399.95 ($249.95 by the time the unit was discontinued in 1998, even less for leftover stock thereafter). The DX-394 was widely modified by many of its sizeable band of dedicated users.

The last of the receivers from returnee manufacturers was from ICOM. Its R71A, introduced in 1984, had been manufactured for more than a decade. An **R72** followed in 1990, but it was a simpler receiver, lacking many features of the R71A, and it was released late in the United States and never gained much of a following here. Almost ten years later, in 1999, ICOM brought out the **R75.** Not quite in the same league as the super sets, at an affordable $799 it was a good value, with coverage up to 60 MHz., keypad tuning, 100 memories, digital readout to three decimal places, two bandwidths suitable for AM, and twin passband controls that offered partial passband tuning and partial variable bandwidth. Noise reduction, an audio notch filter, and other bandwidths were available as extra-cost options. Synchronous detection worked poorly (and was dropped in later production runs), but its solid SSB capabilities produced nearly as good results.

New Brands   Three firms generally unknown to the SWL community entered the communications receiver market in the post–1984 years— Lowe Electronics Ltd. (U.K.), AOR Ltd. (Japan), and Ten-Tec (Sevierville, Tennessee).

Lowe introduced its **HF-125** in 1987. It was a small receiver, well-built, with four bandwidths, 30 memories, 1 kHz. digital readout, and a telescopic antenna. Although it lacked passband tuning and a notch, and both synchronous detection and keypad tuning were extra-cost options, it was a solid, low-noise design, easy to operate — a good performer with excellent audio, of interest mainly to the broadcast listener rather than the DXer. Its price was $599. In 1989 it was succeeded by the slightly improved **HF-225** at $749. Four years later, Lowe released the **HF-225 "Europa"** edition which boasted a number of modifications sought by European DXers to make it more DX-worthy. These included improved bandwidth filters, synchronous detection, and an outboard keypad, all as standard equipment. Its cost in the United States was $899.[76]

Although the Lowe models were interesting receivers, American SWLs had relatively little experience with them because they had to be ordered direct from England. This changed in 1992 when several American suppliers started offering the **HF-150.** It was small, about seven inches on a side and three inches tall, nominally a desktop receiver but operable off eight "AA" batteries as well, and thus easily portable. Ruggedly built, the radio had a main tuning dial, a volume/on-off control and three pushbuttons. While it lacked pass-band tuning, keypad entry (available as an outboard option) and other DX features, it had digital readout to 1 kHz., fast-slow tuning, 60 memories, two well-selected bandwidths, sideband and AM operation with excellent synchronous detection available in each, and very good audio. The result, at $599, was a fun-to-use receiver that compared favorably with far more costly competitors in all but the most demanding DX situations. It became very popular among both program listeners and DXers seeking an escape from complexity.[77] A **PR-150** outboard preselector, designed to compliment the HF-150 and reduce its tendency to overload, became available in 1993, and an upgraded 150, the **150E** ("**Europa**"), appeared in 1998.

In 1995, Lowe introduced the **HF-250,** a major upgrade to the HF-225, followed in 1998 by the improved **HF-250E.** They were good performers, if somewhat under-featured for $1,299, but they never enjoyed the popularity of the HF-150. Starting in 1999, Lowe marketed in Europe and Japan the **HF-350,** a version of a receiver known in America as the Palstar **R30,** made by Palstar, Inc. of Ohio. The R30 was a good performer for non–DX purposes, and the company was friendly to customers. However, even at $495 it gained only a limited following, although those who bought it usually liked it.

Ten-Tec, a manufacturer of ham equipment since 1969, introduced its first communications receiver, the **RX-325,** in 1986 for $699. It was poorly designed and was out of production the next year. Its next offering was the **RX-340,** a high-performance, commer-

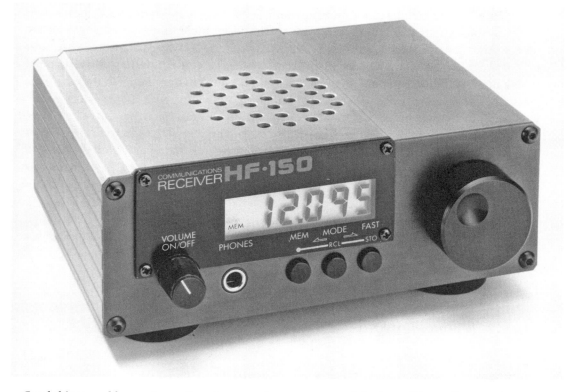

Good things could come in small packages, as the many users of the Lowe HF-150 discovered. Not a DX receiver, this simple but high-performing set could hold its own.

cial-grade dream receiver brought to market in 2000 for an unaffordable $3,950. Thus it was not until 2002, the year the company introduced the **RX-350,** that most SWLs became familiar with Ten-Tec. This $1,199 receiver was loaded with valuable features: two VFOs, 34 bandwidths, over 1,000 memories, an optical encoder-based tuning dial with 1 Hz. resolution, passband tuning, a spectrum display, and an automatic audio notch. It had good audio and few spurious signals. There were some negatives: the synchronous detector lost lock easily, the receiver had more than its share of microprocessor-related anomalies, and keypad tuning, all but a necessity with such a design, was an optional outboard accessory. Being a true DSP (digital signal processing) receiver, the RX-350's firmware was easily upgradeable through internet download, and thus the RX-350 held much promise for future improvement. Unfortunately, although Ten-Tec implemented a few modest changes, it never significantly altered the receiver's firmware, and the hoped-for "rolling improvements" to the RX-350 never materialized.

The third of the new players was AOR, a scanner company located in Japan. Its first general coverage receiver, the **AR3030,** appeared in 1994. At $799 it was a good performer, if lacking in some DX controls, such as passband tuning. AOR's reputation would be made in 1996 with the introduction of the **AR7030,** manufactured at the company's new division in England and designed by John Thorpe, formerly with Lowe. Never had so much receiver been packed into so small a space—10 inches square and three and one-half inches high. It was lightweight and easily transportable, and at $1,149 much more expensive than the AR3030.

The AR7030 maximized the benefits of microprocessing and minimized the number of front panel knobs and buttons, each of which did multiple duty. Familiarity with the menus and submenus displayed on the receiver's small LCD was challenging, at least at first, and some would have no part of such a radical design concept. But most who wanted strong DX capabilities and took the time to learn the AR7030's operation felt the effort was worth it. The receiver had four stock, AM-suitable bandwidths with room for two more, passband tuning, synchronous selectable sideband, and keypad frequency entry by way of an infrared remote control device. A noise blanker and audio notch filter were available as accessories. The quality of the audio was excellent. The receiver was upgraded to the **AR7030 "Plus"** in 1997. In addition to various circuitry improvements, the number of memories was increased from 100 to 400 and now accommodated alpha-numeric labels. Other enhancements followed in later years.

No radio more starkly illustrated the changes over a half century of receiver design than the AR7030. Those who still remembered the "boat anchors" of old could only look with amazement at the capabilities of this receiver, not much bigger than a cigar box.

Higher Technology Still

Since the start of the solid-state era, a number of firms, including Drake, JRC, National and Sony, had marketed many "super receivers," typically for either general or special commercial purposes. Usually costing multiple thousands of dollars, and with features that seldom added much to SWBC DXing, these limited-production models were rarely seen within the DX community. During the 1990s, however, a few of the "supers" caught the eye of DXers with deep pockets and a penchant for advanced technology. The best-known of these were the Watkins-Johnson **WJ-8711** (1990, $4,295) and the later **WJ-8711A**; the Watkins-Johnson **HF-1000** (1993, $3,995);[78] and the Ten-Tec **RX-340** (2000, $3,950).

These receivers boasted features like 1 Hz. tuning (and display), circa 60 bandwidths, tunable IF notch filters, and the like. Many operating parameters, fixed in other receivers, were user adjustable. They were a button-pusher's or knob-twiddler's delight (the HF-1000 had six knobs, 44 buttons and four digital displays). They were not problem free, however. Both Watkins-Johnson receivers suffered from digital hash problems, passband tuning that operated only in CW (or was absent altogether), synchronous detection that was not side-band selectable, and no tone control. The RX-340 had its own quirks and limitations. These receivers excelled in some demanding DX situations, but in others they sometimes could be outperformed by their less expensive consumer counterparts.

A second category of higher-tech radios to whom a few were attracted were the wide-band receivers whose coverage, instead of ending at 30 MHz., extended up to 2,000 MHz. (2 Ghz.) or more. Best known in this category were the ICOM **IC-R9000** (1989, $5,459)[79]; the AOR **AR5000** (1996, $1,899) and several other AOR widebands; the ICOM **IC-R8500** (1996, $1,849); and the Yaesu **VR-5000** (2001, $889). The ICOM **IC-R9500** wideband super set, selling for $13,500, was released in 2007. More enticing to utility DXers, these receivers provided little unique functionality that was directly relevant to SWBC listening, but did offer the capabilities for occasional forays into unfamiliar frequency regions. They had enhanced scanning capabilities, and sometimes a small video display for TV signals or spectrum display, but some were deficient in such DX-vital characteristics as bandwidth, passband tuning, and synchronous detection.

Finally, in 1991 came receivers that operated exclusively with personal computers. Many ordinary communications receivers were equipped with an RS-232 port that permits computer control. The new PC-controlled receivers took the form of a "black box" that was connected to a computer by cable. An antenna was connected to the box, the receiver's "front panel" appeared on screen, and all functions were then controlled exclusively by mouse or keyboard. They were essentially a combination of a receiver and an advanced receiver control program. The first PC-controlled receiver was the McKay-Dymek **DR-333,** which came to market in 1991 at a price of $1,495.[80] A well-known later model was the ICOM **IC-PCR1000** (1998, $499.95).

The PC-controlled receivers contained some high-end features, like spectrum displays and almost unlimited memories, and some had wideband frequency coverage. However, basic DX features were sometimes lacking. In general, they did not perform noticeably better than their non–PC counterparts.

The latest innovation in PC receiver design is the software-defined radio (SDR). While the line between a PC-controlled receiver and SDR is not always clear, in general the SDR offers more flexibility because its basic operating characteristics, rather than just its control, are defined through software rather than hardware. Bandwidths can be made continuously variable, wide swaths of spectrum space can be recorded for later analysis, etc. The Ten-Tec **RX-320,** introduced in 1998 at the bargain price of $295, was the first in this category and set the standard for some time.[81] Since 2003 there have been many **WiNRADiO** models available, including some on PC cards, at prices of $500 and up. Subsequent entries in the field have been, in 2004, the FlexRadio Systems **SDR-1000,** marketed in both transceiver and receiver configurations but no longer available (the receiver was priced at $676); in 2005, the RFspace, Inc. **SDR-14** ($999.95); and in 2007, the RFspace **SDR-IQ** ($499.95).[82] FlexRadio was said to be considering a 2008 release of a receive-only version of the FlexRadio **FLEX-5000,** its latest software-defined transceiver.

While SDR technology is proving useful for later review of recordings of wide swaths of spectrum, in general it is more useful for utility listeners who have multiple transmission modes to deal with. Although choices among traditional communications receivers are now fewer than they were a decade ago, time will tell whether SDR has any more than niche appeal among SWBC listeners.

Portables

In broadcasting's earliest days, all receivers were portable, or at least transportable, because all were battery-powered. The size and weight of the batteries, and the need for headphones and an antenna, made stationary operation the rule, however. The more tubes a radio had, the more battery power was needed, and thus, even by the standards of that time, sets advertised for portable use were almost by definition modest performers. Things improved as lower-drain tubes were developed, loop antennas (for the broadcast band) got smaller, and speakers replaced headphones. As alternating current replaced batteries as the source of power for radio receivers generally, portable receivers became more distinguishable from stationary receivers, but they remained largely a novelty.

Radios got smaller during the 1930s, and 1938 saw the introduction of tubes more compatible with battery operation. Even-smaller tubes followed, and although their required plate voltage still demanded batteries of significant size and weight, save for the war years the portable radio industry blossomed during the 1940s and 1950s. This was especially true when portables became operative on either batteries or AC, when subminiature tubes were developed, and when the ferrite rod antenna was introduced (1951). The mid–1950s introduction of the transistor, with its low battery drain, was a major breakthrough in radio miniaturization. Within five years, tube portables were almost gone.

The Postwar Years

Although not compatible with the needs of the serious shortwave listener, as far back as the early 1940s there were a few mainly BCB portables which had some limited shortwave coverage.[83] These sets weighed circa 15 pounds and covered the 6–18 MHz. range, usually in one band. They sold for around $25. Virtually all other portables were broadcast band–only until portable FM became a significant market in the mid–1960s, by which time portable radios of all sizes and styles were available, shoe box or shirt pocket, speaker or earphone.

Special note should be made of the Zenith **Trans-Oceanic** portable. Its colorful story and those of the company and its president, Commander Eugene F. McDonald, have been comprehensively researched and told by Bryant and Cones in their several volumes.[84] After falling victim to wartime production priorities soon after coming on the market in 1942, manufacturing was resumed after the war, with four vacuum tube Trans-Oceanics released in the years 1946–57. Each was slightly different from its predecessor. The original Trans-Oceanic was priced at $75, the final tube model about twice that. More than three-quarters of a million tube models were made prior to the last production run in 1963.

Over the years the tube complement of the Trans-Oceanic varied between five and eight. Although it lacked almost all the signal handling tools of a communications receiver,

The Zenith Trans-Oceanic was the king of the early shortwave portables. This is the H500, of which almost a quarter million units were sold following its introduction in 1951.

its audio was good, and the Trans-Oceanic had a very big plus for the international short-wave broadcast listener: each of its shortwave bands (16, 19, 25, 31 and 49 meters, changed after 1951 to the first four of these plus 2–4 and 4–8 MHz.) had its own dial scale, meaning that the bands were well spread out and frequency location was comparatively easy. Beyond that, it was a straightforward superhet, and the controls were strictly no-frills—a pushbutton band selector for each band, tuning knob, volume control, and several tone switches. The dial was of the airplane pointer variety until it was replaced by a slide rule design in 1954. The Trans-Oceanic was completely redesigned in 1957. The new **Royal 1000** was fully solid state and housed in an all-new, smaller, modernized metal case with "drum" tuning controlled by a dial on the right side of the case.

The attraction of the Trans-Oceanic was less in its performance than in the romance of owning one.

> With its short-wave bands, the Trans-Oceanics provided access to the world and the appearance of worldliness to its owner. It was used as a prop in fashion layouts, and advertised in upscale magazines, perpetuating its image as a necessity for the socially with-it. An ad in Holiday from 1948 showed a middle-aged man smoking a pipe, listening to his Trans-Oceanic poolside. To underscore this image of success, the ad says: "Of course, it's a Zenith.... The Aristocrat of Portable Radios. The Trans-Oceanic — its owner list reads like 'Who's Who in the World.' The choice of statesmen, executives, leaders in *all* walks of life." The meaning of the Trans-Oceanic had become quite clear. With its upscale image, the set would become especially desirable among Americans of the middle class, whose social aspirations and material cravings then had no limits.[85]

The who's-who was in part a result of Commander McDonald's penchant for promotion, which included making special presentations of a thousand Trans-Oceanics to various well-known personalities and public figures.

There were a number of Trans-Oceanic look alikes in the mid 1950s, including the Sears "Silvertone" 7224 and the "Silvertone Wayfarer" 8224, the RCA "Strato-World," and others.

The Trans-Oceanic was not the only portable of the era. In 1949, Hallicrafters brought out the five-tube, four-band, analog-dial **S-72** for $79.95. It was general coverage up to 30 MHz., had a BFO and bandspread, and came in a brown, luggage-style cabinet. Four years later Hallicrafters introduced the **S-93 "Worldwide,"** a basic five-tube Trans-Oceanic imitator with no DX controls. It covered shortwave up to 18 MHz. and sold for $99.95. It was also known as the **TW 500 "Worldwide,"** one of several similar "TW" receivers sold around the same time.

THE 1960S

The development of the Zenith Trans-Oceanic continued. In 1962, it got a facelift, along with the addition of FM, in the form of the **Royal 3000**. The design was further updated in 1969 with the **Royal 7000**, which for the first time added a BFO and wide-narrow selectivity. During most of this period the price of the solid-state Trans-Oceanic stayed at around $275.

The basic portable shortwave receiver of the mid–1960s market was the "multiband portable," a plastic-and-metal transistor set which typically covered the standard broadcast band, FM and shortwave and sold for $19.95 to $59.95. While a few tuned higher, most limited their shortwave coverage to a portion of the shortwave bands, usually stopping at 12 MHz. These radios had a slide rule dial, a whip antenna, ferrite bar, or both, 3–5" speaker, volume, tone and bandswitch controls, and a handle. A few better sets had dedicated SWBC ranges for easier tuning, a second bandwidth, or a bandspread control, but these were rare. Multibands were promoted for their ability to "pick up amateur, police, aeronautical, foreign broadcasting, marine navigation, and weather," a promise better kept in the advertising than in actual operation. These receivers were made for use on the go, and with few controls, their shortwave capabilities were limited.

A slightly better group of multibands sold for $79.95 to $119.95. Although they were not fundamentally different from their less costly cousins, they were designed to look more like the Trans-Oceanic of the day than the basic carry-around. Their shortwave coverage often extended beyond 12 MHz., or was divided up into more bands for easier tuning, and they were sometimes dressed up with VHF coverage, a "tuning/battery meter," a "fine tuning" control, a "Hi Fi-DX" or "local-distance"switch, a logging scale or world time zone chart, and a leatherette cabinet. However, the emphasis was on speaker size and tone controls, not DX.

Above this group were the several models of the European-style, well-constructed **Nordmende "Globetrotter"** series. At $159.95 to $259.95, one might have expected some DX capabilities in addition to the use of 11 discrete shortwave bands that permitted easier tuning (they did not cover the entire shortwave spectrum but did include all the major international bands *and* the 60 meter band). However, the emphasis was on woofers, tweeters and speaker size (some models did have a BFO).

1970 AND LATER

There were numerous multiband manufacturers in the 1960s, including names like Panasonic, Sanyo, Sharp and Sony. But it was a South African company that no one ever heard of, Barlow Wadley, that would, early in the 1970s, revolutionize the world of shortwave portables with the XCR-30, a cross-over receiver that combined performance with true portability for the first time. The start of an entirely new shortwave receiver market was right around the corner. The result was a plethora of portables, too numerous to recount in detail, some treasures, some trash, with names familiar and unfamiliar to the SWL community — among the latter, Degen, Kaito, Sangean, DAK, Tecsun, and many others. A few of the milestone-setting receivers are described below. They are but the tip of the iceberg, however; there were a multitude of others, good and bad, far too numerous to detail here. Sony was the pace setter.

1972 — Barlow Wadley XCR-30 Mark II   The Barlow Wadley did not look radically different from numerous other multiband portables. With four knobs and a small meter on the front, a telescoping antenna, and powered by six "D" cells, the only distinguishing characteristic of this 8½ pound unit was its tuning mechanism: a two-barrel odometer, one a MHz. scale, the other a 0–100 kHz. scale. But the heart of the receiver was a Wadley Loop circuit, theretofore available only on two high end receivers, the Racal RA17 and the Galaxy R530. On the Barlow Wadley you set the MHz. dial with one thumbwheel, dialed kHz. with the other, adjusted the antenna trimmer, and you were done. Bandspreading was unnecessary. Tuning was linear and drift free on the triple-conversion, 18-transistor radio, and

That a portable shortwave receiver could boast at least 5 kHz. dial accuracy and sell (initially) for under $200 amazed listeners when the Barlow Wadley XCR-30 appeared in the early 1970s.

dial accuracy was better than 5 kHz. It also covered the standard broadcast band, handled sideband reception reasonably well, and there was a 1000 kHz. crystal calibrator to precision-set the kHz. drum. Although construction was below standard and the radio was prone to overload, these complaints paled at the thrill of being able to direct-dial your frequency on a portable radio. The XCR-30 was available in limited quantities from Gilfer starting around 1972 at the initial price of $197.

The introduction of the XCR-30 did not produce an immediate change in the portable receiver market. For several years, most of the available receivers continued to be higher end versions of the basic 1960s multiband design. These included the Sony CRF-160 ($249.50), a double-conversion portable with nine push button-selected SWBC bands, each covering 600 kHz.; the more conventional, single-conversion Panasonic RF-5000 ($299.95); post–1969 versions of the Zenith Royal 7000 (see above); and many similar models. These receivers sometimes had BFOs, two selectivity positions, and some other basic DX controls, but always sported a traditional slide rule dial.

1976 — Sony ICF-5900W This was the first of the new generation of portables that abandoned the traditional 1960s look in favor of something more modern. The 5900W projected a new, "walkie talkie" appearance, although at 8¾" wide × 9¼" high × 4" deep it was not truly a handheld receiver. It was no DX machine — it had only one selectivity position — but it had a BFO for SSB reception, and reception on the whip antenna was surprisingly good, especially for the list price of $149.95. Although not a Wadley Loop receiver, its tuning was somewhat similar if a bit more complicated. Using the receiver's 250 kHz. crystal calibrator, you located the nearest marked 250 kHz. point on the dial, then turned the calibrator off and tuned the linear bandspread dial to the desired frequency, adding or subtracting kHz. from the frequency of the 250 kHz. marker. The dial was entirely analog, but you could read it to about 5 kHz. accuracy.

It was a bit complicated to tune, but the Sony ICF-5900W was a good performer, and reflected the more modern appearance of portables to come.

1977 — RF-2600, RF-2800, RF-2900: Panasonic "Command" Series The half-dozen receivers in Panasonic's "Command" series were released in the United States between 1977 and 1982. The first was the analog-readout RF-2200. Most did equally well as desktop receivers or portables — they had handles, whip antennas, and the option of battery operation, but they were larger than hand-held receivers.

The RF-2600 appeared in 1977, and was soon joined by the better-looking and slightly improved RF-2800, and then the RF-2900. (With some minor differences, they all looked alike.) The receivers had two selectivity positions, "push-pull" two speed tuning (on the 2800 and 2900), variable pitch BFO, and very good audio, and were good all around performers for their often-discounted $249.95 list price. However, the important feature was the LED frequency display. It took all the guesswork out of tuning. One simply tuned to the desired frequency, which was readable to 1 kHz. on the receiver's bright red display (blue on the 2600). This was a huge step forward in shortwave receiver convenience, more accurate than the XCR-30, and accomplishable with a single dial. It is illustrative of the progress in electronics that a digital display was available in these consumer-level portables the same year it was first introduced in an affordable communications receiver, Yaesu's FRG-7000. Previously it had been available only in a limited-production dream receiver, the HRO-600, which cost $2,900 when it was introduced in 1970.

1979 — Zenith Trans-Oceanic   The 1970s were the last full decade in the Trans-Oceanic's life. Zenith brought out the **R-7000** (not to be confused with the Royal 7000) in 1979. It replaced the spread-out international broadcast bands with general coverage over six bands

The main attraction of the Panasonic RF-2800 was its digital display.

(plus citizens band), and added mechanical bandspread to somewhat offset the resulting tuning imprecision. A battery-level meter and an FM tuning meter were added, but no S-meter. More importantly, the R-7000 retained analog slide rule tuning while other receivers were moving to digital displays. If the basic obsolescence of the Trans-Oceanic's solid-state versions had not sealed the radio's fate, the $380 price tag of the R-7000 did. Production ended in 1981.

A last-gasp Hallicrafters portable, the Japanese-made **TW-1200,** also appeared in 1979. However, with an analog slide rule dial it had little chance of success, especially at a price of $399.95.

1980 — Sony ICF-2001 The ICF-2001 was a complete departure in portable receiver design. It had no tuning dial; the desired frequency was entered via a built-in keypad: "6-1-9-5-EXECUTE." It was a space age, "book" design, 11½" wide × 6¾" deep × 2" high, with all controls located either on the top or the sides. There were no knobs, only buttons, slide switches and thumbwheels, and thus the top panel was, like the rest of the surfaces, essentially flat, making the radio, if a little large for traveling, at least easy to pack. The frequency

The no-dials, non-traditional look of Sony's ICF-2001 was unique, a reminder of how the world of shortwave receivers had changed by 1980.

was displayed to 1 kHz. Handy features included eight memories, a variable-pitch BFO, and a built-in delayed timer. Band scanning, accomplished by way of up-down buttons, was a bit cumbersome, and there was only one bandwidth. Battery life was maddeningly short, and long time owners found that the key pad buttons tended to stick as the receiver aged. Although at $329.95 the 2001 was no bargain, it was a novelty, and many DXers bought them as second receivers, especially at the deeply discounted prices sometimes obtainable.

The 2001 was succeeded by the slightly smaller **ICF-2002** in 1983, and the nearly identical **ICF-2003** in 1987, and in 1990 by the redesigned **ICF-SW7600,** all of which were good albeit not ground-breaking modifications of the 2001. The basic design of the 2001 also appeared later in slightly different form under other names (**Uniden CR-2021, Realistic DX-400, ATS-803**).

1985 — Sony ICF-2010   Slightly smaller than the ICF-2001, the ICF-2010 retained the same basic look. That was where the similarity ended, however. Priced at $329.95, the ICF-2010 was a portable that thought it was a communications receiver. It was soon heralded as the best-performing portable on the market. It had all the features of the ICF-2001, but unlike that receiver the 2010 was developed with the DXer in mind. It had two-speed dial tuning as well as keypad frequency entry, two bandwidths, independent upper and lower sideband operation, sideband selectable synchronous detection, and 32 memories. Its sen-

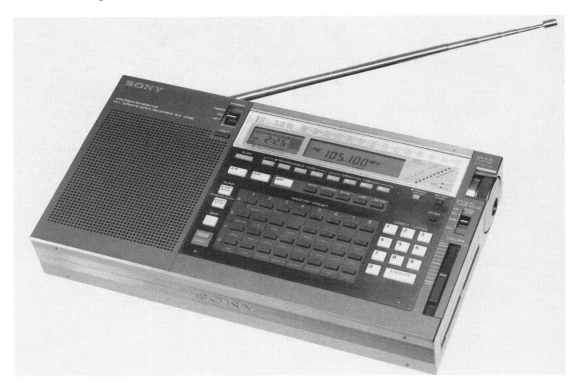

Much more than an upgraded ICF-2001, the Sony ICF-2010 was to portables what the Hammarlund HQ-180 was to vacuum tube sets — the best performing set on the market, a reputation it maintained for two decades.

sitivity was the equal of receivers costing much more. Sticklers may have preferred 10 Hz. tuning to 100 Hz. tuning for better SSB reception, and one of its FETs was prone to damage by static buildup. Overall, however, SWLs soon realized that the 2010's combination of features made it a serious DX receiver that demanded few compromises, especially for a portable. Furthering its reputation, a number of vendors offered aftermarket modifications which improved selectivity and audio response.

In a 1991 survey of ODXA members, more than twice as many ICF-2010s were owned than any other receiver. The results were the same in 1993 and 1995.[86] For 20 years the ICF-2010 would be widely acknowledged as the best shortwave portable. Production declined over time and finally ended in 2002. The receiver is still much sought-after in the used receiver market.

1988 — Sony ICF-SW1S   The ICF-SW1S was noteworthy less for its performance than its size. It was a palm-size receiver, 4¾" wide × 2¾" high × 1" deep, operable on two "AA" batteries, and it sold for $349.95, which included an amplified outboard antenna (it also had a built-in whip). It covered longwave, the standard broadcast band and shortwave, and it was fully synthesized, with digital frequency display and ten memories. Not a design that appealed to everyone — tuning on shortwave was in 5 kHz. steps and by button presses only (no tuning knob), and there was but one bandwidth and no DX controls— the audio quality was decent, especially during earpiece operation (it also had a self-contained speaker). At eight ounces, the SW1 was a triumph of miniaturization, and a technological wonder to those who remembered the days when performance correlated roughly with size and weight. Fifteen years later, receivers of similar size and performance would be available for under $50.

The SW1 was followed in 1994 by the **SW-100,** which was a tiny bit narrower and taller and priced $100 above the SW1's original price. It added synchronous detection, 50 page-based memories, and an alpha display mode, but was slightly under-sensitive when used on the whip.

1991— Sony ICF-SW77   The ICF-2010 was succeeded in 1991 by the ICF-SW77, which got off to a rocky start but finally found its footing in an improved version released in 1992. The top of the Sony line, it retained and improved upon the features of the 2010, but in a complete redesign. The SW77 was a high-performance, technologically advanced receiver, with greater reliance on microprocessor-based operating features than had ever been seen before. Although some found its innovative tuning and memory options a bit too complicated, its biggest drawback was a pronounced synthesizer "chuffing" that could be heard while tuning. This was common in many receivers, but unwelcome at a list price of $624.95. Given the choice, many SWLs opted for the 2010, still available new in limited supply and selling at the time for some $200 less than the SW77.

Sony marketed a number of other successful portables around the time of the SW77, including the **ICF-SW55** (1992, $429.95), innovative for its page-based memory system, annoying for its factory-set mode-bandwidth defaults; and the **ICF-7600G** (1994), at $229.95 the least expensive portable with synchronous selectable sideband.

1993 — Grundig Yacht Boy 400   The Yacht Boy 400 was an excellent portable, but important at least as much as a mark of Grundig success in the United States as for the receiver's performance. The Grundig name was well-known in America for its high quality, concert-style receivers of the 1950s (many of which had shortwave bands). It had been absent from the American market for decades, but continued producing a large variety of shortwave portables in Europe, including the boom box-size Satellit[87] receivers and the smaller Yacht Boy line, among others. Grundig usually targeted the general shortwave listening market rather than those interested in DX performance.

Grundig re-entered the American market in 1987, and achieved its greatest success with the Satellit 650 and Satellit 800 (see below), and the Yacht Boy 400. Compact in size and more modern in appearance than most of the Grundig portables, the 400 was an excellent performer, with keypad tuning in 1 or 5 kHz. steps (no tuning dial), digital display, 40 memories, two well-chosen bandwidths, adjustable BFO for precise SSB tuning, BCB and FM coverage, a novel microprocessor reset for easy clearing of internal glitches, and excellent audio. Although lacking synchronous detection, at $249.95 this China-made radio was probably the best portable in its size and price class at the time. Grundig made the most of its success, widely advertising the Yacht Boy 400 and its slightly modified 1998 successor, the Yacht Boy 400PE.

By the mid–1990s, the period of innovation in portable shortwave receiver design had largely ended. Occasional new models came to market, but they were old wine in new bottles. Slowly, most of the big names, including Sony, retired their portable shortwave lines.

There were two exceptions. One was the **Baygen Freeplay** windup radio, some versions of which had shortwave coverage. The Freeplay first appeared in 1994. It was the brainchild of British inventor Trevor Baylis and marketed by South Africa's BayGen Power Group. Many versions of the Freeplay have been introduced over the years under licensing agreements with other companies. Originally intended mainly for third-world use, and definitely no DX machine, the Freeplay is an interesting novelty that has found a niche in today's "emergencies radio" market.

The second exception, one with more importance to DXers, was Grundig, which continued introducing and advertising new portables. In 1996 it advertised the **Satellit 900,** which was due to be released that year. Supposed heir to the **Satellit 700,** it had a look completely different from that model, its distinctive, ultra-modern style obviously borrowed from two smaller radios in the Yacht Boy line, the Yacht Boy 360 and the Yacht Boy 500.

The Satellit 900 never appeared, however. Introduced in its place, and not until 2000, was the **Satellit 800,** a 15-pound, retro-looking receiver that resembled not the Satellit 700 but the heftier, 20-pound **Satellit 650,** the top-of-the-line Grundig which the 700 had *replaced* in 1992. Made in China, by now the source of most electronic equipment, for $499.95 it was a very good radio, if not quite up to Grundig's claim of "best shortwave receiver on the planet." Surprisingly, the Satellite 800 could be purchased from, among others, R. L. Drake Co., with the same 15-day money-back guarantee that Drake offered on its own R8. Knowledgeable observers noted the similarity in circuitry between the Satellit 800 and the Drake SW8.

2005 — Etón E1-XM   The Satellit 900 remained a mystery radio. In 2005, a new receiver, the Etón E1-XM, appeared. Etón was the former Lextronix, the Grundig American licensee which had re-introduced Grundig receivers in the United States in the 1980s. It was still handling Grundigs, and since 2003 also producing receivers under its own name, with technical help from Drake. The result of this three-way arrangement was the resurrection of the Satellit 900, now known as the Etón E1-XM.

Wary at first of a distinctive-looking receiver made in India and available at, among other non-shortwave venues, The Sharper Image, SWLs were relieved to learn that E1 repairs would be done by Drake, which was well-known for its service. (Drake also repaired the Satellit 800.) As user reports came in, it was clear that the performance of the E1 was unique among portables and the equal of many communications receivers. It had both keypad and two-speed dial tuning, a large digital display, three selectivity positions, passband tuning, selectable synchronous detection that was the best of any receiver at any price, and 1,700

The ICF-2010 met its match in the Etón E1-XM. Introduced in 2005, it was soon in wide use by serious DXers, many of whom favored it over their more expensive communications receivers.

memories. It also had a preamplifier, full scanning functions, and numerous other features dear to the hearts of DXers, and unlike many portables, it accommodated strong signals and long outdoor antennas well. And true to its Grundig provenance, it had excellent audio. It also covered the standard broadcast and FM bands, and it could receive "XM" satellite signals (with an optional XM module). The ICF-2010 had finally met its match, and at a reasonable $499.95, with occasional bargains available at $50–150 off.

The E1 was prone to a small amount of drifting, and a batch of E1s required recall. In addition, more than a few units needed repair for failed displays or microprocessor defects of one kind or another. Problematic as it could be, the receiver gained popularity. Of some two dozen experienced DXers present at a meeting in 2007, all but two had purchased an E1, and many preferred it to their more expensive receivers. However, at the end of 2007 the E1's future was uncertain. Some retailers were selling early stock for $225–279. There was a new version of the E1 without XM capability, and there were conflicting reports that the E1 would either be upgraded or discontinued.

Since the mid–2000s, Etón has marketed, and widely advertised, a host of other shortwave portables. Some have borne the Grundig name, others Etón, and still others have been marketed under both names. With Sony offering no new receivers, Etón was soon the market leader in quality shortwave portables.

In the past few years, numerous other low- and mid-level portables have come to market under a multitude of brand names. One of the most prolific was Sangean, a Taiwan company which had produced many radios for other companies before marketing under its own label.[88] (Of late Sangean appears to be focusing on radio markets other than shortwave.) In the 2008 *Passport to World Band Radio,* 42 portables were rated, 82 if all the names under which they are marketed are counted. Nearly all (including most Etóns) are made in China. While none have offered any technological breakthroughs, and many are wholly forgettable, others have shown themselves surprisingly good performers for their price.

Kits

Many people became acquainted with shortwave through kit building, a popular pastime that required only basic skills. In the mid–1940s, shortwave receiver kits were available from both local electronics stores and the big mail order houses. In the latter category, Allied Radio offered kits under the Knight label, its house brand since the company's founding in 1928, and Lafayette Radio sold kits under its own name. Although the leading kit producer would turn out to be the Heath Company, producer of the Heathkit line, Allied was marketing Knight kits before 1940, long before Heath got into the business. Other shortwave kit makers who sold generic shortwave receiver kits included Meissner, Philmore, Ameco, and Eico.

Allied and Lafayette did not promote their kits extensively until the mid–1950s, when they expanded their lines and heightened the visibility of kits in their catalogs. A wide variety of kits became available, including, in addition to receivers of various kinds, stereo equipment, wireless broadcasters, test equipment, remote control devices for model builders, and experimenters' kits.[89]

For those seeking a low-end kit introduction to shortwave, Allied offered several small regenerative Knight-Kit models, including the two-tube "**Span Master**" (1959, $24.95), the three-tube "**Space Spanner**" (1956, $13.95), and the "**Ocean Hopper**," a three-tube set which Allied had been selling in various incarnations since at least 1940 and which cost $12.95 in 1946. In 1962 it came out with a three-transistor, battery-powered regenerative kit, the "**DXer**," which it sold for $19.95. It covered the broadcast band and 7.5 to 17.5 MHz. Among basic superhets was the four-tube Knight-Kit "**Star Roamer**" (1963, $39.95), a modestly performing but very popular receiver, and its solid-state successor, the "**Star Roamer II**" (1972, $70). Knight-Kit's five-tuber was the "**Span Master II**" (1970, $29.95), which was produced in Japan.

Starting in 1959, Lafayette sold a two-tube regenerative shortwave radio kit, the **KT-**

THE knight-kit OCEAN HOPPER 83Y740

Generations of young kit builders were introduced to shortwave through the Allied Knight-Kit Ocean Hopper, a three-tube regenerative set. The hatch in the top was for changing the plug-in band coils, of which there were six (covering 155 kHz. to 35 MHz.).

135 "Explor-Air," for $18.50 (cabinet extra). Lafayette would also sell kit versions of several of its ready-made "HE" superhet receivers. However, these were called semi-kits, "with all major components and front end premounted and pre-calibrated." They reduced the price by $15–20.

Allied's Knight-Kit **R-100** (1960) and **R-100A** (1963) led the "R"series of receivers, which had more tubes and more features. The R-100 originally came to market in 1958 with just an Allied part number and no "R" designation. It was a well-featured, nine-tube receiver with good sensitivity, bandspread, BFO with pitch control, and a built-in Q-multiplier. Although only single conversion, it was a good performer at its 1958 price of $104.50 (optional S-meter, crystal calibrator and speaker added an additional $28).

A lesser performer in the "R" series was the six-tube **R-55** (1962, $67.50) and its restyled successor, the **R-55A** (1964, $64.95). The last "R" receiver, and one of the last of the Knight-Kit shortwave receivers, was the solid-state **R-195** (1970, $89.95). Despite good features for the price, the R-195 never caught on among shortwave listeners.

Hallicrafters also entered the kit business briefly in 1961, with the **S-119K,** a $39.95 kit version of the three-tube S-119 that tuned up to 16 MHz., and the **SX-140K,** the kit version of the SX-140 ham band receiver. That kit sold for $104.95.

The favorite among kit builders was Heathkit. Heath did not start out in the electronics kit business. In its first incarnation it was known as the Heath Aeroplane Company and its product was an airplane kit. The company went bankrupt after founder Edward Heath died in a plane crash in 1931. An engineer named Howard Anthony purchased the company at auction and kept the name. After the war, Heath was principally a seller of surplus electronic gear and electronic components for tinkerers and hobbyists. The concept of selling electronic kits grew from the availability of a large stock of parts and the notion that prices could be kept low by saving the consumer the largest single element of an electronic product's price, the cost of assembly. Heath began its kit business in 1947 with a 5" oscilloscope. It stuck to test equipment for a long time, entering the ham radio market with caution in 1951 but racking up a string of successes along the way and after.

Over the years Heath offered a number of entry-level, decently performing, general coverage shortwave receiver kits, including the regenerative **GR-81** (1961, $29.95); the five- and six-tube **AR-1, -2** and **-3** superhets (introduced in 1949, 1952 and 1956 respectively, $23.50–27.95 plus cabinet); the solid-state, portable **GC-1 "Mohican"** (1960, $99.50) and its successor, the **GR-78** (1969, $141.95); the no-frills, four-tube superhet, **GR-91** (1961, $39.95), its successor the **GR-64** (1964, $39.95), and the solid-state version of the GR-64, the **SW-717** (1971, $139.95); and the 6-tube, 6-diode **GR-54** (1965, $84.95). Their moderate prices made these receivers popular among SWLs, many of whom enjoyed the added fun of building their own radio. The GR-78 was noteworthy in that, during construction, the user could choose to install either a bandspread dial covering the ham bands or one that covered the major international SWBC bands.

A Heathkit accessory of importance was the **QF-1 Q-multiplier,** first marketed in 1956 for $9.95. Originally designed for the AR series of receivers, it was useable with any receiver having a 450–460 kHz. IF. It was very effective, and over the years it became a popular add-on for SWLs seeking better selectivity. The QF-1 was succeeded by the **HD-11** (1961) and the **GD-125** (1966), both virtually identical to the QF-1 but with the convenience of a built-in power supply (the QF-1 took its power from the receiver).

HEATHKIT COMMUNICATIONS TYPE RECEIVER
MODEL AR-3

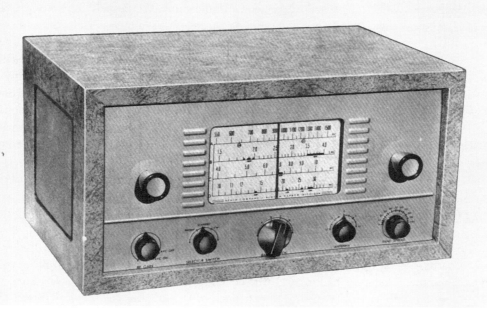

An upgrade from the regeneratives but still a basic set, the Heathkit AR-3 superhet was a favorite of kit builders. It was a good performer and a fun set to use, despite its small bandspread dial and its tendency to go into oscillation, a malady which usually could be remedied by a firm knock on the wooden cabinet.

The top of the line Heath SWL receiver was the **SB-310** "International Broadcast Band Receiver" kit, which came to market in 1967 for $267.95, and its solid-state successor, the **SB-313** (1972, $339.95). The SB-310 was a spinoff of Heath's highly successful SB-300 and SB-301 Collins look-alike, ham band–only receivers. Not a general coverage receiver, it covered nine specific shortwave bands in 500 kHz. chunks, specifically the 75, 49, 41, 31, 25, 19, and 16 meter broadcast bands, plus the 20 meter ham band and the 11 meter citizens band. The main attraction of the SB-310, besides the several bandwidth filters available as options, was the combination 0–100 rotary dial and 0–5 slide rule scale which permitted the listener to dial directly to within 1 kHz. or better of the desired frequency. Enterprising users soon discovered that by lifting the hinged cover, swapping plug-in crystals and tweaking a coil, the "stock" bands could be replaced by almost any other 500 kHz. frequency segment, including the important 90 and 60 meter bands. This gave the receiver added clout as a serious DX machine.

In 1984, Heath offered the **SW-7800** general coverage receiver kit. It was expensive ($349.95), and far behind the technological curve, and it enjoyed little success. By that time the company was in decline, its departure from the kit-building business in part the result of the increasing difficulty, in a rapidly advancing technological world, of designing full-featured kits that could be built by non-technical people and compete in price and quality with the expanding array of ready made (often Japanese) equipment. More importantly, in 1979 Heath had been bought by Zenith, which was interested only in Heath's computer equipment. The amateur-related kit business struggled on until the mid–1980s. Save for a few subsequent attempts to put the Heath name on products manufactured by others, it was the end of an era in which Heath had introduced a whole generation of hams and SWLs to electronics and shortwave. Although its kit-building heritage is just a memory, Heathkit still exists as a seller of educational materials and equipment for computer and electronic training.[90]

Kit building lost its attraction in the solid-state era, particularly after Heathkit left the business. However, kits have appeared on the market from time to time, usually oriented toward amateurs.[91] One shortwave receiver kit was the MFJ **8100K** World Band Shortwave Receiver, which in the mid–1990s sold for $59.95 in kit form, $79.95 assembled. Although solid-state, it was a regenerative receiver. Ten-Tec also offered several receiver kits, both regenerative models (the **1054** at $39 and the **1253** at $89) and a dual-conversion, frequency-synthesized, direct-readout model, the **1254**, for $195.

Electronic Recording

Although some listeners were recording their DX by the only available method—cutting records—as far back as the mid 1930s,[92] widespread off-air recording of radio signals by non-technical listeners had to await the development of magnetic recording tape. Wire recording came first, with consumer units becoming available circa 1945. However, the wire had a tendency to rotate about its axis, resulting in distortion, and there were the ever-present problems of tangling and uneven spooling. These complexities made wire recording unsuitable for widespread use, and it never caught on.

The problems were resolved when wire was replaced with tape, at first made of steel alloy, then plastic. Reel-to-reel tape recorders were in commercial use in the 1940s. They made their first appearance on the consumer market circa 1948, and quickly displaced both disc and wire recording. Multi-channel tape recording fueled the stereo boom that started in the 1950s, but monaural recording, with its wider recording head and relative absence of head compatibility problems, remained the format of choice for DXers.

Reel-to-reel recording was challenged by cassette recording in the late 1960s. Although thought a little gimmicky at first, with subsequent improvements in cassette tape quality, and with the portability afforded by cassettes, the cassette soon replaced reel-to-reel among all but the diehards and the professionals. Cassettes ruled the roost, largely due to low cost, convenience and wide availability, until challenged by CDs. The introduction of minidisc technology in the late 1990s, with CD audio quality and increased flexibility of use, caused some DXers to move to that technology. Today, computer-based hard-disk recording, and digital WAV and MP3 audio devices, are in wide use.

The arrival of electronic recording on the shortwave scene did not mean much to the casual listener. But for the DXer it carried all sorts of new possibilities.[93] In 1949, even those savvy about shortwave listening technology expressed wonder at electronic recording.

> Every DXer can look back through the years to dozens of instances of good listening which never will be forgotten. But no matter how gratifying may be the recollection of outstanding recep-

tion, think of the satisfaction of having recordings of memorable catches which you could play at any time for your own enjoyment and that of your friends! ... ¶Although recording always has offered the same potential appeal to DXers, certain obstacles prevented widespread acceptance in the past. The equipment itself was expensive, and the recording blanks, whether plain or pre-grooved, were too costly for the average budget. The life of a recording was limited to a few dozen playings at best, and there was no such thing as a "temporary recording." Unless you had two separate and distinct recording mechanisms, you could not edit a recording once the blank disk had been cut.... ¶I was fortunate in being able to borrow two recording units for test purposes [one tape and one wire].... ¶After operating both units for some time, I am completely sold![94]

At the most fundamental level, one could re-listen to an interesting program. From the standpoint of DXing, one could play back and study station identifications of which the listener was uncertain, leading to more accurate reporting. The recordings could also be kept for future reference. This was particularly true in the reel-to-reel days, when a segment of a reel tape could easily be snipped out and spliced into an archive reel.

Tape recording also improved communications among listeners. Hobbyists exchanged recordings among themselves, demonstrating the quality of their reception and obtaining the opinions of others as to the identity of a station. It also provided a means for personalized voice communication among hobbyists, and "tapesponding" became an established practice. Many of today's senior listeners remember the pre-cassette days when they exchanged 3" reels of tape with fellow SWLs.

Tape also provided a new vehicle of communication with stations, especially when it came to QSLing. Whereas before one had to satisfy a station through a written description of what was heard, placing somewhat of a premium on one's writing ability, now one could send along a small reel of tape or a cassette and let the station hear for themselves. This was not without occasional pitfalls, however, as with a station whose small signal produced reception that might be understandable to a DXer's ears but basically inaudible to mere mortals at the station itself. In addition, most experienced listeners would acknowledge that improving one's QSL response rate by sending a recording is more a matter of hope than experience. Usually it didn't make much difference. But occasionally it did, and it always improved one's confidence that a reply would be forthcoming, and gave hope that the station might return the tape, over-recorded with some local programming or even a voice message. Receipt of such tapes were great moments for QSLers.

The advent of tape recording also led to occasional discussions of whether a tape recording should be considered an adequate verification for station- or country-counting purposes. Traditionalists felt that only a paper QSL could play that role. Others argued that there could be no better proof of reception than a recording, especially as QSLs came to be seen more as station public relations vehicles than studied confirmations from the engineering side of the house. The argument can be stoked even today, although at much reduced intensity.

7

QSLing

Logging a station is a wonderful feeling, but the best feeling comes when the postman hands you a verification from some far-off station that you have longed to verify.... [T]he verification is the "pay-off."[1] (1946)

A QSL is a card or letter from a station, confirming that the listener heard them. It is sent in response to a reception report, submitted by a listener, containing details of how the signal was received and what programming was heard. Its origins date back to before 1920, when both amateur radio operators and non-transmitting wireless enthusiasts sent postcards to stations they heard, informing them of their reception. The exchange of what came to be known as QSL-cards among hams as proof of their on-air contact soon followed ("QSL" is the Morse code symbol for "I acknowledge receipt.") QSLing became a standard practice among broadcast band stations in the early 1920s, before there were any shortwave broadcasters. Given the long distance nature of shortwave reception, it was natural for the practice to carry over into the world of shortwave broadcasting.[2]

Some QSL-cards are plain, but most reflect the station or its location in some way. QSLs are something tangible from one's listening experience, and evidence of one's achievement. While some DXers do not collect QSLs, there does seem to be a "collecting" gene in the metabolism of most DXers, whether evidenced by a collection of QSLs or a tally of countries or stations heard.

At their best, QSLs contain an unambiguous confirmation statement, plus the date, time and frequency of the reported reception. Not all QSLs meet this test, however, and DXers have gotten used to general, "no data" verifications or thank you letters. Over time, the particulars of a QSL became less important. The modern practice is that if it appears that the station intended to verify your report, it is a verification — the so-called "intent rule." And the rarer the station, the less the DXer cares. As one was heard to say in 1974 upon receiving a rare, flimsy paper QSL from the shortwave station in Baku, Azerbaijan, "I'd have accepted a single square of Charmin."

In shortwave's early days, QSLing was straightforward: a station had a QSL which it issued for its broadcasts. As time went on, the opportunities for QSLing expanded.[3] Some stations offered different QSLs for different services. In the 1950s, for example, each IBRA Radio language service had its own QSL, issued from the IBRA office in the home country of the language. Some stations issued series of related QSLs. HCJB was known for this, offering many series over the years, e.g. a 50th anniversary "then and now" series showing views of the station in 1931 and 1981, and, in 1996, 12 cards which, when placed next to each other, made a panorama view of Quito. During the four quarters of 1992, in commemoration of the 500th anniversary of European exploration of the new world, Trans World Radio, Bonaire, offered a card on which were placed three QSL stickers (for three reports) which together depicted Columbus's three ships.

Stations offered special QSLs for special occasions such as test transmissions, station anniversaries, the first day on the air, the last day on the air (the most famous being East Germany's last broadcast before reunification), the opening of a new station building, the commencement of a new service, etc.

Often a special QSL would be offered in conjunction with a special program arranged by a DX club, and sometimes the club would do the QSLing. This was a frequent practice of Scandinavian clubs in the 1950s. As leased time broadcasting increased, broadcaster-lessees usually offered their own QSLs. Some individual programs issue their own QSLs. A few stations, such as HCJB and Radio Free Asia, change their QSL designs often, while others offer special QSLs periodically (KNLS, Alaska, offers a limited number of specially designed QSLs at the start of each transmission season).

One's "best" QSL usually coincides with a challenging bit of reception, but not always. Goa was "best" for one DXer whose QSL was mailed from the Portuguese enclave just a

Radio Japan offered this special QSL on the occasion of the start of its relay transmissions from Gabon in 1983.

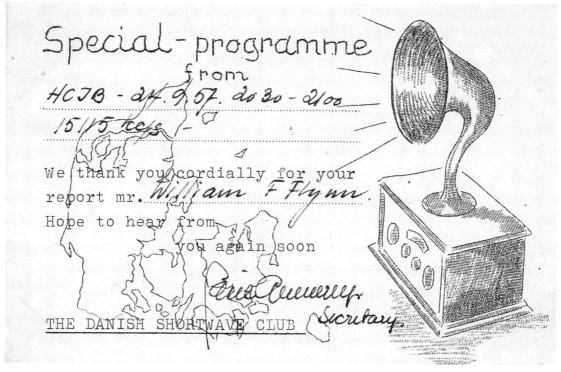

This QSL from a Danish club was offered for a special Danish-language program over HCJB in 1957.

few days before India evicted the Portuguese from the country. Another DXer prized a QSL from a new Venezuelan station that had enthusiastically read his report over the air. For another it was a QSL from a station in Spain that had been on shortwave for only a week while its medium wave transmitter was being repaired. And a San Francisco DXer counted as his favorite a QSL delivered in-hand by the engineer of the Fiji Broadcasting Commission who was in town on vacation.

Perhaps the best-remembered QSL of all time was the "battlefield" QSL received from Radio Biafra by New York City DXer Alan Roth in 1969. Mail to Biafra had been suspended during the Biafran civil war. Roth sent his report via the shortwave station in São Tome, which placed it in an airlift to Biafra. It was delivered to the station engineer, whose reply was hand carried out of the country and given to Biafra's U.N. representative in New York, who mailed it to Roth. Prepared on plain-paper, it had been typed on a typewriter with practically no ink left in the ribbon. Such have been the delights of QSLing.

Nonverifiers

Most stations will verify reports. However, some stations have been "impossible" to verify, even with persistence, although nearly every such station seems to have verified for someone. Occasionally a well-known nonverifier will start verifying. In the late 1970s, Radio Peking began verifying reports for the regional people's broadcasting stations, and soon the stations were verifying direct. Radio Abidjan, a good verifier in the 1950s, then a long-time nonverifier, started QSLing again in 1971. Radio Lesotho and Radio Botswana went from being nonverifiers to verifiers in the 1980s. WWRB, Manchester, Tennessee, which had been on the air since 1995 (originally as WGTG), stopped QSLing reports until 2004, when it reversed itself.

Although Brazilian stations have been decent verifiers, the Spanish-language Latin American stations have long been the most difficult to verify, even when the listener writes in Spanish. Many are small stations, with a local orientation and a shoestring budget, and do not solicit reports. Others do not always understand what QSL seekers want. Some verify for a short period and then stop. At least part of the problem is with the DXers themselves. Station visitors who have observed reports say that many listeners do not know how to write good reports to Latin American stations.[4]

The problem of nonverifiers is an old one. In 1935, Westinghouse announced as follows:

> After the New Year, acknowledgment of shortwave reception reports made to station W8XK, Pittsburgh and W1XK, Boston, will be discontinued, according to a decision made by the company's radio executives. One reason for discontinuing these reports to listeners is due to the fact that many newspapers and magazines in all parts of the world are publishing listings of shortwave transmitters and the time of their programs. Another reason is the increasing volume of mail now being received at the stations.[5]

In 1940, the International Short Wave Club, then still headquartered in the United

Latin American stations have long been the most difficult to verify, but have also produced some of the best-looking QSLs. This card is from Circuito CMQ, Cuba (1958), a station that was well-heard in the United States.

Brazilian rhythms were widely heard on the plethora of Brazilian domestic shortwave stations, of which Radio Globo was one of the largest.

DXers were thrilled when the Galapagos Islands came on shortwave in 1970. This QSL from La Voz de Galapagaos is from 1978.

States, sent letters to many nonverifiers, and over 80 responded and promised to verify. The club also sent many Spanish-language letters to nonverifying Latin American stations, indicating that they would publish the results. Most stations responded positively, while some criticized the club for telling them what to do.[6]

Stations vary a good deal in how long they take to verify. Most SWLs allow at least several months before sending a follow-up. However, there have also been some remarkably fast verifiers. In the 1950s there were many stations from whom immediate replies were reported, including Radio Tahiti (three weeks), Radio Abidjan (18 days), Radio Athens and VE9AI, Edmonton, Alberta (two weeks), the Nigerian Broadcasting Service (10–15 days), TFJ, Iceland (ten days), Radio Sweden (eight days), Radio Japan (seven days), and La Voz Dominicana (less than a week).

"When I first got into this racket back in mid–40s," said one DXer in 1960, "most of the SWBC stations reported were extremely prompt in replying to reports. Now, though, it seems to take quite a bit longer to hear from them; so far, [the] fastest verie I've received was from Radio Norway, about three weeks from my mail to the time I received their's, airmail both ways. I've sent out about 35 reports in [the] past two months, and have received only two as of now."[7] Eight years earlier, in May 1952, DXer Charles McCormick reported verifying Radio Brazzaville in 16 days. He indicated that of 14 reports sent out in 1951, all but one had been QSLed, and that one had not been out long enough.[8]

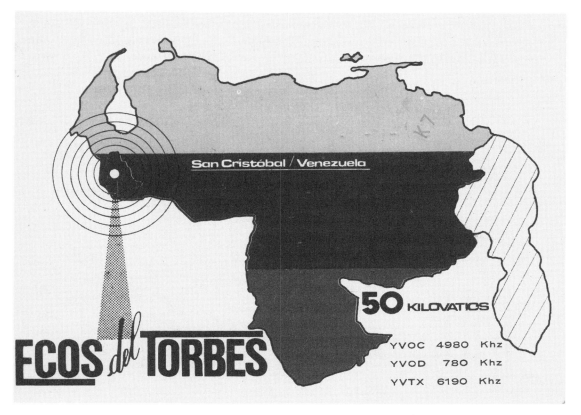

Although no longer on shortwave, for many years one of the most powerful of the many Venezuelan shortwave stations was Ecos del Torbes. This QSL is from 1981.

Stations: Policies, Problems and Solutions

Changes in the QSL practices of stations were common.

POLICIES GOOD AND BAD

To the chagrin of DXers, the BBC had long been a nonverifier on the grounds that the details of BBC programming were reported in numerous publications around the world and that the station was relayed by other stations in many places. But there have been many exceptions to the policy. In the mid–1940s, full-data QSL-cards were issued by the BBC's New York office. At the same time, listeners also began receiving nearly full-data QSL cards from BBC headquarters in London. They were being issued by a DXer who worked in the station's public relations department. The practice came to an end in December 1946 when the chief engineer discovered it.[9]

In 1949, reporters to the British Far Eastern Broadcasting Service, Singapore, were being advised that reports "covering a period long enough to indicate that it is not merely freak reception or deterioration" would always receive a personal reply, but that, since BFEBS was by then part of the BBC, it could not verify reports.[10] For years, listeners received similar messages from the station (which later was known as the BBC Far Eastern Station). Also in 1949, *London Calling* advised readers that while the BBC did not verify reports, they were still welcome, especially if they covered a period of at least a week. In 1959, BBC headquarters started using a "semi-verification" card that said a report was in accordance with the BBC's published schedule. A similar card has been in use ever since. At various times

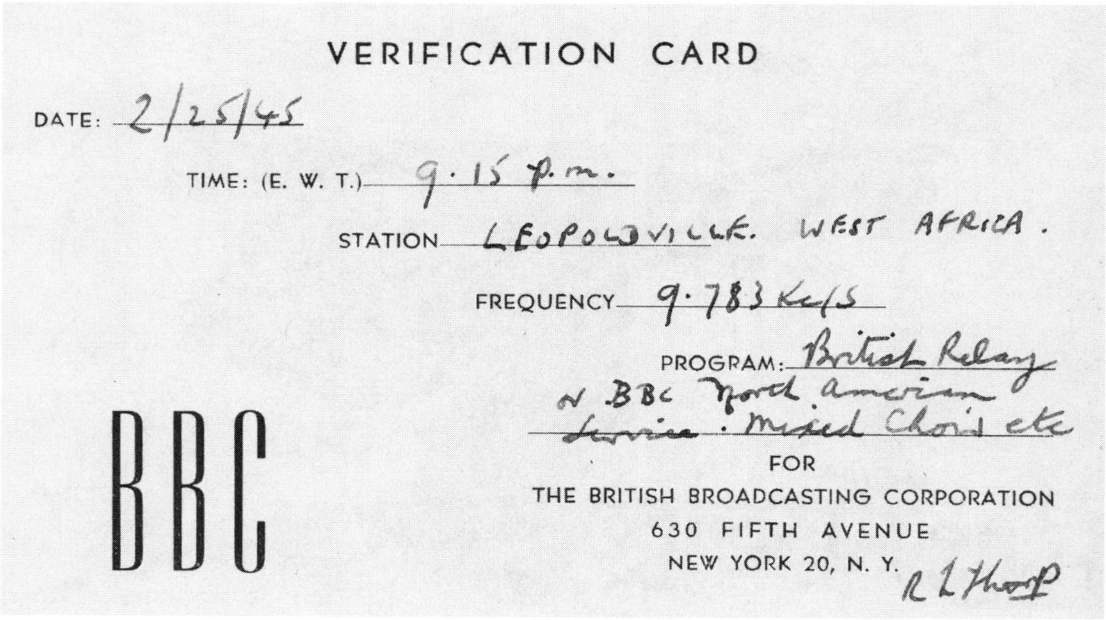

VERIFICATION CARD

DATE: 2/25/45

TIME: (E. W. T.) 9.15 p.m.

STATION LEOPOLDVILLE. WEST AFRICA.

FREQUENCY 9.783 Kc/s

PROGRAM: British Relay
~ BBC North american
Service. Mixed Choir etc

FOR
THE BRITISH BROADCASTING CORPORATION
630 FIFTH AVENUE
NEW YORK 20, N. Y.

BBC

For many years the general rule was that the BBC did not issue verifications. There were, however, many exceptions, including this 1945 full-data card from the BBC's New York office QSLing a relay of BBC programs transmitted from OTC in the Belgian Congo.

B.B.C. FAR EASTERN STATION,
SINGAPORE.

DATE AS POSTMARK

Thank you for your letter of......2nd April, 1958...................... in which you report reception of our transmitting station.

Technical reports, which give details of how our transmissions are received, are very much appreciated here and give valuable information for the improvement of our service. A report covering a period long enough to indicate that it is not merely freak reception or deterioration will always receive a personal reply.

As part of the British Broadcasting Corporation we regret, however, that we have to inform you that it is not our practice to verify reception reports which require details of transmitted programme matter.

We hope you will continue to enjoy good reception of our programmes.

THOMSON ROAD STUDIOS
P. O. BOX 434 SINGAPORE

25.4.58

Resident Engineer

While not a true verification, this carefully-worded card from the BBC Far Eastern Station, Singapore, was for many years used to acknowledge reception reports sent directly to the station.

BBC auf DEUTSCH

The BBC German-language service often issued full-data verifications for its programs.

in the 1990s, the BBC German service and the BBC Spanish service QSLed reports on certain BBC transmissions. And during the late 1960s and 1970s, the BBC World Radio Club issued bona fide BBC verifications during various "DX Award Competition" periods. At present, most of the BBC relay stations verify reports direct.

Radio Moscow was a good verifier of its own transmissions. When, in 1955, DXers sought to verify local Russian stations through Radio Moscow, the station declined. It reversed this policy in 1965, but reverted to the old policy three years later. Over the years, Radio Moscow policy on placing transmitter sites on QSLs has varied. Sometimes sites were provided only for North American transmissions, other times for all transmissions. The site data was always suspect, however. The QSL personnel did not always have access to the official data in the International Frequency Registration Board lists (which were themselves often inaccurate), and the sites shown on Radio Moscow QSLs sometimes did not square with the realities of propagation.

Radio Republik Indonesia, Jakarta, which for many years had verified reports on its regional stations, advised in 1959 that reports could be sent directly to the stations. DXers devised Indonesian report forms, and were soon rewarded with QSLs from a multitude of exotic places. A similar policy was adopted by Radio Peking years later. Although the experience with Radio Peking was not uniformly positive — in 1959 it ceased verifying reports on its home service transmissions because most DXers did not understand Chinese and thus could not, it felt, submit accurate reports — a breakthrough came in 1979 when the station began verifying reports for the provincial people's broadcasting stations. Soon thereafter the stations started verifying direct with Chinese-language cards and letters.[11] It was a happy day when one arrived in the mail.

Indonesia has produced some of the rarest DX and some of the most sought-after verifications.

In the United States, the Armed Forces Radio and Television Service stopped verifying reports for listeners in the U.S. in 1954, a policy it later changed, but resumed in 1982. It was an echo of a practice which the VOA had followed for its overseas relay stations; only in 1953 did it begin verifying them for U.S. listeners. Until 1955, VOA QSLs were full-data. In 1956 the frequency was dropped as a time-saving measure, a six-month backlog of unanswered reports having developed. In 1967, owing to the volume of requests and the limited resources available, VOA announced that it was stopping QSLing for listeners in the United States. This course was reversed almost immediately, however. In 1976, when the VOA stopped identifying transmitter sites on air, the station relented on a threatened end to giving sites on QSLs, agreeing to include them if they were correctly identified in the listener's report. The VOA is, in general, a good verifier, although in recent years there have been long delays in getting replies. In 1991, the VOA station in Bethany, now closed, started verifying direct, and in subsequent years some reports sent directly to the relay stations have brought direct QSLs.

In the past, reception reports have sometimes gotten low priority at VOA. A GAO report observed that in mid–1991 VOA had 100,000 unanswered letters. The station had ceased responding to its mail, most of which was discarded without being opened. (This may have contributed to the 21 percent decline in VOA mail between 1988 and 1991.) Similar problems dogged other stations. Radio Nepal had 3,500 unanswered reports in 1981. In 1989, Radio Greenland had a backlog of 800 unanswered reports. In 1996, FEBC had an eight month backlog.

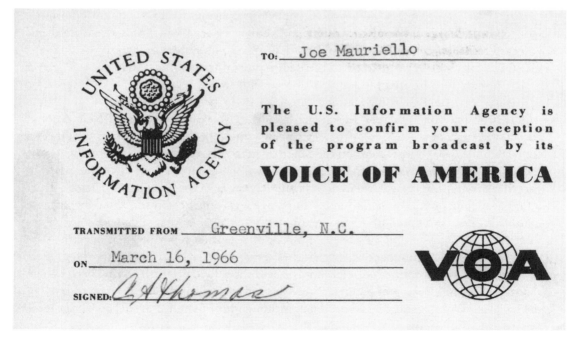

The Voice of America has had many different QSL designs over the years, not all of them showing the transmitter location, a detail which many DXers seek. This card is from 1966.

Retrenchment

As the major international broadcasters have faced the budget constraints of recent decades, some have tried to reduce the burden of QSLing. No-data cards, either verification statements with no mention of date, time or frequency, or acknowledgments with no mention of verification, came into wider use.

Several stations adopted "do it yourself" systems. Radio Canada International pioneered this approach in 1975, enclosing with the year's first program schedule sent to those on the RCI mailing list a blank reporting/QSL-card which the listener filled in with brief program details and returned to the station, which added a "Verified" stamp and returned it. (A study in 1982 showed that about 13 percent of the cards distributed were actually submitted for verification.) Four years later, Radio Sweden introduced a similar plan using a two-part card, one for the usual reception details, the other containing four questions about the station's programs. The questionnaire part was removed by the station; a no-data verification statement was stamped on the other part and returned to the listener. The practice was abandoned within a few years.

The system adopted by the Christian Science Monitor in 1990 replaced the no-data cards which the station had used since commencing shortwave broadcasting in 1987. Three station-specific, do-it-yourself cards, one for each of the station's three transmitter sites, were sent to listeners on the station's mailing list. Listeners sent the cards to the station's Boston headquarters, which forwarded them to the transmitter sites where they were signed and returned to the listener. Most QSLers agreed that because each of the stations had its own distinct card, and because they were returned from the station site rather than Christian Science headquarters in Boston, this was a big improvement over the old system.

In 1978, Radio Finland adopted the policy of requiring comments on programming with a reception report in order to obtain an "Audience Card," which was a no-data thank you card rather than a verification. In 1997 it dropped this system in favor of issuing regular QSLs for reports sent directly to the Radio Finland transmitter site in Pori.

A few stations limited the number of reports it would accept from each listener, or verified reports only from the station's target areas. In 1990, Radio Australia said that it would issue only one QSL per season per listener. Two years later, Radio New Zealand adopted a policy of one QSL per frequency per season. More frequent reports would get a rubber stamp QSL endorsement on the report, which was returned to the sender.

Many SWLs believed that the decline in the willingness of stations to QSL was tied in part to a boom in shortwave listening in Japan that began in the early 1970s. Fueled in part by the availability of consumer-friendly shortwave sets being heavily marketed by Japanese manufacturers, it was a youth phenomenon, with many 13–18-year-olds taking up shortwave listening and sending out huge numbers of reception reports which overwhelmed some broadcasters. A number of glossy shortwave magazines appeared in Japan, and numerous small clubs were formed, often school-based, shortwave listening having been shown as a good way to improve one's English. By the end of the decade, the interest of most of the Japanese SWL generation had declined. Although the boom had produced a large temporary expansion in the membership of the venerable Japan Short Wave Club, and had led to the establishment of a second national shortwave club, the Japan BCL [Broadcast Listener] Federation, it had occurred largely outside the view of the rest of the international shortwave community, due mainly to the language problem and the local orientation of the Japanese listeners.

THIRD-PARTY ASSISTANCE

DXers were willing to do what they could to alleviate the pressure on stations. In 1991, the Ontario DX Association took over the QSLing duties for shortwave station CFRX. The Canadian International DX Club did the same for RCI in 1992 (the do-it-yourself cards were dropped), as did the Australian Radio DX Club for Radio Australia in 1998 when the station stopped QSLing.

And personal initiative sometimes mattered. A volunteer DXer in Nepal had been visiting Radio Nepal to QSL reports. He gave up the duty in 1985, whereupon it became very difficult to QSL the station until the following year when a new person, keen to have good relations with DXers, came on the scene and resumed QSLing. In 1989, Radio Cook Islands ran out of QSL-cards and had no budget to print new ones, whereupon ODXA had 500 cards printed and sent to them. And in 1998, after a request from Radio Tanzania Zanzibar to the German DX community to provide them with QSL-cards, a Belgian DXer took up the challenge and designed and printed 100 colorful QSLs which were subsequently received by some lucky DXers.

A few stations stopped QSLing altogether. As long ago as 1970, in connection with the elimination of its foreign-language broadcasts, Radio Denmark announced that it would reply only to correspondence in Danish (a policy which it soon relaxed, however). The Red Cross Broadcasting Service stopped QSLing in 1996 (and left the air soon thereafter). Tight budgets and reductions in staff led Radio Norway International to stop verifying reports in 1999. Channel Africa and Israel followed in 2002, and Radio Exterior de España in 2004. But there always seemed to be some exceptions, and perseverance sometimes paid off even in the face of a "no QSL" policy.

Technique

Years ago, many of the larger stations had reception report forms on which listeners could submit reports if they wished. In most cases, however, a report was rendered in the form of a letter.

Serious DXers go to great lengths to coax QSLs from stations, but the basic "rules" have always been the same: provide solid program details over at least a half-hour of reception; in the case of Latin America and a few other places, write in the language of the country if possible; include return postage; comment on the station's programming; and be clear but polite in your request (and, for the occasional uninitiated recipient, explain just what a QSL is). Beyond that, DXers bring their individual ingenuity to the task, including with their reports postcards or other trinkets, and preparing a report whose look and content say, "verify me."

Enclosing a recording of the reception has long been a favorite strategy. Only the most dedicated and best equipped could do this in the 1930s and 1940s when disc cutting was required to make a recording. The practice became more prevalent in the 1950s when 3" reel-to-reel recordings could be used, followed in later years by cassette tapes and, today, CDs or e-mail attachments.

In cases where a DXer expects to encounter difficulty, use of a prepared card is commonplace. These are generic cards, supplied by the listener, which it is hoped the station will fill in, sign, stamp, and return. In the age of desktop publishing, prepared cards can take on a highly professional, customized look.

A strategy that has often worked is to involve a country's embassy. In 1974, the Yemeni ambassador in Washington was happy to sign and seal prepared cards for reports sent to him. In 1993, the Bhutan United Nations mission was willing to sign prepared cards on behalf of the Bhutan Broadcasting Service, a notorious nonverifier. While embassy replies were not as satisfying as those received direct, they were better than nothing, even acknowledging that the embassy was not checking the correctness of the reports. In some cases, however, checking was at least implied. The Burma Broadcasting Service was nearly impossible to verify until 1971 when some verification letters began coming through the country's embassy in Washington for reports sent directly to the station.

Hand carrying reports to a station was a powerful if infrequently available tool. Traveling DXers have often taken fellow DXers' reports with them on station visits (usually to Latin America), and returned with their QSLs. In 1973, Roger Stubbe of HCJB hand carried a DXer's reports to Radio Nacional Espejo, obtaining two prepared cards and ending a nine year, 22-report quest.

Although the same complaint was heard 50 years ago, verifications have become more difficult to obtain, probably because, as shortwave reception became more commonplace,

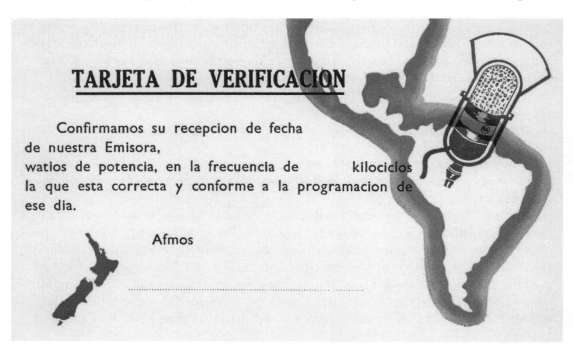

Some DXers have gone to great lengths in the design of prepared cards. This Spanish-language card, showing the islands of New Zealand in the lower left, accompanied reports from a listener in that country to stations in Central and South America.

Radio Republik Indonesia

Ini membenarkan laporan tuan dari Radio Republik Indonesia, Studio __RRI PONTIANAK__ siaran pada gelombang __75__ meter atau frekwensi __3995__ kHz, pada __17 Peb 1993__ 197__, dari __17.59__ sampai __18.17__ waktu Indonesia Bagian __Barat__. Pemancar kuasa studio kami __50__ kw. dengar __Dipole__ antena. Banyak terima kasih untuk laporan tuan.

(Studio Cap)

ACHMAD RUSKAYA,BA
Pimpinan atau Kepala Studio

This is the classic prepared card used for reports to Radio Republik Indonesia local and regional stations. This one was filled out, signed and stamped by a representative of the RRI station in Pontianak on the island of Borneo.

the listener-station QSL connection came to be based less on gratitude for genuinely valuable news of the reception of a station's signal than on ordinary public relations. Today, through technical means, regular monitors and contacts with other broadcasters, international stations can learn how they are being received without relying on reception reports.[12] In such cases, reports may be a nuisance, particularly when the writer has offered no opinion on the program's content.[13] Domestic shortwave stations often have limited interest in reception reports from beyond their local area (except when this evidence of the signal's potency helps promote the station with prospective advertisers). Some stations do not check the accuracy of reception reports at all and issue QSLs as a matter of listener relations. And some listeners have long held that a recording is better proof of one's reception of a station than a QSL anyway.

But such realities do not dampen the spirit of the avid QSLer, who knows in advance that follow-up reports will be needed to QSL some uncooperative stations. And, overall, the QSL yield has been good. In 1989, one experienced DXer reported a year-to-year success rate of about 70–85 percent.[14] In 1994 another reported his success rate, after follow-ups, as about 85 percent.[15] New Zealand DXer Arthur Cushen's success rate varied from year to year but was about 60–85 percent overall.[16] Success rates appear to have declined in recent years, and Latin American stations are even more difficult to verify than they have always been.[17] For many DXers, however, this just adds to the challenge.

Over the long term, the increasing cost of postage has probably inhibited some listeners from writing to stations. In 1945, postage rates varied by country and depended roughly on their distance from the United States. Rates were simplified in 1946, with most countries placed in one of three rate classes: Central and South America, ten cents per half ounce; Europe and Russia, 15 cents; all other countries, 25 cents. And rates increased. Today, the rate to most countries is almost a dollar for the first full ounce, with varying rates above that.

SWLs made frequent use of International Reply Coupons, which could be exchanged in countries of the International Postal Union for stamps sufficient to prepay a letter back to the United States. In 1945 they were nine cents each. Today they are $2.10. The use of IRCs has largely been replaced by the expedient of enclosing with a report a U.S. dollar bill, or mint stamps of the target country. Services offering DXers mint stamps in amounts sufficient to prepay a letter from a foreign country came into operation in the 1950s.[18]

An American postal practice worth remembering if not repeating were the notices received in the 1950s in connection with the arrival from communist countries of newspapers and magazines believed by the post office to be unsolicited propaganda. The mail would be held pending receipt of a written request from the addressee to deliver it. A few SWLs were reluctant to send reception reports to communist countries during the cold war. They were good verifiers, however, and most listeners overcame their concerns, even if it did mean receiving occasional propaganda. Not reporting would mean a country not QSLed. While this problem no longer exists, more than a few present-day DXers have found themselves explaining their hobby to military superiors or security personnel.

Some SWLs were satisfied verifying a station once. Others sent new reports when they learned that the station had a new QSL. For some SWLs in Australia and New Zealand, each new frequency of a broadcaster was a target to be QSLed. And there were always some listeners who bombarded particular stations with repeated reports because the station was a favorite of their's and they enjoyed having regular contact, or just because they liked to send reports. Over time these practices, and the expense of replying, contributed to the increasing unwillingness of stations to QSL.

Although the internet did not change QSLing in a fundamental way, it did create new opportunities. Virtually every station now has a website and an e-mail address, thus permitting e-mail reception reports. Many of the larger stations' websites have forms for this purpose. The internet has also permitted station responses by e-mail, and while these are often very fast by traditional standards, sometimes they consist of only an e-mail message — better than nothing, but not very satisfying as a collectible or a souvenir of one's reception. Some electronic responders have gone further, however, and replied with artful QSL letters or "cards" in the form of graphics files sent as e-mail attachments.

Problem Reports

To some extent, sub-standard reporting has always been a problem. In 1947, a representative of Radio Australia commented on the poor quality of some reports received.

> The following are examples which have been received in recent weeks: "I heard your station VLG4, 25.37 meters on November 18. QSA2. Details of programme — music and song." — From Sweden. "I am collecting verification cards from different radio stations. Will you please be kind

enough to send me your verification cards?"— From America.... Graham D. Hutchins who is in charge of our Correspondence Section, and who is a DXer of longstanding, was instrumental in the formation of the Australian DX Club in 1932. He was surprised to see such poor quality reports coming from overseas countries.... It was decided to notify the various DX clubs through-out the world that a *better class of reports should be submitted* by members; *otherwise we will be unable to continue to issue verification cards.*[19]

Other stations had like experiences. From the Central Broadcasting Administration, Nanking, China, 1948: "In the past we have received many requests for cards with a very brief statement that our station had been heard. We must have detailed information regarding our program contents, the frequency it was heard on, the time (preferably in GMT), reception conditions and station calls before we can issue verification cards. In the past we have received letters from listeners who claim to have heard our station with call letters and frequency that were not being used at the time."[20]

The Australian observations prompted *Short Wave News,* the U.K. magazine that founded and sponsored the International Short Wave League, to start a campaign for better reporting.

[W]e are greatly distressed by the increasing evidence of mass-reporting — the cause of scrappy information — in our midst. Some readers may say that they only send brief reports but get back a fair number of cards. That may be so, as most BC stations either verify or they don't. The fact remains that such reports are still useless. Many stations have now ceased to verify due to the reports received not justifying the expense of providing cards. How many other stations will follow suit? That matter rests entirely with the SWLs.[21]

These comments were picked up and republished by Ken Boord in *Radio News* and by Arne Skoog in a Swedish radio publication, *Roster i Radio.* Skoog deplored the practice and the notion that stations were merely a means to add a QSL to one's collection, but he also wondered why it was that many stations broadcast "programmes of such a poor nature as to repel listeners from consistent listening."[22]

The problem has persisted. NNRC Shortwave Editor Hank Bennett observed in 1960 that he had received a letter from the Far East Network, Japan, accompanied by a report from a west coast DXer indicating that he heard FEN on a date and time when the station was off the air. It turned out that the DXer had copied information from a report in *Popular Electronics.*[23]

Five years later the *New Zealand DX Times* stated the obvious, and the American Short-wave Listeners Club repeated the message in its bulletin.

[W]e would point out that the hobby of DXing depends on the good will of radio stations around the world, and the abuse of this good will by a few unscrupulous "DXers" can and does affect honest DXers seriously. A radio station which has received one obviously dishonest report is not going to be favourably disposed to the honest reports which it receives; these days it seems harder and harder to get stations to reply under normal conditions and it is unfortunate indeed that a handful of unthinking and selfish people should place further obstacles in the way to receiving a verification.[24]

In 1979, Jonathan Marks of Radio Netherlands summed up his experience: "After visiting many European stations, it becomes quite clear that 50–60 percent of incoming reception reports are of little use at all. Either they are incomplete, incorrect, or simply made up using the *WRTVH* and they are written by collectors."[25]

While the abundance of loggings and station audio on the internet provide numerous ways to fake reception, the problem is probably no greater than it ever was. In general, it has been thought more a matter of ignorance, laziness or mistake than dishonesty, and so the principal response has been to educate DXers on proper reporting. This has been done time and again in the pages of club bulletins and other publications. Some other novel strategies have also been tried, including the NASWA "Reception Report Reencounter" of 1966 where members submitted a copy of one of their reception reports, with the three best reports selected as the winners.[26]

The Literature of QSLing

CLUBS

Although how to write good reception reports was a common topic in the North American shortwave press, before the war neither club bulletins nor magazines had sections devoted exclusively to station QSL practices. Reports of verifications received usually appeared mixed in with station loggings or schedule information. From time to time a section of an editor's column might be devoted to recent QSLs.

Things started to change slowly after the war as DXers returned home and resumed listening. Ken Boord started including an occasional "Verifications" section in his monthly *Radio News* column in 1947. Beginning in 1951, Hank Bennett included a page of QSL information in his NNRC shortwave column when there was enough material to warrant it, although the focus of the column remained loggings and other information. Other clubs covered QSLing in a similarly casual fashion.

The American Shortwave Listeners Club was the first major North American club to have a column dedicated to QSLing. It began in 1963 under the editorship of the late Alan Roth of Bridgeport, Connecticut. Other clubs followed suit — NASWA in 1966, the NNRC in 1974 — and soon the QSL column was a standard part of club bulletins. (The NNRC column was edited by Sam Barto, who in 1978 became NASWA's QSL editor, a post he still holds.) The columns would contain reports from members on the appearance of the QSL, other things received with it, enclosures sent with the report, response time, the name and title of the verification signer, etc. NASWA would sometimes aggregate the data and produce "white lists" of good verifiers or "black lists" of recalcitrants.

Before the appearance of the *World Radio Handbook* in 1947, lists of station addresses would sometimes be compiled and published in radio magazines. With the advent of dedicated QSL columns, DXers were able to supplement the *WRTH* data with address information obtained directly from the QSLs themselves.

PUBLICATIONS

In addition to club bulletins, magazines, and anthologies where QSLing was a perennial topic, there were many monographs and small books on the subject.

One of the earliest was the *Proper Reporting Guide* written by ASWLC president Art Glover in 1969[27] (later editions were published in 1974 and 1981). It discussed the do's and don'ts of reporting, and what could be expected from stations in different parts of the world. It also contained pattern reception reports in English, French, Spanish and Portuguese.

Another was *Verification Techniques,* authored in 1978 by Sam Barto with the assistance of well-known DXer Alvin V. Sizer.[28] It was one of the first comprehensive examinations of QSLing. Amply illustrated, it covered writing reports and sending follow-ups, as well as specialized topics such as verifying tentative reception and older reception, and verifying Russian stations.

A comprehensive treatment of the subject was *Secrets of Successful QSL'ing* by Gerry Dexter.[29] First published in 1986, *Secrets* addressed such topics as the format and content of successful reports, postal issues, follow-ups, use of prepared cards, etc. The 1992 edition added a section from the Committee to Preserve Radio Verifications (CPRV) about how to care for QSLs. It also included some 42 pages of "The CPRV Page," an illustrated series about early QSLs which had appeared in various club bulletins.

The most recent treatment of the subject of QSLing is *World QSL Book,* authored by Gayle Van Horn of the *Monitoring Times* editorial staff and distributed in CD form by Grove Enterprises. Its more than 500 illustrated pages provide comprehensive treatment of shortwave broadcast, broadcast band, ham, and utility QSLing, and it contains extensive station contact information.

A popular group of books about QSLing was the "address book" series published by Gilfer. The first of these, the *SWL Address Book,*[30] was published in 1969 and listed all shortwave stations and their addresses, together with a code referencing their verification practices, e.g. whether they sent letters or cards, regularity and promptness of reply, whether reply postage was required, whether tape recordings helped, etc. An *SWL Address Book Updater* was issued that brought the *SWL Address Book* current to the latter part of 1971, and a second edition of the *SWL Address Book,* with Dexter as the author, was published in 1974.[31]

The *SWL Address Book* was followed by Dexter's *QSL Address Book* in 1978, with a subsequent edition published in 1984.[32] These added a discussion of how to write reception reports and replaced the *SWL Address Book* code with a simplified "A," "B," "C" system. This layout was preserved in Dexter's *World Broadcast Station Address Book* which replaced the *QSL Address Book* in 1985.[33]

Some stations, including Deutsche Welle and the Far East Broadcasting Co., issued their own pamphlets describing how to write reception reports. In addition, many publications addressed narrower QSLing issues. An early specialty item was Dexter's *Verification Signers,* a list published by NASWA around 1972.

A number of publications focused on QSLing Latin American stations, which have always been a special challenge. Most of these were published in other countries, with a few finding their way to the United States. In 1969, the Cimbner DX Club in Denmark published *Advertisements in Latin American Radio,* a list of advertisements often heard on Latin American stations. A similar publication, *Latin American Commercials — A Guide to Latin Reporting,* was issued in 1972 by the Swedish DX-Alliansen. Prepared by two Swedish DXers, Nils Ingelström and Henrik Klemetz, it contained the names of products and services known to advertise on the radio, plus information on identifying and sending reception reports to Latin American stations. In 1988, two well-known Argentine DXers, Julian Anderson and Gabriel Ivan Barrera, wrote an eight-page pamphlet of tips, key words and phrases, sample reports, etc. called *The Art of Latin American QSLing.* And in 1989, Klemetz wrote *Latin America By Radio,* a larger work on identifying and reporting Latin American stations, and one of the best treatments of the subject.

In 1970, the Riksförbundet DX-Alliansen in Sweden published the second edition of the *RDXA LA-QSL List* showing all Latin American stations from whom QSLs had been received by Scandinavian DXers between 1944 and 1969, together with the number of QSLs received from each and the years in which they were received.[34]

Many stations, particularly those in Latin America, issued pennants, and these became a favorite collectible among QSLers. A club devoted to the topic, Pennant Collectors International, was promoted in 1980 but never took root. However, that year the St. Louis International DXers (SLIDX) published the 12-page, illustrated *SLIDX Guide for Pennant Collectors* which contained an alphabetical list of stations that sent pennants. A second edition, expanded to 16 pages, was published the following year. In 1984, Sam Barto self-published *Station Pennants of the World,* a virtual encyclopedia of station pennants. Arranged by continent and then by country, it described, and showed images of, each station's pennants. In 2006, the operator of Germany's Pennant Museum, Andreas Schmid, issued a CD, *Radio Pennants of the World,* containing descriptions and pictures of more than 2,600 pennants.

The preparation of reception reports also received much attention, and there were many useful guides to assist DXers in preparing reports in foreign languages. Most often these were foreign-language forms accompanied by words, phrases and sentences that could be inserted as needed. In 1967 and 1968, NASWA offered

Some stations included a station pennant with their QSL. Pennants from Latin American stations were among the best looking. Many Ecuadorean stations had pennants, including Radio Jesus del Gran Poder in Quito.

foreign language report forms in Spanish, Portuguese, French and Indonesian. One DXer used the NASWA format to make available reception report forms in Persian, Chinese, Urdu and Vietnamese. The Australian Radio DX Club issued an *Indonesian Reporting Guide* in 1974. SPEEDX offered the *Foreign language Reporting Guide,* a booklet of relevant words, phrases and sentences in Spanish, Portuguese and English. It also issued a one-page, tri-lingual reporting form (English, French, Spanish). During the 1980s, the Radio Netherlands "Media Network" program offered forms in Spanish and Indonesian. Fom time to time, forms in Italian, Arabic, Russian and other languages were offered by individual DXers.

The best foreign language reporting guides were (and still are) the *Language Lab* series published by Tiare Publications in Spanish, Portuguese, French, and Indonesian editions in 1986 and 1987. These 50-page monographs contained the same general types of form reports, words, phrases, etc., but were more comprehensive, addressing such topics as taped reports, follow ups, use of prepared cards, etc. No longer in print, copies of the *Language Lab* books are still sought after by serious DXers.

In the early 1990s, the Suriname DX Club published several editions of its *QSL Survey,* a booklet containing information about, and a statistical analysis of, the *content* of reports that had brought QSLs, plus a list of verification signers covering a multi-year period. The club also published lists of station addresses and verification signers.

Achievement

Most DXers want to hear as many stations and countries as possible. Typically their success has been reflected through three vehicles: statistics, awards, and contests.

STATISTICS

Many clubs published tables showing the number of stations or countries participating members had heard, their best QSL, etc. It began on the broadcast band where, in the early days of DXing, it was possible for a resourceful listener with good equipment, a good location and persistence to QSL almost every BCB station in North America. In 1933, Earl Roberts of Cambridge, Ohio had QSLed all but 30 of the 602 BCB stations then in operation. The following year, Ray Lewis of Toledo, Ohio reported having QSLed all but two of the 591 active stations, and Carleton Lord had verified 574 out of 589.[35]

In 1946 the NNRC pioneered the use of listings that indicated how many stations and countries participating members had heard. At that time the members with the most countries heard and verified on shortwave were Roger Legge, then of Washington, D.C., with 170 countries heard and 152 verified, and Earl Roberts, with 149 countries heard and 121 verified. However, these totals appeared to include broadcast stations, utility stations and hams, not broadcast stations alone. Although the NNRC started counting shortwave broadcast stations and amateur stations separately in 1947, people's totals suggested that different counting methods were still being used, and when the club got around to better defining terms in 1949, utility stations were still included under "shortwave broadcast."

Other clubs had similar listings using various formats. NASWA began including a SWBC "scoreboard" in 1965 (the ham and BCB columns had their own scoreboards). The publication of such "ladders," as foreign clubs often call them, continues to the present in some clubs, although declining membership has reduced the number of participants.

AWARDS

Club award programs have always been popular. Nearly every club had an awards program which bestowed certificates for various levels of achievement.

Consistent with the club's BCB origins, most of the NNRC awards were for BCB DXers. However, the club did offer two shortwave certificates, the "Traveler" and the "Explorer," of which roughly 20 of each had been issued by 1947. In 1952 the requirements for these certificates were the verification of 30 and 60 countries respectively.[36] They were not SWBC–only awards, however. QSLs could be from either broadcast stations or amateur stations. An award for 50 countries verified on shortwave was added in 1961. In 1964 the awards program was reinvigorated, with new awards established in many categories, including, for SWBC, awards for six continents verified, 10 to 40 zones verified (in five zone steps), and 50 or more countries verified (in 25 country steps). By the 1970s, the awards program had become very popular, particularly in the amateur and SWBC categories.

The ASWLC awards program, operational in the 1960s, issued awards based on total countries verified and countries verified in each continent, and also offered a by-continent

The NNRC offered the "Explorer" award for countries verified on shortwave.

award for countries having SWBC stations. The program was later expanded to include an award for verifying all Russian republics, and a "Tropical Area DX Award" for verifying 10, 25, 50, 75 and 100 countries between the Tropic of Cancer and the Tropic of Capricorn.

Although NASWA had a nominal awards program from the club's inception, it became significant only in 1968. Over the years it underwent many rejuvenations. At various times the bulletin has had columns devoted to recognizing successful awards applicants. By the end of 1982, the NASWA awards program offered 41 different awards recognizing SWBC DXing skills at all levels. The program continues to the present.

Clubs in other countries also had awards programs. In the 1940s, the International Short Wave League had a "Countries Heard" award, including one for "Telephony (Broadcast only)." Grade 1 was for 10 countries heard, Grade 2 for 20 countries, etc. It was actually a "countries verified" award, as the QSLs themselves (or a letter from a club official who had examined them) had to be submitted. In the 1950s, the International Short Wave Club offered the "Heard All Continents" certificate in three categories: shortwave broadcast, amateur phone and amateur CW. The submission of one round of QSLs brought a 1-A certificate, eight rounds brought an 8-A, and so forth, with rounds subsequent to the first earning seals signifying the higher class.

Magazines catering to SWLs also had award programs. *Popular Electronics* set one up in 1963 for those who had registered for WPE calls (see p. 358). Awards were available for 25, 50, 75, 100 or 150 countries verified, with "Letters of Certification" issued above 150 in 10-country steps. There were also awards for states and zones verified, and an "All Canada Award." From 1961 into the 1970s, *Electronics Illustrated* had a "DX Club" which existed

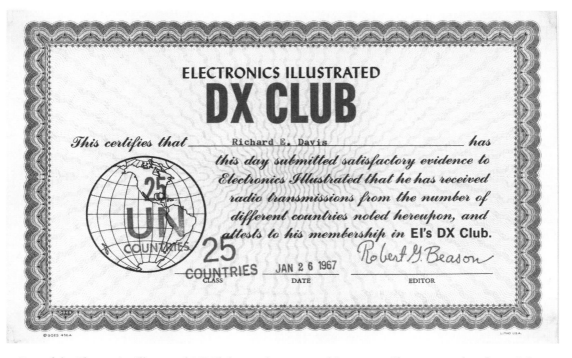

Part of the *Electronics Illustrated* DX Club award program, this 1967 certificate attested to the recipient's confirmation of stations in 25 countries that were U.N. members.

solely to make awards based on verifications received. In the all-band category there were 10-, 50-, and 100-country awards, an all continents award, a 25 U.N.–members award, a 25 most-populated world cities award, and a "six outposts" award, plus, on the broadcast band, a 15 country award and a 25 states or provinces award. There was a similar set of awards for hams.[37] Receipt of an award was necessary for membership, and by 1965 over 2,000 had qualified and had their names listed in the magazine. The "club" had its own countries list, and a monthly column of loggings and news called "Notes from EI's DX Club."

Another source of awards was the Boys' Life Radio Club. Only nominally a club, it was actually a program of awards and contests run in the 1950s and 1960s by the Boy Scouts of America, parent of *Boys' Life* magazine. The awards program ran continuously, and rewarded participants with certificates of achievement for logging all states, all continents and all call areas. There were parallel awards for those submitting QSLs in addition to logs, and a "World Listener" award for those submitting QSLs from 25 countries (including one from each continent). Awards were all-band — any station on any frequency was countable — but, at the applicant's request, they could be endorsed for specific bands, specific types of radio service, or specific emission types. Certificates bore the imprint of the club's donkey mascot, "Pedro," the "Official Ears" of the club.

Each year, *Boys' Life* also ran a bona fide month-long listening contest which paralleled the awards in scope, with receivers, transmitters, books and the like among the prizes. The contests were generally limited to those under age 19, and some years were co-sponsored by the ARRL and the Radio-Electronics-TV Manufacturers Association. The *Boys' Life* contests sometimes attracted a thousand participants.[38]

Some stations also had awards programs of various kinds, usually tied in with their broadcasts. In 1989, Trans World Radio offered a certificate for having confirmed all TWR transmitter sites. The following year, HCJB established the "World By 2000 Confirmed Stations Award" for those with a QSL from at least one partner station in each continent where the World By 2000 project was active. World By 2000 was a joint project of HCJB, FEBC, ELWA and TWR to provide gospel programming to every listener in the world by the year 2000 in a language they could understand.

Other stations have issued awards for reception reports submitted. In 1969, the International Committee of the Red Cross issued a certificate for reporting all six of its 1968 transmissions. Some awards were offered in connection with station-sponsored "clubs." In 1965, Radio Bucharest offered 1,000 diplomas for listeners who sent in 12 reports over three months. In 1994, Radio Bulgaria made available a series of QSL diplomas. There were bronze, silver and gold levels, each requiring increasing numbers of reports to be filed over various time periods. (Gold required three reports the first week, five within the next two weeks, six within the next three, seven within four, nine within five, and ten within six weeks.) Other East European stations had similar QSL programs.

CONTESTS

Although DX contests have never been as popular in the United States as in other countries, there are many examples of contests, some quite novel, that were run by clubs. In the mid–1950s, the section editors of the URDXC's bulletin, *Universalite,* sponsored

This SWL-card was made available to members of the Boys' Life Radio Club.

some informal "logging marathons" where, for a ten-day period, members logged as many stations as possible in a given frequency range. After totaling their points, winners received "a feeling of pride and contentment."

In its early days, the ASWLC held frequent contests for members. They were generally limited to one category of listening — shortwave broadcast, ham or broadcast band —

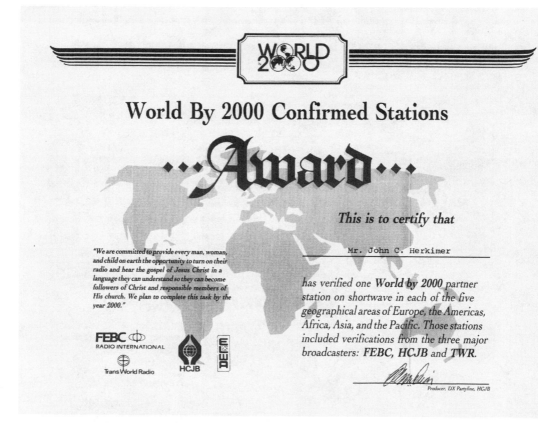

HCJB offered this attractive certificate for listeners who had heard partner stations in the "World By 2000" project which was intended to offer gospel programming in as many different languages as possible.

a particular band, and a particular number of hours or days. Sometimes power and distance of the station were factored in. In the mid–1960s, the ASWLC had annual contests for the most countries verified within two months following 48 hours of listening. It also had contests for the most countries heard within a 48 hour period, and for the most stations and countries heard (once designating separate 10-day periods for monitoring within individual bands, another time covering tropical band DXing during a one-month period, etc.).

SPEEDX had some novel contests. In a 1971 SPEEDX Sweepstakes, the winner was the person whose log of one station from each continent reflected the lowest combined power. SPEEDX Reception Surveys amalgamated reports on a particular station's transmission on a selected date and time, e.g. Radio RSA, Radio Netherlands, and TWR-Bonaire. Participants were rewarded with a special acknowledgment card. In 1976, SPEEDX ran the "Stations by the Hundreds" contest where you had to log as many stations as possible on 20 frequencies that were multiples of 100 kHz.

The NNRC held major, week-long DXing contests during the month of January from 1979 to 1982. In the contest's SWBC section, contestants' listed the details of their 25 best loggings, which were then scored based on transmitter power, location, etc., and a multiplier applied based on the contestant's receiver. Total countries logged during the contest was also a factor. About 35 members participated. Winners in each category — BCB, short-wave, amateur, and utilities — received a plaque, runners up a certificate.

In 1978 and 1979, California DXer James W. Young sponsored a month-long "Tropical Bands Monitoring Contest." Participants could enter as a contestant or a monitor, with the latter submitting loggings and receiving a certificate of merit and a composite list of all loggings. Contestant scoring was: two points for each station on the listener's continent, otherwise five points, plus ten points for a second logging of a station via long path, and one point for an unidentified station. Winners in each band received cash prizes in 1978, and certificates of merit and *WRTHs* in 1979.[39]

The Ontario DX Association conducted the Christmas DX Challenge during the entire month of December in 1986 and 1987, the goal being to hear 100 SWBC countries. All entrants received certificates. Special certificates went to those who achieved the 100 goal, and prize drawings were held for achievers at different levels. The contest was renamed the December DX Challenge, then the DX Challenge, and has been held almost every year since. ODXA sponsored other contests as well. Most of the contests held by the Canadian International DX Club were ham, utility or medium wave affairs. However, in 1990 the club held a "Foreign Language Contest" where, over a two-week period, contestants logged as many stations in as many different languages as possible.

The major North American contest in later years was the North American DX Championships, which took place from 1987 to 1999. It was sponsored by Numero Uno through 1992, when its management was transferred to ANARC where, for all but the first two years, it was administered by a committee led by Harold N. Cones of Newport News, Virginia. Although it ran longer a few times, usually it was a two-day event, with participants competing to log as many stations and countries as possible. In some years, starting in 1991, there was a special tube-radio class. Prizes were books, trophies, plaques and the like, and all participants received certificates. A good year saw the participation of about 20 people.

Listeners in other countries, particularly Europe, were more enthusiastic about contests. The International Short Wave League in the U.K. conducted a series of contests. In 1950 it ran one contest where, for 24 hours, teams competed against each other to see who could log the most countries on shortwave, and another, similar 33-hour contest for individual members. Winners received diplomas. Nine members participated in the contest for individuals, and the winner heard 100 countries.[40] Another contest for individuals was run the following year.

Also in 1950, the ISWL teamed up with the *World Radio Handbook* to sponsor a 40-hour contest in the 9–12 MHz. frequency range, winners being those whose logs O. Lund Johansen felt contained "the most DX stations with regard to the listener's location." Prizes were ISWL memberships and free *WRHs*. The contest was repeated in 1951.[41] Arthur Cushen of New Zealand won both years. In another ISWL-*WRTH* joint venture of the 1960s, listeners had to answer four questions from information broadcast at specified times over 13 cooperating international stations. Prizes were receivers and books. Some 510 entries were received.

Although they drew scant participation from North America, the shortwave listening contests arranged by the Scandinavian countries were the most elaborate. The Nordic DX

Championship is the latest in a line of Scandinavian contests dating back to at least the 1950s. They were conducted several times a year, and often involved arranging special programs or announcements over specific target stations.[42] There were also competitions between clubs, and even between one country's DXers and another's. The NDXC has been run almost annually for over 40 years by clubs in Finland, Norway, Sweden or Denmark on a rotating basis. Another popular European contest has been the DSWCI "Grand Tour With Cancer and Capricorn" which has been conducted annually since 1989. It focuses on tropical band reception.

Swedish DXers were big contest supporters. In 1953, a "Short Wave Game" contest to promote shortwave listening that was sponsored by Teknikens Varid in Stockholm, Sweden yielded 1,451 participants and 2,000 new members for the radio club, which reported having a total of 15,000 club members as of July 1953.[43] Swedish clubs also liked to arrange special Swedish-language programs over various stations, usually larger ones which already had Swedish services. Listeners sent their reports to the sponsoring club, which verified them with a specially designed QSL-card before forwarding them to the station. Some of these special QSL opportunities were part of club-sponsored contests.

A few stations ran their own contests. From 1969 to the mid–1970s, Radio RSA held annual contests wherein listeners had to find a series of unannounced RSA frequencies that were identified on air only by meter band and just before the contest's start. Prizes included tape recorders, wood carvings, books and records. In 1975 the station ran a contest in which

SPECIAL QSL

to Håkan Arvidsson,

For the broadcast over R. Moskva,

the 11 of oct. 1958 at 2200 G.M.T.,

which here is verificated.

Lund the 15 of october 1958

Ordf.　　　　　　　　Tävl.led.

DX - Club Lundensis

DX-CLUB LUNDENSIS
Merkuriusgatan 16
LUND
SWEDEN

DX CL

A club QSL, this 1958 card from a Swedish club verifies a transmission of Radio Moscow.

participants had to verify one Radio RSA or SABC transmission in each shortwave band from 120 to 11 meters. The first 40 people to qualify won a book. And in 1980, AWR-Asia held a 30th anniversary contest requiring the identification of some 20 interval signals and the logging of as many AWR frequencies as possible.

Each year from 1977 to 2004 there was an annual contest sponsored by Adventist World Radio's "Wavescan" program. Usually entrants were required to exhibit some skill in hearing certain stations, identifying interval signsls or the like, or, in later years, describe such things as their five smallest cards, the size of their QSL collection, unique transmissions verified, their best QSL, etc. Prizes included books and AWR mementoes.

The Swiss Shortwave Service ran an interesting 20 meter ham contest back in 1952. On May 5 and 6, each of the nine foreign service transmissions began with information about three Swiss ham stations that were standing by, and the first two amateurs (one within 3,000 km., one beyond) to contact one of the stations would be declared the winner for that transmission.[44] And some stations, such as HCJB, Radio Austria International and Radio Netherlands have held ham band QSO (ham contact) contests.

Today there continue to be occasional shortwave broadcasting contests, usually sponsored by overseas clubs or publications. They are few in number, however, and have but a small following in North America.

COUNTRY LISTS

Some clubs had country lists to guide members in counting countries for purposes of personal statistics and contests. The NNRC adopted its first country list in 1947. Although strongly influenced by the American Radio Relay League ham band countries list, it somewhat broadened the concept of a radio country in deference to the club's membership, many of whom tuned stations other than hams. The list had a total of 296 countries. In 1955, being unsatisfied with the list's haphazard nature, the club adopted the ARRL list for club use until returning to its own, revised list in 1958.[45]

Most lists, including those adopted by foreign clubs, counted "every rock in the ocean" as a separate "country," as preferred by hams and ham band SWLs, or based country counting on fixed dates tied to geopolitical considerations. In 1967, NASWA, by then an all–SWBC club, adopted a new approach. For all SWBC operations after 1945, the rule was "once a country, always a country," regardless of changing political status.[46] The Ontario DX Association adopted the NASWA list in 1976.

There were 212 countries on the original NASWA list. A country list committee has kept the list up-to-date. The original list had 14 inactive countries, included to accommodate loggings of countries that had been active in 1945 or later but were no longer on the air in 1967. It is instructive that in the 2007 revision to the list, 40 years after its creation, the total number of countries had grown to 263 but the number of inactive countries was 76, or 29 percent of the total, reflecting the many countries that had left shortwave broadcasting over the years.

SWL Cards and SWL Calls

SWL Cards

Many SWLs listened to amateur radio contacts, an aspect of shortwave listening that became increasingly popular after the war. The NNRC inaugurated an amateur listening section, "With the Hams," in 1946 and it grew quickly. Regular ham listeners prepared their own "SWL cards," similar to QSL-cards but designed for reporting reception of an amateur's signal rather than confirming a contact. They would typically contain a designator like "W1-SWL," "W2-SWL," etc., indicating the call district in which the listener was located, along with the date, time and frequency heard, the station contacted, a brief signal report, and comments.

In addition to using the cards for their intended purpose, swapping them became popular. Many DX bulletins listed the names and addresses of SWLs who were interested in swapping with other SWLs. A few clubs were established solely to support card swapping. The Grand National SWL Club, founded in 1939 as a club for SWL-card swappers, operated until around 1946 and published a high-quality bulletin for $1 a year. The Short Wave Listener's Registry was founded in Kansas City, Missouri in 1940, also to further SWL-card swapping. Membership was 25 cents a year, and while there was no bulletin, members received a list of SWLs who would respond "100 percent," plus club stickers and a membership card with a "personal club call." The club was reorganized in Pennsylvania after the war, apparently with an expanded mission, and started issuing a monthly bulletin. And in 1954, the International Short Wave Club (U.K.) invited members to send in their cards and offered prizes for the best cards received.

Swapping continued to be popular during the 1960s. The International League of Signal Chasers was formed in Denver, Colorado in 1965, its bulletin, *DX Monitor,* serving mainly card swappers, penpals and tapesponders. In 1969, *Swaps Newsletter* was established by Hal Morgan of Santa Ana, California. Originally appearing as a column in the ASWLC bulletin, it was so popular that the editor spun it off on its own. Circulation diminished over the years, however, and it stopped publishing in 1977. An attempt to revive it was tried in 1983 by John Kapinos, who included such things as shack photos, pennants, equipment — even baseball cards. However, it did not last the year.

Mainstream DX clubs also joined the card swapping action. The NNRC "Short Wave Odds & Ends" column was a major source of names of card swappers. In 1964, the Canadian DX Club offered a "Friendship Certificate" to swappers who had swapped at least 50 cards, been recommended by two well-known swappers, and promised to swap 100 percent. And in 1969, the ASWLC ran a year-long card swappers contest, with points awarded for swaps with each state or province. Results appeared in a swaps scoreboard.

Some felt card swapping had little to do with shortwave listening, and were happy when the card swapping boom ended, which it did by the 1970s.

SWL Call Letters

The United States did not follow the practice in many countries wherein the national amateur radio organization issued numerical "call signs" to SWLs (e.g. EA-6358, JA4-5319, etc.), principally to facilitate the handling of SWL reports to amateur stations and encourage listeners' "advancement" into ham radio.

The use of SWL "calls" in the United States was begun in 1957 by NNRC member Joe P. Morris of Cleveland, Ohio. Morris was an active ham band DXer, and felt that use of a listener identification scheme better than the traditional "W1-SWL," "W2-SWL" would facilitate the work of QSL bureaus in handling SWL-bound ham QSLs. He inquired of the FCC as to whether a system of "licensing" SWLs would be legal. The FCC, while not affirmatively blessing the practice, said: "Call signs are issued for the purpose of identifying radio transmitting stations. There appears to be no objection to the mode of identification of receiving stations contained in your letter." So he started issuing "calls" to any SWL who sent him a blank post card or a stamped, self-addressed envelope (soon replaced by a form card supplied for a one-cent stamp, later a 20-cent fee). The prefix of the calls was WRØ, thus WRØ1AH, WRØ2CF, WRØ3PV, etc. (In zone Ø, the prefix was WR1Ø.)[47]

By March 1958, Morris had issued 81 calls.[48] By September the number had grown to 428, with a few hundred more pending. In October, Morris announced that he could no longer handle the project, and it was taken over by *Popular Electronics.* The prefix was changed to WPE (or the appropriate equivalent in other countries, e.g. VE2PE1HC, G3PE2K, etc.), and certificates were made available for ten cents (no charge if you already had a WRØ call). Tom Kneitel, WPE2AB, was designated Director of Monitoring Station

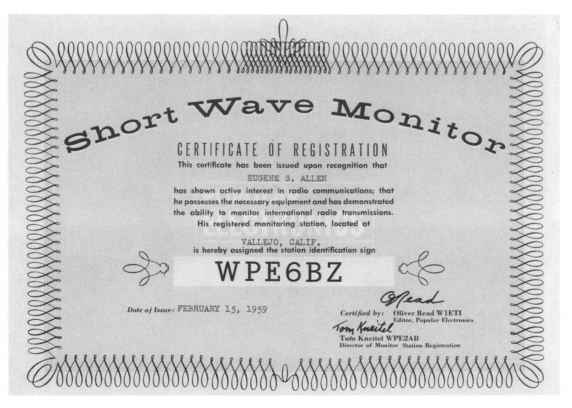

This is the certificate SWLs received from *Popular Electronics* when they registered for a WPE "station identification sign."

Registration. "It will be more than just handing out call letters," said Kneitel. "[I]t will be the first actual international monitoring station registry. Information will be kept on all stations regarding equipment used, etc."[49]

Clubs urged their members to register, and the program was mentioned over many stations, including Radio Australia, VOA, Radio Sweden, the Voice of Denmark, Deutsche Welle, and 4VEH. Soon requests were coming in at the rate of 150 a day. Three thousand SWLs had registered by May 1959, 10,000 by July.[50] Applicants were urged to send a stamped, self-addressed envelope for faster service.

Although nothing appears to have come from the monitoring registry side of the project, save whatever commercial benefit might have accrued to *Popular Electronics,* the WPE calls soon gained some cachet among listeners. A WPE call often accompanied the name of a member in club bulletins, and in the NNRC bulletin and *Popular Electronics,* Hank Bennett included the contributor's WPE call letters when naming contributors of DX items.

Not everyone agreed with the program. Predictably, Arthur E. Bear, head of the International Short Wave Club and anti-jamming maven, editorialized against it. Although he was not against registration if it was to determine how many SWLs there were, he felt that call letters meant nothing to stations and were misleading, and that use of names would preserve the personal nature of the hobby.

Administration of the program was transferred from Kneitel to Bennett, where it continued during the 1960s. It was about to be closed down in 1970 when Bennett's column was dropped in a *Popular Electronics* restructuring that de-emphasized shortwave. With the permission of the magazine, Bennett continued the program on his own, where-

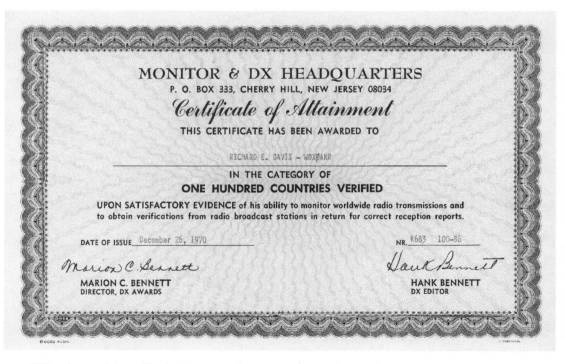

SWLs who participated in the WDX awards program received a certificate like this one, which certifies receipt of SWBC QSLs from 100 countries.

upon it became the WDX Monitor Registration Service. Calls, now "WDX," were $1 each. Special "out of sequence" calls were $2, and one-letter calls were $5 while they lasted. Bennett also continued the WPE awards program (now called the WDX DX Awards Program) and the attendant WDX Honor Roll.[51] It appears that over 40,000 WPE/WDX call signs were issued.

The use of the calls was not without some problems. In 1960, some persons were found transmitting in the amateur bands using WPE calls. Whatever the value of the calls, many years after their heyday they continued to serve as a symbol of the shortwave listening fraternity. In 1982, the Great Circle Shortwave Society was formed to provide a link among those who had been active SWLs in the 1950s and 1960s, particularly WPE call sign holders. Another WPE artifact is a website where WPE call sign holders can give their calls and engage in hobby reminiscing.[52]

Other SWL calls have been issued over the years. Kneitel continued to offer calls as part of a "Monitor Station Registry" which he operated in connection with his business, CRB Research Books, Inc. The format of the calls was, e.g., KPA3CA ("PA" for the state, "3" for the call district, "CA" as the unique identifier).[53] In 1965, a club called SWL Certificate Hunters Club, oriented mainly toward ham band listening, issued "calls" to its members, e.g. "SH-W5-74" ("Shortwave Hunter," call area, club file number). A "World Listening Service" was established in Audubon, New Jersey to issue "WLS" call signs (e.g. WLS5GK) and publish a registry of monitors.[54] And in recent years, a group called the Short Wave Amateur Radio Listeners has established an internet-based, registered call sign system. The group is interested mainly in ham band monitoring, but call signs are available to all radio monitors.[55] The active use of call signs among SWLs ceased long ago, however.

Somewhat related were a few unsuccessful attempts to organize listeners into monitoring networks that would provide an ongoing flow of information to stations about how their signals were received. An early effort was the "International Monitoring Service" of San Carlos, California. Directed by one Edgar W. Parmenter, it appears to have left no traces save for its organizational announcement in 1950.[56] In 1960, Roger Legge proposed a "High Frequency Broadcast Monitoring System" whereby listeners would check reception on various stations at least twice a week and, using prepared forms, submit their observations monthly to Legge, who would assemble them and send them to the stations.[57] Although the project elicited a good deal of interest, neither it, nor a similar Legge effort of the mid–1960s, the "Worldwide Technical Monitoring Service," ever got off the ground.

QSLs as History

In recent years, QSLs have taken on importance as one of the few tangible memorializations of stations now silent. There being in the United States no provision for their archival, a Committee to Preserve Radio Verifications was formed in 1986. Originally a Numero Uno special project, then an ANARC committee, and since 2005 an independent project, the group has archived some 40,000 QSLs, mostly of shortwave and medium wave broadcast stations. The collection, which includes many QSLs going back to the 1920s and

1930s, is located at the Library of American Broadcasting on the campus of the University of Maryland, and continues to grow.[58]

The arrival of the eBay auction site also sparked a market for the sale of QSL-cards. What had been a highly personal memento of a DX experience now became a collectible to satisfy another's nostalgia. A few QSLs have fetched hundreds of dollars.

8

Computers

I have resisted to the best of my ability, but computers keep sneaking into our hobby.[1]

Larry Brookwell, President
International DXer's Club of San Diego, 1983

Although the first personal computer was the Altair 8800, a kit that gained much of its popularity from coverage in the January 1975 issue of *Popular Electronics,* it was Apple that sparked the personal computing revolution. The limited-production Apple I, with its video interface and keyboard (innovations which were added to the Altair only after its initial release), appeared in 1976. It was followed in 1977 by the popular Apple II. Also among the early personal computers were the Commodore PET and the Radio Shack TRS-80. Tandy had hoped to sell 3,000 TRS-80s the first year. It sold 10,000 the first month. Some five to six million Apple II–series computers were sold before the line was retired in 1993. Late models of the Apple II and the TRS-80 were the first computers to use disk drives rather than cassette tapes for storage. A spreadsheet program (VisiCalc) and a word processing program (WordStar) first appeared in 1979, greatly increasing the popularity of personal computers.

In 1981, IBM introduced the "PC," which came bundled with MS-DOS, the first Microsoft operating system. Apple introduced the graphical user interface (GUI) in 1983 on the ill-fated Lisa and in 1984 on the more successful Macintosh. IBM went GUI in 1985 with the introduction of the Microsoft Windows 1.0 operating system. Desktop publishing software became available the same year and improved quickly.

Other companies whose personal computers were popular during the 1979–85 period included Atari; Osborne; Sinclair; Commodore, whose $300 Commodore 64 became the best-selling computer ever (22 million sold over 11 years); and Compaq, which produced the first IBM–compatible unit. IBM compatibility became the standard.

SWLs looked first to their clubs for information about computers and DXing. Probably the first treatment of the subject by an American shortwave club was in 1979 when the preparation of simple BASIC-language programs for "micro-computers" was discussed in the ASWLC's bulletin. A regular column appeared from 1980 to 1982, and was reinstated in 1986.

The corresponding NASWA column, "Computer Corner," was introduced in 1984.

Edited by Bill Cole of North Cape May, New Jersey, it featured instructions for writing programs dealing with darkness patterns, dipole measurements, propagation prediction, receiver control and so forth. Cole was succeeded by Tom McElvy of Norfolk, Virginia, who edited the column from 1986 until it was dropped in 1989. Other clubs with computer columns included CIDX (during the latter half of the 1980s, and again in the mid–1990s) and ODXA, which had a short-lived computer column in 1985 and returned to the subject in 2000 and again in 2006.

In 1983, ANARC appointed Minnesota DX Club member Bill Krause of Anoka, Minnesota as its Computer Information Coordinator. A Computer Information Committee followed. Krause was instrumental in starting the ANARC bulletin board system in 1985. In mid–1988 the job of Computer Information Coordinator passed to David Snyder of Staten Island, New York. Computer-related discussion began appearing occasionally in the *ANARC Newsletter,* which in November 1988 inaugurated a section called "CompuTools." The Computer Information Committee ended its work in 1992, by which time computer bulletin boards were in wide use.

There was treatment of computers in other club bulletins as well. However, considering the impact of computers on shortwave listening, their coverage in club bulletins was modest, even after the internet made computers ubiquitous in shortwave circles. In part this was due to inertia and the scarcity of editorial talent in a relatively unexplored area. It also may have been influenced by fears, subconscious or otherwise, that computers were a pastime all their own and a potential distraction, or even a source of competition, to shortwave listening.

Most SWL computer users were self-taught. The computer bulletin board systems (see below) were a good place to learn, as were the ham radio bulletin boards and ham and computer magazines. Two popular SWL publications that brought together the existing knowledge in the mid–1980s were *The DXers Guide to Computing*, a booklet by George Wood of Radio Sweden which appeared in several editions until 1990, and the Radio Netherlands' *InfoDutch* booklet. ("InfoDutch" stood for "Information for Direct Use to Computer Hobbyists.") Both focused on SWL–related programs, information sources, and suppliers.

Pre-Internet

In the period before 1993, the year the internet became widely available through commercial on-line services, most SWL–related computer activities took the form of word processing and desktop publishing; writing programs for the early computers (especially the Commodore 64); establishing and using bulletin board systems (BBS's); and using online service providers such as CompuServe.

Word processing was the earliest and most useful computer function. It greatly facilitated correspondence and writing reception reports, especially foreign-language reports, whose preparation was usually tedious. In addition, increasing numbers of bulletin and newsletter editors produced their material using word processing, saving time and improving appearance. Word processing was followed in 1985 by desktop publishing, which permitted better page layout, the addition of graphics, and the use of laser printers for high quality printing. Over time, word processors incorporated many desktop publishing features.

At first, the more technically inclined SWLs learned the BASIC computer language and wrote their own programs. These included such things as databases for sorting schedule and program information by time, frequency, language, etc.; propagation programs that calculated sunrise-sunset times or maximum and minimum useable frequencies; and SWL record-keeping programs (stations logged, verified, etc.). Line-by-line code for simple programs sometimes appeared in club bulletins.

As time went on, programs became more sophisticated and often combined multiple functions. Among the more innovative programs of the mid–1980s were the computer control programs which permitted keyboard operation of various functions of the growing number of receivers equipped with an RS-232 port. Some of these programs combined station databases and computer control so that the user could select a station from the database and the receiver would automatically tune to it. These included "Seeker," "ERGO," "Shortwave Navigator" for the Macintosh, and "Watson" (short for "What's On?"). Programs tended to be of more interest to program listeners than DXers because the RF noise generated by the computer would usually cause interference, especially of weak signals. Grounding, cable chokes and shielding lessened the problem but seldom eliminated it. Over time, database programs, and combined database-computer control programs— some generic, some designed for particular receivers— added other features and became much more sophisticated.[2]

Although there grew up a cottage industry of SWL program developers who sold their products commercially, many SWLs who were adept at writing programs or assembling data into files placed them in the public domain for use free of charge (often as downloads from BBS's), or offered them as "shareware," where the user got the program free but was expected to pay for it if he or she liked it. The ANARC Computer Information Committee made a library of free programs and other SWL–related files available on a series of disks.

Shortwave listeners got their first taste of electronic connectivity through bulletin board systems. A BBS was basically a collection of data in a hub computer that could be accessed via telephone dial-up from any other computer equipped with a modem and software that permitted it to act like a terminal of the hub unit. The operator of the BBS was called the system operator, or sysop. The purpose of BBS's was information sharing. They were usually free, and available 24 hours a day. You dialed the BBS phone number, the computers performed a brief connection ritual, and you were "on line."

Once connected, you could read material stored on the BBS, download and save it for reading later, or upload your own items for use by others. This was all before graphical user interfaces, so the BBS screen was all text. Some screens were intuitive and easy to navigate, others cumbersome. The speed with which you could navigate and transfer files depended on the speed of your modem and the capabilities of the hub unit. Early dial-up modems operated at 300 bits per second, then 1,200 and 2,400 (eventually they reached 56 kilobits per second). These speeds were slow compared to broadband, and limited the ability to transfer large files. In addition, if one was paying long distance charges to connect to a BBS, there was a premium in getting on and off line quickly.

The BBS main menu led to more specific sub-menus. BBS content consisted of messages, either to and from individual users or general messages to all users, and files of various kinds, often compressed for faster transfer. Typical of the files were station schedules; logging and QSL reports from listeners; databases containing the times and frequencies of shortwave broadcasts; propagation forecasts; buy-and-sell ads; and "libraries" containing all manner of lists, articles, electronic newsletters and similar items which theretofore were obtainable only by postal mail. On many BBS's one could also download software programs written by other SWLs. Some of the larger shortwave stations, like Radio Netherlands and Radio Canada International, posted program information on the more popular BBS's, providing a valuable new link between station and listener. Radio Sweden posted the "Sweden Calling DXers" bulletin. Some equipment manufacturers had a presence on the BBS's as well. The main benefits of the BBS were that it assembled all of this material in one place, and provided it on demand.

There were thousands of BBS's across the country and around the world. The high point of their popularity was 1994 when there were some 45,000 BBS's in operation in the United States alone. Some BBS's were general, with sub-parts devoted to many different subjects, while others were dedicated to particular topics. Topic-specific BBS's were particularly informative, and were usually supported by small but earnest bands of uploaders who kept the content current.

The pioneer of the major SWL bulletin boards was the ACE BBS, established in November 1984 by the Association of Clandestine Radio Enthusiasts. It was located in Kansas City, Kansas, and it was operated by Kirk Baxter, the ACE loggings editor who would soon become the club's president. The ACE board covered shortwave topics beyond the pirates and clandestines which were the club's focus.

The ACE BBS was followed in September 1985 by the ANARC Shortwave BBS. It was located in Greenfield, Illinois, and it was designed from homemade software originally written for a planned NASWA BBS by Bill Cole, editor of NASWA's "Computer Corner," and Bill Krause of the ANARC Computer Information Committee. Krause was the sysop. After a year, the ANARC board was receiving about 150 calls per month, 200 after 18 months. In June 1988, the ANARC board closed and its functions were transferred to the ACE BBS, which was renamed the ANARC BBS.

The ANARC BBS was the best shortwave BBS. (It won a WRTH Industry Award in 1991.) In addition to the usual BBS features, it had separate sections for individual ANARC clubs. Some editors of ANARC clubs who prepared their columns on word processors uploaded them to the ANARC BBS for advance viewing. The BBS also facilitated intra-club communication. The publisher of the *World Radio TV Handbook* opened a section on the ANARC BBS in 1989.

There were numerous other radio-related BBS's, usually operated by individuals, but sometimes by clubs, publishers, manufacturers, suppliers, stations, even the FCC. Among the better-known were the Pinelands BBS, operated by Tom Sundstrom of Vincentown, New Jersey; UBIX (Universal Bulletin Board & Information Exchange), sponsored by Universal Amateur Radio, Inc., the ham–SWL supplier; and the Grove Systems BBS run by Grove Enterprises, home of *Monitoring Times*.

An important advancement in BBS technology was the creation in 1984 of the FidoNet, a network of BBS's that used Fido software. Participating BBS's could exchange messages and files within FidoNet. Thousands of BBS's, focusing on hundreds of subjects, belonged to the network. An important feature of FidoNet was that a user whose local BBS was part of the network could access network information by a phone call to the local BBS, avoiding long distance charges.

In 1986, the sysop of the "Neverboard" BBS in Pittsburgh, along with a number of other BBS's, established the FidoNet SWL Echo Conference. Shortwave-related messages and files were automatically exchanged among participating BBS's every night. This made the full array of messages and files available at each participating BBS. A valuable part of the network was Echomail, essentially a public discussion forum similar to the Usenet newsgroups (see below). A user of one FidoNet BBS could also send a private message to a user of another FidoNet BBS. FidoNet eventually dropped the requirement that participating BBS's use Fido software, and at its height there were several hundred BBS's in the SWL Echo Conference (not all BBS's carried all Echo Conference material). The ACE BBS was a member of the SWL Echo Conference, and remained so after becoming the ANARC BBS. ANARC BBS sysop Kirk Baxter became the SWL Echo Conference moderator in 1989.

The ANARC BBS closed in 1996, by which time BBS's everywhere were being eclipsed by the internet. The number of BBS's declined to 7,000 by 1998.

The next stage of electronic connectivity for SWLs was provided by the online service providers, mainly CompuServe (once the largest), America On Line, Delphi, GEnie, and Prodigy. The height of their popularity was the late 1980s and early 1990s. These were private computer systems to which users subscribed. Conceptually they were not unlike BBS's, but with graphical interfaces and more extensive, better organized content. Although most of the online services were accessible through local telephone numbers, there was usually a time-based connectivity charge, calculated by the minute or the hour, and monthly bills reminded subscribers how habit forming, and expensive, on-line life could be.

The content of the online services was broader in scope than the BBS's— Compuserve covered almost every conceivable subject — and the graphical interface, with its more professional look and feel, made navigation easier. The services offered all that the BBS's had supplied — messaging, readable and downloadable files, etc.— and more, including "forums," "conferences," "roundtables," and "special interest groups" focused on numerous individual topics, including shortwave listening (which was usually a sub-section of the service's ham area). This was in addition to the news, weather, games, encyclopedias, on-line product support from various companies, shopping, etc. that formed the foundation of the online services. Many BBS operators soon established a presence on these services.

Unlike the internet that was still to come, the online services were proprietary, self-contained communities. In the mid–1990s, as the internet became widely available to the general public by way of independent internet service providers, the popularity of the online services declined. Some closed, while others stayed competitive by replacing their time-based charges with flat monthly fees and supplementing their proprietary content with full internet access and non-proprietary e-mail.

One of the most popular aspects of the online services was the access it provided to Usenet, an independent computer network once described as an "information cooperative." Begun around 1980, by 1990 it consisted of some 16,000 participating sites and over a half-million users. The Usenet was divided into topical groups called newsgroups, of which there were about 600 in 1990. Subscribers posted messages (called articles) to the group, and replied to messages posted by others. Messages were organized into "threads." The servers on which individual articles were posted communicated with each other, so all subscribers to a newsgroup eventually saw all the messages posted by others. Although Usenet messages had been available via some BBS's, access through the online services was much easier.

The most popular of the SWL–related newsgroups was rec.radio.shortwave, "rec" denoting Usenet's "recreation" section. It was begun in 1989. Also popular were rec.radio.info and alt.radio.pirate ("alt" meaning "alternative"). The Usenet is still in operation, and is now accessible directly from the internet. However, the growth of internet alternatives, and the wholesale spamming from which the newsgroups suffer, has reduced their popularity, and today they are seldom used in shortwave circles.

A lower-tech method of electronic connectivity was the telephone-based voice storage-and-retrieval system. Anyone could call, at his or her own expense, the Washington, D.C., number of the "DX Newsline" and receive a public message, updated daily, containing the latest DX log reports. A caller could also contribute information via a 30-second message. For an annual subscription of $10, callers could get immediate, direct access to all messages without waiting for the daily update. The "DX Newsline" was established in 1984. The Ontario DX Association operated a similar system, called the "DX-Change," from 1989 to 1999. Located in Scarborough, Ontario, it was free to both club members and non-members, save for the cost of the call. In 1994, Glenn Hauser experimented briefly with a one-way telephone DX news service called "DX-Daily." Soon, however, the internet would become the main electronic vehicle for distributing DX news, and telephone-based systems would drop from sight.

The Internet

Previously the preserve of academic and industrial research organizations, the internet caught the eye of the general public in 1993 when the Mosaic web browser was released. Mosaic was relatively easy to install and use, and its imbedded graphics greatly simplified internet navigation. A few years later it was succeeded by Netscape Navigator, which remained the consumer standard until eclipsed later in the decade by Microsoft's Internet Explorer which was packaged with the Windows operating system and thus pre-installed on almost every PC sold.

The internet's impact on shortwave listening, as on so many activities, was profound. A 1986 survey of HCJB's listener club, ANDEX, showed that 38 percent of responding members owned a computer, and most did not use it in connection with their radio activities.[3] A survey taken by *Fine Tuning* the following year also showed that most FTers did not use computers in the hobby.[4] By the 2004 Winter SWL Fest, however, 86 percent of attendees had access to the internet, nearly all at home.[5]

Since 1995, nearly every shortwave station, club and vendor, plus countless individuals, have established a web presence. In the internet's early days, the existence of every website was news, and many sites contained links to other shortwave-related websites as part of their content. By far the most comprehensive index to radio on the internet was the "Shortwave/Radio Catalog" website created in 1994 by Pete Costello of Aberdeen, New Jersey. The shortwave broadcasting section contained, among many other things, links to all

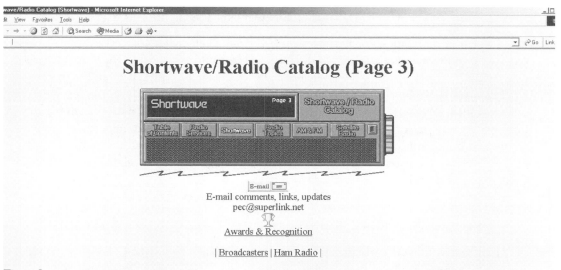

Shortwave/Radio Catalog (Page 3)

E-mail comments, links, updates
pec@superlink.net

Awards & Recognition

| Broadcasters | Ham Radio |

Broadcasters:

o The WWW Shortwave Listening Guide

If you would like to know what programs are on the shortwave bands at the current hour along with their frequencies and target areas, then this interactive web form is for you. Perhaps you have a favorite type of program, say DX show, news documentary, or local folk music, etc., with this form you can get a listing of shortwave programs, frequencies and program times on a selected day, or over the weekend. This site searches John Figliozzi's excellent database of shortwave radio programs in English and other languages. Included in the NASWA hosted site you will find information on how to obtain John's *The Worldwide Shortwave Listening Guide* book. I think all those tuning the shortwave bands should check this site out, but, then again, I have a somewhat biased view since I had a hand in creating the site.

o What's on Shortwave?

An interactive on-line Shortwave Database query WWW form (retired until a database maintainer is found).

Pete Costello's "Shortwave/Radio Catalog" was the first stop in keeping up to date on radio and the internet.

shortwave station websites, arranged by continent. The site also covered amateur radio, AM and FM, utility stations, satellites and scanning, as well as radio-related hardware and software, clubs, propagation, publications, vendors, early internet broadcasting, radio history, and other radio-related topics. The catalog was more than a list of URLs; it contained brief descriptions of each site (early search engines returned only URLs). If it related to radio, it was in the Costello catalog, which soon became the starting point for finding radio-related internet resources. Although the explosion of internet websites made it impossible to update the catalog after 1999, it remained on the web for three more years.

Also noteworthy was "The Internet Guide to International Broadcasters," which was based in Germany and operated by Thorsten Koch. It was on-line from 1992 to 1997, and contained URLs and e-mail addresses for all international broadcasters, plus those of clubs and other organizations, individuals, and vendors. A paper internet index, *Internet Radio Guide,* was tried in 1996 and 1997 by Joerg Klingenfuss of Tuebingen, Germany. The 400+ page manual included screen shots of numerous radio-related sites, most of them oriented to utility listening, but some relevant to shortwave broadcasting. However, it soon became apparent that a paper publication could not keep up with the pace of the internet.

Although modern web searching has made dedicated indices and search facilities less essential, the www.radio-portal.org searchable database of over 42,000 URLs still provides such a service, not just for shortwave but for many other radio-related topics as well.

The internet has impacted many areas.

COMMUNICATION AMONG LISTENERS

Electronic mail quickly replaced letters as the main vehicle for communicating. A close second were the various electronic mailing lists or "reflectors" (e.g. Yahoogroups) whereby conversation occurs through member messages which are distributed to all list members. Such lists are typically run by clubs or other groups of listeners with an interest in some aspect of shortwave, e.g. pirate broadcasting, radio history, shortwave programming, particular receivers or software, transmitter sites, Digital Radio Mondiale, etc. These electronic communication vehicles have also permitted the rapid transfer of audio samples, graphics, etc. Some listeners also stay in contact through the radio-related Internet Relay Chat channels, instant messaging, and blogging.

INFORMATION ABOUT, AND COMMUNICATION WITH, STATIONS

Radio stations were early users of the internet.[6] As they established web sites, the long-standing problem of obtaining program information eased greatly. Programs were typically one of the major content areas of station websites, and while not all stations have provided good website maintenance on program information, many now offer extensive program information, including advance program alerts by e-mail. Station websites also contain time and frequency information, station history, news, contact information, and much else. Electronic mail, and station website forms, have also increased direct contact between listener and station. Broadcasters often announce short-term special transmissions, or changes in schedule, by e-mails that are republished in various shortwave circles.

DXING

After improvements in receivers, the biggest influences on DXing have been the internet and the development of new software tools.

Of great importance have been the new sources for up-to-date time, frequency or contact information. Many electronic mailing lists, websites and databases are available, and most are free. Some are professionally produced, e.g. the periodic on-line updates to the *WRTH* and the HFCC lists, while others are sponsored by clubs or individual listeners, e.g. the Eike Bierwirth lists,[7] the Danish Shortwave Club International *Domestic Broadcasting Survey*,[8] and the several station guides (Africa, India, Middle East, U.K.) offered by the British DX Club,[9] to name just a few. Some web sources are highly specialized, e.g. Mark Mohrmann's Latin American station lists,[10] "ClandestineRadio.com" and "Clandestine Radio Watch,"[11] "DXAsia,"[12] "Asiawaves"[13] and the "Pirate Radio Address List."[14] Those that made their mark in earlier years but are no longer available include the *Tropical Band List*, the FineWare SWBC Schedules, "International Listening Guide" (ILG), and "Jembatan DX" (Indonesian stations). Many DX bulletins and newsletters are archived on the internet and are electronically searchable. Official station lists of national telecommunications authorities in various countries (especially in Latin America) are often available on the internet as well, and while not always current, they sometimes contain useful information.

The internet has spawned numerous new tools and techniques to facilitate DXing.[15] The position of the sunset terminator, and the exact state of propagation, can be obtained

Although not computer-tethered like the software-defined receivers to come, the menus and sub-menus of the high-performance AOR AR7030 reminded users of how computer technology could impact traditional shortwave receiver design, even in 1996.

in real time. Although tape recorders (and minidisc recorders) remain in wide use, reception can be recorded directly to a computer hard disk or other memory device, and enhanced with available audio software. CDs or other electronic media rather than tape are now used for archiving audio, and CDs have replaced cassettes as the medium of choice for sending recordings with reception reports. One could even check one's frequency on dozens of receivers around the world that were remotely tuneable over the internet.

Because listeners around the world can be quickly mobilized to check reception in their areas, the internet is an important tool for identifying stations. A recording of the station might be posted in an electronic mailing list and the assistance of members sought. Groups of DXers sometimes share their experiences in real time via an internet relay chat. The programming of a tentatively identified station might be checked with the streaming audio that is available from an increasing number of stations. An interval signal might be compared against several on-line interval signal archives; likewise a national anthem. And DXers who want to brush up on their knowledge of Latin American music as an aid to identifying stations or writing better reception reports have an SWL website dedicated to that purpose.[16]

PROGRAM LISTENING

In addition to the extensive program and schedule information that can be found on individual station websites, databases available on the internet now aggregate this information in various ways. Current English-language schedules can be accessed on "Prime Time Shortwave"[17] and sorted by time, country, frequency and target area. The NASWA "WWW Shortwave Listening Guide"[18] goes even further. An electronic version of John Figliozzi's paper *Worldwide Shortwave Listening Guide,* one can search the Guide for particular program types, or obtain a list of the programs that are on the air at the current hour.

Since 1999, shortwave program listeners have also enjoyed a "swprograms" internet mailing list. The internet has also made it possible for some of the larger stations to archive individual programs for later listener access.

QSLING

Facilitating the preparation of reception reports through word processing was one the earliest SWL uses for computers. The internet brought additional benefits.

While reception reports are still sent, and replied to, by postal mail, many stations now also accept reception reports by e-mail. Some have established special forms on their websites for this purpose. Electronic mail is an attractive alternative, for while it does not allow for the sending of return postage — a standard listener courtesy when writing to smaller stations — it eliminates the need to produce and mail a letter and await its delivery, contact with the station being virtually immediate. Most pirate broadcasters announce e-mail addresses, and it is not unusual to hear an on-air greeting a few minutes after e-mailing a European pirate.

Reports by e-mail invite e-mail replies. Usually this takes the form of a return e-mail of standard appearance, which, while no less a verification, lacks the personality and collectibility of even the simplest card or letter received by postal mail. Some DXers print out e-mail verifications and decorate them with a logo or other graphic from a station's website in order to make them more presentable. Electronic replies from more DXer-savvy stations take the form of a QSL card or letter graphic file that is sent as an attachment and is more pleasing to the eye when printed out.

At least one website, "QSL Information Pages,"[19] tracks QSLs received by listeners worldwide, compiling the information so that a station's QSL practices can be discerned from the aggregation of the data. There are many sites that display images of QSLs. Other sites have facilitated various aspects of QSLing, such as pirate station addresses, Russian station addresses,[20] or Latin American QSLing.[21] Sites offering foreign language translation, national anthems, and the time at any point around the world can also be helpful, as can the U.S. Postal Service website.

CLUBS ET AL.

Nowhere have both the benefits and burdens of the internet been more evident than in the radio clubs.[22]

Word processing was the first aspect of computers from which clubs benefited. Editors produced better looking columns faster, and soon they were transmitting them to club headquarters electronically. Bulletin managers did their own electronic composition rather than rely on commercial printers.

Clubs and other organized groups were also the source of various on-line electronic initiatives, including electronic mailing lists and e-mail newsletters, e.g. *DX-Window* (DSWCI), *BC-DX* (Worldwide DX Club, Germany), and *Cumbre DX*. Virtually all clubs began accepting member contributions by e-mail.

In the 1990s, declines in club membership, together with ever-increasing printing and mailing costs, made the production of paper bulletins increasingly uneconomic. Some clubs found the answer in the internet. With bulletins already being produced electronically in many respects, it was a small step to provide the completed bulletin in electronic form. This eliminated printing and postage costs, provided immediate delivery, permitted the use of color graphics, and cut the cost of membership by as much as half. ODXA offered an electronic delivery option in 2000, dropping its print version altogether in 2006 (those without internet access received a photocopy of the bulletin). CIDX went to all-electronic delivery in 2005. Overseas clubs took similar steps. In 2000, DSWCI offered an alternative electronic membership, as did ARDXC and NZRDXL the following year. (The percentage of DSWCI members with internet access was 76 percent in 2000, twice the number in 1998.) In an interesting cooperative venture only possible through electronics, ODXA, ARDXC and NZRDXL provide an electronic member of any one club with the bulletins of all three.

But the internet also generated competition for the clubs. With so many internet resources available at no cost, clubs were becoming redundant. Nearly all information in club bulletins, paper or electronic, had appeared earlier on the internet. Often it had been regurgitated multiple times. Monthly bulletins featured less and less original content, and became archival vehicles rather than news sources, save for those who were not on-line.

In addition, the number of DX stations, and DX opportunities, had declined. The shortwave news was now more prosaic — loggings of common stations, schedules of larger ones, new leased time arrangements, time and frequency anomalies, etc. Clubs have struggled to provide unique and interesting bulletin content in a sea of unremarkable internet information.

NASWA, along with some popular foreign clubs such as the World DX Club and the British DX Club, have bucked the trend and continued to publish their bulletins on paper only, but they have supplemented their bulletins with electronic offerings of various kinds. NASWA restarted its weekly "Flash Sheet" in 2001, this time by e-mail, and established a NASWA electronic mailing list in 2007. WDXC and BDXC have like lists, and WDXC distributes some of its columns in advance to members with e-mail.

OTHER AREAS

There are websites on numerous other subjects related to shortwave broadcasting, including shortwave history,[23] Latin American shortwave,[24] shortwave-related technology,[25] receivers and other equipment, shortwave and public diplomacy,[26] even humor.[27] Industry groups,[28] virtually all vendors, and some individual DXers also have websites, and a few "shortwave malls" have ably provided coverage of multiple shortwave-related subjects.[29]

The internet revolution has been a huge boon to the listener, but it has had costs. While most shortwave listeners find computers a valuable complement to their shortwave listening, and the on-line world an interesting if sometimes confounding place, the computer can be a jealous mistress. For those who use them extensively at work or in other activities, computer fatigue can set in, and computers can demand time that might otherwise be spent listening. For those who enjoy computing, it can rival shortwave listening as an enjoyable pastime. Perhaps most significantly, some stations, like the BBC and Deutsche Welle, have replaced shortwave transmissions with an expanded online presence, particularly to target areas like the United States where internet access is widespread.

Listeners who did not make the transition to personal computing lost much of the ability to actively participate in the SWL community, which now meets mainly on-line, and to stay abreast of new activities, which are invariably announced first on the internet. Most breaking developments in the world of shortwave — new stations, special DX targets, one-time listening opportunities, etc.— are eventually chronicled in print, but sometimes after their real-time importance has been lost. For the serious shortwave listener, the computer and the internet have become indispensable.

It is ironic that at the same time that growth of the internet, along with innovation in software development and advances in receiver design, were facilitating shortwave listening in a plethora of ways, the number of shortwave stations, and shortwave listeners, was declining. Future radio historians may encounter another irony, namely, that at the same time that electronic archiving of huge amounts of data has become feasible, the information becomes less visible as archives on small, unlabeled media get lost or discarded and websites disappear. The information explosion notwithstanding, shortwave's internet era may well be its least-well remembered.

9

Conclusion

Come with me, away from the cares and worries of everyday living, on a trip of high adventure. Follow me down the electronic pathway of shortwave radio to far-off lands, to ships at sea, to explorers in the freezing outposts of civilization, and to radio amateurs conversing with the far corners of the earth. Hear the measured, stately tones of Big Ben in London marking the hours, or listen to the Kookaburra bird of Australia. Hear the satellites of Communism nervously defending themselves against the Voice of America. Tune across the airline channels, and the chattering radio teletypes pouring out the news of the day. Tarry awhile with a missionary in the Andes, speaking to another radio ham in Durban, South Africa, in cosmopolitan San Francisco, or in Dakar. You will find the world at your fingertips with a shortwave radio receiver![1]

1957

Such has been the promise of shortwave. The kernel of it was still true in 1995 when Ian McFarland observed that "what attracts many listeners to international broadcasting on the shortwave bands is the exotic nature of the experience, as well as the challenge of tuning in to radio from all those far off places."[2] It is still true today.

The future of shortwave has been the subject of increased discussion in recent years. Issues such as the impact of third-party gate keeping, the true reach of satellite and internet broadcasting vs. shortwave, the advantages of shortwave in times of emergency, DRM, etc., are debated, as are the causes of the changed shortwave environment — advances in technology, post–cold war broadcasting deregulation, budgetary pressures, the absence of a shortwave constituency, and so forth. Former Swiss Radio International personality Bob Zanotti has observed that decision-making on the future of shortwave is in the hands of theoreticians and technocrats rather than broadcasting professionals, and that has led to a denigration of shortwave and the premature promotion of alternatives that cannot serve the large audiences that are routinely within shortwave's reach.[3]

Whatever the future may bring, to listeners it is clear that shortwave broadcasting has been on the wane for some time. The activity on the shortwave bands, year by year, is detailed in the companion volume, *Broadcasting on the Short Waves: 1945 to Today*. The stations, large and small, that have dropped shortwave or gone to leased time, are enumerated there. By the NASWA countries list, there were 198 radio countries actively broadcasting on shortwave in 1945. That number grew by dozens during the decades after, but today stands at less than 190, erasing all the gain and then some.

The size and shape of the SWL community is ultimately a product of the state of shortwave broadcasting itself, and the changes in the broadcasting scene — increased power; jamming (and its near elimination); the expansion, then contraction, of tropical and international band broadcasting; and the leased-time revolution — all have had an obvious impact on shortwave listening. But the listening culture has been affected by other factors as well.

Improvements in receivers have made shortwave listening easier. Digital readout has made the bane of the shortwave listener — knowing the correct frequency — a non-issue. Today's receivers produce better reception, and high performance portable operation is now commonplace.

Listener knowledge about all aspects of shortwave broadcasting has increased exponentially since the war years, especially since the 1960s. It has manifested itself in many ways. Listeners have become much more attuned to differences in musical styles and languages. There has been a large increase in knowledge about propagation. There has also been much DXing specialization. The specialists have contributed much to the knowledge base in areas such as clandestine and pirate broadcasting, and shortwave broadcasting in particular parts of the world, e.g. Indonesia, Latin America, and East Asia. Listeners also know a great deal more about the inside of the shortwave broadcasting industry, as it has organized to confront political challenges and competition from other media.

Although they remain shortwave listening's *raison d'être* for those who take their shortwave seriously, the lure of DXing and QSLing has, in general, lessened. Interest in shortwave program content appears to have increased, the result, perhaps, of both improvements in the content itself and the broadening of interests (and the mellowing of competitive urges) among an increasingly "senior" SWL population.

The change in the shortwave listening community that has been most apparent is its diminution in size. In 1975 there were seven major shortwave clubs in North America. Today there are three. The membership of NASWA, the largest, and the only one in the United States, is only slightly more than a third what it once was. The drop in size of the two Canadian clubs has been even greater. Although there is much information about shortwave listening on the internet, professional writing on the subject is at a low ebb, further reducing the visibility of an already low-visibility medium. The number of affordable, high-performance receivers coming to market is much reduced, and few new listeners are taking up shortwave. Even in the fast-moving world of the internet, where communication is now easier than ever, the SWL fraternity is a comparatively small one.

Other factors, not directly related to the medium, also conspire against shortwave listening. One is the increasingly noisy RF environment. Once the main culprit was TV interference, easily defeatable by turning off the TV. Today, even shortwave listeners in rural areas know that practically every electronic device that they, or their neighbors, plug in seems to emit some level of RF noise, sometimes even while it is turned off. In this regard the consumer electronics revolution, while achieving major improvements in receiver performance, has in other ways been a disincentive to shortwave listening.

So have the loud, sweeping sounds of CODAR transmissions, the ocean current-mapping signals now common in the 60 meter band. Also, the realities of urban living have made the erection of outdoor antennas increasingly problematic.

Diehards may hope for a shortwave boom such as occurred in Japan during the 1970s, or a worldwide epiphany on shortwave broadcasting that gives it a place among the mainstream media. Even if DRM is successful, however, only an optimist would predict a major improvement in the overall shortwave broadcasting picture.

One of the newest and potentially greatest threats to shortwave listening is broadcasting over power lines (BPL), the sending of high speed broadband internet signals over the home electrical system. It is analogous to the once-popular home intercoms and carrier current broadcasting on college campuses, but with a much wider reach. It has the potential to solve the "last mile" problem — getting broadband signals to rural areas not easily served by cable or DSL.

Unfortunately, digital signals sent over unshielded power lines produce noise that blanks out virtually the entire shortwave spectrum. The interference is introduced either through the home electrical line; or through the receiver's antenna, picking up the radiations from power lines acting as giant transmitting antennas; or both. Virtually no shortwave signals would be able to override the strong, continuous interference of BPL.

There are many technological problems associated with BPL. Power lines are not always hospitable to digital transmission, and in the United States the configuration of powerline systems to accommodate BPL requires more new equipment than in many other countries. Some countries have disallowed BPL. The FCC supports it. The FCC rules on BPL do require that it not cause "harmful interference" to licensed users, that it not operate in certain critical public safety frequency ranges, that providers be able to reduce power and remotely notch out frequencies from BPL operation, that a BPL interference database be maintained, etc. However, these provisions are part of an interference-mitigation model rather than an interference-prevention model, leaving it up to the victim to take the steps necessary to solve his BPL problem.

Even the amateurs, over a half million strong in the United States, are not specifically protected under the FCC's BPL regulations, although many BPL operators routinely "notch" out ham frequencies. For SWLs and most other users of the shortwave spectrum, marshalling the facts, arguments and persistence necessary to successfully prosecute a complaint of BPL interference against the local power company would likely be daunting. And given the dynamic time-, target- and season-dependent nature of frequency selection in shortwave broadcasting, mitigation of interference on a channel-by-channel basis would be of no practical use. The notching of all the shortwave broadcasting bands would seem to be the only strategy to preserve an SWL–friendly listening environment. But even then, out-of-band stations would suffer interference, and, in any event, other radio services have sought such treatment during FCC BPL rule making, unsuccessfully.

The ARRL has been active in representing ham interests in the BPL debate. NASWA took a lead on behalf of SWLs when, on several occasions in 2003 and 2004, it submitted comments to the FCC on the impact of BPL on shortwave listening, urging that BPL be allowed only in the 30–47 MHz. range. The recommendation was not accepted.

Whether BPL will prove a major challenge remains to be seen. Some power companies have ended BPL experiments due to interference and other issues. Others have gone forward in buried-line neighborhoods, and some are offering limited BPL service, seemingly without significant complaints. Right now, the BPL user base is still very small.

Gloom about the state of DXing is nothing new. In 1957, well-known broadcast band

DXer David F. Thomas of Ohio declared that branch of DXing on the way out due to the increasing unwillingness of stations to verify reports.[4] On the shortwave side, in 1971, Perry Ferrell, editor of *Popular Electronics,* observed that shortwave listening had probably peaked in the late 1950s and early 1960s.[5] The actual peak was years later, but it is clear that we are now beyond it.

Organized shortwave listening today is not what it once was, but it is neither uninteresting nor without some hitherto unimaginable advantages. Thanks to the internet, the members of the SWL community know each other better, and are able to communicate faster and help each other in ways never before possible. New receivers are few, but better than ever. The depth of knowledge about stations is also greater than it has ever been, even to the point of now being able to "Google Earth" a station's transmitter site. Electronic communication with stations has replaced postal communication in many instances, permitting real-time reception reporting and electronic QSLing, and substituting the anticipation of an electronic message for the eagerness that preceded the postman's arrival.

The impact of the internet has been profound. Internet literacy is essential to a full enjoyment of shortwave listening today, and we will never know how many more SWLs there might have been had the world not gone digital. Whatever negative impact the internet had on clubs and other proprietary, paper-based information channels, and however much the same news now gets repeated endlessly around the internet, it is probably the economies of internet publishing that have saved organized shortwave listening.

By any definition, shortwave listeners are a unique breed. It should come as no surprise that getting enjoyment out of weak and crackling signals in a language you don't understand is not for everyone. As one observer noted after considering the unpredictable and unreliable nature of shortwave, "a lot of things we as hobbyists love about shortwave are a lot of things that the average public hates."[6]

The failure of most attempts to interest others in shortwave listening proves the point. In 1977, Victor Goonetilleke summed it up. "After 10 years of efforts to make DXers and SWLs out of the man on the street," he said, "we learned that DXers and SWLs are people who have it in their blood, and however much we try, it is useless unless you are born a radio enthusiast. Show them your QSLs, the gifts you have gotten, the prizes you have won, tell them the great thrills that SW brings—it's useless. Don't waste your time. But meet a guy who already listens to SW casually—ah, that's where to start!..."[7]

A few years later, a NASWA member told a story that many SWLs will recognize.

> [I]t was ... the spring of 1973. Radio Shack stores were appearing in the suburbs and I went to a local one and bought an "Astronaut 6" multi-band. Primarily it was for its excellent tonal quality, but also it was the VHF-Hi PSB [public service band] I am also interested in. I barely noticed that it also has SW. A few trips later to this store and I noticed a stack of *Elementary Electronics* magazines on the counter. Out of curiosity, I bought one to thumb through that rainy evening. Among its articles was one by Don Jensen and another on the world's international shortwave broadcasters' DX programs. Both features told when and where to tune on the SW band. The next morning, just for the heck of it, I turned to the SW band and tuned halfway between the 9 and 10 on the dial as one article had suggested. Immediately, Radio Australia boomed out of the speaker. A few hours later and a slight twist on the knob, and some man was stating: "This is the Voice of the Andes." I was hooked on shortwave from that moment on.[8]

Time will tell what the fate of shortwave broadcasting will be. One thing is for certain, however: the new technologies will never match the thrill of the old, for it is not just what you hear on shortwave that is important, but how it reaches you. A listener put it well. "[S]omething is lost when I'm listening online. It's too easy. Just about all the work has been done for me, and there is little to fire the imagination. From time to time, I still like to turn a dial, gently and patiently, in search of an ephemeral, transoceanic signal that may transport me to a place far, far away."[9]

Chapter Notes

Preface

1. *NNRC Bulletin,* June 1959, p. BCB.

Chapter 1

1. *Radio News,* January 1947, p. 56.
2. For a good description of the early days of short-wave broadcasting in the United States, see Steven P. Phipps, "The Commercial Development of Short Wave Radio in the United States, 1920–1926," *Historical Journal of Film, Radio and Television,* Vol. 11, No. 3 (1991), p. 215.
3. W.W. Rodgers, "Is Short-Wave Relaying a Step Toward National Broadcasting Stations?" *Radio Broadcast,* June 1923, p. 119.
4. Hugo Gernsback, "The Short-Wave Era," *Radio News,* September 1928, p. 201.
5. Hugo Gernsback, "The Short-Wave Fan," *Radio News,* February 1929, p. 715.
6. C.W. Horn, "Is International Broadcasting Just Around the Corner?" *Radio News,* January 1930, p. 608; see also Lt. H. F. Breckel, "In the Future — Intermediate- or Short-Wave Broadcasting?" *Radio News,* November 1927, p. 461.
7. Charles A. Morrison, "The Future of International Short-Wave Reception, Part III," *Radio News,* June 1935, pp. 745, 780.
8. Arthur J. Green, "Short-Wave Broadcast Listening Grows," *Radio Craft,* March 1931, p. 542.
9. H.P. Davis, "Short-Wave Broadcast Pioneering," *Radio Craft,* March 1931, pp. 543, 566.
10. J.B.L. Hinds, "J.B.L. Hinds Picks the Ten Best Foreign Short-Wave Stations," *Short Wave Radio,* July 1934, p. 10.
11. M. Harvey Gernsback, "Best Heard S-W Stations," *How to Get Best Short Wave Reception* (New York, NY: Short Wave Craft, 1935), p. 45.
12. Hugo Gernsback "The Short Wave Industry" [editorial], *Radio Craft,* January 1936, p. 388B.
13. Charles A. Morrison, "The Future of International Short-Wave Reception, Part I," *Radio News,* April 1935, pp. 598, 599; see also Part II, *Radio News,* May 1935, p. 674; Part III, *Radio News,* June 1935, p. 745; and Part IV, *Radio News,* July 1935, p. 27.
14. Dr. Frank Conrad, "Short Wave Broadcasting — As a Pioneer Sees It," *Short Wave & Television,* August 1938, pp. 197, 256.

15. F. Barrows Colton, "Winged Words— New Weapon of War," *National Geographic,* November 1942, p. 663.
16. *NNRC Bulletin,* July 1, 1943, p. 2.
17. *NNRC Bulletin,* September 1, 1943, p. 2.
18. *International Short Wave Radio,* April 1941, pp. 5–6.
19. *NNRC Bulletin,* January 1, 1944, p. 2.
20. *NNRC Bulletin,* January 1, 1944, p. 4.
21. *International Short Wave Radio,* August 1941, p. 5.
22. *NNRC Bulletin,* March 1, 1945, p. 11. Young Russell Kent French was then five months old.
23. *NNRC Bulletin,* July 1, 1947, p. 1.
24. *NNRC Bulletin,* August 1, 1943, p. 5.
25. *NNRC Bulletin,* December 1, 1943, p. 8.
26. *Radio & Television News,* August 1951, p. 56. For a detailed look at how the New Zealand messaging service worked, see Frank Glen, "A Passion with a Purpose — The Prisoner of War Message Service, 1951–1952," *New Zealand DX Times,* September 2003, p. 39. In 1953, Arthur T. Cushen was awarded the Coronation Medal by Queen Elizabeth II for the POW work he had done over the years. *Radio & Television News,* October 1953, p. 140.
27. *Radio News,* November 1945, p. 45.
28. *NNRC Bulletin,* October 1, 1945, p. 1.

Chapter 2

1. "The Shortwave Center Interview: John Campbell," *FRENDX,* February 1985, p. 1.
2. Simon Spanswick, in Elzbieta Olechowska and Howard Aster, eds., *Challenges for International Broadcasting V — New Tools, New Skills, New Horizons* (Oakville, Ontario: Mosaic, 1999), p. 182.
3. *International Short Wave Radio,* October 1941, p. 4–5.
4. *RADEX,* April 1939, p. 10, republished in Kenneth R. Boord, *The World at a Twirl* (Morgantown, WV: Kenneth R. Boord, 1956), p. 2, and *FRENDX,* August 1964, p. SWBC-2. Fort de France (Martinique), SPD (Poland), COCQ (Cuba), Daventry (U.K.), JZJ (Japan), HH3W (Haiti), Delhi and Bombay (India), ZIK2 (British Honduras) were shortwave stations of the day. QRM is interference; VAC is "Verified All Continents."
5. Oliver "Perry" Ferrell, "The SWL DXer — What Others Think of Us," *SPEEDX,* presented in seven parts, October 1971–January 1972 and March–May 1972.
6. Harry L. Helms, South Carolina, writing in *Review of International Broadcasting,* July 1978, p. 21.

7. Robert Silvey, *Who's Listening? The Story of BBC Audience Research* (London, England: George Allen & Unwin, 1974).

8. Kim Andrew Elliott, "An Overview of International Broadcasting Audience Research," Broadcast Education Association, 2002; and Oliver Zöllner, "International Broadcasters, Audience Research and a Conference: An Overview of Methods and Functions," and Graham Mytton, "Audience Research for International Radio Broadcasters: A Toolkit for Small to Medium-Sized Stations," both in Oliver Zöllner, ed., *An Essential Link with Audiences Worldwide — Research for International Broadcasting* (Cologne, Germany: Deutsche Welle/VISTAS, 2002), p. 13 and 25 respectively. For samples of surveys conducted for the BBC, and discussion of shortwave audience research in general, see Bernard Bumpus, "Broadcasting and Audience Research in the Middle East," *Bulletin (British Society for Middle Eastern Studies),* Vol. 6, No. 9 (1979), p. 13; Graham Mytton, "BBC Audience Research," *The Shortwave Book* (Media, PA: Lawrence Miller, 1984); Graham L. Mytton, "Keeping Tabs On the BBC's 100 Million Listeners," *1987 Radio Database International,* p. 13; R. Eugene Parta, "A New Generation of Audience Research At RFE-RL," presented at the 43rd Annual Conference of the International Communication Association, Washington, D.C., May 27, 1993; Graham Mytton, ed., *Global Audiences: Research for Worldwide Broadcasting — 1993* (London, England: John Libbey, 1993); and Graham Mytton, "New Technology and International Broadcasting," *1996 WRTH,* p. 599. Mytton provides a detailed description of audience research techniques in Graham Mytton, *Handbook on Radio and Television Audience Research* (London, England: UNICEF, UNESCO & BBC World Service Training Trust, 2d ed., 1999). The first edition was published by UNICEF-UNESCO in Paris in 1993 and is available online at unesdoc.unesco.org/images/0012/001242/124231Eo.pdf.

9. Colin M. Wilding, "How Do Surveys Deal With Multiple Forms of Access?" and Mark Rhodes, "Multi-Media Measurement — Measuring a Broadcaster's Total Impact," both in Oliver Zöllner, ed., *Reaching Audiences Worldwide — Perspectives of International Broadcasting and Audience Research, 2001/2002* (Bonn, Germany: CIBAR, 2003), p. 34 and 39 respectively.

10. Graham Mytton, "But Don't Overlook Shortwave," in Zöllner, *Reaching Audiences Worldwide, supra* note 9 at p. 97.

11. Colin M. Wilding, "151 Million Listeners — But What Does It Mean?" and Hélène Robillard-Frayne, "The Process of Estimating the Global Audience at Radio Canada International," both in Zöllner, *An Essential Link, supra* note 8 at p. 61 and 71 respectively.

12. Gene Parta, "Mass Audiences, Elite Audiences and Target Audiences," in Zöllner, *An Essential Link, supra* note 8 at p. 85. Students, educators and government officials have been identified as "better-than-average potential listeners." "Few other groups seem to be of much importance." Don R. Browne, "The Limits of the Limitless Medium — International Broadcasting," *Journalism Quarterly,* Vol. 42, No. 2 (Winter 1965), p. 82.

13. Graham Mytton, "Audience Research For Shortwave Broadcasters," presented at the NASB Annual Meeting, Washington, D.C., May 5, 2000, available at www.shortwave.org/Audience/Audience.htm.

14. Mytton, *supra* note 10.

15. Mytton, *supra* note 13. For discussion of the development of radio and other mass communications vehicles in Africa, see Graham Mytton, *Mass Communication in Africa* (London, England: Edward Arnold, 1983).

16. Graham Mytton & Carol Forrester, "Audiences for International Radio Broadcasts," *European Journal of Communication,* Vol. 3, No. 4 (1988), p. 457.

17. Leonard Carlton, "Voice of America: The Overseas Radio Bureau," *Public Opinion Quarterly,* Vol. 7, No. 1 (Spring 1943), pp. 46–54. For a discussion of the potential and pitfalls of VOA listener mail surveys, see Herta Herzog, "Listener Mail to the Voice of America," *Public Opinion Quarterly,* Vol. 16, No. 4 [Special Issue on International Communications Research] (Winter 1952-1953), p. 607.

18. See Peter Herrmann, "The World-Wide Audience and Their Attitudes," *How to Listen to the World, 8th ed.* (Hvidovre, Denmark: World Radio-TV Handbook, 1974), p. 137, and Peter Herrmann, "Audience and Their Attitudes," *World DX Guide* (London, England: Billboard, 1978), p. 176 (same article).

19. Donald R. Browne, *International Broadcasting: The Limits of the Limitless Medium* (New York, NY: Praeger, 1982), p. 234.

20. *NNRC Bulletin,* April 15, 1942, p. 7.

21. H. Charles Woodruff, *Short-Wave Listener's Guide* (Indianapolis, IN: Howard W. Sams, 4th ed. 1970).

22. Edgar T. Martin & George Jacobs, "Shortwave Broadcasting in the 1970s," *How to Listen to the World 1971* (Hvidovre, Denmark: World Radio-TV Handbook, 1971), p. 4.

23. James L. Hall, *Radio Canada International — Voice of a Middle Power* (Lansing, MI: Michigan State University Press, 1997), p. 167; and *Shortwave Radio Listening in the United States* (Gallup Organization, 1973), referenced in *Report of the Radio Canada International Task Force,* App. H (Montreal, Quebec: CBC, 1973).

24. Kim Andrew Elliott, *International Radio Broadcast Listening in the United States: A Survey of the North American Short Wave Association* (Minneapolis, MN: University of Minnesota, 1977) p. 51 [master's thesis]; *Review of International Broadcasting,* May 1978, p. 26.

25. *ANARC Newsletter,* May 1980, p. 7.

26. Oliver P. Ferrell, "Twilight of the Shortwave Listener," *Popular Electronics,* March 1971, p. 7.

27. Kim Andrew Elliott, in *Review of International Broadcasting,* No. 135 (1990), p. 9.

28. Childs was the founder and first editor of *Public Opinion Quarterly,* and, according to George Gallup in 1972, "the greatest scholar in the general field of public opinion." John B. Whitton, "In Memorium: Harwood L. Childs, 1898–1972," *Public Opinion Quarterly,* Vol. 36, No. 3 (Autumn 1972), p. 475. Childs helped found the Princeton Listening Center, a joint project of Princeton University and the Rockefeller Foundation. The center, which operated from 1938 to 1941, was devoted to the monitoring and analysis of wartime shortwave propaganda broadcasts. It was succeeded by the FCC's Foreign Broadcast Monitoring Service. For more details on the work of the Princeton Listening Center, see Harwood L. Childs & John B. Whitton, eds., *Propaganda by Shortwave* (Princeton, NJ: Princeton University Press, 1942).

29. Harwood L. Childs, "Short-Wave Listening in the United States," *Public Opinion Quarterly,* Vol. 5, No. 2 (June 1941), pp. 210–226. These studies, as well as several others, are more fully described and analyzed in Harwood L. Childs, "America's Short-Wave Audience," *Propaganda by Shortwave, supra* note 28 at p. 303.

30. While broad conclusions can often be reached through sampling, it should be kept in mind that the raw number of shortwave listeners that was identified through these surveys was always very small. The raw number of "shortwave listeners" in the two studies reviewed by Childs where the figures are reported (albeit defined somewhat differently in each study), were: in the American Institute

of Public Opinion nationwide study, 337, a figure Childs thought suspiciously high; and in the Baton Rouge study, 87. In the 1970 Smith study, out of an original sample of 2,126 persons, Smith found only 62 "international political broadcast listeners," i.e. listeners who were not hams or "shortwave listeners" and did not listen principally as a result of their connection with a particular country.

31. Edgar A. Schuler & Wayne C. Eubank, "Sampling Listener Reactions to Short-Wave Broadcasts," *Public Opinion Quarterly,* Vol. 5, No. 2 (June 1941), pp. 260, 266.

32. Don Smith, "Is There a U.S. Audience for International Broadcasts?" *Journalism Quarterly,* Vol. 39 (1962), p. 86.

33. Jerome S. Bruner & Jeanette Sayre, "Shortwave Listening in an Italian Community," *Public Opinion Quarterly,* Vol. 5, No. 4 (Winter 1941), p. 640.

34. Bruner & Sayer, *supra* note 33 at p. 651.

35. Don D. Smith, "America's Short-Wave Audience: Twenty-Five Years Later," *Public Opinion Quarterly,* Vol. 33, No. 4 (Winter 1969-1970), p. 537, and Don D. Smith, "The U.S. Audience for International Broadcasts," *Journalism Quarterly,* Vol. 47 (1970), p. 364.

36. James L. Hall & Drew O. McDaniel, "The Regular Shortwave Listener in the U.S.," *Journal of Broadcasting,* Vol. 19, No. 3 (Summer 1975), p. 363. Although not mentioned in the article, the research for this study was apparently sponsored by Radio Canada International. See James L. Hall, *Radio Canada International — Voice of a Middle Power, supra* note 23 at p. 167.

37. *CQ,* February 1957, p. 104.

38. *NNRC Bulletin,* July 1957, p. SWBC-1.

39. Wayne Green, "de W2NSD," *CQ,* August 1959, p. 12. In his book, *The Fascinating World of Radio Communications,* published in 1971, Green covered ham radio, CB, utilities, the broadcast band, TV and FM DXing and the tropical bands, but not international shortwave broadcasting.

40. Green, *supra* note 39 at p. 14.

41. *QST,* October 1978, p. 47.

42. Lawrence Magne, "The S.R.I. Report Debunked," *Review of International Broadcasting,* November 1978, p. 3; "WARC 79," *FRENDX,* January 1979, p. 2.

43. Robert J. Halprin, "Rediscovering Shortwave Listening," *QST,* May 1991, p. 52.

44. *Short Wave News* [ISWL], May 1949, p. 105; July 1949, p. 162.

45. Figures are not always comparable from survey to survey due to differences in methodologies. The small number of participants in some surveys may also contribute to unrepresentative results. The indicated number of participants ("No.") is the number who participated in the survey as a whole, and this may not be the same as the number who indicated their station preferences. There were some 30,000 "votes" cast in some of the ISWC polls. However, since each participant had multiple votes—for example, five for first place, four for second, etc.—the number of votes was many times larger than the number of actual poll participants. The sources for the data in the table, in chronological order, are: 1950—*Radio & Television News,* September 1950, p. 52; 1953—*Radio & Hobbies* [Australia], October 1953, p. 98; 1956—*International Short Wave Radio,* February 1956, p. 1; 1959—*International Short Wave Radio,* February 1959, p. 1; 1962—*International Short Wave Radio,* February 1962, p. 1; 1965—*International Short Wave Radio,* February 1965, p. 1; 1968—*SWL,* June 1968, p. 7; 1971—*International Short Wave Radio,* April 1976, p. 1; 1974—*International Short Wave Radio,* April 1976, p. 1; 1976—*Complete Results of the 1976 Survey of the North American Shortwave Association* (Liberty, IN: NASWA,

1977 [NASWA Special Pub. 4], App. G, and *FRENDX,* September 1976, p. 27; 1977—*International Short Wave Radio,* May 7, 1977, p. 1; 1979—"R.I.B. Poll Results," *Review of International Broadcasting,* June 1979, p. 6; 1980—"RIB Awards For Excellence," *Review of International Broadcasting,* July 1980, p. 2; 1982—"Results of the 1982 R.I.B. Survey," *Review of International Broadcasting,* July 1982, p. 2; 1983—"1983 R.I.B. Survey Results," *Review of International Broadcasting,* August 1983, p. 3; 1983—"1983 CIDX Survey Results," *Messenger,* November 1983, p. 5; 1984—"1984 R.I.B. Survey Results," *Review of International Broadcasting,* May 1984, p. 3; 1984—*SPEEDX,* November 1984, p. 3; 1985—"1985 Review of International Broadcasting Survey," *Review of International Broadcasting,* No. 106 (1986), p. 3; 1986—*ANDEX,* October-November 1986; 1987—"'Jet Sets,' SWLing Gain in Popularity," *SWL,* August 1987, p. 42; 1989—"1989 CIDX Survey Results," *Messenger,* May 1989, p. 6; 1992—"The 1992 CIDX Survey Results," *Messenger,* September 1992, p. 5; 1994—"1994 Winter Shortwave Fest First Annual Listener Profile," *NASWA Journal,* January 1995, p. 20; 2000—*Messenger,* March 2000, p. 57; 2001—*Messenger,* March 2001, p. 53; 2004—*NASB Newsletter,* April 2004 (survey conducted at the Winter SWL Fest).

46. Similar results were reported in a 1975 study of the NNRC. Hall & McDaniel, *supra* note 36 at p. 363.

47. "I.S.W.C. and Radio Portugal — Storm in a Golden Caravel," *FRENDX,* August 1968, p. 18.

48. "Radio Netherlands Miscellanea," *DX Reporter,* July 1984.

49. *Voices,* February 1981, p. 3.

50. Browne, *supra* note 19 at p. 10.

51. Don D. Smith, "Radio Moscow's North American Broadcasts: An Exploratory Study," *Journalism Quarterly,* Vol. 42, No. 4 (Autumn 1965), p. 643; Don D. Smith, "Some Effects of Radio Moscow's North American Broadcasts," *Public Opinion Quarterly,* Vol. 34, No. 1 (Winter 1970-71), p. 539; Liqing Zhang & Joseph R. Dominick, "Penetrating the Great Wall: The Ideological Impact of Voice of America Newscasts on Young Chinese Intellectuals of the 1980s," *Journal of Radio Studies,* Vol. 5, No. 1 (1998), p. 82.

52. Much has been written on this subject. See, e.g., Charles J. Rolo, *Radio Goes to War* (New York, NY: G. P. Putnam's Sons, 1940); Harwood L. Childs & John B. Whitton, eds., *Propaganda by Shortwave, supra* note 28; Harold Ettlinger, *Axis on the Air* (New York, NY: Bobbs-Merrill, 1943); E. Tangye Lean, *Voices in the Darkness* (London, England: Secker & Warburg, 1943); L.D. Meo, *Japan's Radio War on Australia, 1941–1945* (Melbourne, Australia: Melbourne University Press, 1968); M.A. Doherty, *Nazi Wireless Propaganda* (Edinburgh, Scotland: Edinburgh University Press, 2000); Jane M.J. Robbins, *Tokyo Calling: Japanese Overseas Radio Broadcasting, 1937–1945* (Firenze, Italy: European Press Academic Publishing, 2001).

53. James H. Oliver, "A Comparison of Four Western Russian-Language Broadcasters," *Journalism Quarterly,* Vol. 54, No. 1 (Spring 1977), p. 126 [VOA, BBC, Radio Liberty, Deutsche Welle]; Nien-sheng Lin and Michael Presnell, "Shortwave News: An Exploratory Study," *Journal of Broadcasting and Electronic Media,* Vol. 26, No. 3 (Summer 1982), p. 717 [VOA, BBC, Radio Moscow, Radio Peking]; Donald R. Browne, "The International Newsroom: A Study of Practices at the Voice of America, BBC and Deutsche Welle," *Journal of Broadcasting,* Vol. 27, No. 3 (Summer 1983), p. 205; Philo C. Wasburn, "Voice of America and Radio Moscow Newscasts to the Third World," *Journal of Broadcasting and Electronic Media,* Vol. 32, No. 2 (Spring 1988), p. 197; Natalie Elizabeth Doyle-Hennin, *The World According to International Radio* (Buf-

falo, NY: State University of New York at Buffalo, 1991) [Ph.D. dissertation]; and Ajit K. Daniel, *Voice of Nations—A Comparative Analysis of Three International Radio News Editorial Orientations: Voice of America, All India Radio and British Broadcasting Corporation* (Cincinnati, OH: Union Institute, International Communication, 1995) [Ph.D. dissertation]. See also Joseph T. Klapper, "Content Analysis for the Voice of America: A Symposium," *Public Opinion Quarterly*, Vol. 16, No. 4 (Winter 1952), p. 605. This symposium includes Herta Herzog, "Listener Mail to the Voice of America," p. 607; Alex Inkeles, "Soviet Reactions to the Voice of America," p. 612; Paul W. Massing, "Communist References to the Voice of America," p. 618; Marie Jahoda and Joseph T. Klapper, "From Social Book Keeping to Social Research," p. 623; and Siegfried Kracauer, "The Challenge of Qualitative Content Analysis," p. 631. An interesting exercise in producing student radio "programs" based on foreign shortwave news reports is described in Pat Cranston, "Project 'Listening Post': An Experiment in Education for Broadcasting," *Journal of Broadcasting*, Vol. 2, No. 2 (Spring 1958), p. 149.

54. Harry Helms, "Are Shortwave Broadcasters Failing Listeners?" *Monitoring Times*, November 2002, p. 92, quoted with the permission of *MT*.

55. E-mail from Jonathan Marks to the author, November 25, 2007.

56. Kim Andrew Elliott, "An Alternative Approach to International Broadcast Programming," *Review of International Broadcasting*, January 1980, p. 3; Kim Andrew Elliott, "The Meaningless Messages of International Radio," *Review of International Broadcasting*, August 1980, p. 3; and Kim Andrew Elliott, "An Alternative Programming Prototype," *Review of International Broadcasting*, October 1981, p. 5. Reader discussions of Elliott's ideas appeared in other editions of *RIB*: February 1980, p. 4; March 1980, p. 22; and November 1980, p. 8. Elliott's ideas were the product of research for his Ph.D. dissertation, *An Alternative Programming Strategy for International Radio Broadcasting* (Minneapolis, MN: University of Minnesota, 1979).

Chapter 3

1. Eugene C. Bataille, "The Birth of the Newark News Radio Club," *NNRC Bulletin*, December 1975, p. 3.

2. Irving R. Potts, "The Newark News Radio Club," *Keller's Radio Call Book and Log*, January-February 1934, p. 23.

3. James L. Hall and Drew O. McDaniel, "The Regular Shortwave Listener in the U.S.," *Journal of Broadcasting*, Vol. 19, No. 3 (Summer 1975).

4. *NNRC Bulletin*, January 1965, p. 1.

5. Biddle Arthurs, Pittsburgh, Pennsylvania, writing in the *Universalite*, August 1, 1956, p. 3.

6. *NNRC Bulletin*, March 1, 1942, p. 1; March 31, 1942, p. 1. Courtesy programs resumed after the war. The NNRC arranged 266 such programs in the 1949-50 DX season.

7. *NNRC Bulletin*, June 1, 1947, p. 36.

8. *NNRC Bulletin*, June 1, 1944, p. 1.

9. Technically, the editorship was held jointly by Hank and his then-recent bride, Amelia, although Hank was the face of the team in most respects. Amelia resigned the editorship in 1955.

10. *NNRC Bulletin*, August 1950, p. 1.

11. *NNRC Bulletin*, May 1953, p. 2.

12. For a lengthy discussion of the pros and cons of SWL-card swapping, see *NNRC Bulletin*, April 1962, p. SW-3A; May 1962, p. SW-2A; and June 1962, p. SW-3A.

13. *NNRC Bulletin*, July 1958, p. 4.

14. Actually Potts was the club's second president. He had held the post since 1928 when he succeeded the NNRC's first leader, L.S.J. Cranse, who had been president from 1927 to 1928. Cranse died in 1969.

15. *NNRC Bulletin*, February 1981, p. 3.

16. *NNRC Bulletin*, April 1982, p. 3.

17. "The Jack R. Poppele Transmitting Station of the Voice of America," *Proceedings of the Radio Club of America*, November 1992, p. 4.

18. For a summary history of the club's first 25 years, see Kenneth V. Zichi, "A Look Back at 25 Years of Service," *SWL*, December 1984, p. 4.

19. In actuality, his shortwave colleagues saw him at his least eccentric. See C.M. Stanbury, *Anti-Matter* (Paradise, CA: Dustbooks, 1977).

20. These included John Fischer, Jr., of Glenolden, Pennsylvania; Harold Frodge, Midland, Michigan; Arthur S. Getzel, Flushing, New York; Chris Hansen, New York, New York; John L. Kolb, San Diego, California; Sandra Manning, Garland, Texas; Spence Naylor, Ventura, California; Lani Pettit, Sioux City, Iowa; Woodrow W. Seymour, Jr., Sanford, North Carolina; Jim Thornton, Venice, California; Richard Varron, Wayne, New Jersey; Kenneth V. Zichi of Ann Arbor Michigan; and others.

21. In April 1978 it was separated into "DX Showcase" and "SW Review," both of which continued to be edited by Kenneth V. Zichi of Ann Arbor, Michigan, who had edited "DX-Showcase/Shortwave Review" since May 1977. "DX Showcase" continued to emphasize schedules and miscellaneous items, while "SW Review" focused on programming. Responsibility for these two columns was shifted to other editors in December 1986. (The two were combined and reduced in size in September 1987, and in May 1988 the resulting column became simply "Shortwave Review," with an emphasis on schedules.)

22. *ANARC Newsletter*, January 1989, p. 8; *Monitoring Times*, March 1989, p. 5.

23. The history of NASWA through 1990 was recounted in Don Jensen, *The History of the North American Shortwave Association* (NASWA Editorial Committee, 1990), a monograph which revised slightly a three-part series that appeared in the "Shortwave Center" column of the club's bulletin, *FRENDX*, in June, August and November 1981. There are two other NASWA histories: Richard A. D'Angelo, "An Early History of NASWA," *NASWA Journal*, July 1999, p. 13 (written from material compiled by Jensen); and Don Jensen, "40 Years of the North American Shortwave Association," *NASWA Journal*, September 2001, p. 13, in which Jensen updated his earlier work.

24. Pike became interested in ham radio, and in the years after 1963 he operated as VO1EH, VE1EJ and VO1GW. He died in 1994. See Jerry Berg, "Sterling Pike, Lost and Found," *NASWA Journal*, September 2006, p. 19.

25. Richard Roll was featured in one of Hank Bennett's *Popular Electronics* columns in 1960. See Hank Bennett, "Short-Wave Report," *Popular Electronics*, December 1960, p. 80.

26. *FRENDX*, October 1964, p. HDQ-1.

27. The plaque was to be in the shape of an eagle, 33 inches wide by 15½ inches high, finished in antique gold. The Canadian DX Club made a similar award for the top scorer in Canada. An American chapter of the U.K.-based Certificate Hunters Club was established in 1965. The CHC was a club for hams and SWLs who worked for various ham band certificates which were qualified for CHC credit. CHC had a ham division and an SWL division, hence the common abbreviation "SWL-CHC." The awarding of the

CHC award was assumed by the NNRC when NASWA eliminated ham band coverage in 1966.

28. Participation in the scoreboard dropped somewhat after August 1968 when the list was divided by continent and when member information on stations, in addition to countries, was sought. However, it was back up to 156 by June 1969.

29. *NNRC Bulletin,* September 1967, p. SWBC-5.

30. "Mail Call," a Shaw column which answered members' questions, was added to the bulletin in June 1972. Later it was called "Mailbag." Another column featuring letters from members was "QSO/NASWA," begun in July 1976.

31. *FRENDX,* September 1963, p. Hdq-3.

32. Once on the NASWA list no country could be deleted, unlike other country lists, which usually were governed by fixed dates and fixed political and geographical criteria. Countries that may have ceased to exist as political entities remained as distinct targets as long as there was a SWBC transmitter within their boundaries.

33. Hauser had introduced club members to harmonics in three *FRENDX* articles in 1969. Glenn Hauser, "How to Hear Harmonics," *FRENDX,* January and February 1969; and Glenn Hauser, "Broadcast Harmonics Monitored," *FRENDX,* March 1969.

34. *FRENDX* had focused on English-language broadcasts before. In December 1973, a comprehensive country-by-country list of English programs to North America, compiled by Scott Reeves, was published, followed by similar listings, compiled by Dan Ferguson, in July 1974 and January 1975. From July 1975, when Glenn Hauser assumed the editing of "Listener's Notebook," to December 1980, when he adopted a different format, he denoted a separate section for English-language transmissions. Ferguson resumed the separate publication of a list of English-language broadcasts heard in North America in 1977. This list was eventually produced with the aid of a computer, and, called the *Guide to English Shortwave Broadcasts,* was offered on a six-issue annual subscription basis from 1981 to 1986.

35. Subsequent recipients were: in 2001, Sam Barto, the club's QSL editor since 1978; 2002, member Ed Mauger of Earlville, Pennsylvania; 2003, Wallace C. Treibel of Shoreline, Washington, "Log Report" co-editor since 1984; 2004, Joe Buch of Virginia Beach, Virginia, "Technical Topics" editor since 1992; 2005, Sheryl Paszkiewicz of Manitowoc, Wisconsin, who had been a "Log Report" co-editor for 14 years and the "Flash Sheet" editor at its electronic rebirth in 2001; 2006, member Scott Barbour of Intervale, New Hampshire; and 2007, member Marlin A. Field of Hillsdale, Michigan.

36. *CADEX,* June 1963, p. 7.

37. The part of this discussion covering the 1960s and the 1970s is based on a more detailed history of the club from 1962 to 1987 which was published in installments in the Canadian International DX Club bulletin, the *Messenger,* during 1987, the club's 25th anniversary year. The material in that history covering 1962 to 1972 was authored by R. Lorne Jennings and was reprinted from the July 1972 *Messenger,* where it originally appeared. The material covering 1972 to 1987 was written by Mickey W. Delmage, presently a member of the club's board of directors.

38. Loggings were listed by frequency. Starting in 1984 there was also a monthly time index in each loggings column. The eastern section carried it continuously, but it appeared only occasionally in the western section until it was reinstated in January 1998 when Bob Poirier took over the section. It appeared throughout Poirier's term as editor. In January 1994, loggings from members outside North America were placed in a separate "International Logbook"

column which was edited at first by Harvey, and then by Patrick Perreault of St. Hubert, Quebec. "International Logbook" continued until June 1999, by which time the number of international logs was not large enough to warrant separate treatment.

39. Initially he expanded the list to include specific programs, but he dropped the list altogether after September 1997 because, by then, like information could be found on the internet. He added more personal commentary on shortwave programming, however, and "Prime Time SW" continued to be one of the best pieces of original writing in the *Messenger.* The column returned to a more traditional "schedules" format when Hankison relinquished it to Sheldon Harvey in November 1998. It changed format again in April 1999 when Daniel Sampson of Arcadia, Wisconsin, assumed the editorial duties. Sampson changed the focus to English broadcasts, a topic he knew well through his "Prime Time Shortwave" website, which tracked such broadcasts on an ongoing basis. The column was essentially a printed version of the website database.

40. For more details on this project, see Bill Westenhaver, "Behind the Scenes at the QSL Verification Department," *Messenger,* December 1992, p. 19.

41. There were three board of director positions and an executive secretary position that were subject to election by the membership. Seldom did members place more names in nomination than there were existing incumbents. As a result, the same members tended to be re-elected. Those elected would then decide who would be president and executive secretary. Sheldon Harvey was designated president continuously from 1986 on. The position of executive secretary was replaced by a fourth board position in 2004.

42. Stephen Canney, "CFRB/CFRX: ODXA's Role as QSL Manager," *DX Ontario,* February 1994, p. 72.

43. Tom Williamson, *Across Time — and Space: Listening for Sixty Years from Four Continents* (distributed by ODXA, 1998).

44. *Listening In,* September 2005, p. 2.

45. Because pirate radio broadcasting is illegal, from time to time radio publications and other groups had to consider the propriety of covering the pirate radio scene. At one point the European DX Council opposed giving coverage to pirates. A *Monitoring Times* plan to assist the FCC in spectrum enforcement through a network of registered monitors who would help identify intruders, including pirate stations, was scrapped following the objections of supporters of pirate radio, a topic to which *MT* was already devoting a column. *MT* said it dropped the project due to lack of support in the listening community. See *Monitoring Times,* July 1985, p. 2; *The ACE,* October 1985, p. 8.

46. *The ACE,* March 1984, p. 1.

47. http://membres.lycos.fr/worldbandradio/.

48. DX Inter-Nationale was absorbed by the Canadian DX Club in 1967.

49. The DX Unlimited Shortwave Club later merged with the Five Star DX Club.

50. Folcroft Radio Club merged with DX Inter-Nationale in June 1964.

51. The Midwest DX Shortwave Radio Club was intended for DXers in the "9" call area.

52. United World DXers Club merged into DX Inter-Nationale in January 1967.

53. In 1982, the Atlantic States DX Association, the Southern DX Association, and the Association of Illinois DXers merged to become the United States DX Association. The USDXA ceased operation the same year.

54. The Brooklyn DX Club was founded in 1975. It be-

came the Atlantic States DX Association in 1979. See also *supra* note 53.

55. Formerly the South Jersey DX Association.

56. The Middle Tennessee DX Association was renamed the Southern DX Association in 1980. See also *supra* note 53.

57. Riverside Radio Club merged with SWL International in 1973.

58. The Rocky Mountain DXers Association merged with the Batesville Shortwave Association in May 1972.

59. See *supra* note 53.

60. See *supra* note 53.

61. For a detailed history of ANARC's early years, see Richard A. D'Angelo, "35 Years Ago — The Beginnings: A Retrospective on ANARC's Early Years," *NASWA Journal*, Pt. 1, August 1999, p. 10; Pt. 2, October 1999, p. 11; and Pt. 3, December 1999, p. 10. For a more general description of ANARC and its European counterpart, see Richard A. D'Angelo, "What is an ANARC or an EDXC?" *NASWA Journal*, March 2001, p. 10.

62. Those persons who served as ANARC executive secretary are: Donald N. Jensen, 1964–66; Gerry L. Dexter, 1966–68; Gray Scrimgeour, 1968–70; Wendel Craighead, 1970–72; Allen H. (Al) Reynolds, 1972–74; David B. Browne, 1974–82; Richard T. (Terry) Colgan, 1982–86; Don Hosmer, 1986–87; Robert Horvitz, 1987–90; Sheldon Harvey, 1990–91; Richard A. D'Angelo, 1991–96; Mark Meece, 1997–2004; and Harold Cones, 2004–05.

63. See Jan Tunér, "The DX-Alliansen — A National Union of Swedish Clubs," *How to Listen to the World, 1965/66* (Hellerup, Denmark: World Pub., 1965), p. 162.

64. Hank Bennett, "DX-Alliansen," *Popular Electronics,* October 1962, p. 83; Hank Bennett, "North American Alliance of SWL Clubs," *Popular Electronics,* October 1963, p. 77.

65. This was not the first amateur net oriented toward shortwave listening. NNRC shortwave section editor Hank Bennett, W2PNA, attempted to set up an NNRC net in 1956. It lasted less than a year. A similar net was tried the same year by some Universal Radio DX Club members, and the NNRC tried another in 1958. Both appear to have been unsuccessful.

66. For a detailed description of the ECPA, see Fred Jay Meyer, "Don't Touch That Dial: Radio Listening Under the Electronic Communications Privacy Act of 1986," *NYU Law Review,* May 1988, p. 416. The topic was covered as well in Kenneth Vito Zichi, "Radio Listening and the Law," *Monitoring Times,* April 1988, p. 34; and Frank Terranella, "Lawful Listening," *Monitoring Times,* June 1988, p. 93. A comprehensive work on the subject, including references to applicable state laws, is Frank Terranella, *The Listener's Lawbook* (Brasstown, NC: Grove Enterprises, 1991). It was updated in 1995 and is available on the internet as the *Guide to U.S. Monitoring Laws* at www.grove-ent.com/LLawbook.html.

67. The Man of the Year award was renamed the North American Shortwave DXer of the Year award in 1976 and the North American DXer of the Year award in 1979. In 1982 it was divided into two categories: the (again) North American Shortwave DXer of the Year award, and the North American Specialty Band DXer of the Year award. In 1995 the two awards were recombined into the North American DXer of the Year award, called the DXer of the Year Award from 1996 to 1998. Another award, the International DXer of the Year Award, was created in 1976 and issued through 1995. In 1999, the top ANARC award became the Don Jensen Distinguished Service award. It was issued through 2004. When ANARC disbanded in 2005, it was suggested that the award be renamed the North American

Radio Clubs Distinguished Service award and issued annually on the nomination of individual clubs. However, this plan never came to fruition.

68. *NNRC Bulletin,* June 1973, p. 1.

69. For details on the early days of the EDXC, see Claës-W. Englund, "The European DX Council," *How to Listen to the World 1969/70* (Hvidovre, Denmark: World Radio-TV Handbook, 5th ed. 1969), p. 188; and Jyrki K. Talvitie, "The European DX-Council," *How to Listen to the World 1971* (Hvidovre, Denmark: World Radio-TV Handbook, 6th ed. 1971), p. 145.

70. For a group of articles about the Handicapped Aid Program and the experiences of handicapped listeners, see "Radio and the Handicapped," *1981 WRTH,* p. 560.

71. "HAPpenings," *ANARC Newsletter,* August 1982, p. 2.

72. *Monitoring Times,* April 1986, p. 24; *ANARC Newsletter,* March 1986, p. 10.

73. Chuck Rippel, "Report from the 7.240 SWBC Convention," *FRENDX,* April 1988, p. 5. Another source gives attendance as 40. Harold Cones, "Winter SWL Festival For 20 Years Now," *Monitoring Times,* March 2007, p. 16.

74. *SPEEDX,* January 1978, p. 10.

75. This discussion is based in part on Bob Padula, *The History of DXing in Australia,* originally published by the Australian Radio DX Club in August 1967 and reprinted in 1969; Bob Padula, *The First 20 Years — The Development of the ARDXC: 1965–1985,* a supplement to the June 1985 *Australian DX News;* and information provided by John Wright, ARDXC secretary.

76. See Erik Koie, "The Danish Short-Wave Club 60 Years," *Short Wave News,* August-September 2006, p. 25; and *Radio News,* January 1947, p. 125.

77. For a detailed history of the club, see *Danish Short Wave Club International: 50 Year's Anniversary, 1956–2006* (Greve, Denmark: Danish Short Wave Club International, 2006).

78. The author is grateful to Barry C. Williams of New Zealand for his research on New Zealand DX history.

79. *Radio Record's* shortwave coverage appears to have been the impetus for another club, the New Zealand Shortwave Radio Club, which operated during the 1930s. Its publication was the *New Zealand Shortwaver.*

80. *Radio News,* August 1946, p. 105.

81. "N.Z. DX Radio Association Inc.," *How to Listen to the World, 1965/66* (Hellerup, Denmark: World Pub., 1965), p. 165.

82. For more about Victor, see Colin Miller, "Victor Goonetilleke — Portrait of an Extraordinary DXer," *Monitoring Times,* September 1995, p. 25.

83. *Radio News,* November 1946, p. 45.

84. *Radio News,* February 1947, p. 72.

85. "Sveriges DX-Riksklubb," *How to Listen to the World, 1965/66* (Hellerup, Denmark: World Pub., 1965), p. 164.

86. "The DX-Alliansen — A National Union of Swedish DX-clubs" in O. Lund Johansen, ed., *How to Listen to the World, 1965/66* (Hellerup, Denmark: World Pub., 1965), p. 162.

87. *Short Wave Listener* magazine was itself a companion publication to *Short Wave Magazine,* which focused on radio experimenters and amateur radio operators. Save for the war years, *Short Wave Magazine* published continuously from 1937. In December 2005 it combined with another magazine, *Radio Active,* to form a new title, *Radio User.* In January 2006 an independent magazine called *Monitoring Monthly* was also begun. It featured many of the *Short Wave Magazine* columnists.

88. *CONTACT,* October 2003, p. 17.

89. *CONTACT,* May 1988, p. 4.

90. 10th Anniversary WDXC program pamphlet, 1978.

91. For more about the bulletin's content and the club's activities over the years, see Mark Savage, "A Double Century of *Communication*," *Communication*, July 1991, p. 6; and Mark Savage, "25 Years and Three Hundred Editions of *Communication*," *Communication*, November 1999, p. 5.

Chapter 4

1. *Radio News and The Short-Wave*, October 1933, p. 198.
2. For an interesting tour of the FBIS monitoring facility in Silver Hill, Maryland, see Oliver Read, "Foreign Broadcast Intelligence Service," *Radio News*, January 1945, p. 25. For other descriptions of FBIS activities, see *Radio News*, December 1945, p. 52; Don Jensen, "The Inside Story of FBIS," *FRENDX*, Pt. I, February 1973, p. SWC-1, and Pt. II, March 1973, SWC-11; and Benjamin D. Meyer, "Shhhhhh! The F.B.I.S. is Listening!" *Monitoring Times*, January 1995, p. 30.
3. *NNRC Bulletin*, April 1, 1944, p. 3.
4. *NNRC Bulletin*, December 1, 1943, p. 12.
5. *NNRC Bulletin*, January 1, 1944, p. 3.
6. Formation of the BBC Monitoring Service is addressed in Asa Briggs, *The War of Words — The History of Broadcasting in the United Kingdom, Part III* (London: Oxford University Press, 1970), pp. 187–191. A good summary of BBCM activities in the mid–1970s is John G. T. Sheringham, "BBC Monitoring Service: The Ears of Britain," *How to Listen to the World* (Hvidovre, Denmark: World Radio-TV Handbook, 8th ed., 1974), p. 51.
7. Lesley Chamberlain, "War Across the Airwaves," *The Financial Times*, May 1, 2003. For an interesting account of monitoring during the Evesham period, see Olive Renier & Vladimir Rubinstein, *Assigned to Listen* (BBC External Services, 1986 [reprinted in 1987]).
8. SLS became the World Schedules Section in the 1970s and the Foreign Media Unit in 1994.
9. Alan Thompson, "BBC Monitoring Service — A Layman Looks at Caversham Park" (British Association of DXers, May 1972). Over the years the operations of BBCM have become better known, e.g. Sheringham, *supra* note 6; Henry Pavlovich, "The BBC Monitoring Service," *1987 WRTH*, p. 561; and "Inside BBC Monitoring — Listening to the World," *2001 WRTH*, p. 58. For reports from Caversham visitors and other DX students of BBCM, see Gordon Weston, "Professional DXing," *Short Wave News*, August 1976, p. 20; Anker Petersen, "A Visit to Crowsley and Caversham Park," *Short Wave News*, September 1983, p. 20; James G. Clifford, "A Visit to the BBC Monitoring Service," *DX Reporter*, September 1983, p. Mailbag-2; Ruth Hesch, "The End-Products of the BBC Monitoring Service," *DX Reporter*, January 1984; and Colin Miller, "The 'Ears' of Britain," *Monitoring Times*, September 1996, p. 12. Gerry L. Dexter, "Tuning In With the Pros," *Popular Communications*, February 1987, p. 10, covers FBIS and the BBC Monitoring Service as well as monitoring by Canada and Germany. The history of the monitoring service in Japan is described in Hideharu Torli, "A Short History of Japan's Monitoring Services," *Popular Communications*, October 2002, p. 10.
10. The contributions of Hugo Gernsback and his family of publications to the development of radio are detailed in Keith Massie & Stephen D. Perry, "Hugo Gernsback and Radio Magazines: An Influential Intersection in Broadcast History," *Journal of Radio Studies*, Vol. 9, No. 2 (2002), p. 264. The early issues of *Radio News* (and *Radio Broadcast* magazine) were the subject of a content analysis and historical review in Michael Brown, "Radio Magazines and the Development of Broadcasting: *Radio Broadcast* and *Radio News*, 1922–1930," *Journal of Radio Studies*, Vol. 5, No. 1 (1998), p. 68. The text of what appears to be Gernsback's autobiography, written in the 1950s, was found in 2002 and recently published. See Larry Steckler, ed., *Hugo Gernsback — A Man Well Ahead of His Time* (Marana, AZ: Poptronix, 2007).
11. The magazine changed its name to *Radio & Television News* in August 1948 and *Electronics World* in May 1959. *Electronics World* merged into *Popular Electronics* in January 1972.
12. Coincident with this passing of the baton, a separate column was begun for ham band listeners. Edited by Roger Legge, it bore the name "Ama-Touring," the same name as a similar column edited by Legge in the International DXers Alliance *Globe Circler* bulletin in the early 1940s. "Ama-Touring" in *Popular Electronics* lasted only until January 1956.
13. Harold Cones, "'Tuning the Short-Wave Bands' Revisited: A 1991 Interview with Hank Bennett," *Proceedings* (Stillwater, OK: Fine Tuning Special Publications, Vol. 4, 1991), p. F23.1.
14. *SWL*, October 1961, Ken Boord attachment.
15. *Universalite*, June 1956, p. 3.
16. Wayne Green, *SWL*, July 1969, p. 2.
17. http://home.houston.rr.com/edmayberry/.
18. "The Shortwave Center Interview: Kristina Karjalainen, Editor of *Voices*," *FRENDX*, December 1980, p. 2.
19. In other months, starting in 1992, "Selected Programs" also included a brief summary of "hot" DX news and news of world events relative to shortwave listening. This was later replaced by a brief monthly "Program Highlights" section until "Selected Programs" was dropped altogether in 2005.
20. For a 20-year retrospective referencing many of the past *MT* columns and editors, see Rachel Baughn, "*Monitoring Times* Celebrates 20 Years," *Monitoring Times*, December 2001, p. 22. Other coverage of *MT's* history can be found in: Bob Grove, "From the Editors" and Larry Miller, "Welcome to Our New Home," *Monitoring Times*, July 1986, p. 3; "Lean, Mean, and 15! The Evolution of *Monitoring Times*," *Monitoring Times*, January 1997, p. 8; Bob Grove, "Monitoring Times: The First 15 Years!" *Monitoring Times*," January 1997, p. 104; Bob Grove, "Happy Anniversary!" *Monitoring Times*, January 2001, p. 92; "Monitoring Times Celebrates 25 Years," *Monitoring Times*, December 2006, p. 9; and "A Visit With Bob Grove WJ8HD," *Monitoring Times*, December 2006, p. 10. A history of the magazine has also appeared on the *Monitoring Times* website.
21. Some felt Alice was a pseudonym used by Kneitel. See "It Gets Curiouser and Curiouser," *Popular Communications*, August 1988, p. 4.
22. See Bob Grove, "The End of an Era ... Musings with an Old Friend," *Monitoring Times*, July 1995, p. 120.
23. "The Foundation for International Broadcasting," *Monitoring Times*, June 1987, p. 8.
24. *Radio-TV Experimenter*, December 1966-January 1967, p. 124.
25. A more complete history of *White's Radio Log* may be found in Don Jensen, "The Legend and the Legacy," *White's Radio Log*, Vol. 1, No. 1 (North Branch, NJ: Worldwide Pubs., 1985), p. 1. Note that the 1985 issue was designated Vol. 1, No. 1, even though *White's* had been published in one form or another in most years since 1924. This was likely done in anticipation of the continued publication of a "new series" of *White's*.

26. The reference by O. Lund Johansen to "this first edition" in the 1947 volume would seem to be conclusive as to whether that edition was the first. However, in 1982, and again in 1984, Jens M. Frost, having been editor of the *WRTH* since 1964, stated that the first edition of the *WRTH* was published in 1946, had 80 pages, and featured "big typesetting" and was "very easy to read." "The Shortwave Center Interview: Jens M. Frost and Andrew G. Sennitt," *FRENDX,* April 1982; speech by Jens Frost at the EDXC Conference, Stockholm, Sweden, June 9, 1984. None of those descriptors fit the 1947 edition. However, comments by Frost in 1967 to the effect that the first *WRTH* was published in 1947 suggest that his 1982 and 1984 remarks most likely were the result of faulty memory. *Short Wave News,* February 1967, p. 16. In any event, there is no evidence in Danish libraries of a pre-1947 publication meeting the Frost description. Erik Køie, "The Danish Short-Wave Club 60 Years," *Short Wave News,* August-September 2006, p. 25. (A 1945 Johansen publication, *Kortbolge-Haandbog — Vejledning for Kortbølgelytterre [Shortwave-Handbook — Guide to Short Wave Listening],* some of whose parts the *WRTH* clearly resembled, had 40 pages and small type.)

27. A description of Johansen's early radio-related publishing is contained in Køie, *supra* note 26 at p. 25; see also Richard F. Shepard, "Expert On Short-Wave," *The New York Times,* May 31, 1959, Sec. 2, p. X-9, and Hank Bennett, "Meet O. Lund Johansen," *Popular Electronics,* March 1964, p. 81. For a history of the *WRTH* through 1995, see Andy Sennitt, "The Way It Was— A Brief History of the WRTH," *1996 WRTH,* p. 14. For an interesting analysis and comparison of the contents of the 1948, 1956, 1966 and 1976 editions of the *WRTH,* see "Is There Still Life at Forty," *1986 WRTH,* p. 571.

28. Some of the country listings of the first edition are interesting from an historical viewpoint. Radio Brazzaville was listed under French Equatorial Africa. Ghana was shown as the Gold Coast Colony. Indonesia was "East India (Dutch)" and was subdivided into Java, Celebes, Sumatra and Borneo. Thailand was called Siam, and Newfoundland was shown as a separate country.

29. Shepard, *supra* note 27.

30. Although there are occasional references to one or more Danish-language editions, their existence is open to question. Some of the early English editions did have a Danish-language cover and some Danish-language pages.

31. The publishing structure of the *WRTH* is complex. The earliest issues were published by O. Lund Johansen under his own name. The World Publications imprint (sometimes called World Radio Publications or World Radio-TV Publications) appeared in 1958, with World Publications serving as the parent of the *WRTH,* the *Summer Supplement,* the *World Radio Handbook Newsletter,* and several technical handbooks. In 1964, O. Lund Johansen retired. The structure divided, and the World Radio-TV Handbook Co., Ltd., was established. It produced the *WRTH* and the *Summer Supplement,* while World Publications handled the *World Radio Bulletin, How to Listen to the World,* and other publications. (World Publications issued *News from Around the World* in 1963–67 and *World Medium-Wave Guide* in 1964–68. Neither publication attracted much attention in North America.) In 1968, the World Radio-TV Handbook Co., Ltd., bought *How to Listen to the World* from World Publications and began publishing it under the World Radio-TV Handbook Co., Ltd., imprint, finally incorporating it into the *WRTH* proper in 1976. Save for a few latter years, the *World Radio Bulletin* was published by World Publications until 1976, when it came under the control of the *WRTH* which published it until 1991, first as the *WRTH Newsletter,* and then as *WRTH*

Downlink. Soon after the sale of the *WRTH* to Billboard publications, Inc., circa 1967, printing of the *WRTH* was done in either the United Kingdom or the United States rather than Denmark.

32. From January 1973 until around September 1974, the *World Radio Bulletin* was edited and published privately by David McGarva of Edinburgh, Scotland. McGarva was succeeded by Richard Ginbey of Vanderbijlpark, South Africa.

33. The difficulties in collecting information on the Latin American broadcasting scene are recounted in Torre Ekblom, "Let's Face It," *FRENDX,* December 1972, p. SWC-7 [reprinted from the Finnish *DXclusive*]. Almost 100 persons helped Ekblom with the Latin American information over the years. See Tor-Henrik Ekblom, "World Radio TV Handbook 50 Years Anniversary and Latin America," *Short Wave News,* May 1997, p. 25.

34. For more on Frost, see Anker Petersen, "A Mighty Kauri Has Fallen in the Forest — Jens Frost, 1919–1999," *Short Wave News,* November 1999, p. 22. O. Lund Johansen died in 1975.

35. Speech by Gustav-Georg Thiele at the EDXC Conference, Deutsche Welle, Cologne, Germany, May 28–31, 1982. The number of copies "in print" for each year between 1976 and 1981, as shown on the *WRTH* front cover, is as follows: 1976, 40,000; 1977, over 40,000; 1978, over 43,000; 1979, over 55,000; 1980, over 55,000; 1981, over 60,000.

36. Jens M. Frost & Andrew G. Sennitt, "Message from the Editors," *1982 WRTH,* p. 9.

37. Jens M. Frost & Andrew G. Sennitt, "Message from the Editors," *1984 WRTH,* p. 11. For more on the compilation of the *WRTH,* see Andrew Sennitt, "Not in the WRTH," *1980 WRTH,* p. 574; and "The WRTH-81 Story — Jens Frost Interviewed by Anker Petersen," *Short Wave News,* February 1981, p. 27. Local monitors were always needed to check some of the official data. "The big problem is Latin America. About 1,000 Latin American stations are written to each year via registered letter, but the result is similar to that of the listener who expects a QSL from that part of the world. Therefore, WRTH is compelled to organize regular monitors in Brazil, Chile, Colombia, Dominican Republic, Haiti, etc." *Short Wave News,* February 1967, p. 16.

38. Andy Sennitt, "Jens Frost In Memorium," *2000 WRTH,* p. 613.

39. Andrew G. Sennitt, "Message from the Editor," *1989 WRTH,* p. 5.

40. Andrew G. Sennitt, "Message from the Editor," *1991 WRTH,* p. 5.

41. Hermod Pedersen, "The Decline of the WRTH Empire," Nordic Shortwave Center website, www.nordicdx.com, 1998.

42. For a behind-the-scenes look at the production of the *WRTH* today, see "Diamond Anniversary — It's 60 Years of World Radio TV Handbook," *Monitoring Monthly* [U.K.], March 2006, p. 39.

43. The editions of the *WRTH* from 1947 to 1970 are now available on two CDs.

44. *2006 Passport,* p. 363.

45. Nicholas Hardyman, *The Shortwave Guide* (Milton Keynes, England: WRTH Publications, 2002); Sean Gilbert, ed., *The Shortwave Guide — Volume 2* (Milton Keynes, England: WRTH Publications, 2003).

46. E.g. Wireless World, *Guide to Broadcasting Stations* (London, England: ILIFFEE & Sons, 4th ed. 1948); Wireless World, *Guide to Broadcasting Stations* (London, England: Butterworth & Co., 16th ed. 1970); Philip Darrington, *Guide to Broadcasting Stations* (London, England:

William Heinemann, 19th ed., 1987); Simon Spanswick, *Guide to Broadcasting Stations: 21st Century Edition* (London, England: Butterworth-Heinemann, 21st ed. 2001).

47. Robert B. Grove, *Shortwave Frequency Directory* (Brasstown, NC: Grove Enterprises, 1982, and subsequent editions [titled *Shortwave Directory*]).

48. Fred Osterman, *Shortwave Log* (Reynoldsburg, OH: Universal Shortwave Radio Research, 1st ed. 1983, 2d ed. 1984).

49. E.g. Sony Corp. Radio Division, *Wave Handbook* (Tokyo, Japan: Sony Corp. Radio Division, 1981, 1989, c. 1994).

50. James D. Pickard, *North American Shortwave Frequency Guide* (Burbank, CA: Artsci, 1992, 1994, 1995). The 1992 and 1995 editions added introductory material on shortwave listening and maps, both prepared by Bill Smith.

51. Kenneth R. Boord, *The World at a Twirl* (Morgantown, WV: Kenneth R. Boord, 1956).

52. William I. Orr, *Better Shortwave Reception* (Wilton, CT: Radio Publications, 1st ed. 1957).

53. J. Vastenhoud, *Short Wave Listening* (Eindhoven, Netherlands: Philips, 1966).

54. Richard E. Wood, *Shortwave Voices of the World* (Park Ridge, NJ: Gilfer Associates, 1969).

55. Hank Bennett, *The Complete Shortwave Listener's Handbook* (Blue Ridge Summit, PA: Tab, 1974).

56. Hank Bennett & Harry L. Helms, *The Complete Shortwave Listener's Handbook — 2d ed.* (Blue Ridge Summit, PA: Tab, 1980).

57. Hank Bennett, Harry L. Helms and David T. Hardy, *The Complete Shortwave Listener's Handbook — 3d ed.* (Blue Ridge Summit, PA: Tab, 1986).

58. Hank Bennett, David T. Hardy and Andrew Yoder, *The Complete Shortwave Listener's Handbook — 4th ed.* (Blue Ridge Summit, PA: Tab, 1994).

59. Andrew Yoder, *The Complete Shortwave Listener's Handbook — 5th ed.* (New York, NY: McGraw-Hill, 1997).

60. Edward C. Shaw, *DXing According to NASWA* (Liberty, IN: NASWA, 1975). Approximately six months after publication, the need for a basic "handout" for NASWA members led to the publication of a pamphlet, the 16-page *Welcome to the World of Short Wave Listening!*

61. Arthur T. Cushen, *The World in My Ears* (Invercargill, New Zealand: Arthur T. Cushen, 1979).

62. Arthur Cushen, *Radio Listeners Guide* (Invercargill, New Zealand: Arthur Cushen Publications, 2d ed. 1990).

63. Cushen also contributed to a beginners-level book on radio listening: Ashok Nallawalla, Arthur T. Cushen and Bryan D. Clark, *Better Radio/TV Reception — A Non-Technical Approach* (Werribee, Victoria, Australia: Ashley, 1987, and published the following year by Gilfer Associates).

64. Harry L. Helms, *Shortwave Listening Handbook* (Englewood Cliffs, NJ: Prentice-Hall, 1987).

65. Harry L. Helms, *Shortwave Listening Guidebook* (Solana Beach, CA: High Text, 1991).

66. Len Buckwalter, *abc's of Short-Wave Listening* (Indianapolis, IN: Bobbs-Merrill, 1962).

67. Len Buckwalter, *The Fun of Short-Wave Radio Listening* (New Augusta, IN: Editors and Engineers, 1965).

68. Len Buckwalter, *99 Ways to Improve Your Shortwave Listening* (Indianapolis, IN: Howard W. Sams, 1977).

69. H. Charles Woodruff, *Short-Wave Listener's Guide* (Indianapolis, IN: Howard W. Sams, 1964). Other editions were published at least through 1980. *Short-Wave Listener's Guide* was principally a list of stations arranged by country, frequency and broadcast times, along with some basic material on bands, receivers, antennas, propagation, and QSL addresses. Some editions bore the Radio Shack label,

and an edition called *Short-Wave Listener's Radio Guide* was published for Lafayette Radio Electronics at least once (in 1976).

70. H. Charles Woodruff, *Questions & Answers on Short-Wave Listening* (Indianapolis, IN: Howard W. Sams, 1970).

71. Wayne Green, *The Fascinating World of Radio Communications* (Blue Ridge Summit, PA: Tab, 1971).

72. Norman Fallon, *Shortwave Listener's Handbook* (Rochelle Park, NJ: Hayden, 1972).

73. Forest H. Belt, *Easy-Guide to Shortwave Listening* (Indianapolis, IN: Howard W. Sams, 1973).

74. Louis M. Dezettel, *Introduction to Short-Wave Listening* (Fort Worth, TX: Radio Shack, 1972).

75. John Schultz, *Short Wave Listener's Handbook* (Port Washington, NY: Cowan, 1975).

76. Van Waterford, *Hear All the Action* (Indianapolis, IN: Howard W. Sams, 1978).

77. *Directory of World Band Radio* (New York, NY: Sony Corp. of America, 1981 and 1983).

78. Robert J. Traister, *The Shortwave Listener's Handbook* (Blue Ridge Summit, PA: 1982).

79. Samuel Alcorn, *The World Is Yours on Shortwave Radio* (Park Ridge, NJ: Gilfer Associates, 1984).

80. Bob Grove, *The Listener's Handbook* (Brasstown, NC: Grove Enterprises, 1985). *The Listener's Handbook* was a combination of two earlier books by Grove, *Behind the Dial* and *Communications Monitoring*. The 64-page *Behind the Dial* was first published in 1977 by *73* magazine and became available in the mid–1980s from Grove. It contained information on the segments of the radio spectrum, spectrum users, surveillance, antennas, types of interference, receivers, clubs, frequency allocations, etc. It was primarily of interest to beginners. *Communications Monitoring*, 113 pages, was published in 1979, and covered spectrum use, selecting equipment and designing antennas, tips for successful monitoring, construction projects, etc. At 94 pages, *The Listener's Handbook* covered similar topics, including frequency band classification, transmission modes, security and surveillance, selecting equipment and setting up a listening post, antennas, interference, propagation, and small construction projects.

81. William Barden, Jr., *Shortwave Listening Guide* (Fort Worth, TX: Radio Shack, 1987).

82. Kenneth D. MacHarg, *Introducing International Radio* (Jeffersonville, IN: Global Village, 1987).

83. Gerry L. Dexter, *So You Bought a Shortwave Radio!* (Lake Geneva, WI: Tiare, 1987).

84. Bob Grove, *Bob Grove's Scanner and Shortwave Answer Book* (Brasstown, NC: Grove Enterprises, 1990). This was a 160-page book, divided into scanner and general listening sections, and covering receivers, antennas, accessories, the organization of the bands, and listening tips.

85. Bob Martin, *Quick-N-Easy Shortwave Listening* (Burbank, CA: Artsci, 1993).

86. Anita Louise McCormick, *Shortwave Radio Listening for Beginners* (Blue Ridge Summit, PA: Tab, 1993).

87. Anita Louise McCormick, *The Shortwave Listener's Q&A Book* (Blue Ridge Summit, PA: Tab, 1994).

88. T. J. "Skip" Arey, *Radio Monitoring — The How-To Guide* (San Diego, CA: Index Publishing Group, 1997; Boulder, CO: Paladin, 1997 and 2003). In this 337-page book, "Skip" Arey, a long-time columnist for *Monitoring Times*, past editor of several columns in the NASWA bulletin, and a frequent presenter at the Winter SWL Fest, covers medium wave, shortwave and VHF/UHF monitoring. For each area the author discusses equipment, propagation, what can be heard, resources, and related topics. Intended for beginners, this book contained a good basic description of the areas it covers.

89. This booklet was written by Carl Mann of Omaha, Nebraska, in cooperation with Universal Radio, Reynoldsburg, Ohio. Its contents were broadcast over the air, and a cassette copy of *DX Tips for Beginners* was also available

90. Gerry L. Dexter, *Voices from Home* (Lake Geneva, WI: Tiare, 1988).

91. Andy Sennitt, ed., *Traveler's Guide to World Radio* (New York, NY: Billboard, 1991).

92. Andrew R. Yoder, *Shortwave Listening on the Road — The World Traveler's Guide* (Blue Ridge Summit, PA: Tab, 1995).

93. *Radio & Television News,* May 1950, p. 126; September 1950, p. 165.

94. O. Lund Johansen, ed., *How to Listen to the World* (Hellerup, Denmark: World Radio Handbook [or World-Radio Television Handbook], 1st ed., undated, c. 1950; 2d & 3d eds., undated, c. 1952; 4th & 5th eds., undated, c. 1953).

95. O. Lund Johansen, ed., *How to Listen to the World* (Hellerup, Denmark: World Radio-Television Handbook, 6th ed. 1956).

96. O. Lund Johansen, ed., *How to Listen to the World* (Hellerup, Denmark: O. Lund Johansen, 7th ed., 1959 and 1960). There were two versions of this edition, identical except for the year of publication.

97. To complicate matters, they carried seemingly unrelated numbers on their covers: "5," "6," "11" and "12" respectively. (These may have referred to their place in a broader series of *WRTH* publications.) As with the first five editions, the 1962 volume contained no date, but can be reliably dated in 1962 based on the advertising. All *HTLs* after 1962 were dated. O. Lund Johansen, ed., *How to Listen to the World* "5" (Hellerup, Denmark: O. Lund Johansen, undated, c. 1962); O. Lund Johansen, ed., *How to Listen to the World, 1963/64* "6" (Hellerup, Denmark: O. Lund Johansen, 1963); O. Lund Johansen, ed., *How to Listen to the World, 1965/66* "11" (Hellerup, Denmark: World, 1965); O. Lund Johansen, ed., *How to Listen to the World, 1966/67* "12" (Hellerup, Denmark: World, 1967).

98. J.M. Frost, ed., *How to Listen to the World 1969/70* (Hvidovre, Denmark: World Radio-Television Handbook, 5th ed. 1969); J.M. Frost, ed., *How to Listen to the World 1971* (Hvidovre, Denmark: World Radio-TV Handbook, 6th ed. 1971); J.M. Frost, ed., *How to Listen to the World* (Hvidovre, Denmark: World Radio-TV Handbook, 7th ed. 1973); J.M. Frost, ed., *How to Listen to the World* (Hvidovre, Denmark: World Radio-TV Handbook, 8th ed. 1974).

99. Jim Vastenhoud (arranger and compiler) and Jens M. Frost (ed.), *World DX Guide* (London, England: Billboard, 1978).

100. *The Shortwave Book* (Media, PA: Lawrence Miller, 1984).

101. Gerry L. Dexter, ed., *Shortwave Radio Listening with the Experts* (Indianapolis, IN: Howard W. Sams, 1986), p. vii.

102. Guy Atkins, Kevin Atkins, John H. Bryant, Elton Byington, David M. Clark and Fritz Mellberg, eds., *Proceedings* (Stillwater, OK: Fine Tuning Special Publications, Vol. 1, 1988; Vol. 2, 1989; Vol., 3, 1990; Vol. 4, 1991; Vol. 5, 1992–93; Vol. 6, 1994–95).

103. Rainer Lichte, *Radio Receiver — Chance or Choice* (Park Ridge, NJ: Gilfer Shortwave, 1985); Rainer Lichte, *More Radio Receiver — Chance or Choice* (Park Ridge, NJ: Gilfer Shortwave, 1987).

104. Willem Bos and Jonathan Marks, *WRTH Equipment Buyers Guide* (New York, NY: Billboard, 1993).

105. John Schroder, *How to Improve Short Wave Reception* (Hellerup, Denmark: World Pub., undated, c. 1961).

106. Raymond S. Moore, *Communications Receivers, The Vacuum Tube Era: 1932–1981* (La Belle, FL: RSM Communications, 4th ed. 1997 [also 3d ed. 1993; 2d ed. 1991; 1st ed. 1987]).

107. Fred Osterman, *Shortwave Receivers Past and Present — Communications Receivers 1942–1997* (Reynoldsburg, OH: Universal Radio Research, 3d ed. 1998 [also 2d ed. 1997 ("1945–1996"); 1st ed. 1987 (no time period specified)]).

108. Ted Benson, *Inside Your Shortwave Radio* (Lake Geneva, WI: Tiare, 1992).

109. E.g., Fred Osterman, *Buying a Used Shortwave Receiver: A Market Guide to Used Shortwave Receivers* (Reynoldsburg, OH: Universal Radio Research, 4th ed. 1998). This guide to the 100 most popular solid-state receivers was much simpler than *Shortwave Receivers Past and Present* and included portables. The fourth edition covered many more receivers than the first three.

110. *CQ Shortwave Listener Handbook* (Port Washington, NY: CQ Magazine, 1975).

111. Edward M. Noll, *SWL Antenna Construction Projects* (Indianapolis, IN: Howard W. Sams, 1970).

112. Edward M. Noll, *25 Simple Tropical and M.W. Band Aerials* (London, England: Babani, 1984).

113. Edward M. Noll, *25 Simple Shortwave Broadcast Band Aerials* (London, England: Babani, 1984).

114. Edward M. Noll, *25 Simple Indoor and Window Aerials* (London, England: Babani, 1984).

115. Edward M. Noll, *Easy-Up Antennas for Radio Listeners and Hams* (Indianapolis, IN: Howard W. Sams, 1988 and later editions).

116. Edward M. Noll, *Shortwave Listener's Guide for Apartment/Condo Dwellers* (Mississippi State, MS: MFJ Enterprises, 1991). The book promoted only MFJ products.

117. Edward M. Noll, *73 Dipole and Long-Wire Antennas* (Mississippi State, MS: MFJ Enterprises, 1992 and earlier editions).

118. Wilfred Caron, *Antennas for Receiving* (Brasstown, NC: Grove Enterprises, 1985).

119. Ted Benson, *The SWL Antenna Survey* (Lake Geneva, WI: Tiare, 1989).

120. Robert J. Traister, *The Shortwave Listener's Antenna Handbook* (Blue Ridge Summit, PA: Tab, 1982).

121. Dave Ingram, *Easy Wire Antenna Handbook* (Reynoldsburg, OH: Universal Electronics, 1992).

122. W. Clem Small, *The Antenna Handbook: A Guide to Understanding and Designing Antenna Systems* (Brasstown, NC: Grove Enterprises, 1993). See next footnote.

123. Bob Grove, *The Antenna Factbook* (Brasstown, NC: Grove Enterprises, 1995). *The Antenna Handbook* and *The Antenna Factbook* were later made available by *Monitoring Times* in a joint CD release titled *Antennas for Radio Communications.*

124. Frank P. Hughes, *Limited Space Shortwave Antenna Solutions* (Lake Geneva, WI: Tiare, 1988).

125. Frank P. Hughes, *Easy Shortwave Antennas* (Lake Geneva, WI: Tiare, 1992).

126. Frank P. Hughes, *Long Wire Antennas* (Lake Geneva, WI: Tiare, 1994).

127. Joseph J. Carr, *Practical Antenna Handbook* (Blue Ridge Summit, PA: Tab, 4th ed. 2001; 3d ed. 1998; 2d ed. 1994; 1st ed. 1989).

128. Joseph J. Carr, *Joe Carr's Receiving Antenna Handbook* (Solana Beach, CA: High Text, 1993).

129. Joseph J. Carr, *Joe Carr's Loop Antenna Handbook* (Reynoldsburg, OH: Universal Radio Research, 1999).

130. Andrew R. Yoder, *Build Your Own Shortwave Antennas* (Blue Ridge Summit, PA: Tab, 2d ed. 1994).

131. William R. Nelson, *Interference Handbook* (Wilton,

CT: Radio Publications, 1981; and Lakewood, NJ: Radio Amateur Callbook, 1993).

132. Stanley Leinwoll, *Shortwave Propagation* (New York, NY: John F. Rider, 1959).

133. J.A. Ratcliffe, *Sun, Earth and Radio* (New York, NY: McGraw-Hill, 1970).

134. George Jacobs & Theodore J. Cohen, *The Shortwave Propagation Handbook* (Port Washington, NY: Cowan, 1979).

135. George Jacobs, Theodore J. Cohen and Robert B. Rose, *The NEW Shortwave Propagation Handbook* (Hicksville, NY: CQ Communications, 1995).

136. Jacques d'Avignon, *Propagation Programs — A Review of Current Forecasting Software* (Brasstown, NC: Grove Enterprises, 1993). See also Russell Biggar, "Propagation Forecasting Can Make Life Easier for DXers ... an Interview with Jacques d'Avignon," *DX Ontario,* November 1993, p. 8.

137. Andrew R. Yoder, *Pirate Radio Stations: Tuning in to Underground Broadcasts* (Blue Ridge Summit, PA: Tab, 1990).

138. Andrew R. Yoder, *Pirate Radio: The Incredible Saga of America's Underground, Illegal Broadcasters* (Solana Beach, CA: High Text, 1995).

139. Andrew R. Yoder, *Pirate Radio Stations: Tuning in to Underground Broadcasts in the Air and Online* (New York, NY: McGraw-Hill, 2002).

140. Andrew R. Yoder and Earl T. Gray, *Pirate Radio Operations* (Port Townsend, WA: Breakout Productions/ Loompanics Unlimited, 1997).

141. George Zeller, *The Pirate Radio Directory* (Lake Geneva, WI: Tiare, 1st ed. 1989; 2d ed. 1990; 3d ed. 1991; 4th ed. 1992); George Zeller and Andrew Yoder, *The Pirate Radio Directory* (Lake Geneva, WI: Tiare, 5th ed. 1993; 6th ed. 1994; 7th ed. 1995; 8th ed. 1996, 9th ed. 1997).

142. Andrew Yoder, *1993 Worldwide Pirate Radio Logbook* (Springs, PA: Snallygaster, 1992).

143. Lawrence Soley, *Free Radio — Electronic Civil Disobedience* (Boulder, CO: Westview, 1999).

144. Ron Sakolsky and Stephen Dunifer, eds., *Seizing the Airwaves: A Free Radio Handbook* (San Francisco, CA: AK, 1968).

145. E.g., Paul Harris, *When Pirates Ruled the Waves* (London, England: Impulse, 4th ed. 1970).

146. Paul Harris, *To Be a Pirate King* (Aberdeen, Scotland: Impulse, 1971).

147. Paul Harris, *Broadcasting from the High Seas: The History of Offshore Radio in Europe, 1958–76* (Edinburgh, Scotland: Paul Harris, 1977).

148. John Hind & Stephen Mosco, *Rebel Radio — The Full Story of British Pirate Radio* (London, England: Pluto, 1985).

149. Lawrence C. Soley and John S. Nichols, *Clandestine Radio Broadcasting — A Study of Revolutionary and Counterrevolutionary Electronic Communication* (New York, NY: Praeger, 1987).

150. Lawrence C. Soley, *Radio Warfare: OSS and CIA Subversive Propaganda* (New York, NY: Praeger, 1989).

151. Harry L. Helms, *How to Tune the Secret Shortwave Radio Spectrum* (Blue Ridge Summit, PA: Tab, 1981).

152. Gerry L. Dexter, *Clandestine Confidential* (Columbus, OH: Universal Electronics, 1984).

153. Mathias Kropf, *The Clandestine Broadcasting Directory* (Lake Geneva, WI: Tiare, 1994).

154. Kenneth D. MacHarg, *Tune in the World: The Listener's Guide to International Shortwave Radio* (Media, PA: Lawrence Miller, 1983).

155. Gerry L. Dexter, *Muzzled Media — How to Get the News You've Been Mis*sing! (Lake Geneva, WI: Tiare, 1986).

156. Kraig Krist, *The Shortwave Listener's Program Guide* (Annandale, VA: Krist Computer Consultants, 1989).

157. John A. Figliozzi, *The Shortwave Radioguide* (self-published by author, photocopy format: May 1990 and November 1990).

158. John A. Figliozzi, *The Shortwave Radioguide* (Toronto, Canada: Ontario DX Association & Levittown, PA: North American Shortwave Association: May 1991 and November 1991).

159. John A. Figliozzi, *The Shortwave Radioguide* (Toronto, Ontario: Ontario DX Association & Levittown, PA: North American Shortwave Association, 1993, 1994, 1995).

160. John A. Figliozzi, *The Worldwide Shortwave Listening Guide* (Richardson, Texas: Master, 1st ed. 1996; 2d ed. 1998; and Halfmoon, New York: John A. Figliozzi, 3d ed. 2000).

161. http://www.naswa.net/swlguide/.

162. Kannon Shanmugam, *Guide to Shortwave Programs* (Brasstown, NC: Grove Enterprises, 1992, 1993 and 1994).

163. *LA DXing* (Tokyo, Japan: Radio Nuevo Mundo, No. 1, 1980; No. 2, 1982; No. 3, 1985; No. 4, 1987; No. 5, 1992, No. 6, 1997).

164. Henrik Klemetz, *Latin America by Radio* (Espoo, Finland: Tietoteos, 1989).

165. Carl Huffaker, *Carl Huffaker's Latin Notebooks* (DuBois, PA: SPEEDX, 1992).

166. John Cereghin and Carl Huffaker, *The SPEEDX Guide to Latin American DXing* (Tucson, AZ: SPEEDX, 1985).

167. Hiroyuki Nagase, *Broadcasting Stations in the People's Republic of China* (Tokyo, Japan: Hiroyuki Nagase, 1973).

168. Victor Goonetilleke, "Indonesia," *FRENDX,* August 1974 ["Shortwave Center"]; Dan Henderson and Charles A. Wooten, "Short Wave Stations of Indonesia," *NASWA,* March 1975, p. SWC-3.

169. Robert Butterfield and Doug Snyder, *A Guide to Soviet Broadcasting* (Elsinore, CA: SPEEDX, 1976).

170. Michael Nowicki, *Guide to Soviet Radio* (Tucson, AZ: SPEEDX, 1985).

171. DXpeditions had also been addressed in an earlier 20-page pamphlet: Harold Frodge, *The Guide to DXpeditioning* (Dearborn, MI: Michigan Area Radio Enthusiasts, 1986).

172. *NNRC Bulletin,* October 15, 1946, p. 15.

173. P. Barratt, "One Band Listening," *Short Wave News* [ISWL], September 1950, p. 193.

174. *Shortwave Messenger,* May 21, 1961, p. 3.

175. The birth of *Numero Uno* was influenced by Ken Boord's *World at a Twirl Flash Sheet* of a decade before, and perhaps as well by C.M. Stanbury's *Short Wave Newsletter,* which was in existence when *Numero Uno* was formed and which included Jensen as a member. *Numero Uno* was similar in general approach, although focused almost completely on members' loggings and without the mysteries and conspiracies that burdened Stanbury's work.

176. For insight into Hauser and his many shortwave projects, see "The Shortwave Center Interview: Glenn Hauser," *FRENDX,* January 1981; Sheldon Harvey, "CIDX Interview — Glenn Hauser," *CIDX Messenger,* November 1986, p. 4; and "CIDX Special Feature — The Glenn Hauser Interview" *CIDX Messenger,* September 1999, p. 6.

177. *Review of International Broadcasting,* April 7, 1977 (No. 2), p. 1.

178. *RIB* ceased carrying monthly designations after December 1984, issues thereafter being denominated only by number (the first 1985 issue was No. 96).

179. The last *FT* before the merger was *FT* No. 309,

dated November 4, 1986. The next issue, published on November 17, 1986, was denominated No. 421 to accommodate the insertion of the 112 issues of *The DX Log* into the *FT* chronology.

180. This concept was also embodied in *SWL Forms* published by Tiare Publications in 1990.

181. *DX Listening Digest,* No. 1, October 13, 1985, p. ER0061.

182. *The Wavelength* newsletter should not be confused with the U.K.-based *Wavelength*, a European pirate radio magazine of 1976–78.

183. [Unnamed editor], *Free Radio Handbook* (Brick Through Your Window Publications, 1st ed. undated, distributed by John T. Arthur, then located in Hilo, Hawaii); and John T. Arthur, ed., *Free Radio Handbook* (Wellsville, NY: DVS Communications, 2d ed. undated).

184. Zeke Teflon, *The Complete Manual of Pirate Radio* (publisher unknown: 1st ed. undated; and Tucson, AZ: See Sharp, 2d ed. 1993). The *Manual* was reviewed in *Popular Communications,* July 1988, p. 30.

185. *The Pirate Radio Survival Guide* can be found on the internet at http://www.frn.net/special/prsg/.

186. All issues of the *Hollow State Newsletter* are available on line at http://www.hollowstatenews.com/.

187. Myles Mustoe, *Shortwave Goes to School* (Lake Geneva, WI: Tiare, 1989). Mustoe also formed a group called the International Monitoring Association for Students and Teachers (IMAST), which was designed to promote the educational side of shortwave listening. *Monitoring Times,* May 1988, p. 4. For more on Mustoe's classroom experiences, see Myles Mustoe, "Back to School," *Monitoring Times,* September 1988, p. 22.

188. For more on this topic, see LeRoy R. Shelton, "Shortwave, An Untapped Resource," *Hispania,* March 1964, p. 123; Chuck Yarbrough, "Shortwave Radio as a Teaching Tool," *Proceedings* (Stillwater, OK: Fine Tuning Special Publications, Vol. 4, 1991), p. F24.1; Arthur Edwards, "Shortwave Listening in the Spanish Classroom," *Monitoring Times,* September 1991, p. 20; Neil Carleton, "Shortwave Listening and Amateur Radio in the Classroom," *DX Ontario,* March 1996, p. 7; Neil Carleton, "Shortwave Listening in the Classroom," *DX Ontario,* December 1998, p. 38; Brian Smith, "Neil Carleton's Shortwave Listening Class 2000/2001," *Listening In,* August 2001, p. 33; Neil Carleton, "A Great Year for Radio in the Classroom," *Listening In,* September 2001, p. 33; and Saul Chernos, "Radio Brings the World — and Outer Space — Into the Classroom," *Education Today [Ontario Public School Boards' Assn.],* Fall 2004, p. 23.

189. The recordings were obtained either directly from the stations or recorded at the RCI Receiving Station near Ottawa in 1974. For a later description of the monitoring station, see Jacques d'Avignon, "The RCI Monitoring Station," *Monitoring Times,* April 1994, p. 22.

190. One of Moore's more humorous non-*BLANDX* productions described the rise and fall of the edible QSL. "It seems there are so few of us around now who can still remember the days when you couldn't eat your QSL cards." Don Moore, "You Can't Have Your QSL and Eat It Too," *Review of International Broadcasting,* No. 97, 1985, p. 3.

Chapter 5

1. Apparently from a Philips promotional piece run in several Dutch newspapers in 1936 and quoted in Robert D. Haslach, *Netherlands World Broadcasting* (Media, PA: Lawrence Miller, 1983), p. 29.

2. From time to time, comparisons of the various programs would appear in the DX press, e.g. Gerry L. Dexter, "Calling All SWL DXers," *Popular Electronics,* September 1963, p. 39; Glenn Hauser, "A Survey of DX Programs," *Popular Electronics,* September 1979, p. 92; Roger N. Peterson, "DX Programming," *Elementary Electronics,* September-October 1979, p. 73; Jerome S. Berg, "DX Programs — Treasure or Trash?" *FRENDX,* April 1981, p. 2; Phillip M. Dempier, "DX Programs," *SPEEDX,* October 1983, p. 51; Alex Batman, "DX Programs: A Review," *FRENDX,* Pt. I, October 1988, p. 5; Pt. II, November 1988, p. 3; Pt. III, February 1989, p. 5; Ron Tamburello, "Programs for Radio Enthusiasts," *Monitoring Times,* November 1992, p. 14; Arthur Cushen, "The Medium IS the Message: A Historical Look At DX Programs," *Monitoring Times,* February 1994, p. 24; and John Figliozzi, "Back to Basics: The Best SW Media Programs," *Monitoring Times,* February 2000, p. 69.

3. For an account of Padula's involvement in Australian DX programs, see Bob Padula, "Australian DX Programs," *NASWA Journal,* August 2002, p. 14.

4. See "Arne Skoog of Radio Sweden — The DXer's 'Grand Old Man,'" *1977 WRTH,* p. 454.

5. Claës Englund, "Arne Skoog in Memorium," *2000 WRTH,* p. 613.

6. See "Tribute to a Friend," *1975 WRTH,* p. 375.

7. For a description of the Marks philosophy and the transition from "DX Juke Box" to "Media Network," see "Jonathan Marks! An Interview with Radio Netherland's Media Network Host," *The Shortwave Guide,* March 1985, p. 10; Jeff Chanowitz, "Going Dutch with Radio Netherlands International: Shortwave's True Pioneering Service," *Monitoring Times,* June 1992, p. 14; and "Network Celebrations," *NASWA Journal,* April 1996, p. 10.

8. Reminiscences of Clayton Howard's years at HCJB may be found in Clayton Howard, "The Lord Has Been Good to Me at HCJB," and "HCJB's Clayton Howard to Retire This Summer," both in *The International Shortwave Listener's Program Guide,* April 1984, p. 1.

9. Years earlier, Hardy Hayes had arranged for Ohio powerhouse WLW to turn off its transmitter for one minute so that some of his friends in Cincinnati could hear HCJB on their common frequency of 700 kHz. Several did. Interview with Hardy Hayes, "DX Partyline," June 1, 2003.

10. "Radio Canada Short Wave Club," *How to Listen to the World,* 1965/66 (Hellerup, Denmark: World, 1965), p. 161.

11. Wojtek Gwiazda, "Shortwave's Favorite Canadian — A Profile of Ian McFarland," *Monitoring Times,* September 1990, p. 6.

12. *DX Listening Digest,* No. 39, March 9, 1993, p. 1–16.

13. O. Lund Johansen, ed., *How to Listen to the World,* 1965/66 (Hellerup, Denmark: World, 1965), p. 55.

14. See John G. Pitman, "BBC World Radio Club," *1969 WRTH Summer Supplement,* p. 17; and Joy Boatman, "BBC World Radio Club," *How to Listen to the World* (Hvidovre, Denmark: World Radio-TV Handbook, 7th ed. 1973), p. 98. Henry Hatch had joined the BBC in 1925 as a messenger boy. For 26 years he compiled the transmitter schedules for the Overseas Engineering Information Department. See "A Life Time with Radio," *Voices,* August 1980, p. 19.

15. *UADX Bulletin,* No. 50/51, December 15, 1977, p. 1.

16. For summary descriptions of some of the DX programs on the air in 2003 and 2007 respectively, see John Figliozzi, "DX, SWL, Media & IT Programs," *Monitoring Times,* May 2003, p. 41; Fred Waterer, "DX Programs (Part 1)," *Monitoring Times,* April 2007, p. 36; and Fred Waterer, "More DX Shows," *Monitoring Times,* July 2007, p. 36.

17. John Lackmann, "Stamp Collecting with a Shortwave Radio," *FRENDX,* July 1976, p. 9.

18. For Startz's story on how it all began, see "40 Years Happy Station," *1969 WRTH*, p. 35.

19. *NNRC Bulletin*, May 1976, p. SWBC-1.

20. For a description of his early shortwave listening experiences, see Ray Briem, "Short Wave — Periscope of the World — Personal Reflections on Why SW Listening," *1970 Communications Handbook*, p. 7.

Chapter 6

1. A Chicago member of the Universal Radio DX Club, in the *Universalite*, January 1, 1958, p. 1.

2. *Radio News*, July 1945, p. 124.

3. For a comprehensive history of Collins, see Ken C. Braband, *The First 50 Years — A History of Collins Radio Company and the Collins Division of Rockwell International* (Cedar Rapids, IA: Rockwell International, 1983). For the story of the company's formation and its activities through 1933, see F. Parker Heinemann, "The Collins Radio Company — Ingredients of Success," *AWA Review* (Bloomfield, NY: Antique Wireless Association, Vol. 10, 1996), p. 222. Additional Collins history may be found in Jay H. Miller, *A Pictorial History of Collins Amateur Radio Equipment* (Dallas, TX: Trinity Graphics Systems, 1999). Company founder Arthur A. Collins died in 1987.

4. The Hallicrafters line, and the company, are treated in depth in two publications: Max de Henseler, *The Hallicrafters Story, 1933–1975* (Charleston, WV: Antique Radio Club of America, 1991); and Chuck Dachis, *Radios by Hallicrafters* (Atglen, PA: Schiffer Publishing, 2d. ed. 1999). See also Chuck Dachis, "Hallicrafters: The Man and the Company," *Electric Radio,* January 1990, p. 3.

5. For a brief history of Hammarlund, see Stuart Meyer, "Hammarlund Radio," *AWA Review* (Holcomb, NY: Antique Wireless Association, Vol. 2, 1987), p. 95. See also "The Hammarlund Story," *Old Timer's Bulletin*, Winter 1968, p. 22.

6. For a history of the National Company, see John J. Nagle, "A Brief History of the National Company, Inc." *AWA Review* (Holcomb, NY: Antique Wireless Association, Vol. 1, 1986), p. 65.

7. For more on RME, see Ben Nock, "The RME Receivers," *Radio Bygones,* August-September 2003, p. 22.

8. For a near-encyclopedic review of communications receivers, including descriptions, prices and pictures, see Fred Osterman, *Shortwave Receivers Past and Present — Communications Receivers 1942–1997* (Reynoldsburg, OH: Universal Radio Research, 3d ed. 1998). Another authoritative work on the topic, which includes a detailed history of the development of the communications receiver, is Raymond S. Moore, *Communications Receivers, The Vacuum Tube Era: 1932–1981* (La Belle, FL: RSM Communications, 4th ed. 1997). A briefer examination of the topic is Elton Byington, "Communications Receiver History," *DX Ontario,* Pt. I, April 1993, p. 67; Pt. II, May 1993, p. 69.

9. In 1970, Allied Radio was bought by Radio Shack's then-parent company, Tandy Corp. Allied Radio and Radio Shack merged soon thereafter and the Allied name was retired. For reminiscences about the Allied store in Chicago, see Ken Greenberg, "Memories of Allied Radio," *Electric Radio,* January 1997, p. 36.

10. Lafayette Radio merged with Circuit City in 1981, and the last 15 stores still bearing the Lafayette name, all in the New York area, closed in 1986.

11. For more on the SX-28, see John Bryant, "The Hallicrafters SX-28: THE Classic Shortwave Receiver," *Proceedings* (Stillwater, OK: Fine Tuning Special Publications,

Vol. 5, 1992–93), p. F30.1; and Jim Hanlon, "The Hallicrafters SX-28 Super Skyrider," *Old Timer's Bulletin*, November 2002, p. 54.

12. "The Hammarlund '40' Catalog," 2d ed., p. 19.

13. The development of the AR-88 is reviewed in Bob Haworth, "A Biographical History of the AR88, RCA's 'World Class' Communications Receiver," *Old Timer's Bulletin*, May 1991, p. 11. See also Sam Thompson, "RCA AR-88 — A Classic in Radio Design," *Electric Radio,* July 1990, p. 12; and "RCA Model AR-88 — A Classic In Radio Design" *Old Timer's Bulletin*, June 1977, p. 35.

14. "Annual NNRC Receiving Equipment Listing," *NNRC Bulletin*, September 1957, p. ST-1.

15. *Popular Electronics*, July 1960, p. 6.

16. See Jim Hanlon, "The Echophone Commercial Model EC-1," *Old Timer's Bulletin*, February 2003, p. 38.

17. For a review of the S-38 line, see Chuck Dachis, "The Hallicrafters S-38: 1935 to 1962," *Proceedings* (Vol. 5, 1992–93), p. F29.1.

18. *Universalite*, July 1, 1954, p. 5. Two receivers that were nearly identical to the S-40B, the S-77 and S-77A, were sold from 1950 to 1955.

19. Edward M. Gable, "Restoring the National NC-46," *Old Timer's Bulletin*, May 1990, p. 29.

20. Wayne Childress, "The National Company's NC-57," *Old Timer's Bulletin,* May 1995, p. 11. For a review of the NC-57, see Jim Hanlon, "The National NC-57, or A Tale of Three Receivers," *Electric Radio*, August 1999, p. 20.

21. For a review of the RME-84, see Chuck Teeters, "The RME-84 Receiver," *Electric Radio*, December 2003, p. 2.

22. Jim Hanlon, "The RME-45," *Electric Radio,* July 1996, p. 24. For some interesting history of earlier RME receivers and the company's founder, E.G. Shalkauser, see Jim Hanlon, "The RME-69," *Electric Radio,* May 1992, p. 8; and Jim Hanlon, "The RME-70 — Bringing A Boatanchor Back to Life," *Electric Radio*, November 2000, p. 26.

23. The operation of the sliding-coil receiver is explained in detail in William Fizette, "The National Sliding Coil Tray Receivers," *Old Timer's Bulletin*, February 1988, p. 20. The NC-100 also benefited from an early design of National's "PW" micrometer dial, which would become better known in the HRO series. The NC-200, NC-2-40C and NC-2-40D had more conventional pointer dials. For more on the coil catacomb receivers, see Lawrence R. Ware, "The National Company, Inc. — The Coil-Catacomb Radios and Variations on a Theme," *AWA Review* (Bloomfield, NY: Antique Wireless Association, Vol. 11, 1998), p. 166; and Jim Hanlon, "The National NC-100 Receiver Family," *Electric Radio*, November 1995, p. 20. The NC-100 is also covered in William Fizette, "Restoring A Classic National NC-100 Receiver — Part 1," *Old Timer's Bulletin*, November 2000, p. 11; and "Part 2," *Old Timer's Bulletin*, February 2001, p. 37. A good review of the NC-200 is Bob Haworth, "A World War II Relic Revisited — National's NC-200," *Old Timer's Bulletin*, November 1993, p. 31.

24. A review of the NC-173 and a comparison with the Hammarlund HQ-129X and the Hallicrafters SX-43 may be found in Jim Hanlon, "Radio Kon-Tiki," *Electric Radio*, October 1995, p. 4.

25. The origins and development of the HRO are comprehensively reviewed in Barry Williams, "The Evolution of the National HRO and Its Contribution to Winning World War II," *AWA Review* (Breesport, NY: Antique Wireless Association, Vol. 17, 2004), p. 145. For other discussion of the HRO, see Jim Musgrove, "The Classic HRO and its Evolution," *Electric Radio*, November 1990, p. 22; Elton Byington, "HRO: Portrait of a Classic," *Proceedings*

(Vol. 5, 1992–93), p. F31.1; Jim Hanlon, "An HRO Story —
Part I," *Electric Radio,* June 1994, p. 10; "Part II," *Electric
Radio,* July 1994, p. 10; and Ben Nock, "The HRO Re-
ceiver," *Radio Bygones,* June-July 2003, p. 14.

26. For more on the HRO-7, see William Fizette, "The
National Company HRO-7," *Old Timer's Bulletin,* May
2001, p. 27.

27. Michael D. Tannenbaum, "An HRO Story," *Old
Timer's Bulletin,* August 1991, p. 44.

28. See Jim Hanlon, "The Hammarlund HQ-129-X ... or,
How I Almost Slid Off the Church Roof," *Electric Radio,* Oc-
tober 1994, p. 4; and Jim Hanlon, "The 1945 Hammarlund
HQ-129X," *Old Timer's Bulletin,* May 2003, p. 32.

29. *NNRC Bulletin,* October 15, 1946, p. 16.

30. See Jim Hanlon, "The Hallicrafters SX-43," *Old
Timer's Bulletin,* January 2005, p. 42.

31. Carleton Lord, "The NNRC Radio & Television
Show," *NNRC Bulletin,* December 1951, p. SF-5. The SX-
42 and SX-62 are reviewed in John Bryant and James
Goodwin, "The SX-42 and SX-62: Two Hallicrafters Fa-
vorites," *Proceedings,* (Vol. 6, 1994–95), p. R32.1.

32. For a detailed review, see Dallas Lankford, "The
Collins 51J4," *Proceedings* (Vol. 2, 1989), p. R7.1.

33. *NNRC Bulletin,* December 1, 1947, p. SW-1; see also
A. R. Niblack, "The BC-221 Heterodyne Frequency Meter,"
FRENDX, May 1968, p. 7; and Paul Mayo, "The BC-221
Frequency Meter," *SPEEDX,* November 1971, p. 9.

34. *NNRC Bulletin,* March 2, 1953, p. A-9.

35. For more on the DB preselectors, see Al Brogdon,
"The RME DB-22A Preselector and Me — A Reminis-
cence," *Electric Radio,* May 1997, p. 38; and Chuck Teeters,
"The Last RME Preselector," *Electric Radio,* October 1999,
p. 34.

36. Hallicrafters Catalog P-38, c. 1947; see also Bill
Kleronomos, "The Hallicrafters SP-44 Panadapter," *Electric Radio,* May 1992, p. 32.

37. See Edward J. Insinger, "*FRENDX* Visits Gilfer —
For a Q and A With Perry Ferrell," *FRENDX,* April 1973,
p. SWC-4.

38. See Chuck Teeters, "The Hallicrafters S-53 and S-
53A Receivers," *Electric Radio,* May 2002, p. 12.

39. *Universalite,* July 1, 1954, p. 5.

40. *NNRC Bulletin,* February 1957, p. A-4.

41. Jim Hanlon, "The National Select-O-Ject," *Electric
Radio,* August 1995, p. 10. In an operational comparison, the
author rated the popular Autek QF-1, which came to market circa 1976, a better filter.

42. *QST,* February 1950, p. 61.

43. In the URDXC survey, the SX-71 was the most
wanted receiver in the $200 to $300 class.

44. See Jim Hanlon, "The Hallicrafters SX-73," *Electric
Radio,* March 1999, p. 8.

45. A good review of the technical and operational performance of the SX-88 is Mike O'Brien, "SX-88: Hallicrafters' Highwater Mark," *Electric Radio,* April 1992, p.
30.

46. Ray Osterwald, "National Company's NC-183D Receiver," *Electric Radio,* August 1992, p. 10.

47. Carleton Lord, "The NNRC Radio & Television
Show," *NNRC Bulletin,* January 1953, p. SF-3.

48. *Universalite,* July 1, 1954, p. 5.

49. In 1938 the concept had been incorporated in the
Hallicrafters DD-1 diversity receiver. See Joel Levine, "The
Hallicrafters DD-1 Receiver: Masterpiece or Mistake?"
Electric Radio, November 1993, p. 20.

50. P. Jay Spivack, "The NC-400 Communications Receiver," *Electric Radio,* August 1995, p. 20.

51. See Bill Kleronomos, "Hammarlund Super-Pro: SP-
600JX-17," *Electric Radio,* May 1990, p. 10; and Phil Bythe-

way, "The Hammarlund SP-600," *Proceedings* (Vol. 6,
1994–95), p. R33.1.

52. See Mitchell A. Sams, "Sizing Up the Hammarlund
HQ-180A in the 1980s," *Proceedings* (Vol. 1, 1988), p. R2.1.

53. See David Clark, "The Collins 51S1: An 'S' Line Classic," *Proceedings* (Vol. 5, 1992–93), p. R24.1.

54. For a detailed review of the electrical and mechanical design of the R-390A, see Ray W. Osterwald, "The
R390A Receiver: A Milestone in HF Communications —
Part I," *Electric Radio,* April 1991, p. 22; "Part II," May 1991,
p. 22; and "Part III," June 1991, p. 22. An operational eval-
uation is John R. Tow, "A Review of the R-390A Receiver,"
Proceedings (Vol. 1, 1988), p. R3.1.

55. See Bruce Portzer, "The RACAL RA-17 Communications Receiver," *Proceedings* (Vol. 4, 1991), p. R19.1.

56. Drake's history, and information on troubleshoot-
ing Drake equipment, is recounted in John Loughmiller,
A Family Affair — The R. L. Drake Story (published by John
Loughmiller, 2007).

57. Richard A. Seifert, "Broadcasting Bedfellows — Re-
calling the Unusual Alliance Between Radio New York
Worldwide and the Drake SW-4(A)," *Monitoring Times,*
August 1996, p. 25.

58. Jon Williams, "The Drake R4B and R4C — Two Receivers from the Past," *Proceedings* (Vol. 2, 1989), p. R-11.1.

59. "Annual Receiver Survey," *SWL,* August 1975, p. 45.

60. See Tom Hoitenga, "Squires-Sanders (SS-1R, SS-1V,
SS-1S, SS-1RS)," *Electric Radio,* July 1994, p. 26.

61. See John D. Drummond, "Is the HRO-500 the
Greatest Receiver Ever Made?" *Popular Electronics,* August
1965, p. 45 (Answer: "*We don't know ... but it could very
well be*"); and David Kuraner, "The National HRO-500:
Last of the Dinosaurs or Technological Innovation?" *Electric Radio,* August 1996, p. 30.

62. Dave Ishmael, "The Hallicrafters SX-122," *Electric
Radio,* December 1994, p. 24.

63. Steve Kennedy, "Receiver Survey," *SPEEDX,* September 1979, p. 51.

64. See Kirk Allen and David Clark, "The Japan Radio
Company's NRD-515 — A DXer's Perspective," *Proceedings*
(Vol. 1, 1988), p. R-6.1.

65. See Jerry Strawman, "The Drake R7 Communications Receiver Revisited," *Proceedings* (Vol. 1, 1988), p. R-
5.1; and David Clark and Jon Williams, "The Drake R-7/R-
7A," *Proceedings* (Vol. 6, 1994–95), p. R34.1.

66. See Kevin Atkins, John Tow and Jerry Strawman,
"The ICOM IC-R70 — A DXer's Review, Technical De-
scription and Modifications," *Proceedings* (Vol. 2, 1989), p.
R10.1.

67. See David Clark, "The ICOM IC-R71A Receiver —
A 1991 Appraisal," *Proceedings* (Vol. 4, 1991), p. R20.1.

68. Many are contained in Donald E. Moman, *ICOM
R71 Performance Manual* (Edmonton, Alberta: Shortwave
Horizons, 1985).

69. See John H. Bryant, "Wastegunner On A 525 — A
Dirty Elbows DXer's View of Japan's Current Best DX Machine," *Proceedings* (Vol. 2, 1989), p. R12.1; and John Tow,
"Modifications for the NRD-525," *NASWA Journal,* June
1990, p. 14.

70. Craig Siegenthaler, "Synchronous Detection," *Proceedings* (Vol. 3, 1990), p. R15.1.

71. See Bob Evans, "The Japanese Radio Corporation
NRD-535," *Proceedings* (Vol. 5, 1992–93), p. R27.1; and
John Bryant, "Comparing the Drake R8 and the JRC NRD-
535D," *Proceedings* (Vol. 6, 1994–95), p. R37.1.

72. See Don Moore, "A Review of the Kenwood R-
5000," *Proceedings* (Vol. 3, 1990), p. R16.1.

73. Drake's pre-R8 history is described by Bill Frost, the
company's service department manager, in a paper, "The

R.L. Drake Co., 45 Years Young," January 1988, reprinted in part in the *ANARC Newsletter,* December 1988, p. 8.

74. See Elton Byington, "The Drake R-8 Receiver: Through DXers' Eyes and Ears," *Proceedings* (Vol. 5, 1992–93), p. R28.1.

75. See Bob Grove, "R.L. Drake Co.— The Passing of a Legend," *Monitoring Times,* August 2005, p. 16.

76. See Chuck Mitchell, "The Lowe Europa HF-225," *Proceedings* (Vol. 6, 1994–95), p. R36.1.

77. See Elton Byington, "Two British Boxes: Lowe's HF-225 and HF-150 Receivers," *Proceedings* (Vol. 5, 1992–93), p. R26.1.

78. See David Clark and James Goodwin, "The Watkins-Johnson HF-1000 Digital Receiver," *Proceedings* (Vol. 6, 1994–95), p. R29.1.

79. George Zeller, "The ICOM R-9000: The Rolls Royce of Receivers, Or A Pentagon Toilet Seat?" *Proceedings* (Vol. 4, 1991), p. R23.1.

80. John Bryant, "A First Look at the Mckay-Dymek DR-333 HF Receiver," *Proceedings* (Vol. 4, 1991), p. R22.1.

81. See Alan Johnson, "The Ten-Tec Model RX-320 PC Radio," *NASWA Journal,* December 1998, p. 12; Joe Buch, "Collaborative Receiver Design— The RX-320 Saga," *NASWA Journal,* February 2000, p. 20; Joe Cooper, "The 'Black Box' Radio— Ten-Tec's RX320. And 240+ Loggings," *Popular Communications,* November 2001, p. 62; and Joe Cooper, "'Black Box' Radio Part II— Software Definition and Control," *Popular Communications,* December 2001, p. 66.

82. See John Catalano, "Three Software Radios," *Monitoring Times,* January 2005, p. 69.

83. It is instructive that in the entertaining and informative history of the portable radio, Michael Brian Schiffer, *The Portable Radio in American Life* (Tucson, AZ: University of Arizona Press, 1991), save for the Zenith Trans-Oceanic, shortwave is barely mentioned.

84. John H. Bryant and Harold N. Cones, *The Zenith Trans-Oceanic— The Royalty of Radios* (Atglen, PA: Schiffer Publishing, 1995); Bryant & Cones, *Zenith Radio— The Early Years: 1919–1935* (Schiffer, 1997); Bryant, Cones & Martin Blankinship, *Zenith Radio— The Glory Years, 1936–1945 (History and Products)* (Schiffer, 2003); and Bryant, Cones & Blankinship, *Zenith Radio— The Glory Years, 1936–1945 (Illustrated Catalog & Database)* (Schiffer, 2003).

85. Schiffer, *supra* note 83 at p. 151.

86. "Receiver Survey," *DX Ontario,* July 1995, p. 73.

87. See Thomas Baier, *Grundig Satellit— All Models in Word and Picture* (Homburg, Germany: Thomas Baier, 2d. ed. 2001).

88. See James Careless, "Cheap and Wonderful World Band Radio Contest, Round One," *Listening In,* March 2003, p. 36, and "Round Two," *Listening In,* July 2003, p. 5.

89. See "Kits for the Short-wave Listener," *Electronics Illustrated,* November 1959, p. 79; and Herb Friedman, "Receiver Kits for Beginning SWLs," *Radio-TV Experimenter,* October-November 1967, p. 61. During the late 1950s and early 1960s, *Popular Electronics* published several editions of an annual called the *Electronic Kits Directory.*

90. Chuck Penson, *Heathkit— A Guide to the Amateur Radio Products* (Hicksville, NY: CQ Communications, Inc., 2003), is a comprehensive history of the Heath Company. An abbreviated version is Randy Kaeding, "Heathkit— From Airplanes to Educational Courses," *Popular Communications,* October 2007, p. 17. For an interesting and informative collection of reminiscences by Heath employees, see Terry Perdue, *Heath Nostalgia* (Lynnwood, WA: Terry A. Perdue, 1992).

91. For a review of various kits, see Rich Arland, "Kits Keep It Simple," *Monitoring Times,* July 1999, p. 80; and Ken Reitz, "Beginner's Corner," *Monitoring Times,* March 2002, p. 24.

92. M. Harvey Gernsback, "Record Your Foreign Programs," *Short Wave Craft,* February 1935, p. 586.

93. See Jerry Berg, "Tape Recording and Shortwave" in Gerry L. Dexter, ed., *Shortwave Radio Listening with the Experts* (Indianapolis, IN: Howard W. Sams, 1986), p. 97.

94. Carleton Lord, "Leaves From A DX-er's Scrapbook," *NNRC Bulletin,* February 21, 1949, p. SF-2.

Chapter 7

1. *Radio News,* January 1946, p. 43, quoting a Virginia DXer in the *Universalite,* bulletin of the Universal Radio DX Club.

2. The early years of broadcast station QSLing are covered in greater depth in Jerome S. Berg, *On the Short Waves, 1923–1945: Broadcast Listening in the Pioneer Days of Radio* (Jefferson, NC: McFarland, 1999), pp. 178–199.

3. Jerry Berg, "Keeping the QSL Fires Burning," *Monitoring Times,* February 1996, p. 20.

4. E.g. Don Moore, "Fiesta QSL: Verifying the Latins," *Monitoring Times,* January 1990, p. 10.

5. Page Taylor, "Short Waves— Past, Present and Future," *RADEX,* February 1936, p. 16, quoting a Westinghouse press release.

6. *International Short Wave Radio,* June 1941, p. 7.

7. Drayton Cooper, *NNRC Bulletin,* April 1960, p. SW-4A.

8. *Universalite,* May 1, 1952, p. 12.

9. *NNRC Bulletin,* March 1, 1947, p. 31.

10. *Radio & Television News,* March 1949, p. 121.

11. Jerry Berg, "Verifying Chinese Regional Stations," *FRENDX,* December 1983, p. 1, and "P.S.," January 1984, p. 11.

12. See Denis Casey, "Radio Canada International Shortwave Monitoring Station," *DX Ontario,* September 1993, p. 69; Jacques d'Avignon, "The RCI Monitoring Station," *Monitoring Times,* April 1994, p. 22; and Victor Goonetilleke, "Monitoring for the International Broadcasting Bureau," *Monitoring Times,* September 2003, p. 10.

13. Many stations have kept track of reception reports as part of the process of monitoring listener mail. See Richard E. Wood, "Austria Evaluates a Year's Reports," *FRENDX,* February 1976, p. SWC-7. Respondents to a survey of the QSL practices of international broadcasters conducted by the European DX Council in 1980 rated the quality of most reception reports as moderate to good. Most stations QSLed by way of the correspondence department and then sent reports to the technical department. "QSL Survey" (St. Ives, Huntingdon, England: European DX Council, 1980).

14. "Member's Newsletter," Michigan Area Radio Enthusiasts, April-May 1989, p. 4.

15. *FRENDX,* April 1994, p. 37.

16. Arthur Cushen, "Trends in QSLing," *Monitoring Times,* February 1997, p. 86.

17. An experienced DXer in Brazil reported an overall return rate, after follow-ups, of 62 percent from 425 Latin American stations to whom he reported from 1963 to 1967. Giacomo "Jack" Perolo, "Latin American Veries: Personal Statistics," *FRENDX,* March 1968, p. 18. An intensive program of Latin American reporting from 1981 to 1985 brought another DXer a return rate, after follow-ups, of 78 percent for 172 stations reported. Jerry Berg, "Latin Amer-

ican QSLing Made Easy (?)" *FRENDX*, September 1985, p. 9.

18. In 1997, William J. Plum of Flemington, New Jersey, who has sold mint stamps to SWLs and hams for over 25 years, wrote *Foreign Airmail Postage for Successful QSLing— The SASE [Self-Addressed Stamped Envelope] Method*. Mainly for ham operators, the 20-page booklet contained many practical tips on the use of IRCs, mint stamps and U.S. currency to cover the cost of return postage.

19. Robin Wood, Program Manager, Radio Australia, *Radio News*, April 1947, p. 107.

20. T.Y. Woo, director general, Central Broadcasting Administration, Nanking, China, quoted in *Radio & Television News*, October 1948, p. 190.

21. Editorial, *Short Wave News* [ISWL], May 1947, p. 113.

22. Editorial, *Short Wave News* [ISWL], December 1947, p. 309.

23. Hank Bennett, *NNRC Bulletin*, February 1960, p. SWBC-1.

24. *New Zealand DX Times*, September 1965, p. 1, reprinted in *SWL*, November 1965.

25. Jonathan Marks, "Radio Finland's Attitude Is *Not* Repugnant," *Review of International Broadcasting*, May 1979 (No. 27), p. 10, responding to criticisms of Radio Finland's decision to stop QSLing and to respond to reports with an "Audience Card."

26. "Reception Report Reencounter," *FRENDX*, February 1966, and Don Jensen and Gerry Dexter, "The Results of Reception Report Re-Encounter," *FRENDX*, May 1966.

27. Art Glover, *Proper Reporting Guide* (Huntington Beach, CA: ASWLC, 1969).

28. Sam Barto, *Verification Techniques* (self-published, 1978).

29. Gerry L. Dexter, *Secrets of Successful QSL'ing* (Lake Geneva, WI: Tiare, 1986).

30. J.C. Gillespie, *SWL Address Book* (Park Ridge, NJ: Gilfer Associates, 1st ed. 1969).

31. Gerry L. Dexter, *SWL Address Book* (Park Ridge, NJ: Gilfer Associates, 2d ed. 1974).

32. Gerry L. Dexter, *QSL Address Book* (Park Ridge, NJ: Gilfer Associates, 3rd ed. 1978, 4th ed. 1984).

33. Gerry L. Dexter, *World Broadcast Station Address Book* (Park Ridge, NJ: Gilfer Associates, 5th ed. 1985).

34. Zet Jacobsson, *RDXA LA-QSL List* (Boras, Sweden: Riksförbundet DX-Alliansen, 1970).

35. *NNRC Bulletin*, April 1, 1948, p. SF-1.

36. *NNRC Bulletin*, December 1951, p. 4 (unpaginated).

37. "DX Club Awards for Hams, SWLs , BCBers!" *Electronics Illustrated*, January 1965, p. 53.

38. *Boys' Life* occasionally ran articles about shortwave listening and included some SWL questions in its "Hobby Hows" column. It carried some shortwave-related advertising as well. In addition, from time to time *Boys' Life* issued reprints of shortwave material originally published in the magazine. These typically included an article by Ken Boord about shortwave listening (with a list of stations to try for), plus information on ham radio, antennas, kit building, other available publications, the awards program, and a Boys' Life Radio Club membership certificate.

39. *Fine Tuning*, December 9, 1977 (No. 21), p. 3; *FRENDX*, January 1978, p. LN-15; *SPEEDX*, November 1978, p. 3.

40. *Short Wave News* [ISWL], February 1951, p. 355.

41. *Short Wave News* [ISWL], October-November 1950, p. 246; *NNRC Bulletin*, November 1951, p. SW-8.

42. Jan Tunér, "Competitions— Scandinavian Style," *FRENDX*, February 1968, p. 19.

43. *Radio & Television News*, October 1953, p. 82.

44. *NNRC Bulletin*, April 1952, p. 2-B.

45. In 1966, James J. Hart, who had served as the NNRC's shortwave editor from 1943 to 1949, claimed the honor of being the first ever to have verified every country then-recognized by the ARRL.

46. Don Jensen, "Counting Countries— An In Depth Look," *Proceedings* (Stillwater, OK: Fine Tuning Special Publications, Vol. 2, 1989), p. F10.1; and *NASWA Journal*, January 1995, p. 8.

47. *NNRC Bulletin*, May 1957, p. A-14; *NNRC Bulletin*, October 1957, p. A-6.

48. *NNRC Bulletin*, March 1958, p. SW-4A.

49. *NNRC Bulletin*, March 1959, p. SW-1A.

50. *Popular Electronics*, May 1959, p. 105; *Popular Electronics*, July 1959, p. 127.

51. Hank Bennett, "WDX — A Service to and for the SWL," *Monitoring Times*, September 1985, p. 24.

52. "Were *You* A 'WPE'?" http://www.qsl.net/wb1gfh/swl.html.

53. *Popular Communications*, November 1990, p. 74.

54. In 1979 and 1980, the WLS claimed to have started in 1976. However, in 1985 it announced its 25th anniversary. See *CANDX*, June 1979, p. 8; *FRENDX*, April 1980, p. 1; and *FRENDX*, July 1985, p. 53.

55. See http://www.members.shaw.ca/swarl/.

56. *Radio & Television News*, April 1950, p. 62.

57. *NNRC Bulletin*, May 1960, p. SWBC-9.

58. See Jerry Berg, "QSLs From Attic to Archive," *Popular Communications*, May 2001, p. 6. For more information on the committee, see its website at http://www.ontheshortwaves.com A QSL archival project also exists in Austria. See *http://www.qsl.at/english/welcome.html*.

Chapter 8

1. *Bulletin of the International DXer's Club of San Diego*, December 1983, p. 11.

2. E.g. Tom Sundstrom's 1989 WRTH Industry Award-winning "Shortwave Database" program; Radio Listener's Database (no longer available); and SWLog, http://www.shortwavelog.com/. See Todd D. Dokey, *Computerized Radio Monitoring* (Lake Geneva, WI: Tiare, 1992).

3. *FRENDX*, August 1986, p. 12.

4. *Fine Tuning*, No. 456, January 7, 1988.

5. *NASB Newsletter*, April 2004.

6. Gayle Van Horn, "Sleepless in Brasstown — The Insomniac's Guide to International Broadcast Sites on the Internet," *Monitoring Times*, November 1996, p. 18.

7. http://www.eibi.de.vu/.

8. http://www.dswci.org/.

9. http: //www.bdxc.org.uk/.

10. http://home.tele2.it/MCDXT/LASWLOGS.html.

11. http://www.clandestineradio.com, http://www.schoechi.de/crw.html.

12. http://www.dxasia.info/index.html.

13. http://www.asiawaves.net/.

14. http://www.schoechi.de/pwdb.html.

15. The extent to which the DXing world changed is captured in a series of articles, "Emerging Techniques of High-Tech DXpeditioning " by Guy Atkins, John H. Bryant, Nick Hall-Patch and Don Nelson, in the *NASWA Journal*, Pt. I, February 2003, p. 10; Pt. II, March 2003, p. 9; Pt. III, April 2003, p. 24; and in *Listening In*, Pt. I, May 2003, p. 33; Pt. II, June 2003, p. 35; Pt. III, July 2003, p. 35; Pt. IV, August 2003, p. 33. The authors explain how laptop computers can be used to control a receiver (some-

times from a remote location), access databases and record reception. They also describe commonly available software programs for keeping a log, viewing the sunrise-sunset status, slowing down audio while preserving correct pitch, etc.

16. "Latin American Music Styles," http://home.swip net.se/gersnaes/henriks/lamusic.html.

17. http://www.primetimeshortwave.com/.

18. http://www.naswa.net/swlguide/.

19. http://www.schoechi.de/qip.html.

20. http://www.irkutsk.com/radio/russia.htm.

21. "Eldorado for LA DXers," archived e-mail addresses and verification signer names for Latin American stations. It is available at http://web.comhem.se/mwm/eldorado/ but has not been updated since 2003.

22. See Anker Petersen, "The Influence of the Internet on DX Clubs," *Short Wave News,* October 2000, p. 14.

23. E.g., "On the short waves.com," http://www.onthe shortwaves.com; "Radio Heritage Foundation," http://www.radioheritage.net/.

24. E.g., "Patepluma Radio," http://www.pateplumara dio.com/.

25. E.g., "Digital Radio Mondiale," http://www.drm. org/; "HCJB Global Technology Center," http://www.hcjb. org/index.php?option=com_content&task=view&id=83&I tem=0.

26. "Kim Andrew Elliott," http://www.kimandrewel liott.com/.

27. "BLANDX," http://www.blandx.com/.

28. "Association for International Broadcasting," http://www.aib.org.uk/; "National Association of Short-wave Broadcasters," http://www.shortwave.org/.

29. E.g., "hard-core-dx.com," http://www.hard-core-dx.com; "DXing.com," http://www.dxing.info/; and "Nordic Shortwave Center" (no longer online).

Chapter 9

1. William I. Orr, *Better Shortwave Reception* (Wilton, CT: Radio Publications, 1957), p. 5.

2. Ian McFarland, "The Course of International Broadcasting (A View from the Inside)," *Monitoring Times,* January 1995, p. 14.

3. Robert Zanotti, "The Future of International Broadcasting: A Critical Point of View," August 31, 2007, http://www.undpi.org/index.php?name=News&file=article&sid=181.

4. David F. Thomas, *NNRC Bulletin,* February 1957, p. FM-2.

5. Oliver P. Ferrell, "Twilight of the Shortwave Listener," *Popular Electronics,* March 1971, p. 7.

6. Harry Helms, "The Future of Radio," *Monitoring Times,* April 2007, p. 16.

7. Victor Goonetilleke, *UADX,* No. 50/51, Dec. 15, 1977, p. 2.

8. Robert Fraser, "Shortwave — Boom or Thud?" *FRENDX,* April 1980, p. SWC-5.

9. Stephen J. Morgan, "A Signal Heard Round the World," *The Christian Science Monitor,* February 2, 2007, p. 19.

Selected Bibliography

Books and Other Major Works

Baier, Thomas. *Grundig Satellit — All Models in Word and Picture.* Homburg, Germany: Thomas Baier, 2d. ed. 2001.

Bennett, Hank. *The Complete Shortwave Listener's Handbook.* Blue Ridge Summit, PA: Tab, 1974.

_____ and Harry L. Helms. *The Complete Shortwave Listener's Handbook — 2d ed.* Blue Ridge Summit, PA: Tab, 1980.

_____, Harry L. Helms and David T. Hardy. *The Complete Shortwave Listener's Handbook — 3d ed.* Blue Ridge Summit, PA: Tab, 1986.

_____, David T. Hardy and Andrew Yoder. *The Complete Shortwave Listener's Handbook — 4th ed.* Blue Ridge Summit, PA: Tab, 1994.

_____ and Andrew Yoder. *The Complete Shortwave Listener's Handbook — 5th ed.* New York, NY: McGraw-Hill, 1997.

Berg, Jerome S. *On the Short Waves, 1923–1945: Broadcast Listening in the Pioneer Days of Radio.* Jefferson, NC: McFarland, 1999.

Boord, Kenneth R. *The World at a Twirl.* Morgantown, WV: Kenneth R. Boord, 1956.

Bos, Willem, and Jonthan Marks. *WRTH Equipment Buyers Guide.* New York, NY: Billboard, 1993.

Braband, Ken C. *The First 50 Years — A History of Collins Radio Company and the Collins Division of Rockwell International.* Cedar Rapids, IA: Rockwell International, 1983.

Broadcasting Stations of the World. Washington, D.C.: Foreign Broadcast Information Service, published periodically, 1945–74.

Browne, Donald R. *International Broadcasting: The Limits of the Limitless Medium.* New York, NY: Praeger, 1982.

Bryant, John H., Guy Atkins, Kevin Atkins, Elton Byington, David M. Clark and Fritz Mellberg, eds. *Proceedings.* Stillwater, OK: Fine Tuning Special Publications, Vol. 1, 1988; Vol. 2, 1989; Vol., 3, 1990; Vol. 4, 1991; Vol. 5, 1992–93; Vol. 6, 1994–95.

Bryant, John H., and Harold N. Cones. *The Zenith Trans-Oceanic — The Royalty of Radios.* Atglen, PA: Schiffer, 1995.

_____. *Zenith Radio — The Early Years: 1919–1935.* Schiffer, 1997.

_____ and Martin Blankinship. *Zenith Radio — The Glory Years, 1936–1945. History and Products.* Schiffer, 2003.

_____. *Zenith Radio — The Glory Years, 1936–1945: Illustrated Catalog & Database.* Schiffer, 2003.

Childs, Harwood L., and John B. Whitton, eds. *Propaganda by Shortwave.* Princeton, NJ: Princeton University Press, 1942.

Cushen, Arthur. *Radio Listeners Guide.* Invercargill, New Zealand: Arthur Cushen, 2d ed. 1990.

Cushen, Arthur T. *The World in My Ears.* Invercargill, New Zealand: Arthur T. Cushen, 1979.

Dachis, Chuck. *Radios by Hallicrafters.* Atglen, PA: Schiffer, 1st ed. 1996; 2d. ed. 1999.

Darrington, Philip. *Guide to Broadcasting Stations.* London, England: William Heinemann, 19th ed. 1987.

Dexter, Gerry L. *Clandestine Confidential.* Columbus, OH: Universal Electronics, 1984.

_____, ed. *Shortwave Radio Listening with the Experts.* Indianapolis, IN: Howard W. Sams, 1986.

Elliott, Kim Andrew. *An Alternative Programming Strategy for International Radio Broadcasting,* Ph.D. dissertation. Minneapolis, MN: University of Minnesota, 1979.

_____. *International Radio Broadcast Listening in the United States: A Survey of the North American Short Wave Association,* master's thesis. Minneapolis, MN: University of Minnesota, 1977.

Farmerie, William T., ed. *National Radio Club 50th Anniversary, 1933–1983 — Commemorating the History of Broadcast Band DXing.* Cambridge, WI: National Radio Club, 1983.

Figliozzi, John A. *The Shortwave Radioguide.* Self-published, May 1990 and November 1990; *The Shortwave Radioguide.* Toronto, Canada: Ontario DX Assn., and Levittown, PA: North American Shortwave Assn.: May 1991 and November 1991; *The Shortwave Radioguide.* Toronto, Canada: Ontario DX Assn., and Levittown, PA: North American Shortwave Assn.: 1993, 1994, 1995; *The Worldwide Shortwave Listening Guide.* Richardson, TX: Master, 1st ed. 1996; 2d ed. 1998; and Halfmoon, NY: John A. Figliozzi, 3d ed. 2000.

Frost, J.M., ed. *How to Listen to the World 1969/70.* Hvidovre, Denmark: World Radio-Television Handbook, 5th ed. 1969; *How to Listen to the World 1971.* Hvidovre, Denmark: World Radio-TV Handbook, 6th ed. 1971; *How to Listen to the World.* Hvidovre, Denmark: World Radio-TV Handbook, 7th ed. 1973; *How to Listen to the World.* Hvidovre, Denmark: World Radio-TV Handbook, 8th ed. 1974. This series, which began under editor O. Lund Johansen, is numbered irregularly.

Gilbert, Sean, ed. *The Shortwave Guide — Volume 2.* Milton Keynes, England: WRTH, 2003.

Guha, Manosij. *India BroadBase.* Toronto, Ontario: Ontario DX Assn., various eds., 1993–97.

Hall, James L. *Radio Canada International — Voice of a Middle Power.* Lansing, MI: Michigan State University Press, 1997.

Hardyman, Nicholas. *The Shortwave Guide.* Milton Keynes, England: WRTH, 2002.

Haring, Kristen. *Ham Radio's Technical Culture.* Cambridge, MA: MIT Press, 2007.

Haslach, Robert D. *Netherlands World Broadcasting.* Media, PA: Lawrence Miller, 1983.

Helms, Harry L. *Shortwave Listening Guidebook.* Solana Beach, CA: High Text, 1991.

_____. *Shortwave Listening Handbook.* Englewood Cliffs, NJ: Prentice-Hall, 1987.

de Henseler, Max. *The Hallicrafters Story, 1933–1975.* Charleston, WV: Antique Radio Club of America, 1991.

Jacobs, George, and Theodore J. Cohen. *The Shortwave Propagation Handbook.* Port Washington, NY: Cowan, 1979.

_____ and Robert B. Rose. *The NEW Shortwave Propagation Handbook.* Hicksville, NY: CQ Communications, 1995.

Johansen, O. Lund, ed. *How to Listen to the World.* Hellerup, Denmark: World Radio Handbook (or World Radio-Television Handbook), 1st ed., undated, c. 1950; 2d and 3d eds., undated, c. 1952; 4th and 5th eds., undated, c. 1953; *How to Listen to the World.* Hellerup, Denmark: World Radio-Television Handbook, 6th ed. 1956; *How to Listen to the World.* Hellerup, Denmark: O. Lund Johansen, 7th ed., 1959 and 1960; *How to Listen to the World "5."* Hellerup, Denmark: O. Lund Johansen, undated, c. 1962; *How to Listen to the World, 1963/64 "6."* Hellerup, Denmark: O. Lund Johansen, 1963; *How to Listen to the World, 1965/66 "11."* Hellerup, Denmark: World, 1965; *How to Listen to the World, 1966/67 "12."* Hellerup, Denmark: World, 1967. This series, which continued under editor J.M. Frost, is numbered irregularly.

Klemetz, Henrik. *Latin America by Radio.* Espoo, Finland: Tietoteos, 1989.

Krone, Finn. *Clandestine Stations List.* Greve, Denmark: Danish Shortwave Club Intl., various eds., 1985–98.

LA DXing. Tokyo, Japan: Radio Nuevo Mundo, No. 1,

1980; No. 2, 1982; No. 3, 1985; No. 4, 1987; No. 5, 1992, No. 6, 1997.

Leinwoll, Stanley. *Shortwave Propagation.* New York, NY: John F. Rider, 1959.

Lichte, Rainer. *Radio Receiver — Chance or Choice.* Park Ridge, NJ: Gilfer Shortwave, 1985.

_____. *More Radio Receiver — Chance or Choice.* Park Ridge, NJ: Gilfer Shortwave, 1987.

Loughmiller, John. *A Family Affair — The R. L. Drake Story.* Self-published, 2007.

MacHarg, Kenneth D. *Tune in the World: The Listener's Guide to International Shortwave Radio.* Media, PA: Lawrence Miller, 1983.

Maes, Ludo. *Transmitter Documentation Project.* Rijkevorsel, Belgium: Ludo Maes/TDP, various eds., 1994–98.

Magne, Lawrence, ed. *Passport to World Band Radio.* Penn's Park, PA: International Broadcasting Services, various annual eds., 1988–2008.

_____. *Radio Database International.* Penn's Park, PA: International Broadcasting Services, various eds., 1985–87.

Meo, L.D. *Japan's Radio War on Australia, 1941–1945.* Melbourne, Australia: Melbourne University Press, 1968.

Moore, Raymond S. *Communications Receivers, the Vacuum Tube Era: 1932–1981.* La Belle, FL: RSM Communications, 1st ed. 1987; 2d ed. 1991; 3d ed. 1993; 4th ed. 1997.

Mytton, Graham, ed. *Global Audiences: Research for Worldwide Broadcasting — 1993.* London, England: John Libbey, 1993.

Orr, William I. *Better Shortwave Reception.* Wilton, CT: Radio Publications, 1st ed. 1957.

Osterman, Fred. *Shortwave Receivers Past and Present — Communications Receivers 1942–1997.* Reynoldsburg, OH: Universal Radio Research, 1st ed. 1987 (no time period specified); 2d ed. 1997 ("1945–1996"); 3d ed. 1998.

Passmann, Willi H. *Tropical Band List.* Mülheim an der Ruhr, Germany: Willi H. Passmann Media Consulting, various eds., 1992–2006.

Penson, Chuck. *Heathkit — A Guide to the Amateur Radio Products.* Durango, CO: Electric Radio Press, 1st ed. 1995; Hicksville, NY: CQ Communications, 2d ed. 2003.

Perdue, Terry. *Heath Nostalgia.* Lynnwood, WA: Terry A. Perdue, 1992.

Petersen, Anker. *Domestic Broadcasting Survey.* Greve, Denmark: Danish Shortwave Club Intl., various eds., 1999–2007.

Petersen, Anker, et al. *Tropical Bands Survey.* (Greve, Denmark: Danish Shortwave Club Intl., various eds., 1973–98.

"Radio in My Life" Essay Collection. Toronto, Ontario: Ontario DX Assn., 1999.

Robbins, Jane M.J. *Tokyo Calling: Japanese Overseas Radio Broadcasting, 1937–1945.* Firenze, Italy: European Press Academic Publishing, 2001.

Rolo, Charles J. *Radio Goes to War.* New York, NY: G.P. Putnam's Sons, 1940.

Schiffer, Michael Brian. *The Portable Radio in American Life.* Tucson, AZ: University of Arizona Press, 1991.

Shaw, Edward C. *DXing According to NASWA.* Liberty, IN: North American Shortwave Assn., 1975.

The Shortwave Book. Media, PA: Lawrence Miller, 1984.

Shortwave Radio Listening In the United States. The Gallup Organization, 1973, referenced in *Report of the Radio Canada International Task Force,* App. H. Montreal, Quebec: CBC, 1973.

Soley, Lawrence. *Free Radio — Electronic Civil Disobedience.* Boulder, CO: Westview, 1999.

Soley, Lawrence C. *Radio Warfare: OSS and CIA Subversive Propaganda.* New York, NY: Praeger, 1989.

_____ and John S. Nichols. *Clandestine Radio Broadcasting — A Study of Revolutionary and Counterrevolutionary Electronic Communication.* New York, NY: Praeger, 1987.

Spanswick, Simon. *Guide to Broadcasting Stations: 21st Century Edition.* London, England: Butterworth-Heinemann, 21st ed. 2001.

Stearns, Ben W. *Arthur Collins — Radio Wizard.* Marion, IA: Ben W. Stearns, 2002.

Thompson, Alan. "BBC Monitoring Service — A Layman Looks at Caversham Park." British Association of DXers, May 1972.

Vastenhoud, J. *Short Wave Listening.* Eindhoven, Netherlands: Philips, 1966.

Vastenhoud, Jim, arr. and comp., and Jens M. Frost, ed. *World DX Guide.* London, England: Billboard, 1978.

Williams, Barry C. *The Origins of DXing in New Zealand.* Auckland, New Zealand: Barry C. Williams, 2007, draft.

Williamson, Tom. *Across Time — and Space: Listening for Sixty Years from Four Continents.* Distributed by Ontario DX Assn., 1998.

Wireless World. *Guide to Broadcasting Stations.* London, England: ILIFFEE & Sons, 4th ed. 1948; *Guide to Broadcasting Stations.* London, England: Butterworth, 16th ed. 1970.

Wood, Richard E. *Shortwave Voices of the World.* Park Ridge, NJ: Gilfer Associates, 1969.

World Radio TV Handbook, published annually from 1947 to 2008; originally edited and published by O. Lund Johansen in Hellerup, Denmark, later edited by others and published in varying venues; also *WRTH Summer Supplement,* annually, 1959–66, and *WRTH Summer Edition,* annually, 1967–71.

The Zenith Story: A History from 1918 to 1954. Glenview, IL: Zenith Electronics, 1955.

Articles

Atkins, Guy, John H. Bryant, Nick Hall-Patch and Don Nelson. "Emerging Techniques of High-Tech DX-peditioning." *NASWA Journal,* Pt. I, February 2003, p. 10; Pt. II, March 2003, p. 9; Pt. III, April 2003, p. 24; and *Listening In,* Pt. I, May 2003, p. 33; Pt. II, June 2003, p. 35; Pt. III, July 2003, p. 35; Pt. IV, August 2003, p. 33.

Bataille, Eugene C. "The Birth of the Newark News Radio Club." *NNRC Bulletin,* December 1975, p. 3.

Breckel, Lt. H.F. "In the Future — Intermediate- or Short-Wave Broadcasting?" *Radio News,* November 1927, p. 461.

Browne, Don R. "The Limits of the Limitless Medium — International Broadcasting." *Journalism Quarterly,* Vol. 42, No. 2 (Winter 1965), p. 82.

Childs, Harwood L. "America's Short-Wave Audience," in Harwood L. Childs and John B. Whitton, eds., *Propaganda by Shortwave.* Princeton, NJ: Princeton University Press, 1942, p. 303.

_____. "Short-Wave Listening in the United States." *Public Opinion Quarterly,* Vol. 5, No. 2 (June 1941), p. 210.

Cones, Harold. "'Tuning the Short-Wave Bands' Revisited: A 1991 Interview with Hank Bennett." *Proceedings* (Vol. 4, 1991), p. F23.1.

Conrad, Dr. Frank. "Short Wave Broadcasting — As a Pioneer Sees It." *Short Wave & Television,* August 1938, p. 197.

Dachis, Chuck. "The Hallicrafters S-38: 1935 to 1962." *Proceedings* (Vol. 5, 1992–93), p. F29.1.

D'Angelo, Richard A. "An Early History of NASWA." *NASWA Journal,* July 1999, p. 13.

_____. "35 Years Ago — The Beginnings: A Retrospective on ANARC's Early Years." *NASWA Journal,* Pt. 1, August 1999, p. 10; Pt. 2, October 1999, p. 11; and Pt. 3, December 1999, p. 10.

_____. "What is an ANARC or an EDXC?" *NASWA Journal,* March 2001, p. 10.

Davis, H. P. "Short-Wave Broadcast Pioneering." *Radio Craft,* March 1931, p. 543.

"Diamond Anniversary — It's 60 Years of World Radio TV Handbook." *Monitoring Monthly* (U.K.), March 2006, p. 39.

Elliott, Kim Andrew. "An Alternative Approach to International Broadcast Programming." *Review of International Broadcasting,* January 1980, p. 3.

_____. "An Alternative Programming Prototype." *Review of International Broadcasting,* October 1981, p. 5.

_____. "The Meaningless Messages of International Radio." *Review of International Broadcasting,* August 1980, p. 3.

_____. "An Overview of International Broadcasting Audience Research." Presentation to the Broadcast Education Association, April 2002.

Ferrell, Oliver P. "Dream Receivers for the SWL." *Popular Electronics,* October 1968, p. 53.

Fizette, William. "The National Sliding Coil Tray Receivers." *Old Timer's Bulletin,* February 1988, p. 20.

Gernsback, Hugo. "The Short-Wave Era." *Radio News,* September 1928, p. 201.

_____. "The Short-Wave Fan." *Radio News,* February 1929, p. 715.

Glen, Frank. "A Passion with A Purpose — The Prisoner of War Message Service, 1951–1952." *New Zealand DX Times,* September 2003, p. 39.

Hall, James L., and Drew O. McDaniel. "The Regular Shortwave Listener in the U.S." *Journal of Broadcasting,* Vol. 19, No. 3 (Summer 1975), p. 363.

Headrick, Daniel R. "Shortwave Radio and Its Impact on International Telecommunications Between the Wars." *History and Technology,* Vol. 11, p. 21 (1994).

Heinemann, F. Parker. "The Collins Radio Company — Ingredients of Success." *AWA Review* (Bloomfield, NY: Antique Wireless Association, Vol. 10, 1996), p. 222.

Herrmann, Peter. "The World-Wide Audience and Their Attitudes," *How to Listen to the World, 8th ed.* (Hvidovre, Denmark: World Radio-TV Handbook, 1974), p. 137, and Peter Herrmann, "Audience and Their Attitudes," *World DX Guide* (London, England: Billboard, 1978), p. 176. Same article.

Horn, C.W. "Is International Broadcasting Just Around the Corner?" *Radio News,* January 1930, p. 608.

Insinger, Edward J. "*FRENDX* Visits Gilfer — For a Q and A with Perry Ferrell." *FRENDX,* April 1973, p. SWC-4.

"The Jack R. Poppele Transmitting Station of the Voice of America." *Proceedings of the Radio Club of America,* November 1992, p. 4.

Jensen, Don. *The History of the North American Shortwave Association* (NASWA Editorial Committee, 1990), a monograph appearing in revised form in a three-part series in *FRENDX,* June, August and November 1981, and updated in Don Jensen, "40 Years of the North American Shortwave Association," *NASWA Journal,* September 2001, p. 13.

Magne, Lawrence. "The S.R.I. Report Debunked." *Review of International Broadcasting,* November 1978, p. 3.

Martin, Edgar T., and George Jacobs. "Shortwave Broadcasting in the 1970s," *How to Listen to the World 1971* (Hvidovre, Denmark: World Radio-TV Handbook, 1971), p. 4.

Massie, Keith, and Stephen D. Perry. "Hugo Gernsback and Radio Magazines: An Influential Intersection in Broadcast History." *Journal of Radio Studies,* Vol. 9, No. 2 (2002), p. 264.

McFarland, Ian. "The Course of International Broadcasting (A View from the Inside)." *Monitoring Times,* January 1995, p. 14.

Meyer, Stuart. "Hammarlund Radio." *AWA Review* (Holcomb, NY: Antique Wireless Association, Vol. 2, 1987), p. 95.

Morgan, Stephen J. "A Signal Heard Round the World." *The Christian Science Monitor,* February 2, 2007, p. 19.

Morrison, Charles A. "The Future of International Short-Wave Reception, Part I." *Radio News,* April 1935, p. 598; Part II, May 1935, p. 674; Part III, June 1935, p. 745; Part IV, July 1935, p. 27.

Mytton, Graham. "Audience Research for International Radio Broadcasters: A Toolkit for Small to Medium-Sized Stations," in Oliver Zöllner, ed., *An Essential Link With Audiences Worldwide — Research for International Broadcasting* (Cologne, Germany: Deutsche Welle/VISTAS, 2002), p. 25.

_____. "Audience Research for Shortwave Broadcasters." Presented at NASB Annual Meeting, Washington, D.C., May 5, 2000, http://www.shortwave.org/Audience/Audience.htm.

_____. "...But Don't Overlook Shortwave," in Oliver

Zöllner, *Reaching Audiences Worldwide — Perspectives of International Broadcasting and Audience Research, 2001/2002* (Bonn, Germany: CIBAR, 2003), p. 97.

_____. "New Technology and International Broadcasting," in Andrew G. Sennitt, ed., *1996 World Radio TV Handbook* (Amsterdam, Netherlands: Billboard, 1996), p. 599.

_____ and Carol Forrester. "Audiences for International Radio Broadcasts." *European Journal of Communication,* Vol. 3, No. 4 (1988), p. 457.

Nagle, John J. "A Brief History of the National Company, Inc." *AWA Review* (Holcomb, NY: Antique Wireless Association, Vol. 1, 1986), p. 65.

Padula, Bob. *The First 20 Years — The Development of the ARDXC: 1965–1985.* Supplement to the June 1985 *Australian DX News.*

_____. *The History of DXing in Australia.* Published by the Australian Radio DX Club in August 1967, reprinted in 1969.

Parta, Gene. "Mass Audiences, Elite Audiences and Target Audiences," in *An Essential Link with Audiences Worldwide — Research for International Broadcasting* (Cologne, Germany: Deutsche Welle/VISTAS, 2002), p. 85.

Petersen, Anker. "The Influence of the Internet on DX Clubs." *Short Wave News,* October 2000, p. 14.

Potts, Irving R. "The Newark News Radio Club." *Keller's Radio Call Book and Log,* January-February 1934, p. 23.

Read, Oliver. "Foreign Broadcast Intelligence Service." *Radio News,* January 1945, p. 25.

Rodgers, W.W. "Is Short-Wave Relaying a Step Toward National Broadcasting Stations?" *Radio Broadcast,* June 1923, p. 119.

Sams, Mitch. "Shortwave Broadcasting in the 1980s: A Hobby Perspective." *Proceedings* (Vol. 4, 1991), p. F25.1

Savage, Mark. "A Double Century of *Communication.*" *Communication,* July 1991, p. 6.

_____. "25 Years and Three Hundred Editions of *Communication.*" *Communication,* November 1999, p. 5.

Schuler, Edgar A., and Wayne C. Eubank. "Sampling Listener Reactions to Short-Wave Broadcasts." *Public Opinion Quarterly,* Vol. 5, No. 2 (June 1941), p. 260.

Seifert, Richard A. "Broadcasting Bedfellows — Recalling the Unusual Alliance between Radio New York Worldwide and the Drake SW-4(A)." *Monitoring Times,* August 1996, p. 25.

Sheringham, John G.T. "BBC Monitoring Service: The Ears of Britain," in *How to Listen to the World* (Hvidovre, Denmark: World Radio-TV Handbook, 8th ed., 1974), p. 51.

Smith, Don. "Is There a U.S. Audience for International Broadcasts?" *Journalism Quarterly,* Vol. 39 (1962), p. 86.

Smith, Don D. "America's Short-Wave Audience: Twenty-Five Years Later." *Public Opinion Quarterly,* Vol. 33, No. 4 (Winter 1969–1970), p. 537.

_____. "The U.S. Audience for International Broadcasts," *Journalism Quarterly,* Vol. 47 (1970), p. 364.

Startz, Edward. "40 Years Happy Station," in J.M. Frost,

ed., *1969 World Radio TV Handbook* (Hellerup, Denmark: World Radio-Television Handbook, 1969), p. 35.

Ware, Lawrence R. "The National Company, Inc.— The Coil-Catacomb Radios and Variations on a Theme." *AWA Review* (Bloomfield, NY: Antique Wireless Association, Vol. 11, 1998), p. 166.

Williams, Barry. "The Evolution of the National HRO and Its Contribution to Winning World War II." *AWA Review* (Breesport, NY: Antique Wireless Association, Vol. 17, 2004), p. 145.

Zanotti, Robert. "The Future of International Broadcasting: A Critical Point of View." August 31, 2007, http://www.undpi.org/index.php?name=News&file=article&sid=181.

Zichi, Kenneth V. "A Look Back At 25 Years of Service." *SWL,* December 1984, p. 4.

Zöllner, Oliver. "International Broadcasters, Audience Research and a Conference: An Overview of Methods and Functions," in Oliver Zöllner, ed., *An Essential Link with Audiences Worldwide— Research for International Broadcasting* (Cologne, Germany: Deutsche Welle/VISTAS, 2002), p. 13.

Websites

Association for International Broadcasting, http://www.aib.org.uk/.

Clandestineradio.com, http://www.clandestineradio.com.

Digital Radio Mondiale, http://www.drm.org/.

Kim Andrew Elliott, http://www.kimandrewelliott.com/.

National Association of Shortwave Broadcasters, http://www.shortwave.org/.

On the Short Waves, http://www.ontheshortwaves.com.

QSL Collection, http://www.qsl.at/english/welcome.html.

Radio Heritage Foundation, http://www.radioheritage.net/.

"Were *You* A 'WPE'?" http://www.qsl.net/wb1gfh/swl.html.

Periodicals

RADIO- AND DX-RELATED MAGAZINES

Antique Radio Classified
The Antique Radio Gazette (Antique Radio Club of America)
DX Listening Digest
DXing Horizons
Electric Radio
Electronics Illustrated
Elementary Electronics
International Listener
The International Shortwave Listener's Program Guide, The Shortwave Guide, International Radio
Monitoring Times
Popular Communications
Popular Electronics
QST
Radio News, Radio & Television News
Radio-TV Experimenter
Review of International Broadcasting
S9
73
Voices
World Radio Report

CLUB PUBLICATIONS

The ACE, The Monthly ACE (Assn. of Clandestine Radio Enthusiasts)
ANARC Newsletter (Assn. of North American Radio Clubs)
Bulletin of the International DXer's Club of San Diego
CADEX (Canadian DX Club)
Communication (British DX Club)
CONTACT (World DX Club)
DX Ontario, Listening In (Ontario DX Assn.)
DX Reporter (Assn. of DX Reporters)
FRENDX, NASWA Journal (North American Shortwave Assn.)
International Short Wave Radio (International Short Wave Club)
The Messenger (Canadian International DX Club)
NNRC Bulletin (Newark News Radio Club)
Old Timer's Bulletin, AWA Journal (Antique Wireless Assn.)
Short Wave News (Danish Short Wave Club International)
SPEEDX (Society for the Preservation of the Engrossing Enjoyment of DXing)
SWL (American Shortwave Listeners Club)
Union of Asian DXers
Universalite (Universal Radio DX Club)
WPE Call Letter (Great Circle Shortwave Society)

NEWSLETTERS

Bandspread (British Assn. of DXers)
Clandestine Confidential Newsletter
"Down Under" DX Survey ("Down Under" DX Circle)
DX South Florida, Mosquito Coast News
Fine Tuning
NASB Newsletter (Natl. Assn. of Shortwave Broadcasters)
Numero Uno, DXplorer
NEDS News (New England DX Society)
OZDX [Australia]
R390 Users Group Newsletter, Hollow State Newsletter
Sweden Calling DXers
Tropical DX Newsletter
USSR High Frequency Broadcast Newsletter

Catalogs

Allied Radio Corp.
Gilfer Associates

Heathkit
Lafayette Radio Corp.
Radio Shack
Universal Radio
World Radio Laboratories

Other Publications

AWA Review (Antique Wireless Assn.)
Communications Handbook
Communications World

Index

Numbers in **bold italics** indicate pages with illustrations.